LONDON MATHEMATICAL SOCIETY LECTURE NOTE SERIES

Managing Editor: Professor M. Reid, Mathematics Institute,
University of Warwick, Coventry CV4 7AL, United Kingdom

The titles below are available from booksellers, or from Cambri~ ~~ ~~~~~~~v Press at
http://www.cambridge.org/mathematics

London Mathematical Society Lecture Note Series: 441

Polynomials and the mod 2 Steenrod Algebra

Volume 1: The Peterson Hit Problem

GRANT WALKER
University of Manchester

REGINALD M. W. WOOD
University of Manchester

CAMBRIDGE
UNIVERSITY PRESS

CAMBRIDGE
UNIVERSITY PRESS

University Printing House, Cambridge CB2 8BS, United Kingdom

One Liberty Plaza, 20th Floor, New York, NY 10006, USA

477 Williamstown Road, Port Melbourne, VIC 3207, Australia

4843/24, 2nd Floor, Ansari Road, Daryaganj, Delhi – 110002, India

79 Anson Road, #06-04/06, Singapore 079906

Cambridge University Press is part of the University of Cambridge.

It furthers the University's mission by disseminating knowledge in the pursuit of education, learning, and research at the highest international levels of excellence.

www.cambridge.org
Information on this title: www.cambridge.org/9781108414487
DOI: 10.1017/9781108333368

First published 2018

Printed in the United Kingdom by Clays, St Ives plc

A catalogue record for this publication is available from the British Library.

ISBN – 2 Volume Set 978-1-108-41406-7 Paperback
ISBN – Volume 1 978-1-108-41448-7 Paperback
ISBN – Volume 2 978-1-108-41445-6 Paperback

Contents of Volume 1

Contents of Volume 2

Preface

This book is about the mod 2 Steenrod algebra A_2 and its action on the polynomial algebra $P(n) = \mathbb{F}_2[x_1, \ldots, x_n]$ in n variables, where $\mathbb{F}_2$ is the field of two elements. Polynomials are graded by degree, so that $P^d(n)$ is the set of homogeneous polynomials of degree d. Although our subject has its origin in the work of Norman E. Steenrod in algebraic topology, we have taken an algebraic point of view. We have tried as far as possible to provide a self-contained treatment based on linear algebra and representations of finite matrix groups. In other words, the reader does not require knowledge of algebraic topology, although the subject has been developed by topologists and is motivated by problems in topology.

There are many bonuses for working with the prime $p = 2$. There are no coefficients to worry about, so that every polynomial can be written simply as a sum of monomials. We use a matrix-like array of 0s and 1s, which we call a 'block', to represent a monomial in $P(n)$, where the rows of the block are formed by the reverse binary expansions of its exponents. Thus a polynomial is a set of blocks, and the sum of two polynomials is the symmetric difference of the corresponding sets. Using block notation, the action of A_2 on $P(n)$ can be encoded in computer algebra programs using standard routines on sets, lists and arrays. In addition, much of the literature on the Steenrod algebra and its applications in topology concentrates on the case $p = 2$. Often a result for $p = 2$ has later been extended to all primes, but there are some results where no odd prime analogue is known.

We begin in Chapter 1 with the algebra map $\mathsf{Sq} : P(n) \to P(n)$ defined on the generators by $\mathsf{Sq}(x_i) = x_i + x_i^2$. The map Sq is the total Steenrod squaring operation, and the Steenrod squares $Sq^k : P^d(n) \to P^{d+k}(n)$ are its graded parts. The linear operations Sq^k can be calculated using induction on degree and the Cartan formula $Sq^k(fg) = \sum_{i+j=k} Sq^i(f)Sq^j(g)$, which is equivalent to the multiplicative property of Sq. A general Steenrod operation is a sum of compositions of Steenrod squares.

The multiplicative monoid $M(n)$ of $n \times n$ matrices over $\mathbb{F}_2$ acts on the right of $P(n)$ by linear substitution of the variables. Thus $P^d(n)$ gives a representation over $\mathbb{F}_2$ of $M(n)$ and of $GL(n)$, the general linear group of invertible matrices. This matrix action commutes with the action of the Steenrod squares, and the interplay between the two gives rise to a host of interesting algebraic problems.

One of these, the 'hit' problem, is a constant theme here. A polynomial f is 'hit' if there are polynomials f_k such that $f = \sum_{k>0} Sq^k(f_k)$. The hit polynomials form a graded subspace $H(n)$ of $P(n)$, and the basic problem is to find the dimension of the quotient space $Q^d(n) = P^d(n)/H^d(n)$. We call $Q(n)$ the space of 'cohits'. Since $P(n)$ is spanned by monomials, $Q(n)$ is spanned by their equivalence classes, which we refer to simply as 'monomials in $Q(n)$'. A monomial whose exponents are integers of the form $2^j - 1$ is called a 'spike', and cannot appear as a term in a hit polynomial. It follows that a monomial basis for $Q(n)$ must include all the spikes.

At a deeper level, the hit problem concerns the structure of $Q(n)$ as a representation of $GL(n)$ or $M(n)$. We develop the tools needed to answer this in the 1- and 2-variable cases in Chapter 1. These include the 2-variable version of the maps introduced by Masaki Kameko to solve the 3-variable case, and a map which we call the duplication map. We hope that this opening chapter is accessible to graduate students and mathematicians with little or no background in algebraic topology, and that it will serve as an appetizer for the rest of the book.

Chapter 2 introduces a second family of Steenrod operations, the conjugate Steenrod squares $Xq^k : P^d(n) \to P^{d+k}(n)$. These are useful in the hit problem because of a device known as the 'χ-trick'. This states that the product of f and $Xq^k(g)$ is hit if and only if the product of $Xq^k(f)$ and g is hit. We use the χ-trick to prove that $Q^d(n) = 0$ if and only if $\mu(d) > n$, where $\mu(d)$ is the smallest number of integers of the form $2^j - 1$ (with repetitions allowed) whose sum is d. This establishes the 1986 conjecture of Franklin P. Peterson which first stimulated interest in the hit problem.

Here is a rough guide to the structure of the rest of the book, in terms of three main themes: the Steenrod algebra A_2, the Peterson hit problem, and matrix representations. Volume 1 contains Chapters 1 to 15, and Volume 2 contains Chapters 16 to 30.

Chapters 3 to 5 develop A_2 from an algebraic viewpoint.

Chapters 6 to 10 provide general results on the hit problem, together with a detailed solution for the 3-variable case.

Chapters 11 to 14 introduce the Hopf algebra structure of A_2 and study its structure in greater depth.

Chapter 15 introduces the theme of modular representations by relating the hit problem to invariants and the Dickson algebra.

Chapters 16 to 20 develop the representation theory of $Gl(n, \mathbb{F}_2)$ via its action on 'flags', or increasing sequences of subspaces, in an n-dimensional vector space $V(n)$ over $\mathbb{F}_2$.

Chapter 21 explores a fundamental relation between linear maps $V(m) \rightarrow V(n)$ and Steenrod operations, leading to a maximal splitting of $P(n)$ as a direct sum of A_2-modules.

Chapter 22 studies the A_2-summands of $P(n)$ corresponding to the Steinberg representation of $GL(n, \mathbb{F}_2)$.

Chapters 23 and 24 develop the relation between flag modules and the dual hit problem.

Chapters 25 and 26 study the hit problem for symmetric polynomials over $\mathbb{F}_2$.

Chapters 27 and 28 study the splitting of $P(n)$ as an A_2-module obtained using a cyclic subgroup of order $2^n - 1$ in $GL(n, \mathbb{F}_2)$.

Chapters 29 and 30 return to Peterson's original problem, with a partial solution of the 4-variable case.

The contents of Chapters 3 to 30 are summarized below in more detail.

In Chapter 3 we interpret the operations Sq^k as generators of a graded algebra A_2, subject to a set of relations called the Adem relations. The algebra A_2 is the mod 2 Steenrod algebra, and the operations Sq^k of Chapter 1 provide $P(n)$ with the structure of a left A_2-module. If $f = \sum_{k>0} Sq^k(f_k)$, then f can be reduced to a set of polynomials of lower degree modulo the action of the positively graded part A_2^+ of A_2. A monomial basis for $Q(n)$ gives a minimal generating set for $P(n)$ as an A_2-module. Thus the hit problem is an example of the general question of finding a minimal generating set for a module over a ring. The structure of A_2 itself is completely determined by its action on polynomials, in the sense that two expressions in the generators Sq^k are equal in A_2 if and only if the corresponding operations on $P(n)$ are equal for all n. For example, the results $Sq^1 Sq^{2k}(f) = Sq^{2k+1}(f)$ and $Sq^1 Sq^{2k+1}(f) = 0$ of Chapter 1 imply the Adem relations $Sq^1 Sq^{2k} = Sq^{2k+1}$ and $Sq^1 Sq^{2k+1} = 0$.

In Chapter 3 we also establish the two most important bases for A_2 as a vector space over $\mathbb{F}_2$. These are the admissible monomials in the generators Sq^k, due to Henri Cartan and Jean-Pierre Serre, and the basis introduced by John W. Milnor by treating A_2 as a Hopf algebra. As mentioned above, we represent a monomial by a 'block' whose rows are the reversed binary expansions of its exponents, and whose entries are integers 0 or 1. We use blocks to keep track of Steenrod operations on monomials. This 'block technology' and 'digital engineering' works well for the prime 2, and greatly facilitates our understanding of techniques which can appear opaque when expressed in more standard notation.

Chapter 4 begins with the multiplication formula for elements of the Milnor basis. This combinatorial formula helps to explain the ubiquity of the Milnor basis in the literature, as a product formula is not available for other bases of A_2. We also discuss the compact formulation of the Adem relations due to Shaun R. Bullett and Ian G. Macdonald. We use this to construct the conjugation χ of A_2, which interchanges Sq^k and Xq^k.

Chapter 5 provides combinatorial background for the algebra A_2, the hit problem and the representation theory of $GL(n)$ over $\mathbb{F}_2$. Sequences of non-negative integers appear in various forms, and we distinguish 'finite sequences' from 'sequences'. A 'finite sequence' has a fixed number of entries, called its 'size', while a 'sequence' is an infinite sequence $R = (r_1, r_2, \ldots)$ with only a finite number of nonzero terms, whose 'length' is the largest ℓ for which $r_\ell > 0$. However, a sequence R is usually written as a finite sequence $(r_1, \ldots, r_n)$, where $n \geq \ell$, by suppressing some or all of the trailing 0s. The modulus of a sequence or a finite sequence is the sum of its terms. For example, the degree of a monomial is the modulus of its sequence of exponents. The set of all sequences indexes the Milnor basis of A_2.

For brevity, we call a sequence R 'decreasing' if $r_i \geq r_{i+1}$ for all i, i.e. if it is non-increasing or weakly decreasing. Thus a decreasing sequence of modulus d is a partition of d. Such a partition can alternatively be regarded as a multiset of positive integers with sum d. We discuss two special types of partition; 'binary' partitions, whose parts are integers of the form 2^j, and 'spike' partitions, whose parts are integers of the form $2^j - 1$.

We introduce two total order relations on sequences, the left (lexicographic) order and the right (reversed lexicographic) order, and two partial order relations, dominance and 2-dominance. The ω-sequence $\omega(f) = (\omega_1, \omega_2, \ldots, \omega_k)$ of a monomial f is defined by writing f as a product $f_1 f_2^2 \cdots f_k^{2^{k-1}}$, where f_i is a product of ω_i distinct variables. In terms of blocks, $\omega(f)$ is the sequence of column sums of the block representing f, and the degree of f is $\omega_1 + 2\omega_2 + 4\omega_3 + \cdots + 2^{k-1}\omega_k$, the '2-degree' of $\omega(f)$. The set of decreasing sequences of 2-degree d has a minimum element $\omega^{\min}(d)$, which plays an important part in the hit problem, and is the same for the left, right and 2-dominance orders. We end Chapter 5 by relating this combinatorial material to the admissible and Milnor bases of A_2.

In Chapter 6 we return to the hit problem and introduce 'local' cohit spaces $Q^\omega(n)$. A total order relation on ω-sequences of monomials gives a filtration on $P^d(n)$ with quotients $P^\omega(n)$. For the left and right orders, this passes to a filtration on $Q^d(n)$ with quotients $Q^\omega(n)$. A polynomial f in $P^\omega(n)$ is 'left reducible' if it is the sum of a hit polynomial and monomials with ω-sequences $< \omega$ in the left order, and similarly for the right order. We prove the theorem of William M. Singer that $Q^\omega(n) = 0$ if $\omega < \omega^{\min}(d)$ in the left order. We introduce the 'splicing' technique for manufacturing hit equations, extend the Kameko and duplication maps of Chapter 1 to the n-variable case, and determine $Q^\omega(n)$ for 'head' sequences $\omega = (n-1, \ldots, n-1)$ and 'tail' sequences $\omega = (1, \ldots, 1)$.

We begin Chapter 7 by proving that $\dim Q^d(n)$ is bounded by a function of n independent of d. Thus only finitely many isomorphism classes of $\mathbb{F}_2 \mathrm{GL}(n)$-modules can be realized as cohit modules $Q^d(n)$. We extend splicing techniques and show that $Q^\omega(n) = 0$ if ω is greater than every decreasing ω-sequence in the left order. A correspondence between blocks with decreasing ω-sequences and Young tableaux is used to define 'semi-standard' blocks (or monomials), and we show that $Q^\omega(n)$ is spanned by such blocks when $\omega = \omega^{\min}(d)$.

In Chapter 8, we obtain reduction theorems for $Q^\omega(n)$ when the sequence ω has a 'head' of length $\geq n-1$ or a 'tail' of length $\geq n$. It follows that $\dim Q^d(n) = \prod_{i=1}^{n}(2^i - 1)$ for degrees $d = \sum_{i=1}^{n}(2^{a_i} - 1)$, when $a_i - a_{i+1} \geq i+1$ for $i < n$ and when $a_i - a_{i+1} \geq n-i+1$ for $i < n$. We complete a solution of the 3-variable hit problem by giving bases for $Q^\omega(3)$ in the remaining cases.

The techniques introduced so far are useful for obtaining upper bounds for $\dim Q^d(n)$, but are less efficient for obtaining lower bounds, where we may wish to prove that no linear combination of a certain set of monomials is hit. Chapter 9 introduces the dual problem of finding $K^d(n)$, the simultaneous kernel of the linear operations $Sq_k : \mathrm{DP}^d(n) \to \mathrm{DP}^{d-k}(n)$ dual to Sq^k for $k > 0$. Here $\mathrm{DP}(n)$ is a 'divided power algebra' over $\mathbb{F}_2$, whose elements are sums of dual or 'd-monomials' $v_1^{(d_1)} \cdots v_n^{(d_n)}$. As a $\mathbb{F}_2 \mathrm{GL}(n)$-module, $K^d(n)$ is the dual of $Q^d(n)$ defined by matrix transposition, and so $\dim K^d(n) = \dim Q^d(n)$. Thus we aim to find upper bounds for $\dim Q^d(n)$ by using spanning sets in $Q^d(n)$, and lower bounds by using linearly independent elements in $K^d(n)$.

An advantage of working in the dual situation is that $K(n)$ is a subalgebra of $\mathrm{DP}(n)$. Since the dual spikes are in $K(n)$, they generate a subalgebra $J(n)$ of $K(n)$ which is amenable to calculation. In the cases $n = 1$ and 2, $J(n) = K(n)$, and when $n = 3$, $K^d(n)/J^d(n)$ has dimension 0 or 1. We explain how to construct the dual $K^\omega(n)$ of $Q^\omega(n)$ with respect to an order relation. We study the duals of the Kameko and duplication maps, and solve the dual hit problem for $n \leq 3$. In Chapter 10 we extend these results by determining $K^d(3)$ and

$Q^d(3)$ as modules over $\mathbb{F}_2 GL(3)$. Here the flag module $FL(3)$, given by the permutation action of $GL(3)$ on subspaces of the defining module $V(3)$, plays an important part. We describe tail and head modules in terms of the exterior powers of $V(3)$.

Hopf algebras are introduced in Chapter 11. A Hopf algebra A has a 'coproduct' $A \to A \otimes A$ compatible with the product $A \otimes A \to A$, and an 'antipode' $A \to A$. We show that the coproduct $Sq^k \mapsto \sum_{i+j=k} Sq^i \otimes Sq^j$ and the conjugation χ provide the mod 2 Steenrod algebra A_2 with the structure of a Hopf algebra. For a graded Hopf algebra A of finite dimension in each degree, the graded dual A^* is also a Hopf algebra. In this sense, the divided power algebra $DP(n)$ is dual to the polynomial algebra $P(n)$. We show that the graded dual A_2^* of A_2 is a polynomial algebra on generators ξ_j of degree $2^j - 1$ for $j \geq 1$, and determine its structure maps. We conclude this chapter with the formula of Zaiqing Li for conjugation in A_2, which complements Milnor's product formula of Chapter 4.

Chapters 12 and 13 give more detail on the internal structure of A_2. In Chapter 12 we focus on two important families of Hopf subalgebras of A_2, namely the subalgebras of 'Steenrod qth powers' A_q, where q is a power of 2, and the finite subalgebras $A_2(n)$ generated by Sq^k for $k < 2^{n+1}$. We also introduce some more additive bases of A_2. We continue in Chapter 13 by introducing a 'cap product' action of the dual algebra A_2^* on A_2, which can be used to obtain relations in A_2 by a process which we call 'stripping'. We use the 'halving' map (or Verschiebung) of A_2 to explain why its action on $P(n)$ reproduces itself by doubling exponents of monomials Sq^A and squaring polynomials. This map sends Sq^k to 0 if k is odd and to $Sq^{k/2}$ if k is even. Since it is the union of the finite subalgebras $A_2(n)$, the algebra A_2 is nilpotent. We apply the stripping technique to obtain the nilpotence order of certain elements of A_2.

Chapter 14 is devoted to a proof of the 2-dominance theorem of Judith H. Silverman and Dagmar M. Meyer. This deep result states that a monomial f in $P^d(n)$ is hit if $\omega(f)$ is not greater than $\omega^{\min}(d)$ in the 2-dominance order. This strengthens the Peterson conjecture of Chapter 2 and the theorem of Singer from Chapter 6. One consequence is the Silverman–Singer criterion, which states that if g and h are homogeneous polynomials such that $\deg g < (2^k - 1)\mu(\deg h)$, where μ is the numerical function of Chapter 2, then $f = gh^{2^k}$ is hit.

In Chapter 15, we consider the Dickson algebra $D(n)$ of $GL(n)$-invariants in $P(n)$. Following Nguyen H. V. Hung and Tran Ngoc Nam, we show that all Dickson invariants of positive degree are hit in $P(n)$ when $n \geq 3$. There is a large class of similar problems: given a subgroup G of $GL(n)$, the subalgebra of G-invariant polynomials $P(n)^G$ is an A_2-module, and the 'relative' hit problem asks for the elements of $P(n)^G$ which are hit in $P(n)$. The corresponding 'absolute' hit problem asks for a minimal generating set for $P(n)^G$. We consider the absolute problem for the Weyl subgroup $G = W(n)$ of permutation matrices in $Gl(n)$ in Chapter 25.

In the chapters which follow, we shift attention to the representation theory of $GL(n)$ over $\mathbb{F}_2$. We begin in Chapter 16 by studying the flag module $FL(n)$, which is defined by the permutation action of $GL(n)$ on the right cosets of the Borel subgroup $B(n)$ of lower triangular matrices. This module is isomorphic to $Q^d(n)$ when the degree d is 'generic' in the sense of Chapter 8. The Bruhat decomposition $A = BWB'$ of a matrix A in $GL(n)$ is used to define certain subspaces of $FL(n)$ which we call 'Schubert cells'. Here $B, B' \in B(n)$ are lower triangular matrices and $W \in W(n)$ is a permutation matrix. We show that $FL(n)$ is the direct sum of 2^{n-1} submodules $FL_I(n)$, where $I \subseteq \{1, 2, \ldots, n-1\}$ is the set of dimensions of the subspaces in the 'partial' flags given by right cosets of parabolic subgroups of $GL(n)$.

The main aim of Chapter 17 is to construct a full set of 2^{n-1} irreducible $\mathbb{F}_2 GL(n)$-modules $L(\lambda)$. Following C. W. Curtis, $L(\lambda)$ is defined as the head of the summand $FL_I(n)$ of $FL(n)$, where λ is the 'column 2-regular' partition corresponding to I, i.e. $\lambda_i - \lambda_{i+1} = 1$ if $i \in I$, 0 if $i \notin I$. The summand of $FL(n)$ corresponding to complete flags is the Steinberg module $St(n)$. We use the Hecke algebra $H_0(n)$ of endomorphisms of $FL(n)$ which commute with the action of $GL(n)$, and follow the methods of R. W. Carter and G. Lusztig.

In Chapter 18, we review the background from modular representation theory that we use to study $P(n)$ and $DP(n)$ as $\mathbb{F}_2 GL(n)$-modules. We explain the role of idempotents in obtaining direct sum decompositions, and introduce the Steinberg idempotent $e(n) = \overline{B}(n)\overline{W}(n) \in \mathbb{F}_2 GL(n)$, the sum of all products BW, where $B \in B(n)$ and $W \in W(n)$. We study $e(n)$ and the conjugate idempotent $e'(n) = \overline{W}(n)\overline{B}(n)$ by means of an embedding of $H_0(n)$ in the group algebra $\mathbb{F}_2 GL(n)$ due to N. J. Kuhn. We also discuss Brauer characters and the representation ring $R_2(GL(n))$.

In Chapter 19 we use idempotents in $\mathbb{F}_2 GL(n)$ to split $P(n)$ as a direct sum of A_2-submodules $P(n, \lambda)$, each occurring $\dim L(\lambda)$ times. We discuss the problem of determining the number of factors isomorphic to $L(\lambda)$ in a composition series for $P^d(n)$. Following Ton That Tri, we use the Mui algebra of $B(n)$-invariants in $P(n)$ to determine the minimum degree d in which $P^d(n)$ has a submodule isomorphic to $L(\lambda)$.

As Weyl modules and their duals are central topics of modular representation theory, it is no surprise that they appear here also. As these modules are defined over infinite coefficient fields, we begin Chapter 20 by reviewing some results on modular representations of the algebraic group $\overline{G}(n)$ of nonsingular $n \times n$ matrices over $\overline{\mathbb{F}}_2$, the algebraic closure of $\mathbb{F}_2$. We then introduce the 'restricted' Weyl module $\Delta(\lambda, n)$ over $\mathbb{F}_2$ and its transpose dual $\nabla(\lambda, n)$, and show that if λ is column 2-regular and if the ordering on ω-sequences is suitably chosen, then $\Delta(\lambda, n) \cong K^\omega(n)$ and $\nabla(\lambda, n) \cong Q^\omega(n)$, where ω is the partition conjugate to λ. We use the theory of polynomial $\overline{G}(n)$-modules to determine the minimum degree d in which $L(\lambda)$ occurs as a composition factor of $P^d(n)$.

Chapter 21 gives a self-contained proof of an important result of J. F. Adams, J. Gunawardena and H. R. Miller. This states that every degree-preserving A_2-module map $P(m) \rightarrow P(n)$ is given by a sum of linear substitutions given by the action of $m \times n$ matrices over $\mathbb{F}_2$. It follows that the A_2-summands in a maximal splitting of $P(n)$ obtained using idempotents in $\mathbb{F}_2 M(n)$ rather than $\mathbb{F}_2 GL(n)$ are indecomposable. Hence such a splitting is a maximal direct sum decomposition of $P(n)$ as an A_2-module.

Chapter 22 is concerned with the A_2-summands of $P(n)$ corresponding to the Steinberg representation $St(n)$ of $GL(n)$. We discuss the 'internal' model $MP(n)$ of the A_2-module $P(St(n))$ defined by Stephen A. Mitchell and Stewart B. Priddy using admissible monomials of length n in A_2 itself. Although the hit problem for $P(n)$ can be split into a corresponding problem for $P(n, \lambda)$ for each λ, the Steinberg summand is the only case where this problem has been solved for all n, and we give minimal generating sets for the summands given by the idempotents $e(n)$ and $e'(n)$ of Chapter 18.

In Chapter 23, we identify the module $J^d(n)$ of $K^d(n)$ generated by the dual spikes in degrees $d = \sum_{i=1}^{n} (2^{a_i} - 1)$, where $a_1 > a_2 > \cdots > a_n$, in terms of the flag module $FL(n)$. We show that $Q^d(n) \cong FL(n)$ in 'generic' degrees d, and extend the method in Chapter 24 to obtain results of Tran Ngoc Nam on $J(n)$ relating cohit modules to partial flag modules. Following Nguyen Sum, we give counterexamples for $n \geq 5$ to Kameko's conjecture that $\dim Q^d(n) \leq \dim FL(n)$ for all d.

In Chapters 25 and 26 we discuss the hit problem for the action of A_2 on the algebra of symmetric polynomials $S(n)$, the invariants in $P(n)$ of the group $W(n)$ of permutations of the variables. More generally, we discuss the 'absolute' hit problem for any subgroup G of $W(n)$, and show that the Peterson conjecture, the Kameko map and Singer's minimal spike theorem have analogues for $P(n)^G$. We solve the symmetric hit problem for $n \leq 3$, using the dual problem to obtain the lower bound in the case $n = 3$. Following Singer, we introduce the 'bigraded Steenrod algebra' $\widetilde{A}_2$, which is obtained by omitting the relation $Sq^0 = 1$ in the definition of A_2, and apply $\widetilde{A}_2$ to the dual problem.

In Chapters 27 and 28 we consider the cyclic subgroup $C(n)$ of order $2^n - 1$ in $GL(n)$. This is obtained by regarding $P^1(n)$ as the underlying vector space of the Galois field $\mathbb{F}_{2^n}$. The action of $C(n)$ on $P(n)$ is diagonalized over $\mathbb{F}_{2^n}$ by a change of variables which 'twists' the action of A_2, in the sense that $Sq^1(t_i) = t_{i-1}$ in the new variables, which are indexed mod n. Following H. E. A. Campbell and P. S. Selick, we show that the polynomial algebra $\widetilde{P}(n) = \mathbb{F}_2[t_1, \ldots, t_n]$ splits as the direct sum of $2^n - 1$ A_2-modules $\widetilde{P}(n,j)$ corresponding to the 1-dimensional representations of $C(n)$. In particular, $\widetilde{P}(n,0)$ can be identified with the ring of $C(n)$-invariants of $P(n)$. In Chapter 28, we solve the dual cyclic hit problem for $n = 3$ by using the twisted analogue $\widetilde{J}(n)$ of the d-spike module $J(n)$.

In Chapters 29 and 30, we collect some results on the hit problem in the 4-variable case as further illustration of our methods. Nguyen Sum has extended the method introduced by Kameko to find a monomial basis for $Q^d(4)$ for all d. We include without proof some of the results of Sum, and also some results which we have verified only by computer using MAPLE. Thus there remain some challenging aspects of the hit problem even in the case $n = 4$.

The Steenrod algebra was originally defined for all primes p, but we have restricted attention to the case $p = 2$. All the problems have analogues for odd primes, but in general much less is known, and a number of difficulties arise in trying to extend our techniques to the odd prime case. The 2-variable hit problem for the action of A_p on the polynomial algebra $\mathbb{F}_p[x, y]$ has been solved by Martin D. Crossley, but little appears to be known even for the 3-variable case. In common with many authors on the Steenrod algebra, we have therefore confined ourselves to the prime 2.

There are several good textbooks on topology which include material on the Steenrod algebra and its applications, such as those by Brayton I. Gray [70] and by Robert E. Mosher and Martin C. Tangora [147], in addition to the classic Annals of Mathematics Study [196], based on lectures by Steenrod himself, and the Cartan seminars [33]. A treatment of the Steenrod algebra from an algebraic viewpoint, including Steenrod operations over an arbitrary finite field, is given in Larry Smith's book [190] on invariant theory. The book of Harvey R. Margolis [129] treats the general theory of modules over the Steenrod algebra. Still other approaches to the Steenrod algebra are possible. The survey article [233] treats Steenrod operations as linear differential operators with polynomial coefficients, and is a precursor for this book.

We sometimes introduce definitions and constructions for a small number of variables and extend them to the general case in later chapters. Although this can involve a certain amount of repetition, it has the advantage of leading to interesting results at an early stage by elementary methods. We hope that our approach will appeal to readers whose main interests are in algebra, especially in the modular representation theory of linear groups, or in the combinatorics related to symmetric polynomials and to the invariant theory of finite groups.

In order to avoid interruptions to the text, citations and background material are collected in the 'Remarks' sections at the end of each chapter. The occasional reference to topology may occur in these, but we have not tried to explain the topology. We have also omitted important topics such as the Singer transfer map and its applications to the homotopy groups of spheres through the Adams spectral sequence. These would require another volume, which we are not qualified to write. For similar reasons, we do not treat the theory of analytic functors and the category $\mathcal{U}/\mathcal{N}il$ of unstable A_2-modules modulo nilpotent objects due to Hans-Werner Henn, Jean Lannes and Lionel Schwartz. Finally, in a subject which crosses several disciplines, notation presents a problem because the traditional symbols of one area may be in conflict with those of another. A list of symbols for the main ingredients of our subject appears at the end of the book, together with an index of the main terms defined in the text.

We should like to offer our sincere thanks to the School of Mathematics of the University of Manchester for providing us with office space and computing facilities during work on this text. We should also like to thank several colleagues at Manchester for mathematical help and support, and in particular Peter Eccles, Nige Ray, Peter Rowley and Bob Sandling. Of colleagues farther afield, we should like to thank Nguyen H. V. Hung, Ali S. Janfada, Bill Singer and Larry Smith for their interest in our project. The first author would also like to thank Stephen R. Doty for teaching him something about modular representation theory. Finally we should like to thank Roger Astley and his colleagues at Cambridge University Press for their encouragement, support and patience. We set out with the modest aim of providing a beginning graduate student in topology and algebra with a basic primer on the Steenrod algebra, illustrated by our favourite application to a problem proposed by Frank Peterson, but the project has expanded substantially in scope over the past eight years.

1

Steenrod squares and the hit problem

1.0 Introduction

In this chapter we introduce our main subject, the algebra of polynomials in n variables over the field of two elements $\mathbb{F}_2$ under the left action of linear operations called **Steenrod squares** and the right action of $n \times n$ matrices. We denote this polynomial algebra by $P(n) = \mathbb{F}_2[x_1, \ldots, x_n]$, in variables x_i for $1 \le i \le n$. For small n it is convenient to use x, y, z for variables. The algebra $P(n)$ is graded by the vector spaces $P^d(n)$ of homogeneous polynomials of degree $d \ge 0$. In particular, the variables x_i have degree 1, and form a basis of the n-dimensional vector space $P^1(n)$.

In Section 1.1 the Steenrod squaring operations $Sq^k : P^d(n) \to P^{d+k}(n)$ are defined for $k \ge 0$, and their basic properties, such as the **Cartan formula**, are established. In Section 1.2 we explain how $P^d(n)$ is a right module over the monoid algebra $\mathbb{F}_2 M(n)$, where $M(n) = M(n, \mathbb{F}_2)$ is the multiplicative monoid of $n \times n$ matrices over $\mathbb{F}_2$. This right action commutes with the left action of the Steenrod squares. By restricting to non-singular matrices, $P^d(n)$ gives a modular representation of the general linear group $GL(n) \subset M(n)$, and the Steenrod squares are maps of $\mathbb{F}_2 GL(n)$-modules. Further properties of the Steenrod squares are developed in Section 1.3.

In Section 1.4 we introduce the **hit problem**. We call a polynomial 'hit' if it is a linear combination of elements in the images of positive Steenrod squares. The hit polynomials form a $\mathbb{F}_2 M(n)$-submodule $H(n)$ of $P(n)$. The corresponding quotient $Q(n) = P(n)/H(n)$ is the **cohit module**, and the hit problem is to determine $Q^d(n)$ for each n and degree d. Although this problem arose in algebraic topology, we treat it simply as a problem in algebra. We shall see later that the cohit modules are also of interest in group representation theory.

In the 1-variable case, the hit problem is a straightforward exercise in handling binomial coefficients mod 2. We give the solution in Section 1.4. For

1

all n, certain monomials called **spikes** cannot appear in the image of a positive Steenrod square. We introduce these in Section 1.5. In the rest of the chapter we focus on the 2-variable case. In Sections 1.6 and 1.7, we introduce the **Kameko** and **duplication** maps, which connect the cohit modules $Q^d(2)$ in different degrees d. Section 1.8 completes the solution of the hit problem for $P(2)$.

1.1 The total square Sq

Most of our calculations will be carried out with polynomials whose coefficients are in the field of two elements $\mathbb{F}_2 = \{0, 1\}$. Such a polynomial can be written as a sum without repetitions of monomials called its **terms**. That is to say, we do not write down monomials with coefficient 0, except in the case of the zero polynomial 0, which has no terms, and we do not write the coefficient 1. The joy in working with these polynomials is that there are no explicit coefficients or signs to worry about. Further, since we are working mod 2, $(x+y)^2 = x^2 + y^2$.

Definition 1.1.1 For $n \geq 1$, $P(n) = \mathbb{F}_2[x_1, \ldots, x_n]$ is the polynomial algebra in n variables $x_1, \ldots, x_n$ over the field $\mathbb{F}_2$. For $n = 0$, $P(0) = \mathbb{F}_2$.

As an algebra over $\mathbb{F}_2$, $P(n)$ is graded by integers $d \geq 0$. That is, $P(n)$ is the direct sum $\sum_{d \geq 0} P^d(n)$, where $P^d(n)$ is the vector space of homogeneous polynomials of degree d. We identify $P^0(n)$ with $\mathbb{F}_2$ for all $n \geq 0$. The monomials $x_1^{d_1} \cdots x_n^{d_n}$ such that $d_1 + \cdots + d_n = d$ and $d_i \geq 0$ for $1 \leq i \leq n$ form a basis for $P^d(n)$ as a vector space over $\mathbb{F}_2$. As usual, $x_i^0 = 1$ and $x_i^1 = x_i$. By considering a monomial as a string of symbols and separators (e.g. $xxx \cdot x \cdot xx$ for $x^3 y z^2$), we see that the dimension of $P^d(n)$ is

$$\dim P^d(n) = \binom{d+n-1}{n-1}. \tag{1.1}$$

Since $P(n)$ is freely generated as a commutative algebra by the variables x_i, an algebra map $\phi : P(n) \to P(n)$ is defined uniquely by assigning a value $\phi(x_i)$ to each variable x_i, $1 \leq i \leq n$. We shall always assume that $\phi(1) = 1$. We use a particular map of this kind to define the Steenrod squaring operations on $P(n)$.

Definition 1.1.2 The **total Steenrod square** $Sq : P(n) \to P(n)$ is the algebra map defined by

$$Sq(1) = 1, \quad Sq(x_i) = x_i + x_i^2, \ 1 \leq i \leq n.$$

The **Steenrod squares** $Sq^k : P^d(n) \to P^{d+k}(n)$ are the linear maps defined for $k, d \geq 0$ by restricting Sq to $P^d(n)$ and projecting on to $P^{d+k}(n)$. Thus $Sq = \sum_{k \geq 0} Sq^k$ is the formal sum of its graded parts.

Proposition 1.1.3 *For all $x \in P^1(n)$, $Sq(x) = x + x^2$. Thus $Sq^0(x) = x$, $Sq^1(x) = x^2$ and $Sq^k(x) = 0$ for all $k > 1$.*

Proof Let $x = \sum_{i=1}^n a_i x_i$, where $a_i \in \mathbb{F}_2$. Then $Sq(x) = \sum_{i=1}^n a_i Sq(x_i) = \sum_{i=1}^n a_i(x_i + x_i^2) = \sum_{i=1}^n a_i x_i + \sum_{i=1}^n (a_i x_i)^2 = x + x^2$. The second statement follows by equating graded parts. $\square$

The most important rule for calculating with Steenrod squares is as follows.

Proposition 1.1.4 (Cartan formula) *For polynomials $f, g \in P(n)$ and $k \geq 0$,*

$$Sq^k(fg) = \sum_{i+j=k} Sq^i(f) Sq^j(g).$$

Proof This follows from the multiplicative property $Sq(fg) = Sq(f)Sq(g)$ by equating terms of degree k. $\square$

Proposition 1.1.5 *Sq^0 is the identity map of $P(n)$.*

Proof Setting $k = 0$ in Proposition 1.1.4, Sq^0 is an algebra map of $P(n)$. Since $Sq^0(1) = 1$ and $Sq^0(x_i) = x_i$ for $1 \leq i \leq n$, $Sq^0(f) = f$ for all $f \in P(n)$. $\square$

Definition 1.1.6 A **Steenrod operation** is a linear map $\theta : P(n) \to P(n)$ which can be obtained from the operations Sq^k by addition and composition. Thus θ is a finite sum of operations of the form $Sq^{k_1} Sq^{k_2} \cdots Sq^{k_s}$.

In principle, any Steenrod operation can be evaluated on a polynomial by means of Propositions 1.1.3 and 1.1.4.

Example 1.1.7 In $P(2) = \mathbb{F}_2[x, y]$ we have

$$Sq^1(xy) = Sq^1(x)Sq^0(y) + Sq^0(x)Sq^1(y) = x^2 y + xy^2.$$

The next two results show how to evaluate a Steenrod square on a monomial. All binomial coefficients which appear in formulae such as the following are understood to be reduced mod 2.

Proposition 1.1.8 *For all $x \in P^1(n)$,*

$$Sq^k(x^d) = \binom{d}{k} x^{d+k}.$$

In particular, $Sq^1(x^d) = x^{d+1}$ if d is odd and $Sq^1(x^d) = 0$ if d is even.

Proof By the multiplicative property of Sq, we have

$$\mathrm{Sq}(x^d) = (\mathrm{Sq}(x))^d = (x + x^2)^d = x^d(1+x)^d = \sum_{k=0}^{d} \binom{d}{k} x^{d+k}.$$

The result follows by equating terms of degree $d + k$. □

Proposition 1.1.9 *Let $f = x_1^{d_1} \cdots x_n^{d_n}$ be a monomial in* P(n). *Then*

$$Sq^k(f) = \sum_{k_1 + \cdots + k_n = k} Sq^{k_1}(x_1^{d_1}) \cdots Sq^{k_n}(x_n^{d_n}).$$

Proof This follows by induction on n using the Cartan formula 1.1.4. □

The next result explains why Sq^k is called a squaring operation.

Proposition 1.1.10 *For $f \in \mathrm{P}^d(n)$, $Sq^k(f) = 0$ for $k > d$ and $Sq^d(f) = f^2$.*

Proof Since Sq^k is linear and $(f + g)^2 = f^2 + g^2$, we may assume that f is a monomial of degree d. We use induction on d. The base case $d = 0$ holds since $\mathrm{Sq}(1) = 1$. For $d > 0$, let $f = xg$, where x is one of the variables x_i and g is a monomial of degree $d - 1$. By the Cartan formula, $Sq^k(f) = xSq^k(g) + x^2 Sq^{k-1}(g)$ for $k > 0$. If $k > d$, then $Sq^k(g) = 0$ and $Sq^{k-1}(g) = 0$ by the inductive hypothesis, so $Sq^k(f) = 0$. If $k = d$, then $Sq^k(g) = 0$ and $Sq^{k-1}(g) = g^2$ by the inductive hypothesis, so $Sq^k(f) = x^2 g^2 = f^2$. This completes the induction. □

Proposition 1.1.11 *If $f \in$ P(n) is a monomial and $k \geq 0$, then every term of $Sq^k(f)$ involves exactly the same variables as f does.*

Proof The monomial $f = x_1^{d_1} \cdots x_n^{d_n}$ involves x_i if and only if $d_i > 0$. If $d_i > 0$, then $k + d_i > 0$ for all k. If $d_i = 0$, then $\mathrm{Sq}(x_i^{d_i}) = \mathrm{Sq}(1) = 1$. The result follows from Proposition 1.1.9. □

1.2 The action of matrices on P(n)

As well as algebra operations in P(n), we can substitute polynomials for the variables x_i in a polynomial $f \in$ P(n). In this section, we show that the Steenrod operations on P(n) commute with linear substitutions of the variables. For $n \geq 1$, we write GL(n) = $GL(n, \mathbb{F}_2)$ for the general linear group of non-singular $n \times n$ matrices over $\mathbb{F}_2$, and $\mathbb{F}_2 GL(n)$ for its group algebra over $\mathbb{F}_2$. We also write M(n) = $M(n, \mathbb{F}_2)$ for the multiplicative monoid of all $n \times n$ matrices over $\mathbb{F}_2$, and $\mathbb{F}_2 M(n)$ for its monoid algebra.

Definition 1.2.1 For $A = (a_{i,j}) \in M(n)$ and $1 \leq i \leq n$, let

$$x_i \cdot A = \sum_{j=1}^{n} a_{i,j} x_j,$$

and extend this action of A to all polynomials $f \in P(n)$ by substitution, so that $f \cdot A$ is the polynomial $f(x_1 \cdot A, \ldots, x_n \cdot A)$. If $f \in P^0(n)$ is constant, then $f \cdot A = f$.

Proposition 1.2.2 *For* $d \geq 0$, $P^d(n)$ *is a right* $\mathbb{F}_2 M(n)$-*module, and so also a right* $\mathbb{F}_2 GL(n)$-*module. For* $A \in M(n)$, *the map* $f \mapsto f \cdot A$ *is an algebra map of* P(n).

Proof Let $f, g \in P(n)$ and $A, B \in M(n)$, and let $I_n \in M(n)$ be the identity matrix. We have $f \cdot I_n = f$, and we check that $f \cdot (AB) = (f \cdot A) \cdot B$. Since linear substitutions preserve degree, this proves the first statement. For the second statement, we check that $1 \cdot A = 1$, $(f + g) \cdot A = f \cdot A + g \cdot A$ and $(fg) \cdot A = (f \cdot A)(g \cdot A)$. $\square$

Example 1.2.3 The group $GL(2)$ is generated by S and U, where

$$S = \begin{pmatrix} 0 & 1 \\ 1 & 0 \end{pmatrix}, \quad U = \begin{pmatrix} 1 & 1 \\ 0 & 1 \end{pmatrix},$$

which act on the variables x, y by $x \cdot S = y$, $y \cdot S = x$ and $x \cdot U = x + y$, $y \cdot U = y$. Evaluating in degree 2, we obtain $x^2 \cdot S = y^2$, $xy \cdot S = xy$, $y^2 \cdot S = x^2$ and $x^2 \cdot U = x^2 + y^2$, $xy \cdot U = xy + y^2$, $y^2 \cdot U = y^2$.

By choosing a basis for $P^d(n)$, we obtain a matrix representation ρ of $GL(n)$, i.e. a homomorphism $\rho : GL(n) \rightarrow GL(m)$, where $m = \dim P^d(n)$. Thus with respect to the basis $\{x^2, xy, y^2\}$ of $P^2(2)$, S and U are represented by

$$\rho(S) = \begin{pmatrix} 0 & 0 & 1 \\ 0 & 1 & 0 \\ 1 & 0 & 0 \end{pmatrix}, \quad \rho(U) = \begin{pmatrix} 1 & 0 & 1 \\ 0 & 1 & 1 \\ 0 & 0 & 1 \end{pmatrix}.$$

The module $P^1(n)$ gives the **defining representation** of $GL(n)$ or $M(n)$. It is equivalent to the representation $V(n)$ given by matrix multiplication $v \cdot A$, where v is a row vector regarded as a element of $\mathbb{F}_2^n$. The defining representation is irreducible both as a $\mathbb{F}_2 GL(n)$-module and as a $\mathbb{F}_2 M(n)$-module, meaning that it has no submodules other than itself and 0. In general, the module structure of $P^d(n)$ is complicated, as the following observation shows.

Proposition 1.2.4 *For* $n \geq 1$ *and* $d \geq 0$, *the squaring map* $f \mapsto f^2$ *embeds* $P^d(n)$ *as a* $\mathbb{F}_2 M(n)$-*submodule in* $P^{2d}(n)$.

Proof Since $(f+g)^2 = f^2 + g^2$ in $P(n)$, the squaring map is linear, and it is clearly injective. It is also an $\mathbb{F}_2 M(n)$-module map, since $f^2 \cdot A = (f \cdot A)^2$. $\square$

We shall usually consider $P^d(n)$ only as a $\mathbb{F}_2 GL(n)$-module. The difference between this and its (even more complicated) $\mathbb{F}_2 M(n)$-module structure can be seen as follows. There are two non-equivalent 1-dimensional representations of $M(n)$, the 'trivial' representation $I(n)$, where all matrices act as the identity, and the determinant representation det. Singular matrices act as the identity map for $I(n)$, and as the zero map for det. The $\mathbb{F}_2 M(1)$-module $P^d(1)$ is $I(1)$ if $d = 0$, det if $d > 0$. In $P(2)$, $GL(2)$ permutes x, y and $x+y$. Hence $f = xy(x+y)$ is an invariant, and generates a $\mathbb{F}_2 GL(2)$-submodule I of $P^3(2)$ which is isomorphic to $I(2)$. Since a singular matrix maps f to 0, I is isomorphic to det as a $\mathbb{F}_2 M(2)$-module. This difference affects higher-dimensional modules as well. For example, the $\mathbb{F}_2 GL(2)$-submodule of $P^4(2)$ with basis $\{xf, yf\}$ is isomorphic to $P^1(2)$. As every singular matrix acts as zero on xf and yf, while only the zero matrix acts as zero on $P^1(2)$, these modules are not isomorphic as $\mathbb{F}_2 M(2)$-modules.

The next result is central to our whole subject.

Proposition 1.2.5 *The right action of $\mathbb{F}_2 M(n)$ on $P(n)$ commutes with the left action of the Steenrod squares Sq^k. That is to say*

(i) $Sq(f) \cdot A = Sq(f \cdot A)$, *for $f \in P(n)$ and $A \in M(n)$,*

(ii) $Sq^k : P^d(n) \to P^{k+d}(n)$ *is a map of $\mathbb{F}_2 M(n)$-modules.*

Proof Since Sq and $f \mapsto f \cdot A$ are algebra maps of $P(n)$, we need only check (i) when f is one of the variables x_i. Let $A = (a_{i,j})$. Then $Sq(x_i) \cdot A = (x_i + x_i^2) \cdot A = x_i \cdot A + (x_i \cdot A)^2$, while $Sq(x_i \cdot A) = Sq(\sum_{j=1}^n a_{i,j} x_j) = \sum_{j=1}^n a_{i,j} Sq(x_j) = \sum_{j=1}^n a_{i,j}(x_j + x_j^2)$. But $\sum_{j=1}^n a_{i,j} x_j^2 = (\sum_{j=1}^n a_{i,j} x_j)^2 = (x_i \cdot A)^2$, so $Sq(x_i) \cdot A = Sq(x_i \cdot A)$. Statement (ii) follows by taking the graded parts of Sq. $\square$

We may also consider the set of rectangular $m \times n$ matrices $M(m,n)$ over $\mathbb{F}_2$, which define linear maps $P(m) \to P(n)$ as in Definition 1.2.1. The proof of (i) shows more generally that $Sq(f) \cdot A = Sq(f \cdot A)$, for $f \in P(m)$ and $A \in M(m,n)$. Although we shall mainly consider $P(n)$ as a representation of $GL(n)$ rather than $M(n)$, it is important to be aware that Steenrod operations commute with specializations of the variables given by singular matrices. For example, since the action of $\left(\begin{smallmatrix} 0 & 0 \\ 0 & 1 \end{smallmatrix}\right)$ sets $x = 0$, and that of $\left(\begin{smallmatrix} 1 & 0 \\ 1 & 0 \end{smallmatrix}\right)$ sets $y = x$, $Sq^1(xy^2 + y^3) = x^2 y^2 + y^4$ implies $Sq^1(x^3) = x^4$, by first setting $x = 0$ and then $y = x$.

It is convenient to fix notation for some matrices in $GL(n)$.

Definition 1.2.6 The **switch matrix** obtained from the identity matrix I_n by exchanging rows i and j will be denoted by $S_{i,j}$, or by S_i in the case $j = i+1$.

The **transvection** obtained from I_n by replacing the entry 0 in position (i,j) by 1 will be denoted by $T_{i,j}$. We also write $U_i = T_{i,i+1}$ and $L_i = T_{i+1,i}$.

Thus S in Example 1.2.3 is also denoted by $S_{1,2}$ or S_1, and U by $T_{1,2}$ or U_1.

1.3 Some properties of Sq^k

The action of Steenrod operations on polynomials is reproduced on squares of polynomials by doubling exponents of the operations.

Proposition 1.3.1 *For $f \in P(n)$ and $k \geq 0$,*

$$Sq^k(f^2) = \begin{cases} (Sq^j(f))^2, & \text{if } k = 2j \text{ is even,} \\ 0, & \text{if } k \text{ is odd.} \end{cases}$$

Proof By the Cartan formula, $Sq^k(f^2) = \sum_{i+j=k} Sq^i(f)Sq^j(f)$. By exchanging i and j, the terms in the sum with $i \neq j$ cancel in pairs. $\square$

Proposition 1.3.2 *For $f \in P(n)$ and $s, k \geq 0$,*

$$Sq^k(f^{2^s}) = \begin{cases} (Sq^j(f))^{2^s}, & \text{if } k = 2^s j, \\ 0, & \text{otherwise.} \end{cases}$$

In particular, $Sq^k(x^{2^s})$ is x^{2^s} if $k = 0$, $x^{2^{s+1}}$ if $k = 2^s$, and is 0 otherwise.

Proof This follows from Proposition 1.3.1 by induction on s. $\square$

Proposition 1.3.3 *For $f, g \in P(n)$ and $s, k \geq 0$,*

(i) $Sq^k(gf^{2^s}) = \displaystyle\sum_{i+2^s j=k} Sq^i(g)(Sq^j(f))^{2^s}$, (ii) $Sq^k(gf^{2^s}) = Sq^k(g)f^{2^s}$ *if $k < 2^s$.*

Proof (i) follows from the Cartan formula and Proposition 1.3.2, and (ii) is immediate from (i) and Proposition 1.1.10. $\square$

Proposition 1.3.4 *For $f \in P(n)$ and $k \geq 0$,*

(i) $Sq^1 Sq^{2k}(f) = Sq^{2k+1}(f)$, (ii) $Sq^1 Sq^{2k+1}(f) = 0$.

Proof By linearity, we may assume that f is a monomial of degree d. We use induction on d. The case $d = 0$ is true since $Sq(1) = 1$. For $d > 0$, let $f = xg$ where $x = x_i$ is one of the variables. By the Cartan formula, $Sq^j(f) = xSq^j(g) + x^2 Sq^{j-1}(g)$ and $Sq^1 Sq^j(f) = xSq^1 Sq^j(g) + x^2 Sq^j(g) + x^2 Sq^1 Sq^{j-1}(g)$. The inductive hypothesis for (i) gives $Sq^1 Sq^{2k}(g) = Sq^{2k+1}(g)$, and the inductive hypothesis for (ii) gives $Sq^1Sq^{2k-1}(g) = 0$ and $Sq^1Sq^{2k+1}(g) = 0$. Hence $Sq^1Sq^{2k}(f) = xSq^{2k+1}(g) + x^2Sq^{2k}(g) = Sq^{2k+1}(f)$ and $Sq^1Sq^{2k+1}(f) = x^2Sq^{2k+1}(g) + x^2Sq^{2k+1}(g) = 0$. This completes the induction. □

Proposition 1.3.5 *Given* $f \in P^d(n)$ *for* $d \geq 1$, $Sq^1(f) = 0$ *if and only if* $f = Sq^1(g)$ *for some* $g \in P^{d-1}(n)$.

Proof Since $Sq^1Sq^1(g) = 0$ for all polynomials g, the 'if' part follows from Proposition 1.3.4. To prove the 'only if' part, we use induction on n. The case $n = 1$ is true by Proposition 1.1.8. Assume the result for $P(n-1)$, and let $f \in P(n)$ with $Sq^1(f) = 0$. Then $f = xf_1 + f_2$, where $x = x_n$, $f_1 \in P^{d-1}(n)$ and $f_2 \in P^d(n-1)$. Hence $0 = Sq^1(f) = Sq^1(xf_1) + Sq^1(f_2) = x^2f_1 + xSq^1(f_1) + Sq^1(f_2)$. Since f_2 is independent of x, setting $x = 0$ and using Proposition 1.1.11 we have $Sq^1(f_2) = 0$. Hence $f_2 = Sq^1(g_2)$ for some $g_2 \in P^{d-1}(n-1)$ by the inductive assumption. Then $x^2f_1 + xSq^1(f_1) = 0$ and so $Sq^1(f_1) = xf_1$. Thus $f = xf_1 + f_2 = Sq^1(f_1 + g_2)$. This completes the induction. □

Since $Sq^1(fg) = Sq^1(f)g + fSq^1(g)$, the operation Sq^1 on $P(n)$ is a differential. In Section 1.9 we explain how the Steenrod squares may be interpreted in terms of differential operators with polynomial coefficients. Here we show that the action of a Steenrod square commutes with partial differentiation.

Proposition 1.3.6 *Let* $f \in P^d(n)$, $x = x_i$, $1 \leq i \leq n$, *and* $k \geq 0$. *Then*

$$Sq^k\left(\frac{\partial f}{\partial x}\right) = \frac{\partial}{\partial x}(Sq^k(f)).$$

Proof By linearity, we may assume that f is a monomial of degree d. We use induction on d. If f does not involve x, then nor does $Sq^k(f)$ by Proposition 1.1.11. Hence we may assume that $f = xg$, where g is a monomial of degree $d - 1$. By the Cartan formula, $Sq^k(\partial(xg)/\partial x) = Sq^k(g + x\partial g/\partial x) = Sq^k(g) + xSq^k(\partial g/\partial x) + x^2Sq^{k-1}(\partial g/\partial x)$ for $k > 0$. On the other hand, $\partial/\partial x(Sq^k(xg)) = \partial/\partial x(xSq^k(g) + x^2Sq^{k-1}(g)) = Sq^k(g) + x\partial/\partial x(Sq^k(g)) + x^2\partial/\partial x(Sq^{k-1}(g))$. By the induction hypothesis, Sq^k and Sq^{k-1} commute with the operator $\partial/\partial x$ on g, and so $Sq^k(\partial f/\partial x) = \partial/\partial x(Sq^k(f))$. This completes the induction. □

1.4 The hit problem

The Steenrod squaring operations allow us to express many elements of $P^d(n)$ in terms of polynomials of lower degree. For example, $x^5 = Sq^2(x^3)$, $x^4y = Sq^2(x^2y)$ and $x^3y^2 = Sq^1(x^3y) + Sq^2(x^2y)$. By exchanging the variables x and y

and taking sums, it is clear that every element of $P^5(2)$ can be written in this form.

Definition 1.4.1 A polynomial $f \in P^d(n)$ is **hit** if it satisfies a **hit equation**

$$f = \sum_{i>0} Sq^i(f_i), \; f_i \in P^{d-i}(n)$$

By Proposition 1.1.10, $Sq^i(f_i) = 0$ if $i > d/2$, so the sum has $\leq d/2$ terms.

Proposition 1.4.2 *For all $f \in P^d(n)$, where $d > 0$, f^2 is hit.*

Proof By Proposition 1.1.10, $Sq^d(f) = f^2$. If $d > 0$, this is a hit equation. $\quad\square$

The hit polynomials in $P^d(n)$, together with the zero polynomial, form a vector subspace $H^d(n)$. For $f, g \in P^d(n)$ we write $f \sim g$ if $f - g$ is hit. (Since we are working mod 2, $f - g = f + g$.) This is an equivalence relation, the equivalence classes being cosets of $H^d(n)$ in $P^d(n)$.

Proposition 1.4.3 *For $n \geq 1$ and $d \geq 0$, $H^d(n)$ is an $\mathbb{F}_2 M(n)$-submodule of $P^d(n)$.*

Proof Let $f = \sum_{i>0} Sq^i(f_i)$ be a hit equation for f, and let $A \in M(n)$. Then $f \cdot A = \sum_{i>0} (Sq^i(f_i) \cdot A) = \sum_{i>0} Sq^i(f_i \cdot A)$ by Proposition 1.2.5. This is a hit equation for $f \cdot A$. $\quad\square$

When d is even, Proposition 1.4.2 shows that $H^d(n)$ contains the submodule whose elements are squares (Proposition 1.2.4). The next result gives a related submodule of $H^d(n)$ when d is odd.

Proposition 1.4.4 *Let $g \in P^d(n)$ be hit and let $x \in P^1(n)$. Then xg^2 is hit.*

Proof Let $g = \sum_{i>0} Sq^i(g_i)$ be a hit equation for g. By Proposition 1.3.2, $Sq^{2i}(xg_i^2) = x(Sq^i(g_i))^2$. Hence $xg^2 = x(\sum_{i>0} Sq^i(g_i))^2 = \sum_{i>0} x(Sq^i(g_i)^2) = \sum_{i>0} Sq^{2i}(xg_i^2)$. $\quad\square$

Definition 1.4.5 The **hit problem** is to determine the **cohit module**

$$Q^d(n) = P^d(n)/H^d(n), \; n \geq 1, \; d \geq 0.$$

The hit problem can be put at different levels. The most basic question is to ask whether or not all homogeneous polynomials of degree d in $P(n)$ are hit. At the next level, we can ask for the dimension of $Q^d(n)$ as a vector space over $\mathbb{F}_2$. Further, we can ask for a basis of $Q^d(n)$. For example, as a quotient space of $P^d(n)$, a basis can be chosen from the equivalence classes of monomials. Finally, we can seek information about $Q^d(n)$ as a $\mathbb{F}_2 GL(n)$-module, or as a $\mathbb{F}_2 M(n)$-module.

Note that we shall normally write (and refer to) elements of $Q^d(n)$ as polynomials, although strictly speaking they are equivalence classes of polynomials.

Example 1.4.6 As $Sq^2(x) = 0$ and $Sq^1(x^2) = 0$, x^3 is not hit in $P(1) = \mathbb{F}_2[x]$. Hence $Q^3(1)$ is 1-dimensional, generated by x^3. In $P(2) = \mathbb{F}_2[x,y]$ we have $Sq^1(x) = x^2$ and $Sq^1(y) = y^2$. Hence x^2 and y^2 are hit, but xy is not hit and so $Q^2(2)$ is 1-dimensional, generated by xy. The discussion at the start of this section shows that $Q^5(2) = 0$.

As far as the vector space structure of $Q(n)$ is concerned, the hit problem can be decomposed into smaller problems. Let $Z[n]$ denote the set $\{1,\ldots,n\}$ of the first n positive integers. For $Y \subseteq Z[n]$, let $P(Y)$ be the subspace of $P(n)$ spanned by monomials which are divisible by x_i if and only if $i \in Y$. If $Y = \emptyset$, then $P(Y) = P(0) = \mathbb{F}_2$. By Proposition 1.1.11, $P(Y)$ is preserved by the action of the Steenrod squares, and so we have a corresponding vector space $H(Y)$ of hit polynomials and quotient space $Q(Y) = P(Y)/H(Y)$. Then there are direct sum decompositions $P(n) = \bigoplus_Y P(Y)$, $H(n) = \bigoplus_Y H(Y)$ and $Q(n) = \bigoplus_Y Q(Y)$ as vector spaces over $\mathbb{F}_2$, where the sums are over all 2^n subsets Y of $Z[n]$. If $|Y| = |Y'|$ then $Q(Y) \cong Q(Y')$, by permuting variables and using Proposition 1.2.5. Since there are $\binom{n}{k}$ k-element subsets Y, this gives the following result.

Proposition 1.4.7 $\dim Q^d(n) = \sum_{k=1}^n \binom{n}{k} \dim Q^d(Z[k])$ *for $d > 0$.* $\qquad \square$

Example 1.4.8 We compute $\dim Q^3(3)$. Since x^3 is not hit in $P(1)$ by Example 1.4.6, $\dim Q^3(Z[1]) = 1$. By Example 1.1.7, $xy^2 \sim x^2y$, but clearly these monomials are not hit, so $\dim Q^3(Z[2]) = 1$. Finally $\dim Q^3(Z[3]) = 1$, since the only monomial in $P^3(Z[3])$ is xyz. Hence $\dim Q^3(3) = 7$.

In general it is not easy to determine whether a monomial is hit. However, the 1-variable case can be solved using Proposition 1.1.8. For this, we need to evaluate binomial coefficients $\binom{a+b}{a}$ mod 2. This is done by comparing the binary expansions of a and b.

Definition 1.4.9 For $d > 0$, $\mathrm{bin}(d) = \{2^{d_1},\ldots,2^{d_r}\}$, where $d = 2^{d_1} + \cdots + 2^{d_r}$ with distinct terms, and $\mathrm{bin}(0) = \emptyset$. For $d \geq 0$, $\alpha(d) = |\mathrm{bin}(d)|$.

Example 1.4.10 $\mathrm{bin}(11) = \{1,2,8\}$, $\alpha(11) = 3$; $\mathrm{bin}(12) = \{4,8\}$, $\alpha(12) = 2$; $\mathrm{bin}(27) = \{1,2,8,16\}$, $\alpha(27) = 4$.

Thus the elements of the set $\mathrm{bin}(d)$ correspond to the 1s in the binary expansion of d, and $\alpha(d)$ is the total number of 1s. By considering binary addition of a and b, it is clear that $\alpha(a+b) \leq \alpha(a) + \alpha(b)$, with equality if and only if no 'carries' occur. The next result shows that this happens precisely when the binomial coefficient $\binom{a+b}{a}$ is odd.

Proposition 1.4.11 *For $a, b \geq 0$, the conditions* (i) $\binom{a+b}{a}$ *is odd,* (ii) $bin(a) \subseteq bin(a+b)$, (iii) $bin(a) \cap bin(b) = \emptyset$ *and* (iv) $\alpha(a+b) = \alpha(a) + \alpha(b)$ *are equivalent.*

Proof Working in $\mathbb{F}_2[x]$, the term x^a appears in the expansion of

$$(1+x)^d = \prod_{i=1}^{r}(1+x)^{2^{d_i}} = \prod_{i=1}^{r}(1+x^{2^{d_i}})$$

if and only if all summands in the binary expansion of a are also summands in the binary expansion of d. Setting $d = a + b$, this proves that (ii) is equivalent to (i). Consideration of binary addition of a and b as above shows that (iii) and (iv) are equivalent to (ii). □

We can now solve the case $n = 1$ of the hit problem.

Theorem 1.4.12 *In $P(1) = \mathbb{F}_2[x]$, x^d is hit if and only if d is not of the form $2^s - 1$, where $s \geq 0$. Hence $Q^d(1) \cong \mathbb{F}_2$ if $d = 2^s - 1$, and $Q^d(1) = 0$ otherwise.*

Proof Clearly, x^d is hit if and only if $Sq^b(x^a) = x^d$, where $a + b = d$ and $a, b > 0$. By Proposition 1.1.8, $Sq^b(x^a) = \binom{a}{b}x^{a+b}$. For $d = 2^s - 1$, let $d = a + b$ be any decomposition of d, where $a, b > 0$. Then $bin(a)$ is a subset of $bin(d) = \{1, 2, \ldots, 2^{s-1}\}$ and $bin(b)$ is the complementary subset. By Proposition 1.4.11, $\binom{a}{b} = 0$ if $b > 0$. Hence x^d is not hit. On the other hand, if d is not of the form $2^s - 1$, then, for some i, 2^{i+1} is a term in the binary expansion of d, but 2^i is not. Let $b = 2^i$ and $a = d - 2^i$. Then $2^i \in bin(a)$, and hence $\binom{a}{b} = 1$ by Proposition 1.4.11. Hence x^d is hit. □

Proposition 1.4.11 can be extended to multinomial coefficients mod 2.

Proposition 1.4.13 *Let $a = a_1 + \cdots + a_s$ where $a_1, \ldots, a_s \geq 0$. Then the multinomial coefficient $\binom{a}{a_1, \ldots, a_s}$ is odd if and only if $bin(a)$ is the disjoint union of the sets $bin(a_i)$, $1 \leq i \leq s$.*

Proof We have $(x_1 + \cdots + x_s)^a = \prod_k (x_1 + \cdots + x_s)^{2^k} = \prod_k (x_1^{2^k} + \cdots + x_s^{2^k})$ mod 2, where the product is taken over all k such that $2^k \in bin(a)$. The monomial $x_1^{a_1} \cdots x_s^{a_s}$ appears in the product if and only if no two integers a_i and a_j have the same power of 2 in their binary expansions. □

1.5 Spikes

Theorem 1.4.12 has implications which reach beyond the case $n = 1$. In this section we apply it in two ways. The first application uses the fact that the hit polynomials form a $\mathbb{F}_2 \mathrm{M}(n)$-module.

Proposition 1.5.1 *For $j \geq 0$, no monomial of degree $2^j - 1$ in $\mathrm{P}(n)$ is hit.*

Proof We specialize all the variables in the monomial f to x_1 by using the matrix $A = (a_{i,j})$, where $a_{i,j} = 1$ if $j = 1$ and $a_{i,j} = 0$ if $j > 1$. Then $f \cdot A = x_1^{2^j - 1}$. By Theorem 1.4.12, $x_1^{2^j - 1}$ is not hit, so the result follows from Proposition 1.4.3.
$\square$

The second application of Theorem 1.4.12 gives a special class of monomials which are of fundamental importance in the hit problem. Not only are they not hit, but they cannot be terms of a polynomial in the image of Sq^k for $k > 0$.

Definition 1.5.2 A monomial in $\mathrm{P}(n)$ is a **spike** if all its exponents are integers of the form $2^j - 1$, where $j \geq 0$.

Proposition 1.5.3 *For $k > 0$ and $f \in \mathrm{P}(n)$, a spike cannot be a term of $Sq^k(f)$. Every monomial basis of $\mathrm{Q}(n)$ contains all the spikes in $\mathrm{P}(n)$.*

Proof Let $x_1^{2^{j_1} - 1} \cdots x_n^{2^{j_n} - 1}$ be a spike. By Proposition 1.4.12, $x_i^{2^{j_i} - 1}$ cannot be the ith factor $Sq^{k_i}(x_i^{d_i})$ of any summand in the expansion of $Sq^k(x_1^{d_1} \cdots x_n^{d_n})$ by the Cartan formula 1.1.9 unless $k_i = 0$. Hence $k_i = 0$ for all i, and so $k = 0$. Hence a spike cannot appear as a term in a hit equation. Consequently the spikes in degree d are linearly independent mod $\mathrm{H}^d(n)$, and so any monomial basis of $\mathrm{Q}^d(n)$ must contain them all.
$\square$

The next example shows that, even when $n = 2$, the spikes are not sufficient to give a basis for $\mathrm{Q}(n)$.

Example 1.5.4 As in Example 1.4.8, the monomials $x^2 y$ and xy^2 are not hit, and one of them, but not both, must be included in a monomial basis for $\mathrm{Q}^3(2)$. Hence $\dim \mathrm{Q}^d(2) = 2, 1, 3$ for $d = 1, 2, 3$ respectively, with generating sets $\{x, y\}$, $\{xy\}$ and $\{x^3, x^2 y, y^3\}$.

1.6 The Kameko maps for $\mathrm{P}(2)$

A homogeneous polynomial $f \in \mathbb{F}_2[x, y]$ can be written uniquely as $f = xyg^2 + h^2$ if $\deg f$ is even, and as $f = xg^2 + yh^2$ if $\deg f$ is odd. Thus the equations

$xyg_1^2 + h_1^2 = xyg_2^2 + h_2^2$ or $xg_1^2 + yh_1^2 = xg_2^2 + yh_2^2$ imply that $g_1 = g_2$ and $h_1 = h_2$, by comparing coefficients of xy and 1, or of x and y. We begin with the even degree case.

Proposition 1.6.1 *For* $d \geq 1$, $xyg^2 \in$ P(2) *is hit if and only if g is hit.*

Proof Let $f = xyg^2$ and suppose that f is hit. Using Proposition 1.3.4 to collect the terms in a hit equation which involve odd Steenrod squares, we can write $f = Sq^1(f_0) + \sum_{i>0} Sq^{2i}(f_i)$ where f_0 has odd degree and f_i has even degree for $i > 0$. Then $f_0 = xg_0^2 + yh_0^2$ and $f_i = xyg_i^2 + h_i^2$. Hence by Proposition 1.3.1

$$f = xyg^2 = x^2g_0^2 + y^2h_0^2 + \sum_{i>0}\left(xy(Sq^i(g_i))^2 + x^2y^2(Sq^{i-1}(g_i))^2 + (Sq^i(h_i))^2\right).$$

Comparing coefficients of xy gives $g = \sum_{i>0} Sq^i(g_i)$. Hence g is hit.

Conversely, if g is hit, let $g = \sum_{i>0} Sq^i(g_i)$. Then $f = \sum_{i>0} xySq^{2i}(g_i^2) = Sq^{2i}(\sum_{i>0} xyg_i^2) + \sum_{i>0} x^2y^2(Sq^{i-1}(g_i))^2$. Since the second term is a square, it is hit by Proposition 1.4.2. Hence f is hit.   □

Definition 1.6.2 For $d \geq 0$, the **down Kameko map** $\kappa : P^{2d+2}(2) \to P^d(2)$ is the surjective linear map defined on monomials by $\kappa(f) = g$ if $f = xyg^2$ and $\kappa(f) = 0$ otherwise. The **up Kameko map** $\upsilon : P^d(2) \to P^{2d+2}(2)$ is the injective linear map defined on monomials by $\upsilon(g) = xyg^2$.

Thus the cokernel of υ is the subspace of polynomials f of the form $f = h^2$, which are hit. Hence Proposition 1.6.1 gives the following result.

Proposition 1.6.3 *For* $d \geq 0$, *the down Kameko map induces a linear isomorphism* $\kappa : Q^{2d+2}(2) \to Q^d(2)$, *with inverse* υ.   □

We next turn to the odd degree case.

Proposition 1.6.4 *For d odd,* $Q^d(2) = 0$ *unless* $d = 2^j - 1$ *for some* $j \geq 1$.

Proof Let f be a monomial in $P^d(2)$. We begin with the case $d = 4k + 1$, where $k > 0$. By exchanging x and y if necessary, we may assume that $f = xg^2$ for some monomial $g \in P^{2k}(2)$. If $g = h^2$, then g is hit by Proposition 1.4.2, and so f is hit by Proposition 1.4.4. If $g = xyh^2$, then $Sq^1(x^3yh^4) = x^4yh^4 + f$. Since x^4yh^4 is hit by Proposition 1.4.4, it follows that f is hit.

Next we consider the case $d = 8k + 3$ where $k > 0$. As before, we may assume that $f = xg^2$ for some monomial $g \in P^{4k+1}(2)$. Then g is hit by the preceding case. By Proposition 1.4.4, it follows that f is hit. Continuing in this way, the result is true by induction on $r \geq 2$ for integers $d = 2^r k + (2^{r-1} - 1)$. This gives all odd numbers not of the form $2^j - 1$.   □

1.7 The duplication map for $P(2)$

The next result gives a necessary and sufficient condition for a polynomial of odd degree in $P(2)$ to be hit, corresponding to Proposition 1.6.1 for even degrees. Note that the converse of Proposition 1.4.4 is false. For example, $g = y^3 + xy^2$ is not hit since it contains a spike, but $xg^2 = xy^6 + x^3y^4 = Sq^1(xy^5 + x^3y^3) + Sq^2(x^2y^3)$ is hit.

Proposition 1.7.1 *For $d > 0$, $f = xg^2 + yh^2 \in P^{2d+1}(2)$ is hit if and only if $g \sim ye$ and $h \sim xe$ for some $e \in P^{d-1}(2)$.*

Proof Let $f = Sq^1(f_0) + \sum_{i>0} Sq^{2^i}(f_i)$. Then f_0 has even degree and f_i has odd degree, so $f_0 = xyg_0^2 + h_0^2$ and $f_i = xg_i^2 + yh_i^2$ for $i > 0$. Hence

$$f = xg^2 + yh^2 = (x^2y + xy^2)g_0^2 + \sum_{i>0}\left(x(Sq^i(g_i))^2 + y(Sq^i(h_i))^2\right).$$

Comparing coefficients of x and y gives $g = yg_0 + \sum_{i>0} Sq^i(g_i)$, $h = xg_0 + \sum_{i>0} Sq^i(h_i)$. So we take $e = g_0$ to prove the result in one direction. The converse follows by reversing the argument. □

Proposition 1.7.2 *For $d \geq 0$, the linear map $\delta : P^{2d+1}(2) \to P^{4d+3}(2)$ defined by $\delta(xg^2) = x^3g^4$ and $\delta(yh^2) = y^3h^4$ sends hit polynomials to hit polynomials.*

Proof Suppose $f = xg^2 + yh^2$ is hit. Then by Proposition 1.7.1, $g = ye + \sum_{i>0} Sq^i(g_i)$, $h = xe + \sum_{i>0} Sq^i(h_i)$ for some $e \in P^{d-1}(2)$. Hence $xg^2 = xy^2e^2 + \sum_{i>0} Sq^{2^i}(xg_i^2) \sim y(xye^2)$ and $yh^2 = yx^2e^2 + \sum_{i>0} Sq^{2^i}(yh_i^2) \sim x(xye^2)$. By Proposition 1.7.1 it follows that $\delta(f)$ is hit. □

Definition 1.7.3 *For $d \geq 0$, the **duplication map** $\delta : Q^{2d+1}(2) \to Q^{4d+3}(2)$ is the linear map induced by δ.*

Proposition 1.7.4 *For $d \geq 0$, the up Kameko map $\upsilon : Q^d(2) \to Q^{2d+2}(2)$ and the duplication map $\delta : Q^{2d+1}(2) \to Q^{4d+3}(2)$ are maps of $\mathbb{F}_2 GL(2)$-modules. Further, δ is a map of $\mathbb{F}_2 M(2)$-modules, as also is υ when d is positive and even.*

Proof Let the matrices S and U be as in Example 1.2.3, and let $E = \left(\begin{smallmatrix} 1 & 0 \\ 0 & 0 \end{smallmatrix}\right)$. Then S and U generate $GL(2)$ and S, U and E generate $M(2)$. Clearly υ and δ commute with the action of S. We show that they also commute with the action of U. First consider υ. For $f \in P^d(2)$ we have $\upsilon(f) = xyf^2$. Then $\upsilon(f) \cdot U = (x+y)y(f \cdot U)^2 = xy(f \cdot U)^2 + y^2(f \cdot U)^2$. Since the second term is hit, $\upsilon(f) \cdot U \sim xy(f \cdot U)^2 = \upsilon(f \cdot U)$, as required. Next consider δ, and let $f = xg^2 + yh^2$. Then $f \cdot U = (x+y)(g \cdot U)^2 + y(h \cdot U)^2$. Hence $\delta(f \cdot U) = (x^3 + y^3)(g \cdot U)^4 + y^3(h \cdot U)^4$, while $\delta(f) \cdot U = (x+y)^3(g \cdot U)^4 + y^3(h \cdot U)^4 = (x^3 + y^3 + x^2y + xy^2)(g \cdot U)^4 + y^3(h \cdot U)^4$. Using the hit equation $Sq^1(xy(g \cdot U)^4) = (x^2y + xy^2)(g \cdot U)^4$, we see that $\delta(f) \cdot U \sim \delta(f \cdot U)$, as required. Hence υ and δ are $\mathbb{F}_2 GL(2)$-module maps.

Next we show that δ commutes with the action of E. If $f = xg^2 + yh^2$ then $f \cdot E = x(g \cdot E)^2$. Hence $\delta(f \cdot E) = r^3(g \cdot E)^4 = r^3g^4 \cdot E$ and $\delta(f) \ E = (x^3g^4 + y^3h^4) \cdot E = x^3g^4 \cdot E$ as required. Next consider υ. Since $\upsilon(f) = xyf^2$ is divisible by y for all f, $\upsilon(f) \cdot E = 0$. If $d > 0$ is even, then $f = xyg^2 + h^2$ and $f \cdot E = (h \cdot E)^2$ is hit.   □

Example 1.7.5 At the level of polynomials, neither $\upsilon : \mathsf{P}^d(2) \to \mathsf{P}^{2d+2}(2)$ nor $\delta : \mathsf{P}^{2d+1}(2) \to \mathsf{P}^{4d+3}(2)$ commutes with U. Also $\upsilon : \mathsf{Q}^d(2) \to \mathsf{Q}^{2d+2}(2)$ does not commute with E when $d = 2^j - 1, j \geq 0$, since we have seen that $\upsilon(f) \cdot E = 0$ for all f, whereas $\upsilon(x^d \cdot E) = \upsilon(x^d) = x^{2d+1}y$ is a spike.

1.8 The hit problem for P(2)

As $\mathsf{Q}^0(2) = \mathsf{P}^0(2)$ is the trivial 1-dimensional module $\mathsf{I}(2)$ and $\mathsf{Q}^1(2) = \mathsf{P}^1(2)$ is the defining 2-dimensional module $\mathsf{V}(2)$, we begin with $\mathsf{Q}^2(2)$. Since $\mathsf{H}^2(2)$ is spanned by x^2 and y^2, Example 1.2.3 gives $(xy) \cdot S = xy$ and $(xy) \cdot U = xy$ in $\mathsf{Q}^2(2)$, so that $\mathsf{Q}^2(2) \cong \mathsf{I}(2)$. A similar calculation in $\mathsf{Q}^3(2)$ gives $x^3 \cdot S = y^3$, $y^3 \cdot S = x^3, x^2y \cdot S = x^2y$ and $x^3 \cdot U = x^3 + y^3, y^3 \cdot U = y^3, x^2y \cdot U = x^2y + y^3$. We see that $g = x^3 + y^3 + x^2y$, viewed as an element of $\mathsf{Q}^3(2)$, is a GL(2)-invariant, while x^3 and y^3 span a complementary 2-dimensional submodule. If we regard GL(2) as the group $\Sigma(3)$ of permutations of a set of three elements, $\mathsf{Q}^3(2)$ is the representation of GL(2) which permutes $x^2y, x^3 + x^2y$ and $y^3 + x^2y$.

By iteration of the Kameko and duplication maps, the GL(2)-module $\mathsf{Q}^d(2)$ can be linked to $\mathsf{Q}^0(2)$, $\mathsf{Q}^1(2)$ or $\mathsf{Q}^3(2)$ for all d. Iteration of the up Kameko map υ shows that the cohit module in every degree is isomorphic to $\mathsf{Q}^0(2)$ or to $\mathsf{Q}^d(2)$ for d odd. Proposition 1.6.4 then reduces the problem to the case $d = 2^j - 1$. We show that in this case $\mathsf{Q}^d(2)$ is isomorphic to $\mathsf{Q}^1(2)$ or to $\mathsf{Q}^3(2)$.

Proposition 1.8.1 *The duplication map* $\delta : \mathsf{Q}^d(2) \to \mathsf{Q}^{2d+1}(2)$ *is an isomorphism for* $d = 2^j - 1$ *and* $j > 1$.

Proof We show first that $\delta : \mathsf{Q}^d(2) \to \mathsf{Q}^{2d+1}(2)$ is surjective. After exchanging x and y if necessary, a monomial in $\mathsf{Q}^{2d+1}(2)$ has the form $f = gh^8$ for some monomial h, where $g = x^7, x^3y^4, x^6y$ or x^5y^2. In the first two cases, f is clearly in the image of δ. Since $Sq^2(x^3y^2h^8) = (x^5y^2 + x^3y^4)h^8$ and $Sq^1(x^5yh^8) = (x^6y + x^5y^2)h^8$, we have $x^3y^4h^8 \sim x^5y^2h^8 \sim x^6yh^8$. This shows that δ is surjective.

For all $j > 1$, $Q^d(2)$ contains two spikes x^{2^j-1} and y^{2^j-1} and monomials divisible by xy. It follows by Propositions 1.5.1 and 1.4.7 that $\dim Q^d(2) \geq 3$. Since $\dim Q^3(2) = 3$ by Example 1.5.4, δ is an isomorphism. □

Thus we can summarize the solution of the hit problem for $P^d(2)$ in terms of $GL(2)$-modules as follows.

Theorem 1.8.2 *The following table gives the dimension of $Q^d(2)$ when $d = (2^{s+t}-1)+(2^t-1)$ and $s,t \geq 0$. For all other values of d, $Q^d(2) = 0$.*

	$s=0$	$s=1$	$s \geq 2$
$t \geq 0$	1	2	3

For $s = 0$, $Q^d(2)$ is isomorphic as a $\mathbb{F}_2 GL(2)$-module to the trivial module $I(2)$, for $s = 1$ to the defining module $V(2)$ and for $s = 2$ to the permutation module on a set of three elements. □

The following diagram shows a monomial basis for $Q^d(2)$ for $d \leq 30$, together with the maps υ and δ. In each case, the last listed monomial generates $Q^d(2)$ as a $\mathbb{F}_2 GL(2)$-module. All maps shown, except the first δ, are isomorphisms of $\mathbb{F}_2 GL(2)$-modules, and all others, except the first row of υ, are also isomorphisms of $\mathbb{F}_2 M(2)$-modules. The down Kameko map $\kappa : Q^{2^{j+1}}(2) \to Q^{2^j-1}(2)$ for $j \geq 2$ is not an isomorphism of $\mathbb{F}_2 M(2)$-modules, because $E = \left(\begin{smallmatrix} 1 & 0 \\ 0 & 0 \end{smallmatrix}\right)$ acts as zero in the first module, but not in the second. However, the composition $\upsilon \circ \delta \circ \kappa : Q^{2^{j+1}}(2) \to Q^{2^{j+2}}(2)$ is an isomorphism, because E acts as zero in both modules. Hence, for all d, $Q^d(2)$ is isomorphic as a $\mathbb{F}_2 M(2)$-module to $Q^{d'}(2)$, where $d' = 0,1,2,3,4$ or 8. These six cases are distinguished by dimension and by the action of singular matrices.

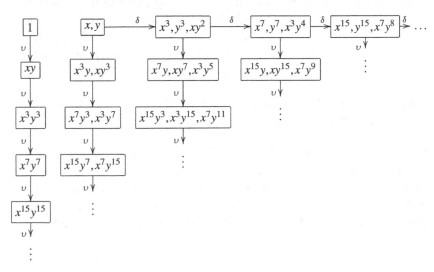

1.9 Remarks

The Steenrod square $Sq^k : P^d(n) \to P^{d+k}(n)$ has its origin in algebraic topology as a cohomology operation. Given a topological space X and an integer $d \geq 0$, the cohomology ring $H^*(X; \mathbb{F}_2) = \sum_{d \geq 0} H^d(X; \mathbb{F}_2)$ is a graded $\mathbb{F}_2$-algebra and $Sq^k : H^d(X, \mathbb{F}_2) \to H^{d+k}(X; \mathbb{F}_2)$ is a linear map. The algebra $P(n)$ is the mod 2 cohomology ring of $X(n) = \mathbb{R}P^\infty \times \cdots \times \mathbb{R}P^\infty$, the product of n copies of infinite real projective space. The basic facts about Steenrod operations can be found in Steenrod–Epstein [196]. The total Steenrod square Sq goes back at least to [11]: our purely algebraic approach is in the spirit of [190, Section 10.3].

The vector space $P(Y)$ introduced in Section 1.4 is the reduced cohomology of the smash product $\mathbb{R}P^\infty \wedge \cdots \wedge \mathbb{R}P^\infty$ of $|Y|$ copies of $\mathbb{R}P^\infty$, the quotient space of the corresponding product obtained by identifying all points where at least one component is the base point. The decomposition $P(n) = \bigoplus_{Y \subseteq Z[n]} P(Y)$ can be realized by a homotopy equivalence of the suspension $S^1 \wedge X(n)$ with a one-point union of spaces. In the 1970s, topologists began to investigate such splittings of suspensions of classifying spaces of Lie groups. These split their cohomology into summands invariant under the action of Steenrod operations. The space $X(n)$ can be regarded as the classifying space of the group $\mathbb{Z}/2 \times \cdots \times \mathbb{Z}/2$ of order 2^n.

The hit problem is motivated by several problems in topology and algebra. In 1987 Frank Peterson determined the degrees d such that $Q^d(2) \neq 0$ [168], and conjectured that $Q^d(n) = 0$ unless $\alpha(d+n) \leq n$. An equivalent statement is that f is hit if there are no spikes with the same degree and in the same variables as f. Following [229, 230], we prove this conjecture in Section 2.5.5. Peterson's motivation was to prove that if M is a smooth manifold of dimension d such that all products of length n of Stiefel–Whitney classes of its normal bundle vanish, then either $\alpha(d) \leq n$ or M is cobordant to zero. This is proved in [169], where Peterson encourages work on the hit problem in order to exploit Singer's transfer map [186] from $\mathrm{Tor}_n^{A_2}(\mathbb{F}_2, \mathbb{F}_2)$ to the $\mathrm{GL}(n)$ invariants of the cohits $Q(n)$. The terminology of spikes and hit polynomials is due to Singer [187]. The Kameko maps κ, υ appear in [106], and the duplication map δ in [221].

The action of Steenrod squares on polynomials can be expressed in terms of differential operators with variable coefficients. Let $D = \sum_{i \geq 1} x_i^2 \, \partial/\partial x_i$. Although D is a formally infinite sum, its action on a polynomial is finite, and it is easy to verify that $Sq^1(f) = D(f)$ mod 2 for any $f \in P(n)$. The wedge product of two differential operators allows the first operator to pass the second without acting, so that all symbols commute, for example

$$\sum_{i \geq 1} x_i^2 \frac{\partial}{\partial x_i} \vee \sum_{i \geq 1} x_i^2 \frac{\partial}{\partial x_i} = \left(\sum_{i \geq 1} x_i^2 \frac{\partial}{\partial x_i} \right)^{\vee 2} = \sum_{i \geq 1, j \geq 1} x_i^2 x_j^2 \frac{\partial^2}{\partial x_i \partial x_j}.$$

As the operator $D \vee D$ is divisible by 2, we can write $(D \vee D)/2$ as an integral operator. More generally, $D^{\vee k}/k!$ is an integral operator, and $Sq^k = D^{\vee k}/k!$ mod 2. The total Steenrod square may be expressed by an exponential generating function $Sq = \exp_\vee(D) = \sum_{k \geq 0} D^{\vee k}/k!$. For example, if $x \in P^1(n)$ then $\exp_\vee(D)(x) = x + x^2$. In this context, the Cartan formula is the Leibniz formula $\exp_\vee(D)(fg) = \exp_\vee(D)(f)\exp_\vee(D)(g)$. This point of view is developed in [232] and [233].

2

Conjugate Steenrod squares

2.0 Introduction

In this chapter we introduce a second family of linear operations on the polynomial algebra $P(n)$, the **conjugate Steenrod squares** $Xq^k : P^d(n) \to P^{d+k}(n)$. By expressing Xq^k in terms of Steenrod squares, we show that it is a Steenrod operation. It follows that the left action of Xq^k on $P(n)$ commutes with the right action of matrices in $M(n)$. As for Sq^k, we begin by defining the **total conjugate Steenrod square** $Xq = \sum_{k \geq 0} Xq^k$. Since evaluation of Xq on a nonzero element $x \in P^1(n)$ yields a power series, and not a polynomial, we begin by reviewing some properties of power series rings in Section 2.1. We define Xq^k in Section 2.2, and establish properties analogous to those of Sq^k. In particular, when $x \in P^1(n)$ and $k = 2^j - 1$, $Xq^k(x) = x^{2^j}$. The formulae $\sum_{i+j=k} Sq^i Xq^j = 0$ and $\sum_{i+j=k} Xq^i Sq^j = 0$ of Section 2.3 enable us to express Xq^k recursively in terms of Steenrod squares.

Decompositions of $k \geq 0$ as a sum of integers of the form $2^j - 1$, which have appeared in Chapter 1 in connection with spike monomials, also play a part in the study of the operations Xq^k. Section 2.4 introduces the important function $\mu(k)$, the minimum number of terms in such a sum. We define the **excess** of a Steenrod operation and show that the excess of Xq^k is $\mu(k)$. In Section 2.5 we apply the operations Xq^k to prove the Peterson conjecture, Theorem 2.5.5. The key result, known as the **conjugation** or χ-**trick**, implies that the product of polynomials f and $Sq^k(g)$ is hit if and only if the product of $Xq^k(f)$ and g is hit. Finally Section 2.6 offers an alternative proof of Proposition 2.3.2 which connects the operations Xq^k with the Catalan numbers.

2.1 Power series

Let R be a commutative ring with identity 1. A (**formal**) **power series** over R in one variable x has the form $f = \sum_{i \geq 0} f_i x^i$, where $f_i \in R$. If $g = \sum_{i \geq 0} g_i x^i$

is another such series, then $f = g$ if and only if $f_i = g_i$ for all $i \geq 0$. The ith term of f is $f_i x^i$. We identify x^0 with the identity $1 \in R$, so that the constant term $f_0 x^0$ of f is identified with $f_0 \in R$ and is also written as $f(0)$. Power series are added and subtracted term by term, and are multiplied using the Cauchy product $h = fg$, where $h_k = \sum_{i+j=k} f_i g_j$ for $k \geq 0$. These operations make the set of power series into a ring $R[[x]]$. Then $\mathbb{F}_2[[x]]$ is an algebra over $\mathbb{F}_2$, and we identify the subalgebra whose elements have a finite number of nonzero terms with the polynomial algebra $P(1)$. In contrast to $P(1)$, the algebra $\mathbb{F}_2[[x]]$ is not graded, since f is not obtained from its terms by ring operations. On the other hand $\mathbb{F}_2[[x]]$ is filtered, the filtration of $f \neq 0$ being the smallest integer i such that the ith term of f is nonzero.

More generally, a power series over R in n variables $x_1, \ldots, x_n$ assigns a coefficient in R to each monomial. We index monomials by listing their exponents in a fixed order, so that $x^I = x_1^{i_1} \cdots x_n^{i_n}$ is indexed by the sequence $I = (i_1, \ldots, i_n)$, and a product of monomials by the sum of their exponent sequences.

Definition 2.1.1 A **power series** in variables $x_1, \ldots, x_n$ over a commutative ring R with 1 is a formal sum $f = \sum_I f_I x^I$, $f_I \in R$, with terms indexed by sequences $I = (i_1, \ldots, i_n)$ of integers ≥ 0. In the ring $R[[x_1, \ldots, x_n]]$, the sum $f + g$ of the power series f and g is defined by $(f + g)_I = f_I + g_I$, and the product fg by $(fg)_I = \sum_{I=J+K} f_J g_K$. For all I, the element $x^I = x_1^{i_1} \cdots x_n^{i_n}$ is a **monomial** and has **degree** $\deg x^I = |I| = i_1 + \cdots + i_n$. The unique monomial of degree 0 is the identity $1 \in R$, and the corresponding coefficient f_0 (or $f(0)$) of a power series f is the **constant term** of f.

We identify the subring of $R[[x_1, \ldots, x_n]]$ generated by $x_1, \ldots, x_n$ with the polynomial ring $R[x_1, \ldots, x_n]$. When R is a field, this is the vector space spanned by the monomials x^I and $R[[x_1, \ldots, x_n]]$ is an algebra over R. Since $\deg fg = \deg f + \deg g$ for polynomials f and g, the only polynomials with multiplicative inverses are the invertible elements of R, regarded as constant polynomials. On the other hand, every polynomial with invertible constant term has a multiplicative inverse in the power series ring, the geometric series $\sum_{k \geq 0} x^k = (1 - x)^{-1}$ being a familiar example.

Proposition 2.1.2 *The power series f is invertible in $R[[x_1, \ldots, x_n]]$ if and only if its constant term f_0 is invertible in R.*

Proof If f is invertible and $g = f^{-1}$, then $f_0 g_0 = 1$, so f_0 is invertible. Conversely, if f_0 is invertible then, by taking a constant multiple, we may assume that $f_0 = 1$. Let $h = 1 - f$, so that $h_0 = 0$. Then for $k > 0$, h^k has no terms of degree $< k$. It follows that the expansion $f^{-1} = (1-h)^{-1} = \sum_{k \geq 0} h^k$ is valid in $R[[x_1, \ldots, x_n]]$, since the sum contains only a finite number of terms in each degree. $\qquad\square$

Example 2.1.3 In $\mathbb{F}_2[[x]]$, let $f = 1 + \sum_{k \geq 0} x^{2^k}$ and $g = \sum_{k \geq 0} x^{2^k - 1}$. Then $f + f^2 = x$ and $f = 1 + xg$. Hence $x/f = 1 + f = xg$ and so $g = f^{-1}$.

Proposition 2.1.4 *Let* $f = 1 + \sum_{k \geq 0} x^{2^k}$ *in* $\mathbb{F}_2[[x]]$. *Then the coefficient of* x^r *in* f^{r+1} *is* 0 *for all* $r \geq 1$.

Proof We write $r = 2^k(2s - 1)$ where $k \geq 0$ and $s \geq 1$, and argue by induction on k. For the base case $k = 0$, $r + 1 = 2s$ and $f^{r+1} = (f^2)^s = (1 + x^2 + x^4 + \cdots)^s$ contains no odd powers of x.

Thus let $k \geq 1$, and assume that the result holds for $k - 1$. Then $f^{r+1} = (1 + x + x^2 + x^4 + \cdots)(1 + x^{2^k} + x^{2^{k+1}} + \cdots)^{2s-1}$. Since r is even and the second factor is a power series in x^2, the term x in the first factor cannot contribute to the coefficient of x^r. Thus the coefficient of x^r in f^{r+1} is also the coefficient of x^r in $f^{r+2} = (1 + x^2 + x^4 + x^8 + \cdots)(1 + x^{2^k} + x^{2^{k+1}} + \cdots)^{2s-1}$. Let $y = x^2$, so that $x^r = y^{r/2}$. Since $r/2 = 2^{k-1}(2s - 1)$, the induction hypothesis on k implies that the coefficient of $y^{r/2}$ in $(1 + y + y^2 + y^4 + \cdots)^{(r/2)+1}$ is 0. But $(1 + y + y^2 + y^4 + \cdots)^{(r/2)+1} = (1 + y + y^2 + y^4 + \cdots)(1 + y + y^2 + y^4 + \cdots)^{r/2} = (1 + y + y^2 + y^4 + \cdots)(1 + y^{2^{k-1}} + y^{2^k} + \cdots)^{2s-1} = f^{r+2}$. This completes the induction. $\qquad\square$

Given polynomials f and g in $R[x]$, we can form the composition $(f \circ g)(x) = f(g(x))$ by substituting $g(x)$ for x in f. In other words, we think of polynomials as functions from R to R, although we cannot literally treat them in this way when R is finite. Composition is an associative operation with two-sided identity x, and $\deg(f \circ g) = \deg f \cdot \deg g$. For example, let $f = x + x^2$ and $g = 1 + x \in \mathbb{F}_2[x]$, then $f \circ g = (1 + x) + (1 + x)^2 = x + x^2$ and $g \circ f = 1 + x + x^2$, $f \circ f = x + x^4$ and $g \circ g = x$. Since composition multiplies degrees, $g \in R[x]$ can have a composition inverse only if $\deg g = 1$.

For power series, on the other hand, the composition inverse is the usual inverse function. The composition $f \circ g$ of two power series is defined if and only if $g_0 = 0$, so that substitution of g for x in f gives an expression with a finite number of terms in each degree. If also $f \circ g = x = g \circ f$ and $f_0 = 0$, then $g = f^{(-1)}$ is the **composition inverse** of f.

Proposition 2.1.5 *Let* $f = x + x^2$ *and* $g = x + x^2 + x^4 + \cdots = \sum_{k \geq 0} x^{2^k}$ *in* $\mathbb{F}_2[[x]]$. *Then* $g = f^{(-1)}$.

Proof We have $f \circ g = (x + x^2 + x^4 + \cdots) + (x + x^2 + x^4 + \cdots)^2 = (x + x^2 + x^4 + \cdots) + (x^2 + x^4 + x^8 + \cdots) = x$, and $g \circ f = (x + x^2) + (x + x^2)^2 + (x + x^2)^4 + \cdots = (x + x^2) + (x^2 + x^4) + (x^4 + x^8) + \cdots = x$. $\qquad\square$

In more familiar language, the quadratic equation $t = x + x^2$ in $\mathbb{F}_2[[x]]$ has the solution $x = \sum_{k \geq 0} t^{2^k}$. This gives an isomorphism between $\mathbb{F}_2[[x]]$ and $\mathbb{F}_2[[t]]$.

Proposition 2.1.6 *Let R be a commutative ring with identity. Then the elements $f = \sum_{i \geq 0} f_i x^i \in R[[x]]$ such that $f_0 = 0$ and f_1 is invertible in R form a group under composition, with identity element x.*

Proof Let $f(x) = f_1 x + f_2 x^2 + f_3 x^3 + \cdots$ and $g(x) = g_1 x + g_2 x^2 + g_3 x^3 + \cdots$ satisfy $f(g(x)) = x$. By substituting $g(x)$ for x in f and equating coefficients, we obtain $f_1 g_1 = 1$, $f_1 g_2 + f_2 g_1^2 = 0$, $f_1 g_3 + 2 f_2 g_1 g_2 + f_3 g_1^3 = 0$, and so on. Thus f_1 and g_1 are invertible, and we may assume that $f_1 = g_1 = 1$ by replacing f and g by f/f_1 and g/g_1. Then the jth equation expresses g_j in terms of $g_1, \ldots, g_{j-1}$ and $f_1, \ldots, f_j$. Hence for all f the equations have a unique solution g, which can be calculated by recursion on j. Similarly, given g, there is a unique solution for f. Thus every power series f such that $f_0 = 0$ and f_1 is invertible has both a right inverse g and a left inverse h. Since $g = (h \circ f) \circ g = h \circ (f \circ g) = h$, f has a two-sided inverse. $\qquad\square$

It follows from Proposition 2.1.2 that $\mathbb{F}_2[[x]]$ is a principal ideal domain with a single descending chain of nonzero ideals (x^r) where $r \geq 0$. Its field of fractions $\mathbb{F}_2((x))$ has elements f/g where $f, g \in \mathbb{F}_2[[x]]$ and $g \neq 0$. If $r \geq 0$ is minimal such that $g_r \neq 0$, so that g has filtration r, then $g = x^r h$ where h is invertible. Thus $f/g = x^{-r} f h^{-1}$ is a (formal) Laurent series in x.

2.2 The total conjugate square Xq

In this section we define the operations Xq^k on $P(n)$ for $k \geq 0$, following the method used in Section 1.1 to define the Steenrod squares Sq^k.

Definition 2.2.1 For $n \geq 1$, $P[n] = \mathbb{F}_2[[x_1, \ldots, x_n]]$ is the power series algebra in n variables $x_1, \ldots, x_n$ over the field $\mathbb{F}_2$. For $n = 0$, $P[0] = \mathbb{F}_2$.

Definition 2.2.2 The **total conjugate Steenrod square** is the algebra map $Xq : P(n) \to P[n]$ defined by $Xq(1) = 1$ and

$$Xq(x) = x + x^2 + x^4 + \cdots = \sum_{j \geq 0} x^{2^j},$$

where $x = x_i$, $1 \leq i \leq n$. The **conjugate Steenrod square** $Xq^k : P^d(n) \rightarrow P^{d+k}(n)$, for $k \geq 0$ and $d \geq 0$, is the linear map defined by restricting Xq to $P^d(n)$ and projecting on to $P^{d+k}(n)$, so that $Xq = \sum_{k \geq 0} Xq^k$ is the formal sum of its graded parts.

Proposition 2.2.3 *For $x \in P^1(n)$, $Xq(x) = \sum_{j \geq 0} x^{2^j}$. Thus $Xq^k(x) = x^{2^j}$ if $k = 2^j - 1$ and $Xq^k(x) = 0$ otherwise. More generally, for $r \geq 0$, $Xq^k(x^{2^r}) = x^{2^{j+r}}$ if $k = (2^j - 1)2^r$, where $j \geq 0$, and $Xq^k(x^{2^r}) = 0$ otherwise.*

Proof Let $x = \sum_{i=1}^{n} a_i x_i$, where $a_i \in \mathbb{F}_2$. Then $Xq(x) = \sum_{i=1}^{n} a_i Xq(x_i) = \sum_{j \geq 0} (\sum_{i=1}^{n} a_i x_i^{2^j}) = \sum_{j \geq 0} (\sum_{i=1}^{n} a_i x_i)^{2^j} = \sum_{j \geq 0} x^{2^j}$. Since Xq is multiplicative, $Xq(x^{2^r}) = (\sum_{j \geq 0} x^{2^j})^{2^r} = \sum_{j \geq 0} x^{2^{j+r}}$. The second statement follows by equating graded parts. $\square$

Proposition 2.2.4 (Cartan formula) *For polynomials $f, g \in P(n)$ and $k \geq 0$,*

$$Xq^k(fg) = \sum_{i+j=k} Xq^i(f)Xq^j(g).$$

Proof This follows from the multiplicative property $Xq(fg) = Xq(f)Xq(g)$ by equating terms of degree k. $\square$

Proposition 2.2.5 $Xq^0 = Sq^0$, *the identity map of* $P(n)$.

Proof The proof is the same as for Sq^0 in Proposition 1.1.5. $\square$

As for Sq^k, in principle these rules allow the evaluation of any conjugate square on a polynomial.

Example 2.2.6 In $P(2) = \mathbb{F}_2[x, y]$ we have for $j \geq 0$

$$Xq^{2^j-1}(xy) = Xq^{2^j-1}(x)y + xXq^{2^j-1}(y) = x^{2^j}y + xy^{2^j}.$$

As for Proposition 1.3.2, the next result follows from the multiplicative property of Xq.

Proposition 2.2.7 *For $f \in P(n)$ and $s, k \geq 0$,*

$$Xq^k(f^{2^s}) = \begin{cases} (Xq^j(f))^{2^s}, & \text{if } k = 2^s j, \\ 0, & \text{otherwise.} \end{cases} \quad \square$$

2.3 Some properties of Xq^k

We begin by reducing the evaluation of a conjugate Steenrod square Xq^k on a monomial to the 1-variable case.

Proposition 2.3.1 *Let* $f = x_1^{d_1} \cdots x_n^{d_n}$ *be a monomial in* $\mathsf{P}(n)$. *Then*

$$Xq^k(f) = \sum_{k_1 + \cdots + k_n = k} Xq^{k_1}(x_1^{d_1}) \cdots Xq^{k_n}(x_n^{d_n}).$$

Proof As for Sq^k, this follows from the Cartan formula 2.2.4 by induction on n. $\qquad\square$

The formula

$$\binom{u}{v} = \frac{u(u-1)\cdots(u-v+1)}{v!} \tag{2.1}$$

defines binomial coefficients for all integers u. For example, $\binom{-1}{v} = (-1)^v$ for all $v \geq 0$. Thus for $a, b \geq 0$

$$\binom{-a}{b} = \frac{-a(-a-1)\cdots(-a-b+1)}{b!} = (-1)^b \binom{a+b-1}{b}. \tag{2.2}$$

The next result is the analogue of Proposition 1.1.8 for Xq^k. We give an alternative proof of it in Section 2.6 using the Catalan numbers.

Proposition 2.3.2 *For* $x \in \mathsf{P}^1(n)$,

$$Xq^k(x^d) = \binom{d+2k}{k} x^{d+k}.$$

Proof By (2.2), $\binom{d+2k}{k} = \binom{-d-k-1}{k}$ mod 2. We use the second form to prove the result. Let $t = x + x^2$. We work in $\mathbb{F}_2[[x]]$, keeping in mind the isomorphism with $\mathbb{F}_2[[t]]$ explained following Proposition 2.1.5. Then $Xq(t) = x$ and $Sq(x) = t$. Hence $Xq(t^d) = x^d$ and $Xq^k(t^d) = c_{d,k} t^{d+k}$, where $c_{d,k}$ is the coefficient of t^{k+d} in x^d. Thus $c_{d,k}$ is the coefficient of t^{-1} in $t^{-d-k-1}x^d$. We have $t^{-d-k-1}x^d = (x+x^2)^{-d-k-1}x^d = x^{-k-1}(1+x)^{-d-k-1} = \sum_{j\geq 0} \binom{-d-k-1}{j} x^{j-k-1}$.

For $r \geq 0$, x^r is a power series in t, and so the coefficient of t^{-1} in x^r is 0. Thus only negative powers of x can contribute to the above sum, so that $c_{d,k}$ is the coefficient of t^{-1} in

$$\sum_{j=0}^{k} \binom{-d-k-1}{j} x^{j-k-1}. \tag{2.3}$$

The term with $j = k$ is $\binom{-d-k-1}{k} x^{-1}$, and we have $x^{-1} = (t + t^2 + t^4 + \cdots)^{-1} = t^{-1}(1 + t + t^3 + t^7 + \cdots)^{-1} = t^{-1}(1 + t + t^2 + t^4 + \cdots)$ by Example 2.1.3. Thus the coefficient of this term in (2.3) is $\binom{-d-k-1}{k}$.

To complete the proof, we show that the coefficient of t^{-1} in x^{-r} is even when $r > 1$. This follows from Proposition 2.1.4, since $x^{-r} = (t + t^2 + t^4 + \cdots)^{-r} = t^{-r}(1 + t + t^3 + t^7 + \cdots)^{-r} = t^{-r}(1 + t + t^2 + t^4 + \cdots)^r$ by

Example 2.1.3, and hence the required coefficient is the coefficient of t^{r-1} in $(1+t+t^2+t^4+\ldots)^r$.   □

The following formulae relate the two families of operations Sq^k and Xq^k.

Proposition 2.3.3 *Let $k \geq 1$ and $f \in \Gamma(n)$. Then*

$$\text{(i)} \sum_{i+j=k} Sq^i \circ Xq^j(f) = 0, \quad \text{(ii)} \sum_{i+j=k} Xq^i \circ Sq^j(f) = 0.$$

Proof Let $\theta^k = \sum_{i+j=k} Sq^i \circ Xq^j$ and $\phi^k = \sum_{i+j=k} Xq^i \circ Sq^j$ for $k \geq 0$. A calculation using Propositions 1.1.4 and 2.2.4 shows that θ^k and ϕ^k are also evaluated on products of polynomials $f, g \in P(n)$ by the Cartan formula

$$\theta^k(fg) = \sum_{i+j=k} \theta^i(f)\theta^j(g), \quad \phi^k(fg) = \sum_{i+j=k} \phi^i(f)\phi^j(g).$$

Thus it suffices to show that $\theta^k(x) = 0$ and $\phi^k(x) = 0$ when $x = x_i$ is one of the variables. Since $Xq \circ Sq(x) = Xq(x+x^2) = Xq(x) + Xq(x)^2 = \sum_{j\geq 0} x^{2^j} + \sum_{j\geq 0} x^{2^{j+1}} = x$, $\phi^k(x) = 0$ when $k > 0$. By Propositions 1.1.3 and 2.2.3, $Sq^i(Xq^j(x)) = 0$ unless there is an integer $r \geq 0$ such that $j = 2^r - 1$ and $i = 0$ or $i = 2^r$. Thus $\theta^k(x) = 0$ if k is not of the form $2^s - 1$. But if $k = 2^s - 1$ with $s \geq 1$, then $\theta^k(x) = Sq^0 Xq^{2^s-1}(x) + Sq^{2^{s-1}} Xq^{2^{s-1}-1}(x) = x^{2^s} + x^{2^s} = 0$. Hence $\theta^k(x) = 0$ for all $k \geq 1$.   □

These formulae can be used to calculate $Xq^k(f)$ recursively, but inefficiently, in terms of $Sq^j(f)$, $1 \leq j \leq k$. In particular, as $Sq^1 Sq^1(f) = 0$ and $Sq^1 Sq^2(f) = Sq^3(f)$ for all $f \in P(n)$ by Proposition 1.3.4, we obtain using (i)

$$Xq^1(f) = Sq^1(f),$$
$$Xq^2(f) = Sq^2(f) + Sq^1 Xq^1(f) = Sq^2(f),$$
$$Xq^3(f) = Sq^3(f) + Sq^2 Xq^1(f) + Sq^1 Xq^2(f) = Sq^2 Sq^1(f).$$

Continuing in this way, we obtain $Xq^4(f) = Sq^4(f) + Sq^2 Sq^2(f)$ and $Xq^5(f) = Sq^4 Sq^1(f) + Sq^2 Sq^2 Sq^1(f)$. (Using (ii) leads to the same formulae.) As we show in Chapter 3, Steenrod squares satisfy many more relations than those of Proposition 1.3.4, for example, $Sq^2 Sq^2 Sq^1(f) = 0$, so that $Xq^5(f) = Sq^4 Sq^1(f)$ for all f. However, Proposition 2.3.3 gives the following result.

Proposition 2.3.4 *For $k \geq 0$, $Xq^k : P(n) \to P(n)$ is a Steenrod operation of degree k. Hence $Xq^k(f)$ is hit for all $f \in P^d(n)$ and $k > 0$.*   □

To illustrate the use of Proposition 2.3.2, we give another proof of Theorem 1.4.12 using the conjugate squaring operations Xq^k.

Example 2.3.5 For $d = 2^j - 1$, let $d = a + b$ where $a, b > 0$. Let 2^i be the minimum element of bin(b), so that $b = 2^i$ mod 2^{i+1}. Then $a = 2^i - 1$ mod 2^{i+1}, and so $2^i \notin$ bin($a + 2b$). Hence $\binom{a+2b}{b} = 0$ mod 2, so $x^d \neq Xq^b(x^a)$ by Proposition 2.3.2. Hence x^d is not hit.

Now assume that $d \neq 2^j - 1$, and choose a and b, as in the proof of Theorem 1.4.12, so that $2^i \notin$ bin(d) and $2^{i+1} \in$ bin(d). Let $b = 2^i$ and $a = d - 2^i$. Then $a + 2b = d + 2^i$ and so $2^i \in$ bin($a + 2b$). Hence $\binom{a+2b}{b}$ is odd, and so x^d is hit.

2.4 The numerical function μ

The binary expansion of an integer $d \geq 0$ as a sum of distinct 2-powers is usually written as a sequence of 0s and 1s in descending order of 2-powers. We shall use ascending order, so that for example 25 is written as $1\,0\,0\,1\,1$. This makes it easier to add trailing 0s, so that $1\,0\,0\,1\,1\,0\,0$ also means 25. We regard this as an infinite sequence of 0s and 1s, with only a finite number of 1s. Recall from Section 1.4 that bin(d) is the set of 2-powers in the binary expansion of d, and that $\alpha(d) = |\text{bin}(d)|$ is the number of 1s.

Definition 2.4.1 The **ω-sequence** $\omega(d) = (\omega_1(d), \omega_2(d), \ldots)$ of an integer $d \geq 0$ is defined by

$$\omega_j(d) = \begin{cases} 1, & \text{if } 2^{j-1} \in \text{bin}(d), \\ 0, & \text{if } 2^{j-1} \notin \text{bin}(d). \end{cases}$$

Thus

$$\alpha(d) = \sum_{j \geq 1} \omega_j(d), \quad d = \sum_{j \geq 1} 2^{j-1} \omega_j(d). \tag{2.4}$$

The results of Chapter 1 show that spikes, or monomials with all exponents of the form $2^j - 1$, play a key role in the hit problem. In particular, it is important to know whether spikes occur in $P^d(n)$, and if so, how many. For example $P^{17}(2)$ contains no spikes, but $P^{17}(3)$ contains the spikes $x_1^7 x_2^7 x_3^3$, $x_1^{15} x_2 x_3$ and four more obtained by permuting the variables. By selecting three of the four variables, these yield 24 spikes in $P^{17}(4)$. For $n \geq 5$, $P^{17}(n)$ contains spikes with other exponent sequences, such as $x_1^7 x_2^3 x_3^3 x_4^3 x_5$.

Definition 2.4.2 For $d > 0$, $\mu(d)$ is the minimum number of terms of the form $2^j - 1$, with repetitions allowed, whose sum is d. We also define $\mu(0) = 0$.

Example 2.4.3 Since $19 = 15 + 3 + 1$ and $\mu(19) > 1$, $\mu(19) = 3$. Since $12 = 7 + 3 + 1 + 1$ and $\mu(12) > 2$, $\mu(12) = 4$.

Clearly $\mu(d) = d \bmod 2$, and $\mu(a+b) \leq \mu(a) + \mu(b)$ for all $a, b \geq 0$. Thus $\mu(d)$ is analogous to $\alpha(d)$, but we consider d as a sum of integers of the form $2^j - 1$. In particular, $\mu(d) = 1$ if and only if $d = 2^s - 1$ where $s \geq 1$, and $\mu(d) = 2$ if and only if $d = (2^{s+t} - 1) + (2^t - 1)$, where $s \geq 0$ and $t \geq 1$. By Theorems 1.4.12 and 1.8.2, these conditions correspond to solutions of the hit problem for $P(1)$ and $P(2)$. Theorem 2.5.5 states that a similar result holds for $P(n)$ for all n.

The next result relates the functions μ and α.

Proposition 2.4.4 *For all $d \geq 0$, $\mu(d) \leq k$ if and only if $\alpha(d+k) \leq k$. Hence $\mu(d)$ is the minimum k such that $\alpha(d+k) \leq k$.*

Proof Suppose that $\mu(d) \leq k$. We can write $d = \sum_{i=1}^{k}(2^{a_i} - 1)$, where $a_i \geq 0$. Then $d + k = \sum_{i=1}^{k} 2^{a_i}$, showing that $\alpha(d+k) \leq k$. Conversely, if $\alpha(d+k) \leq k$ then $d + k = \sum_{i=1}^{m} 2^{a_i}$ for some $m \leq k$. If $m < k$ then not all a_i can be 0. Hence, by splitting a 2-power into two equal parts, we can express $d + k$ as the sum of $m + 1$ 2-powers. By iteration, we may assume $m = k$. Then $d = \sum_{i=1}^{k}(2^{a_i} - 1)$, and so $\mu(d) \leq k$. $\square$

Proposition 2.4.5 *For all $d \geq 0$, $\mu(d) \leq k$ if and only if $\mu(2d+k) \leq k$.*

Proof By Proposition 2.4.4, $\mu(d) \leq k$ if and only if $\alpha(d+k) \leq k$, and similarly $\mu(2d+k) \leq k$ if and only if $\alpha(2d+2k) \leq k$. But $\alpha(2d+2k) = \alpha(d+k)$. $\square$

We can evaluate $\mu(d)$ explicitly in some special cases.

Proposition 2.4.6 $\mu(2^n - k) = k$ *for* $1 \leq k \leq n$.

Proof The decomposition $2^n - k = (2^{n-k+1} - 1) + \sum_{j=1}^{k-1}(2^{n-j} - 1)$ shows that $\mu(2^n - k) \leq k$. If $\mu(2^n - k) \leq k - 1$, then Proposition 2.4.4 gives the contradiction $n = \alpha(2^n - 1) \leq k - 1$. Hence $\mu(2^n - k) = k$. $\square$

Proposition 2.4.7 *Let $d = 2^n - n - 1$ for $n \geq 1$. Then $\mu(d) = n - 1$, and $d = \sum_{j=1}^{n-1}(2^j - 1)$ is the unique expression for d as the sum of $n - 1$ integers of the form $2^j - 1$.*

Proof By Proposition 2.4.4 $\mu(d) \geq n - 1$, since $\mu(d) \leq n - 2$ leads to the contradiction $n - 1 = \alpha(2^n - 3) \leq n - 2$. Thus the stated decomposition implies that $\mu(2^n - n - 1) = n - 1$. The uniqueness with $n - 1$ summands follows from that of the decomposition $2^n - 2 = \sum_{j=1}^{n-1} 2^j$. $\square$

The next result relates the function μ to the hit problem.

Proposition 2.4.8 (i) *If* $\mu(k) > d$, *then* $Xq^k(f) = 0$ *for all* $f \in P^d(n)$, *and* (ii) *if* $\mu(k) = d$, *then* $Xq^k(x_1 \cdots x_d) \neq 0$.

Proof For (i), we may assume by linearity that f is a monomial. Let $f = y_1 \cdots y_d$, where each y_i is one of the variables x_j. Then

$$Xq(f) = \prod_{i=1}^{d} (y_i + y_i^2 + y_i^4 + \cdots + y_i^{2^j} + \cdots).$$

If k is not the sum of d integers of the form $2^j - 1$, then there are no terms of degree $d + k$ in $Xq(f)$, and so $Xq^k(f) = 0$.

For (ii), if k is the sum of d integers of the form $2^j - 1$, then $Xq(f)$ contains terms of degree $d + k$. If $y_1, \ldots, y_d$ are distinct, then $Xq^k(f) \neq 0$ as there can be no cancellation of terms. $\square$

Definition 2.4.9 The **excess** $ex(\theta)$ of a nonzero Steenrod operation θ is the minimum $d \geq 0$ such that $\theta(x_1 x_2 \cdots x_d) \neq 0$, where $x_1 x_2 \cdots x_d = 1$ if $d = 0$.

Propositions 1.1.10 and 2.4.8 determine the excess of the squares and conjugate squares.

Proposition 2.4.10 *For* $k \geq 0$, $ex(Sq^k) = k$ *and* $ex(Xq^k) = \mu(k)$. $\square$

Proposition 2.4.11 *If* $ex(\theta) > d$, *then* $\theta(f) = 0$ *for all* $f \in P^d(n)$.

Proof By Proposition 1.2.5, θ commutes with all linear substitutions in $M(n)$, and in particular with specializations of the variables. The result follows since any monomial of degree $d > 0$ can be obtained from $x_1 \cdots x_d$ in this way. $\square$

2.5 The χ-trick and the Peterson conjecture

The following formula connects the families of operations Sq^k and Xq^k.

Proposition 2.5.1 *For all* $g, h \in P^d(n)$ *and* $k \geq 1$

$$gXq^k(h) + Sq^k(g)h = \sum_{i=1}^{k} Sq^i(gXq^{k-i}(h)).$$

Proof The rows of the following array are the first term on the left of the equation and the expansion by the Cartan formula of the terms on the right.

$gXq^k(h)$

$gSq^1Xq^{k-1}(h)$ $\quad + Sq^1(g)Xq^{k-1}(h)$

$gSq^2Xq^{k-2}(h)$ $\quad + Sq^1(g)Sq^1Xq^{k-2}(h)$ $\quad + Sq^2(g)Xq^{k-2}(h)$

$\qquad \vdots \qquad\qquad\qquad \vdots \qquad\qquad\qquad \vdots \qquad\qquad \ddots$

$gSq^k(h)$ $\qquad + Sq^1(g)Sq^{k-1}(h)$ $\qquad + Sq^2(g)Sq^{k-2}(h) + \quad \cdots \quad + Sq^k(g)h$

By Proposition 2.3.3(i), the sum of each column is 0, except for the last term in the last line, which is the second term on the left of the equation. $\qquad\square$

Since all the terms on the right are of the form $Sq^i(h_i)$ where $i > 0$, the formula of Proposition 2.5.1 is a hit equation, and so we have the following result.

Proposition 2.5.2 (The χ-trick) *For $g, h \in P(n)$, $gSq^k(h) \sim Xq^k(g)h$. In particular, $gSq^k(h)$ is hit if and only if $Xq^k(g)h$ is hit.* $\qquad\square$

The χ-trick provides a direct means of applying the operations Xq^k to the hit problem without having first to evaluate them in terms of Steenrod squares. By combining it with Proposition 2.4.8, we obtain a useful condition for a homogeneous polynomial to be hit.

Proposition 2.5.3 *If $f = gh^2 \in P^d(n)$ where $\deg g < \mu(\deg h)$, then f is hit.*

Proof Let $\deg g = a$ and $\deg h = b$. By Proposition 1.1.10, $h^2 = Sq^b(h)$. By Proposition 2.4.8, $Xq^b(g) = 0$ since $\mu(b) > a$. Then the χ-trick shows that $f = gh^2 = gSq^b(h) \sim Xq^b(g)h = 0$. $\qquad\square$

Example 2.5.4 Let $n = 3$ and let $f = x^7y^2z^2$. Then $f = gh^2$ where $g = x$ has degree 1 and $h = x^3yz$ has degree 5. Since $\mu(5) = 3 > 1$, f is hit. The same argument applies to any monomial of the form $f = x^ay^2z^2$ of odd degree not of the form $2^j - 1$. By Proposition 1.5.1, a monomial of degree $2^j - 1$ cannot be hit.

Peterson conjectured that $Q^d(n) = 0$ if $\alpha(d + n) > n$. By Proposition 2.4.4, this condition is equivalent to $\mu(d) > n$.

Theorem 2.5.5 (Peterson conjecture) $Q^d(n) \neq 0$ *if and only if $\mu(d) \leq n$.*

Proof If $\mu(d) \leq n$, then we can write $d = (2^{j_1} - 1) + \cdots + (2^{j_n} - 1)$. Then $\mathrm{P}^d(n)$ contains the spike $x_1^{2^{j_1}-1} \cdots x_n^{2^{j_n}-1}$, which is not hit (Proposition 1.5.3). Hence $Q^d(n) \neq 0$.

Conversely, suppose $\mu(d) > n$. A monomial $f \in \mathrm{P}^d(n)$ has the form $f = gh^2$, where g is the product of $a \leq n$ distinct variables. Then $d = a + 2b$, where $b = \deg h$. Since $\mu(d) > n \geq a$, $\mu(b) > a$ by Proposition 2.4.5. By Proposition 2.5.3, f is hit. Hence $Q^d(n) = 0$. $\qquad\square$

Theorem 2.5.5 shows that $Q^d(n)$ can only be nonzero in degrees d where spikes exist. Example 2.5.4 shows that Proposition 2.5.3 is actually stronger than Theorem 2.5.5, as it can sometimes be applied to prove that a monomial of degree d is hit in cases where $\mu(d) \leq n$.

2.6 Catalan numbers and Xq^k

In this section we give an alternative proof of Proposition 2.3.2, which links the operations Xq^k to Catalan numbers. It is based on an integral relationship between the power series

$$C = \sum_{k \geq 0} \binom{2k}{k} \frac{x^k}{(k+1)}, \quad F(d) = \sum_{k \geq 0} \binom{d+2k}{k} x^k,$$

where $C = \sum_{k \geq 0} c_k x^k$ is the generating function for the Catalan numbers.

Proposition 2.6.1 $F(d+1) = CF(d)$ *for* $d \geq 0$.

Proof Since $\binom{d+2k}{k} = \binom{d-1+2k}{k} + \binom{d-1+2k}{k-1}$, we have the recurrence relation

$$xF(d+1) = F(d) - F(d-1).$$

We assume as induction hypothesis that the result is proved up to some $d \geq 1$. Then $xF(d+1) = F(d) - F(d-1)$ and multiplying through by C gives $xCF(d+1) = CF(d) - CF(d-1) = F(d+1) - F(d)$ by the inductive assumption. Hence $xCF(d+1) = xF(d+2)$ by the recurrence relation, and $CF(d+1) = F(d+2)$ as required in the inductive step.

It remains to verify the statement for $d = 0, 1$. We use the following property of the series C.

Proposition 2.6.2 *The Catalan series* C *satisfies the identity* $xC^2 = C - 1$.

Multiplying through the identity by $F(0)$ and assuming $F(1) = CF(0)$, we obtain $xCF(1) = xC^2 F(0) = CF(0) - F(0) = F(1) - F(0) = xF(2)$ by the recurrence relation. Hence $F(2) = CF(1)$. It remains to show that $F(1) =$

$CF(0)$. From the definition of $F(d)$,

$$F(0) = C + x\frac{dC}{dx}, \quad F(1) = C + 2x\frac{dC}{dx}.$$

Now we need to show that $C^2 + xC\,dC/dx = C + 2x\,dC/dx$. Differentiating the identity 2.6.2 we have $C^2 + 2xC\,dC/dx = dC/dx$. Solving for dC/dx, the problem reduces to showing that

$$\frac{C^2 - C}{2x - xC} = \frac{C^2}{1 - 2xC}.$$

The result follows by elementary algebra, using Proposition 2.6.2 again. $\square$

By induction on d, Proposition 2.6.1 gives $F(d) = F(0)C^d$. We can now prove Proposition 2.3.2. Recall that Xq is multiplicative and $\mathsf{Xq}(x) = x + x^2 + \cdots + x^{2^k} + \cdots$. It is known that the Catalan numbers c_k are even except when k has the form $2^r - 1$, so that working mod 2 we have

$$C = 1 + x + x^3 + \cdots + x^{2^r - 1} + \cdots.$$

Hence $\mathsf{Xq}(x) = xC$ mod 2. Also the binomial coefficients $\binom{2k}{k}$ are even for $k > 0$. Hence $F(0) = 1$ mod 2 and we obtain

$$\mathsf{Xq}(x^d) = \mathsf{Xq}(x)^d = (xC)^d = x^d C^d = x^d F(d) \text{ mod } 2.$$

Proposition 2.3.2 follows by selecting the terms of grading $k + d$.

2.7 Remarks

Stanley [192] is a good reference for power series: see in particular Volume 2, Section 5.4, where Propositions 2.1.6 and 2.6.2 appear, and Section 6.1 for Laurent series. The conjugate Steenrod operations Xq^k and the relations of Proposition 2.3.3 were first studied by Thom [210] and Wu [238] in the context of duality for characteristic classes of vector bundles. The work of Milnor and Moore on Hopf algebras [135] placed them in a general algebraic context. The function μ has been studied by several authors [49, 187]. Partitions of integers as the sum of numbers of the form $2^j - 1$ are studied in [19] from a combinatorial point of view.

The Peterson conjecture was first proved in [229]. The version presented here appears in [230] alongside Peterson's application to his original problem

[169]. H. E. A. Campbell and P. S. Selick had previously verified the conjecture for $n \leq 5$. The χ-trick was known to Peterson and other topologists since at least the 1970s in the context of bordism theory; it appears for example in [158]. Section 2.6 is based on an exercise in [174, p.154].

3
The Steenrod algebra A$_2$

3.0 Introduction

In this chapter we define the **mod 2 Steenrod algebra** A$_2$. We begin by using the notation Sq^k as a symbol for an element of A$_2$, and show that the action of Sq^k on P(n) defined in Chapter 1 makes P(n) into a left A$_2$-module. Our central object of study is the commuting actions of A$_2$ on the left and of $\mathbb{F}_2$M(n) on the right of P(n). In particular, the hit problem asks for a minimal set of generators of P(n) as an A$_2$-module.

In Section 3.1 we define A$_2$ by generators Sq^k subject to the **Adem relations**. We introduce **admissible monomials** $Sq^A = Sq^{a_1} \cdots Sq^{a_s}$, and show that they span A$_2$ as a vector space over $\mathbb{F}_2$. In Section 3.2 we show that the operations Sq^k on P(n) defined in Chapter 1 satisfy the Adem relations. Thus P(n) is an A$_2$-module. We also show that the elements Sq^{2^j}, $j \geq 0$, form a minimal generating set for A$_2$ as an algebra over $\mathbb{F}_2$.

In Section 3.3 we introduce some combinatorial notation that is helpful in studying the action of A$_2$ on P(n). In particular, we represent a monomial $f \in$ P(n) by a 'block', which is a matrix-like array of 0s and 1s whose rows are the reverse binary expansions of the exponents of f. A polynomial in P(n) is then represented by a formal sum of blocks. We are particularly concerned in this section with the special case of monomials where all exponents are 2-powers, so that each row of the corresponding block contains a single entry 1, and with the corresponding monomial symmetric functions obtained by taking the sum of all distinct monomials obtained by permutations of the variables, or equivalently by permuting the rows of the corresponding blocks. We call these 'Cartan symmetric functions', and we show that for $\theta \in$ A$_2$ and $n \geq 1$, $\theta(x_1 \cdots x_n)$ is a sum of Cartan symmetric functions.

In Section 3.4 we use Cartan symmetric functions to show that Steenrod operations of degree d are faithfully represented by their action on the product

33

$c(n) = x_1 \cdots x_n$ when $n \geq d$. It follows that the admissible monomials form a vector space basis for A_2. The argument leads naturally to another basis for A_2, the important **Milnor basis**, which we introduce in Section 3.5.

3.1 The Adem relations

Definition 3.1.1 The **Steenrod algebra** A_2 is the associative algebra over $\mathbb{F}_2$ generated by symbols Sq^k, $k \geq 0$, subject to the relations $Sq^0 = 1$, the identity element of A_2 and the **Adem relations**

$$Sq^a Sq^b = \sum_{j=0}^{[a/2]} \binom{b-j-1}{a-2j} Sq^{a+b-j} Sq^j, \quad a < 2b, \tag{3.1}$$

where the binomial coefficient is reduced mod 2.

Every term on the right hand side of (3.1) has the form $Sq^{a'} Sq^{b'}$, where $a' > 2b'$, since the conditions $2j \leq a < 2b$ imply $a+b-j > 2j$. Thus the Adem relations express all products of two generators Sq^k as sums of compositions $Sq^a Sq^b$, where $a \geq 2b$. Some examples of the Adem relations are as follows:

(1) $Sq^1 Sq^1 = 0$, and in general $Sq^{2k-1} Sq^k = 0$ for $k \geq 1$.
(2) $Sq^1 Sq^2 = Sq^3$ and $Sq^1 Sq^3 = 0$. In general, $Sq^1 Sq^k = Sq^{k+1}$ if k is even, and $Sq^1 Sq^k = 0$ if k is odd.
(3) $Sq^2 Sq^2 = Sq^3 Sq^1$ and $Sq^2 Sq^3 = Sq^5 + Sq^4 Sq^1$. In general, $Sq^2 Sq^k = Sq^{k+1} Sq^1$ if $k = 1, 2$ mod 4 and $Sq^2 Sq^k = Sq^{k+2} + Sq^{k+1} Sq^1$ if $k = 0, 3$ mod 4.
(4) $Sq^3 Sq^2 = 0$, $Sq^3 Sq^3 = Sq^5 Sq^1$, $Sq^3 Sq^4 = Sq^7$ and $Sq^3 Sq^5 = Sq^7 Sq^1$. In general, $Sq^3 Sq^k = Sq^{k+2} Sq^1$ if k is odd, $Sq^3 Sq^k = Sq^{k+3}$ if $k = 0$ mod 4 and $Sq^3 Sq^k = 0$ if $k = 2$ mod 4.

These examples show that the generating set Sq^k, $k \geq 1$, of A_2 is far from minimal. The Adem relations also contain much redundancy. For example, (4) follows from (2) and (3) since $Sq^1 Sq^2 = Sq^3$.

Although A_2 is not commutative, a product $Sq^A = Sq^{a_1} Sq^{a_2} \cdots Sq^{a_s}$ in A_2 corresponding to a sequence $A = (a_1, a_2, \ldots, a_s)$ of integers ≥ 0 is called a **monomial**. Every element of A_2 is a finite sum of monomials, but this expression is not unique. As it involves an infinite sum, the total Steenrod square Sq is not an element of A_2.

Proposition 3.1.2 *The algebra* A_2 *is graded by giving* Sq^k *degree k for* $k \geq 0$.

Proof All terms in the Adem relation (3.1) have the same exponent sum $a+b$, and this is also unchanged by inserting or omitting factors $Sq^0 = 1$. Thus for $A = (a_1, a_2, \ldots, a_s)$, Sq^A has degree $|A| = a_1 + \cdots + a_s$. $\qquad\square$

Definition 3.1.3 For $d \geq 0$, we denote by A_2^d the vector space over $\mathbb{F}_2$ spanned by monomials Sq^A of degree d. Thus $A_2 = \sum_{d \geq 0} A_2^d$. We write $\deg \theta = d$ if $\theta \in A_2^d$. We also denote by $A_2^+ = \sum_{d > 0} A_2^d$ the two-sided ideal of A_2 generated by the elements Sq^k, $k > 0$.

Since $Sq^0 = 1$, omitting 0s from the sequence A does not change the element Sq^A of A_2. Hence A_2^d is a finite dimensional vector space over $\mathbb{F}_2$. For the same reason, adding trailing 0s to a finite sequence A does not change Sq^A, and so we may regard A as an infinite sequence with only a finite number of nonzero terms.

Definition 3.1.4 We denote by Seq the set of sequences $A = (a_1, a_2, \ldots,)$ of integers $a_i \geq 0$ such that $a_i > 0$ for only a finite number of $i \geq 1$. The **length** len(A) of $A \in$ Seq is ℓ if $a_\ell \neq 0$ and $a_i = 0$ for all $i > \ell$. The zero sequence $(0, 0, \ldots)$ has length 0. If $m \geq$ len(A), we may identify $A \in$ Seq with a finite sequence of size m by omitting trailing 0s. The **modulus** of A is $|A| = \sum_{i \geq 1} a_i$. We denote the set of sequences of modulus d by Seq^d.

We introduce two linear orders on sequences, the **left order** $<_l$ and the **right order** $<_r$.

Definition 3.1.5 Let $A = (a_1, a_2, \ldots)$ and $B = (b_1, b_2, \ldots)$ be sequences in Seq. Then $A <_l B$ if and only if, for some k, $a_j = b_j$ for $1 \leq j < k$ and $a_k < b_k$, and $A <_r B$ if and only if, for some k, $a_j = b_j$ for $j > k$ and $a_k > b_k$.

Thus $<_l$ is the usual left lexicographic order, but $<_r$ is the *reversed* right lexicographic order. Thus len(A) < len(B) implies $A >_r B$.

Definition 3.1.6 The sequence $A = (a_1, a_2, \ldots) \in$ Seq, and the corresponding monomial $Sq^A \in A_2$, are **admissible** if $a_i \geq 2a_{i+1}$ for $i \geq 1$. We denote the set of admissible sequences by Adm and those of modulus d by Adm^d.

In particular, Sq^a is an admissible monomial for all $a \geq 0$. We show that the admissible monomials span A_2.

Proposition 3.1.7 *Every element of* A_2 *is a sum of admissible monomials.*

Proof For $d \geq 0$, the set S^d of sequences $A = (a_1, a_2, \ldots) \in \mathsf{Seq}$ with $a_i > 0$ for $i \leq \mathrm{len}(A)$ and $|A| = d$ is finite, and the corresponding monomials Sq^A span A_2^d. If $A \in S^d$ is not admissible, then $0 < a_k < 2a_{k+1}$ for some k.

Using the Adem relation (3.1) with $a = a_k$ and $b = a_{k+1}$, we may write Sq^A as a sum of monomials Sq^B, where B is obtained from A by replacing (a, b) by $(a + b - j, j)$ for some j such that $0 \leq j \leq [a/2] < b$. In the case $j = 0$, we omit the corresponding term of B to obtain a sequence in S^d: this does not affect Sq^B since $Sq^0 = 1$. Then $B \in S^d$ and $B >_{l,r} A$. Hence Sq^A can be written as a sum of monomials Sq^B which are greater than Sq^A in both the left and right orders. Iteration of this procedure must stop, since S^d is a finite set. Hence Sq^A can be expressed as a sum of admissible monomials.   □

We shall prove in Section 3.4 that the set of admissible monomials is a vector space basis of A_2. The table below shows this basis in degrees $d \leq 9$. In this range the left and right orders coincide, and the basis is listed in descending order. The left and right orders are different when $d \geq 12$, since $(9, 2, 1) >_l (8, 4)$ but $(9, 2, 1) <_r (8, 4)$.

degree d	admissible basis of A_2^d	dim A_2^d
0	$Sq^0 = 1$	1
1	Sq^1	1
2	Sq^2	1
3	$Sq^3, Sq^2 Sq^1$	2
4	$Sq^4, Sq^3 Sq^1$	2
5	$Sq^5, Sq^4 Sq^1$	2
6	$Sq^6, Sq^5 Sq^1, Sq^4 Sq^2$	3
7	$Sq^7, Sq^6 Sq^1, Sq^5 Sq^2, Sq^4 Sq^2 Sq^1$	4
8	$Sq^8, Sq^7 Sq^1, Sq^6 Sq^2, Sq^5 Sq^2 Sq^1$	4
9	$Sq^9, Sq^8 Sq^1, Sq^7 Sq^2, Sq^6 Sq^3, Sq^6 Sq^2 Sq^1$	5

The following example shows how the Adem relations can be used to convert an element of A_2 to admissible form.

Example 3.1.8 $Sq^4 Sq^2 Sq^3 = Sq^4(Sq^5 + Sq^4 Sq^1) = (Sq^9 + Sq^8 Sq^1 + Sq^7 Sq^2) + (Sq^7 Sq^1 + Sq^6 Sq^2)Sq^1 = Sq^9 + Sq^8 Sq^1 + Sq^7 Sq^2 + Sq^6 Sq^2 Sq^1$.

3.2 The action of A_2 on $P(n)$

In this section we prove that the linear operations $Sq^k : P(n) \to P(n)$ defined in Chapter 1 define an action of the Steenrod algebra A_2 on $P(n)$. Since the correspondingly named elements Sq^k generate A_2 as an algebra, for $\theta \in A_2$ a corresponding linear operation $\theta : P(n) \to P(n)$ is defined by addition and composition of the operations Sq^k.

For this action to be well defined, it must be compatible with all relations between the generators Sq^k of A_2. For example, $Sq^0 = 1$ is a relation in A_2, and Proposition 1.1.5 states that Sq^0 is the identity map of $P(n)$. Hence the action is compatible with the relation $Sq^0 = 1$. In the same way, Proposition 1.3.4 states that the action is compatible with the Adem relations $Sq^1 Sq^{2k} = Sq^{2k+1}$ and $Sq^1 Sq^{2k+1} = 0$ for $k \geq 0$. Since the Adem relations are a set of defining relations for A_2, our task is to extend this argument to all the Adem relations (3.1).

We argue by induction on the grading in A_2, and begin by extending this to all integers by defining $Sq^j = 0$ for $j < 0$, so that $A_2^d = 0$ for $d < 0$. Hence relation 3.1 can be written as

$$Sq^a Sq^b = \sum_j \binom{b-j-1}{a-2j} Sq^{a+b-j} Sq^j, \text{ for } a < 2b, \tag{3.2}$$

where the summation is over all integers j, and we define $\binom{u}{v} = 0$ if $v < 0$. When $u < 0$ and $v \geq 0$, we use equation (2.2) to interpret the mod 2 binomial coefficient. With these conventions, the relation $\binom{u}{v} = \binom{u-1}{v} + \binom{u-1}{v-1}$ is valid for all integers u and v.

We shall prove that the operations Sq^k on $P(n)$ are compatible with the relation (3.1) for all integers a and b, and not only for $0 < a < 2b$. For $a + b < 0$ there is nothing to prove, giving the base of the induction. We shall see in Section 3.4 that the extra relations (3.1) obtained by omitting the condition $0 < a < 2b$ are not only properties of the linear operations Sq^k on $P(n)$, but are relations in A_2. Hence they are implied by the Adem relations for $0 < a < 2b$.

If $a < 0$ then the relation (3.1) is again trivial, since the left hand side is zero and there are no terms on the right, but if $a \geq 2b \geq 0$, the right hand side includes terms with $j \geq b$. For example, when $a = 2$ and $b = -1$, (3.1) states that $Sq^2 Sq^{-1} = \binom{-2}{2} Sq^1 Sq^0 + \binom{-3}{0} Sq^0 Sq^1$, which reduces to the relation $Sq^1 + Sq^1 = 0$.

Proposition 3.2.1 *For all integers a, b and all* $f \in P(n)$, $R^{a,b}(f) = 0$, *where* $R^{a,b} \in A_2$ *is the element*

$$R^{a,b} = Sq^a Sq^b + \sum_j \binom{b-j-1}{a-2j} Sq^{a+b-j} Sq^j.$$

Proof As explained above, we may assume that the result is true for all a', b' with $a' + b' < a + b$. By linearity, we may assume that f is a monomial of degree $d \geq 0$. Thus we fix a and b and argue by induction on d. If $d = 0$, then $f = 1$ and so $Sq^k(f) = f$ if $k = 0$, and otherwise $Sq^k(f) = 0$. Hence $R^{a,b}(f) = 0$.

For $d > 0$, let $f = xg$ where $x = x_i$ is a variable. By the Cartan formula 1.1.9

$$Sq^a Sq^b(xg) = Sq^a(xSq^b(g) + x^2 Sq^{b-1}(g))$$
$$= xSq^a Sq^b(g) + x^2(Sq^{a-1} Sq^b + Sq^a Sq^{b-1})(g) + x^4 Sq^{a-2} Sq^{b-1}(g).$$

Applying this expansion to each term in $R^{a,b}(g)$, we have

$$R^{a,b}(xg) = xR^{a,b}(g) + x^2 S^{a,b}(g) + x^4 T^{a,b}(g), \tag{3.3}$$

where

$$S^{a,b} = Sq^{a-1} Sq^b + Sq^a Sq^{b-1} + \sum_j \binom{b-j-1}{a-2j}(Sq^{a+b-j-1} Sq^j + Sq^{a+b-j} Sq^{j-1})$$

and

$$T^{a,b} = Sq^{a-2} Sq^{b-1} + \sum_j \binom{b-j-1}{a-2j} Sq^{a+b-j-2} Sq^{j-1}.$$

Thus $T^{a,b} = R^{a-2,b-1}$, by shifting the summation index. We claim that $S^{a,b} = R^{a-1,b} + R^{a,b-1}$. By shifting the summation index in the second term, we have

$$S^{a,b} = Sq^{a-1} Sq^b + Sq^a Sq^{b-1} + \sum_j \left\{ \binom{b-j-1}{a-2j} + \binom{b-j-2}{a-2j-2} \right\} Sq^{a+b-j-1} Sq^j,$$

while $R^{a-1,b} + R^{a,b-1}$ is a similar sum, where the coefficient of $Sq^{a+b-j-1} Sq^j$ is

$$\binom{b-j-1}{a-1-2j} + \binom{b-j-2}{a-2j}.$$

Writing $c = b - j - 1$, $d = a - 2j$, we have

$$\binom{c}{d} + \binom{c-1}{d-2} = \binom{c-1}{d} + \binom{c-1}{d-1} + \binom{c-1}{d-2} = \binom{c-1}{d} + \binom{c}{d-1},$$

proving the claim.

Since the operations $T^{a,b} = R^{a-2,b-1}$ and $S^{a,b} = R^{a-1,b} + R^{a,b-1}$ correspond to elements of lower degree in A_2, both are the zero operation on $P(n)$. In particular, $T^{a,b}(g) = S^{a,b}(g) = 0$. By (3.3), $R^{a,b}(xg) = xR^{a,b}(g) = 0$ by the induction hypothesis on d. $\qquad\square$

Since the Adem relations and the relation $Sq^0 = 1$ are a set of defining relations for A_2, Proposition 3.2.1 implies the following result.

Theorem 3.2.2 *The operations* Sq^k *on* $P(n)$ *define a left* A_2*-module structure on* $P(n)$. $\qquad\square$

The hit problem asks for a minimal generating set for $P(n)$ as an A_2-module. For example, Theorem 1.4.12 states that $P(1)$ has the minimal generating set $\{1, x, x^3, x^7, \ldots\}$. Thus $f \in P(n)$ is hit if and only if $f \in A_2^+ P(n)$. In principle, we can evaluate the operation of any element of A_2 on $P(n)$ using the results of Chapter 1.

Proposition 3.2.3 *Let* $A = (a_1, a_2, \ldots,) \in$ Seq *with* $a_i > 0$ *for* $1 \leq i \leq len(A)$, *and let* $x \in P^1(n)$. *Then*

$$Sq^A(x^{2^s}) = \begin{cases} x^{2^s}, & \text{if } A = (0), \\ x^{2^r}, & \text{if } r > s \text{ and } A = (2^{r-1}, 2^{r-2}, \ldots, 2^s), \\ 0, & \text{otherwise.} \end{cases}$$

Proof Applying Proposition 1.3.2, we obtain $Sq^k(x^{2^s}) = x^{2^s}$ if $k = 0$, $Sq^k(x^{2^s}) = x^{2^{s+1}}$ if $k = 2^s$, and 0 otherwise. The result follows by induction on r. $\qquad\square$

Proposition 3.2.4 *The elements* Sq^{2^j}, $j \geq 0$, *are a minimal generating set for* A_2 *as an algebra over* $\mathbb{F}_2$.

Proof Since $Sq^0 = 1$, A_2 is generated by Sq^k for $k \geq 1$. If k is not a 2-power, then $k = 2^r(2s + 1)$, where $r \geq 0$ and $s \geq 1$. Let $a = 2^r$ and $b = 2^{r+1}s$ in the Adem relation (3.1). Since $2^r \in bin(b - 1)$, $\binom{b-1}{a}$ is odd by Proposition 1.4.11, and so the last term $Sq^{a+b} = Sq^k$ appears in the Adem relation. Hence Sq^k is in the subalgebra generated by the elements Sq^i for $i < k$. It follows by iterating the argument that A_2 is generated by Sq^{2^j}, $j \geq 0$.

If Sq^{2^j} can be omitted from this generating set, then by considering the grading on A_2 it would follow that Sq^{2^j} is in the subalgebra generated by the elements Sq^i for $i < 2^j$. But this is false, since $Sq^i(x^{2^j}) = 0$ for $0 < i < 2^j$ by Proposition 3.2.3, while $Sq^{2^j}(x^{2^j}) = x^{2^{j+1}}$. Hence the generating set is minimal. $\qquad\square$

3.3 Symmetric functions and blocks

A polynomial in $P(n)$ is symmetric if it is invariant under all permutations of the variables $x_1, x_2, \ldots, x_n$. Standard examples are the elementary symmetric function $e_k(n)$, the sum of all k-fold products of distinct variables, the complete symmetric function $h_k(n)$, the sum of all monomials of degree k and the power sum function $p_k(n) = x_1^k + x_2^k + \cdots + x_n^k$. In particular, the product $x_1 x_2 \cdots x_n$ is a symmetric polynomial.

Proposition 3.3.1 *Let $\theta \in A_2$ and let $f \in P(n)$ be symmetric. Then $\theta(f)$ is symmetric.*

Proof This follows from Proposition 1.2.5, since permutations of the variables $x_1, \ldots, x_n$ are given by the action on $P(n)$ of permutation matrices in $GL(n)$. $\square$

Given a monomial $f \in P(n)$, the sum of all distinct monomials formed from f by permuting the variables is called a monomial symmetric function. The functions $e_k(n)$ and $p_k(n)$ above are examples. Every symmetric polynomial can be uniquely written as a sum of monomial symmetric functions. In particular, a symmetric polynomial with 2-power exponents can be written uniquely as a sum of monomial symmetric functions of the same form.

Recall from Definition 2.4.1 that $\omega(d)$ is the sequence $\omega = (\omega_1, \omega_2, \ldots) \in \mathrm{Seq}$ given by the reversed binary expansion of d. Conversely, if $\omega \in \mathrm{Seq}$ then $\omega = \omega(d)$ where d is given by (2.4). Writing $\omega(d)$ as $\omega(x^d)$, we extend the definition to monomials in $P(n)$ as follows.

Definition 3.3.2 Let $n \geq 1$ and let $f = x_1^{d_1} \cdots x_n^{d_n}$ be a monomial in $P(n)$. The ω-**sequence** $\omega(f)$ of f is the sum $\omega(f) = \sum_{i=1}^{n} \omega(d_i)$ of the ω-sequences of its exponents.

We usually write ω-sequences as finite sequences by omitting trailing 0s. For example, $\omega(x_1^7 x_2^4 x_3) = \omega(7) + \omega(4) + \omega(1) = (1, 1, 1) + (0, 0, 1) + (1, 0, 0) = (2, 1, 2)$. For given n and d, we often consider the set of sequences which arise as ω-sequences of monomials $f \in P^d(n)$. The following definition is made so that $\deg f = \deg_2 \omega(f)$.

Definition 3.3.3 The 2-**degree** of a sequence $R = (r_1, r_2, \ldots)$ of integers ≥ 0, of length ℓ, is

$$\deg_2 R = \sum_{i=1}^{\ell} 2^{i-1} r_i.$$

We denote the set of sequences of 2-degree d by Seq_d. Thus the lower d in Seq_d refers to the 2-degree, while the upper d in Seq^d refers to the modulus.

Since $\omega(f) = \omega(g)$ if g is obtained from f by a permutation of the variables, the sum of all monomials with a given ω-sequence R is a symmetric polynomial. The same is true if we restrict all the exponents to be 2-powers. In this case, r_i is the number of variables with exponent 2^{i-1} in each term of the symmetric polynomial, and $|R| = \sum_i r_i$ is the total number of variables.

Definition 3.3.4 For $R = (r_1, \ldots, r_\ell)$ and $|R| = n$, the **Cartan symmetric function** $c(R) \in P(n)$ is the sum of all monomials whose exponents are 2-powers and whose ω-sequence is R. In particular, if $R = (n, 0, 0, \ldots)$, $c(R)$ is the product of the variables, and we write $c(n) = x_1 x_2 \cdots x_n$. The **leading term** of $c(R)$ is the unique term whose exponents are in decreasing order.

By Definition 3.3.3, the homogeneous polynomial $c(R)$ has degree $\deg_2 R$. For example, for $n = 5$, $c(2, 1, 2) \in P^{12}(n)$ is the sum of all distinct monomials obtained from the leading term $x_1^4 x_2^4 x_3^2 x_4 x_5$ by permuting the variables, and $\deg_2(2, 1, 2) = 2 \cdot 1 + 1 \cdot 2 + 2 \cdot 4 = 12$. In particular, $c(n)$ divides $c(R)$ for all R such that $|R| = n$.

Proposition 3.3.5 *If $c(R)$ is a Cartan symmetric function and Sq^A is a monomial in A_2, then $Sq^A(c(R))$ is a sum of Cartan symmetric functions $c(S)$ such that $|S| = |R|$ and $\deg_2 S = |A| + \deg_2 R$.*

Proof By Proposition 3.2.3 and the Cartan formula, all exponents in $Sq^A(c(R))$ are 2-powers. By Proposition 3.3.1, $Sq^A(c(R))$ is symmetric, and so it can be written uniquely as a sum of Cartan symmetric functions $c(S)$, where $|S| = |R|$ by Proposition 1.1.11. Since $c(R)$ has degree $\deg_2 R$, $\deg_2 S = |A| + \deg_2 R$. $\quad\square$

The following **block** notation for monomials is helpful for practical calculations involving the action of Steenrod operations on $P(n)$. We represent a monomial by writing the ω-sequences of its exponents as the rows of an array. We generally omit trailing 0s in the rows, but show at least one entry as a marker in a zero row. The following example shows some monomials in $P(2)$ and their associated 2-blocks.

Example 3.3.6

$x^{11}y^7$	x^6	$x^{2^k-1}y^{2^k}$
1 1 0 1	0 1 1	$1 \cdots 1$
1 1 1	0	$0 \cdots 0\,1$

Definition 3.3.7 An n-**block** $B = (b_{i,j})$ is an array of integers 0 or 1 defined for $1 \le i \le n$ and $j \ge 1$. For each i there are only finitely many j's such that $b_{i,j} = 1$. The row sums $\alpha_i(B) = \sum_{j \ge 1} b_{i,j}$ form the α-**sequence** $\alpha(B) = (\alpha_1(B), \ldots, \alpha_n(B))$ of B , and the column sums $\omega_j(B) = \sum_{i=1}^{n} b_{i,j}$ form the ω-**sequence** $\omega(B) = (\omega_1(B), \omega_2(B), \ldots)$ of B. Thus $|\alpha(B)| = |\omega(B)|$ is the number of digits 1 in B. The block B has 2-**degree** $\deg_2 B = \deg_2 \omega(B)$.

We associate to a monomial $f = x_1^{d_1} \cdots x_n^{d_n}$ in $P(n)$ the n-block F whose ith row is $\omega(x_i^{d_i})$ for $1 \le i \le n$. In particular, the zero n-block corresponds to the monomial 1, the identity element of $P(n)$. Then $\omega(F) = \omega(f)$, and we define $\alpha(f) = \alpha(F)$. The degree of the monomial f is the 2-degree of the block F or of the sequence $\omega(F)$, whereas it is the modulus of the exponent sequence $(d_1, \ldots, d_n)$.

In Section 3.4, we shall prove that admissible monomials in A_2 are linearly independent by associating to an admissible monomial Sq^A a particular term in $Sq^A(c(n))$ for $n \ge |A|$, which we call its 'leading term'. By Proposition 3.3.5, $Sq^A(c(n))$ has a unique expression as a sum $\sum_S c(S)$ of Cartan symmetric functions. We order the summands $c(S)$ using the right order on the corresponding sequences S.

Definition 3.3.8 Let $Sq^A \in A_2$ be a monomial (which need not be admissible), and let $n \ge |A|$. The **leading term** of $Sq^A(c(n))$ is the leading term of the $<_r$-minimal summand $c(S)$ of $Sq^A(c(n))$, and the **leading block** of $Sq^A(c(n))$ is the corresponding block.

Thus if $R = (r_1, r_2, \ldots)$ with $|R| = n$, the leading block of $c(R)$ is an n-block with one digit 1 in each row, where the last r_1 rows have a 1 in the first column, the preceding r_2 rows have a 1 in the second column, and so on.

Example 3.3.9 Let $A = (2,1)$ and $R = (3)$. Then $c(R) = x_1 x_2 x_3$ and $Sq^A(c(R)) = Sq^2(Sq^1(x_1 x_2 x_3)) = Sq^2(x_1^2 x_2 x_3 + x_1 x_2^2 x_3 + x_1 x_2 x_3^2) = x_1^4 x_2 x_3 +$

$x_1 x_2^4 x_3 + x_1 x_2 x_3^4 + x_1^2 x_2^2 x_3^2$. In block notation, this becomes

$$Sq^{(2,1)} \begin{pmatrix} 1 \\ 1 \\ 1 \end{pmatrix} = Sq^2 \begin{pmatrix} 0\,1 & 1 & 1 \\ 1 & +\,0\,1 & +\,1 \\ 1 & 1 & 0\,1 \end{pmatrix} = 1 \begin{matrix} 0\,0\,1 & 1 & 1 & 0\,1 \\ & +\,0\,0\,1 & +\,1 & +\,0\,1 \\ 1 & 1 & 0\,0\,1 & 0\,1 \end{matrix}$$

where the first block is the leading block of $Sq^{(2,1)}(\mathsf{c}(3))$.

Example 3.3.10 The diagram

shows the leading blocks of $Sq^A(\mathsf{c}(6))$ for $Sq^A = Sq^6$, $Sq^5 Sq^1$ and $Sq^4 Sq^2$. These correspond to the Cartan summands $\mathsf{c}(0,6)$, $\mathsf{c}(2,3,1)$ and $\mathsf{c}(4,0,2)$. More generally, for $Sq^A(\mathsf{c}(n))$, where $n \geq 6$, the Cartan summands are $\mathsf{c}(n-6,6)$, $\mathsf{c}(n-4,3,1)$ and $\mathsf{c}(n-2,0,2)$ respectively, since an element of A_2^d can act non trivially on at most d rows of a block. As they have different leading blocks, the three polynomials $Sq^A(\mathsf{c}(n))$ are linearly independent in $P^{n+6}(n)$, and hence the admissible monomials Sq^6, $Sq^5 Sq^1$ and $Sq^4 Sq^2$ are linearly independent in A_2.

This is justified as follows. By the Cartan formula and Proposition 1.3.2, each operation Sq^k maps a block with a single digit 1 in each row to a set of new blocks by moving some of the 1s one column to the right. Each new block corresponds to a term in the Cartan formula in which k is expressed as a sum of 2-powers. This expression can be recovered from the block by ignoring the first column and treating each row in the rest of the block as the reversed binary expansion of a 2-power.

We observe that the monomials represented by the blocks shown above can arise in only one way from the action of the operations Sq^k on the full polynomials. We shall see that this is a consequence of the fact that the monomials Sq^6, $Sq^5 Sq^1$ and $Sq^4 Sq^2$ are admissible. This fails for non-admissible monomials, as can be seen by trying to prove results such as $Sq^1 Sq^1(\mathsf{c}(2)) = 0$ or $Sq^3 Sq^2(\mathsf{c}(5)) = 0$ in the same way. We also observe that, in the cases of the compositions $Sq^5 Sq^1$ and $Sq^4 Sq^2$, all digits 1 in the leading block which are moved by the first operation are moved again by the second operation. This is possible because $Sq^5 Sq^1$ and $Sq^4 Sq^2$ are admissible. Thus a block formed in this way is $<_r$-minimal.

3.4 The admissible basis of A_2

In this section we show that admissible monomials are linearly independent, and so form a vector space basis for A_2 over $\mathbb{F}_2$. We do this by showing that if $n \geq d$, the polynomials obtained by applying admissible monomials of degree d to the product $x_1 x_2 \cdots x_n$ are linearly independent in $P^{n+d}(n)$.

We associate to an admissible monomial Sq^A the sequence R given by deleting the first entry of $\omega(B)$, where B is the leading block of $Sq^A(c(n))$. The sequence R depends only on A and not on n, and we prove in Proposition 3.4.3 that the R is defined in terms of A as follows. In Example 3.3.10, these sequences are (6), $(3,1)$ and $(0,2)$.

Definition 3.4.1 Let $A = (a_1, \ldots, a_\ell)$ be an admissible sequence of length ℓ. The **Milnor sequence** of A, or of Sq^A, is $R = (r_1, \ldots, r_\ell)$, where $r_j = a_j - 2a_{j+1}$ for $1 \leq j < \ell$ and $r_\ell = a_\ell$.

Proposition 3.4.2 *The map* Adm $\to$ Seq *which sends an admissible sequence* A *to its Milnor sequence* R *is a bijection preserving length and the right order* $<_r$. *If* $A \in \mathrm{Adm}^d$, *then* $|R| = 2a_1 - d$, $\deg_2 R = a_1$ *and* $d = \sum_{j=1}^{\ell} (2^j - 1) r_j$.

Proof Since $r_\ell = a_\ell$, this map preserves length. Given $R = (r_1, \ldots, r_\ell)$ of length ℓ, the linear equations (3.4.1) can be solved recursively for $j = \ell, \ell - 1, \ldots, 1$ to give $a_j = \sum_{i=j}^{\ell} 2^{i-j} r_i$. In particular, $a_1 = \sum_{i=1}^{\ell} 2^{i-1} r_i = \deg_2 R$. Since $A = (a_1, a_2, \ldots)$ is admissible, these equations give an inverse of the given map.

For the right order, it suffices to consider sequences of the same length ℓ, since $\mathrm{len}(A) > \mathrm{len}(B)$ implies that $A <_r B$. Let $A = (a_1, \ldots, a_\ell)$ and $B = (b_1, \ldots, b_\ell)$ be admissible of length ℓ, with Milnor sequences $R = (r_1, \ldots, r_\ell)$ and $S = (s_1, \ldots, s_\ell)$ respectively. If $a_j = b_j$ for $j > k$ and $a_k > b_k$, then $r_j = s_j$ for $j > k$, and $r_k > s_k$. Hence $R <_r S$ if $A <_r B$. The sum of the equations (3.4.1) gives $|R| = a_1 - (a_2 + \cdots + a_s) = 2a_1 - d$. With the jth equation weighted by 2^{j-1}, the sum gives $\deg_2 R = a_1$. Then $d = 2\deg_2 R - |R| = \sum_{j=1}^{s} (2^j - 1) r_j$. Since $a_1 \leq d$, $|R| \leq d$. $\qquad \square$

The correspondence between admissible and Milnor sequences does not preserve the left order. For example, for $A = (6,3)$ and $A' = (6,2,1)$, $A >_l A'$

but $R = (0,3)$ and $R' = (2,0,1)$, so $R <_l R'$. For the right order, we have $A >_r A'$ but $R >_r R'$.

Proposition 3.4.3 *Let* $A = (a_1,\ldots,a_\ell)$ *be an admissible sequence with Milnor sequence* $R = (r_1,\ldots,r_\ell)$, *and let* $n \geq |A|$. *Then* $Sq^A(c(n))$ *has* $<_r$-minimal Cartan summand $c(R^+)$, where $R^+ = (n - |R|, r_1, r_2, \ldots)$.

Proof We evaluate Sq^{a_i} successively for $i = \ell, \ell - 1, \ldots, 1$ using the Cartan formula. For the first step, $Sq^{a_\ell}(c(n)) = c(n - r_\ell, r_\ell)$, with leading block B_1 obtained by applying Sq^1 in each of the first $r_\ell = a_\ell$ rows. Since $a_{\ell-1} = 2a_\ell + r_{\ell-1}$, a block B_2 in $Sq^{a_{\ell-1}}Sq^{a_\ell}(c(n))$ is obtained by applying Sq^2 in the first r_ℓ rows of B_1 and Sq^1 in the next $r_{\ell-1}$ rows. Since B_2 can arise only in this way, the corresponding monomial appears in $Sq^{a_{\ell-1}}Sq^{a_\ell}(c(n))$, and so $c(n - r_\ell - r_{\ell-1}, r_{\ell-1}, r_\ell)$ is a summand in $Sq^{a_{\ell-1}}Sq^{a_\ell}(c(n))$. It is the $<_r$-minimal summand, as all entries 1 in the second column of B_1 have been moved to the third column of B_2.

The contribution to the degree of B_2 from the first $r_\ell + r_{\ell-1}$ variables is $4r_\ell + 2r_{\ell-1} = 2a_{\ell-1}$. Since $a_{\ell-2} = 2a_{\ell-1} + r_{\ell-2}$, a block B_3 in $Sq^{a_{\ell-2}}Sq^{a_{\ell-1}}Sq^{a_\ell}(c(n))$ is obtained by applying Sq^4 in the first r_ℓ rows of B_2, Sq^2 in the next $r_{\ell-1}$ rows and Sq^1 in the next $r_{\ell-2}$ rows. Since the block B_3 can arise only in this way, B_3 represents a monomial which appears in $Sq^{a_{\ell-2}}Sq^{a_{\ell-1}}Sq^{a_\ell}(c(n))$. Since B_2 is the leading block of $c(n - r_\ell - r_{\ell-1} - r_{\ell-2}, r_{\ell-2}, r_{\ell-1}, r_\ell)$, this Cartan symmetric function is a summand in $Sq^{a_{\ell-2}}Sq^{a_{\ell-1}}Sq^{a_\ell}(c(n))$. Again, it is the $<_r$-minimal summand, since all entries 1 in the second and third columns of B_2 have been moved to the right. It follows by iteration that the $<_r$-minimal summand in $Sq^A(c(n))$ is $c(n - |R|, r_1, \ldots, r_\ell)$. $\square$

Theorem 3.4.4 *The set of admissible monomials is a vector space basis for* A_2.

Proof Using Propositions 3.1.2 and 3.1.7, it is sufficient to prove that the admissible monomials of degree d are linearly independent for $d \geq 0$. Let $n \geq d$ and consider the linear map $A_2^d \to P^{n+d}(n)$ defined by $\theta \mapsto \theta(c(n))$. Since an admissible sequence A is uniquely determined by its Milnor sequence R, it follows from Proposition 3.4.3 that the symmetric polynomials obtained by applying admissible monomials Sq^A of degree d to $c(n)$ have different leading terms. Hence the polynomials $Sq^A(c(n))$ are linearly independent, and so the elements Sq^A are also linearly independent. $\square$

It follows that we can use the action of A_2 on $P(n)$ to study the structure of A_2 itself, since elements of A_2^d are faithfully represented by the action of the corresponding operations on $P(n)$ when $n \geq d$. In fact, we have seen that it suffices to consider the universal case provided by the action on the product of variables $c(n) = x_1 \cdots x_n$. Thus we have the following result.

Proposition 3.4.5 *For $0 \leq d \leq n$, the linear map $A_2^d \to P^{n+d}(n)$ defined by $Sq^A \mapsto Sq^A(c(n))$ is injective.* □

As a first example of this method, Proposition 3.2.1 has the following consequence. Recall that $Sq^k = 0$ if $k < 0$.

Proposition 3.4.6 *The Adem relation (3.1) holds in A_2 for all integers a and b.* □

Next we prove an exactness property of Sq^1 related to Proposition 1.3.5.

Proposition 3.4.7 *For $\theta \in A_2$, $\theta Sq^1 = 0$ if and only if $\theta = \phi Sq^1$ for some $\phi \in A_2$.*

Proof The 'only if' part follows from the relation $Sq^1 Sq^1 = 0$. To prove the 'if' part, let $\theta = \sum_i Sq^{A_i}$ in the admissible basis. If the last nonzero term of A_i is > 1, then $Sq^{A_i} Sq^1$ is admissible, and if it is 1 then $Sq^{A_i} Sq^1 = 0$. Hence if $\theta Sq^1 = 0$ then $Sq^{A_i} = Sq^{B_i} Sq^1$ for all i, where Sq^{B_i} is admissible, and so $\theta = \phi Sq^1$, where $\phi = \sum_i Sq^{B_i}$. □

3.5 The Milnor basis of A_2

In this section, we introduce the most widely used basis of the Steenrod algebra A_2, the **Milnor basis**. This consists of elements $Sq(R)$ indexed by sequences $R \in \text{Seq}$. If A is the admissible sequence associated to R by (3.4.1), $Sq(R)$ has the same degree as Sq^A. By Proposition 3.4.2, this means that $Sq(R) \in A_2^d$ where $d = \sum_{j=1}^{\ell} (2^j - 1) r_j$. Since $|R| = \sum_{j \geq 1} r_j$, $|R| \leq d$, with equality only for $R = (d)$. Recall from Definition 3.3.4 that $c(R) \in P(n)$ is the Cartan symmetric function corresponding to a sequence R such that $|R| = n$.

Theorem 3.5.1 *Let $R = (r_1, r_2, \ldots) \in \text{Seq}$, $d = \sum_{j \geq 1} (2^j - 1) r_j$, and $n \geq d$. Then there is a unique element $Sq(R) \in A_2^d$ such that*

$$Sq(R)(c(n)) = c(n - |R|, r_1, r_2, \ldots).$$

*The elements $Sq(R)$ form a $\mathbb{F}_2$-basis for A_2. This basis, the **Milnor basis**, is triangularly related to the admissible basis in the right order $<_r$.*

Proof As above, let A be the admissible sequence associated to R. Then d is the degree of $Sq^A \in \mathsf{A}_2$. By Proposition 3.4.3, $Sq^A(c(n))$ is a sum of Cartan symmetric functions

$$Sq^A(c(n)) = c(R^+) + \sum_i c(R_i) \tag{3.4}$$

where $R^+ = (n - |R|, r_1, r_2, \ldots)$, $c(R_i)$ has degree $n + d$, and $R_i >_r R^+$ for all terms in the sum. Thus we obtain a triangular system of linear equations in A_2^d by listing equations (3.4) in increasing right order for A. For example, for $d = 6$ we obtain the equations

$$Sq^4 Sq^2(c(6)) = c(4,0,2) + a\,c(2,3,1) + b\,c(0,6),$$
$$Sq^5 Sq^1(c(6)) = c(2,3,1) + c\,c(0,6)$$
$$Sq^6(c(6)) = c(0,6),$$

where $a, b, c \in \mathbb{F}_2$, and the corresponding Milnor elements satisfy $Sq(6)(c(6)) = c(0,6)$, $Sq(3,1)(c(6)) = c(2,3,1)$ and $Sq(0,2)(c(6)) = c(4,0,2)$. Thus by using Proposition 3.4.5 and solving the resulting equations in A_2 recursively, we obtain an element

$$Sq(R) = Sq^A + \sum_j Sq^{B_j} \tag{3.5}$$

in A_2^d such that $Sq(R)(c(n)) = c(R^+)$ and each B_j in the sum is admissible with $B_j >_r A$. By Proposition 3.4.5, equation (3.5) determines $Sq(R)$ uniquely.

Again by Proposition 3.4.5, every element of A_2^d is determined uniquely by its action on $c(n)$, since $n \geq d$. Hence equation (3.4) gives a corresponding equation

$$Sq^A = Sq(R) + \sum_i Sq(S_i) \tag{3.6}$$

in A_2^d, where $S_i^+ = R_i$ for all i. Since $R_i >_r R^+$, $S_i > R$. By Theorem 3.4.4, the admissible monomials form a basis for A_2. Hence the elements $Sq(R)$ also form a basis. By (3.6), the two bases are triangularly related in the right order. $\square$

Since $Sq(R) \in \mathsf{A}_2^d$ where $d = \sum_{j \geq 1}(2^j - 1)r_j$, Milnor basis elements in degree d correspond to partitions of d as sums of integers of the form $2^j - 1$, and $|R| = \sum_{j \geq 1} r_j$ is the number of parts in such a partition. By Definition 2.4.2, the minimum value of $|R|$ for Milnor basis elements $Sq(R) \in \mathsf{A}_2^d$ is $\mu(d)$. The table below shows the Milnor basis in degrees ≤ 9.

degree d	Milnor basis of A_2^d	dim A_2^d
0	$Sq(0) = 1$	1
1	$Sq(1)$	1
2	$Sq(2)$	1
3	$Sq(3), Sq(0,1)$	2
4	$Sq(4), Sq(1,1)$	2
5	$Sq(5), Sq(2,1)$	2
6	$Sq(6), Sq(3,1), Sq(0,2)$	3
7	$Sq(7), Sq(4,1), Sq(1,2), Sq(0,0,1)$	4
8	$Sq(8), Sq(5,1), Sq(2,2), Sq(1,0,1)$	4
9	$Sq(9), Sq(6,1), Sq(3,2), Sq(0,3), Sq(2,0,1)$	5

The next result strengthens Theorem 3.5.1 and helps in evaluating Milnor basis elements on polynomials. We show in Proposition 5.8.3 that Theorem 3.5.1 can be further refined by using partial orderings on the corresponding sequences.

Proposition 3.5.2 *For $R = (r_1, r_2, \ldots)$, $Sq(R)(c(n)) = c(n - |R|, r_1, r_2, \ldots)$ if $n \geq |R|$ and $Sq(R)(c(n)) = 0$ if $n < |R|$.*

Proof By Theorem 3.5.1 $Sq(R)(c(n)) = c(R^+)$ when n is sufficiently large, and in particular for $n \geq \sum_{i \geq 1} (2^i - 1) r_i$, where $R^+ = (n - |R|, r_1, r_2, \ldots)$. By Proposition 1.3.6, $Sq(R)$ commutes with the partial differentiation operator $\partial / \partial x_i$ on $P(n)$ for $1 \leq i \leq n$. If all exponents in a monomial $f \in P(n)$ are 2-powers, then $\partial f / \partial x_i = f / x_i$ if the exponent of x_i in f is 1, and $\partial f / \partial x_i = 0$ otherwise. Hence $\partial c(R^+) / \partial x_n = c(n - |R| - 1, r_1, r_2, \ldots)$ if $n - 1 \geq |R|$, and $\partial c(R^+) / \partial x_n = 0$ if $|R| = n$. However $\partial c(n) / \partial x_n = c(n - 1)$ for all $n \geq 1$. It follows that $Sq(R)(c(n)) = c(R^+)$ for all $n \geq |R|$. When $n = |R| - 1$, $Sq(R)(c(n - 1)) = Sq(R) \partial c(n) / \partial x_n = (\partial / \partial x_n) Sq(R)(c(n)) = (\partial / \partial x_n) c(0, r_1, r_2, \ldots) = 0$. Continuing in this way, we have $Sq(R)(c(n)) = 0$ for all $n < |R|$. □

Proposition 3.5.3 *For $x \in P^1(n)$, $Sq(R)(x) = x^{2^k}$ if $R_k = (0, \ldots, 0, 1)$ is the sequence defined by $r_k = 1$ and $r_i = 0$ for $i \neq k$, and $Sq(R)(x) = 0$ otherwise.*

Proof For $n = 1$, this is the case $n = 1$ of Proposition 3.5.2. The general case follows by considering linear maps $P(1) \to P(n)$ as in Section 1.2. □

For $d \geq 0$, Sq^d is the $<_r$-maximal admissible monomial in A_2^d, and $A = (d,0,\ldots)$ has Milnor sequence $R = A$. Hence $Sq(d) = Sq^d$, so that Sq^d is an element of both bases. In particular $Sq(0,0,\ldots) = 1$. In principle, we can express a given admissible monomial $Sq^A \in A_2^d$ in the Milnor basis by evaluating it on $c(d)$ and using (3.4) and (3.6). For example, $Sq^2 Sq^1(c(3)) = Sq^2(c(2,1)) = c(0,3) + c(2,0,1)$, and so $Sq^2 Sq^1 = Sq(3) + Sq(0,1)$.

Example 3.5.4 The table below gives the conversion from the admissible basis to the Milnor basis in degree 9, using the right order. For example, the last row corresponds to the equation $Sq^6 Sq^2 Sq^1(c(n)) = c(n - 3,2,0,1) + c(n - 3,0,3) + c(n - 5,3,2)$. Note that the two bases are not triangularly related in the left order.

	$Sq(9)$	$Sq(6,1)$	$Sq(3,2)$	$Sq(0,3)$	$Sq(2,0,1)$
Sq^9	1	0	0	0	0
$Sq^8 Sq^1$	1	1	0	0	0
$Sq^7 Sq^2$	0	0	1	0	0
$Sq^6 Sq^3$	0	1	1	1	0
$Sq^6 Sq^2 Sq^1$	0	0	1	1	1

The combinatorics of keeping track of the multiplicities mod 2 of the monomials which arise in this type of calculation soon become unmanageable. Milnor's product formula (Theorem 4.1.2) provides a much more efficient method.

Example 3.5.5 We express $Sq^a Sq^b$ in the Milnor basis by evaluating it on $c(a+b)$. Thus $Sq^b(c(a+b)) = c(a,b)$, and $Sq^a(c(a,b))$ is a sum of terms of the form $c(2k, a+b-3k, k)$, where $0 \leq k \leq [a/2]$. We count the number of ways in which the leading term of $c(2k, a+b-3k, k)$ can arise. This depends on the choice of $b - k$ of $a+b-3k$ rows where the move $x_i \mapsto x_i^2$ is effected by Sq^b rather than Sq^a. Thus

$$Sq^a Sq^b(c(d)) = \sum_{k=0}^{[a/2]} \binom{a+b-3k}{b-k} c(2k, a+b-3k, k). \tag{3.7}$$

It follows that $Sq^a Sq^b = \sum_{k=0}^{[a/2]} \binom{a+b-3k}{b-k} Sq(a+b-3k, k)$.

Definition 3.5.6 For $\ell \geq 0$, let Ad_ℓ denote the $\mathbb{F}_2$-subspace of A_2 spanned by admissible monomials Sq^A of length $\text{len}(A) = \ell$, and define $Ad_{\leq \ell}$, $Ad_{\geq \ell}$ similarly.

Since the correspondence between admissible and Milnor sequences preserves length, by Theorem 3.5.1 the Milnor basis elements $Sq(R)$ where $\text{len}(R) \leq \ell$ also form a basis for $\text{Ad}_{\leq \ell}$. If Sq^A is a non-admissible monomial in A$_2$, where $\text{len}(A) \leq \ell$, then we can express Sq^A as a sum of admissible monomials of length $\leq \ell$ by using the Adem relations (3.1), and so $Sq^A \in \text{Ad}_{\leq \ell}$.

The Cartan formula 1.1.4 can be generalized to Milnor basis elements. We illustrate the argument using the special case $Sq(r) = Sq^r$. Writing the variables in $P(m+n)$ as $x_1, \ldots, x_m, y_1, \ldots, y_n$, we have

$$Sq(r)(x_1 \cdots x_m y_1 \cdots y_n) = \sum_{s+t=r} Sq(s)(x_1 \cdots x_m) \cdot Sq(t)(y_1 \cdots y_n),$$

since a choice of r variables on the left hand side corresponds to a choice of s of the x's and t of the y's on the right hand side, where $r = s + t$. By Proposition 1.2.5, the operations $Sq(r)$ commute with specializations of the variables, and so the equation $Sq(r)(fg) = \sum_{s+t=r} Sq(s)(f) \cdot Sq(t)(g)$ holds for monomials f and g, and by linearity it also holds for polynomials.

Proposition 3.5.7 (Cartan formula) *For all Milnor basis elements $Sq(R)$ and $f, g \in P(n)$,*

$$Sq(R)(fg) = \sum_{R=S+T} Sq(S)(f) \cdot Sq(T)(g).$$

Proof Recall that $Sq(R)(c(n)) = c(R^+)$, the sum of all monomials in $x_1, \ldots, x_n$ in which r_j of the variables are raised to the 2^jth power for $j \geq 1$. We rename the variables as $x_1, \ldots, x_m, y_1, \ldots, y_n$ and consider $Sq(R)(c(m+n))$. Each monomial in $Sq(R)(c(m+n))$ corresponds to a choice of s_j of the x's and t_j of the y's to be raised to the power 2^j, where $s_j + t_j = r_j$ for $j \geq 1$. Writing $S = (s_1, s_2, \ldots)$ and $T = (t_1, t_2, \ldots)$ we obtain

$$Sq(R)(x_1 \cdots x_m y_1 \cdots y_n) = \sum_{S+T=R} Sq(S)(x_1 \cdots x_m) \cdot Sq(T)(y_1 \cdots y_n),$$

and the proof is completed as in the special case $Sq(R) = Sq^r$. □

The Cartan formula shows that in principle the evaluation of the operation $Sq(R)$ on $P(n)$ can be reduced to the case $|R| = 1$. Thus the case $|R| = 1$ gives the **primitive** Milnor basis elements (see Proposition 11.2.4).

Definition 3.5.8 For $k \geq 1$, let $R_k = (0, \ldots, 0, 1)$ be the sequence defined by $r_k = 1$ and $r_i = 0$ for $i \neq k$, and let $Q_k = Sq(R_k) \in A_2^{2^k-1}$ be the corresponding Milnor basis element.

Example 3.5.9 By Proposition 3.5.3, $Q_k(x) = x^{2^k}$, and setting $n = 2$ in Proposition 3.5.2 gives $Q_k(xy) = x^{2^k}y + xy^{2^k}$. Comparing with Section 2.2, we see that Xq^{2^k-1} also has these properties. This does not imply that Q_k and Xq^{2^k-1} are equal in A_2 for $k > 1$, because Proposition 3.4.5 requires us to test their action on $c(n)$ for $n \geq 2^k - 1$. For example, $Q_2(xyz) = x^4yz + xy^4z + xyz^4$, but the Cartan formula 2.2.4 gives $Xq^3(xyz) = x^4yz + xy^4z + xyz^4 + x^2y^2z^2$.

Part (i) of the next result states that Q_k acts as a derivation of $P(n)$.

Proposition 3.5.10 (i) *For* $f, g \in P(n)$, $Q_k(fg) = Q_k(f)g + fQ_k(g)$,
(ii) $Q_k(g^2) = 0$ *and* $Q_k(fg^2) = Q_k(f)g^2$.
(iii) *For* $x \in P^1(n)$ *and* $a \geq 0$, $Q_k(x^{2a+1}) = x^{2a+2^k}$.
(iv) $Q_k = \sum_{i=1}^{n} x_i^{2^k} \partial/\partial x_i$.
(v) Q_k *is given recursively by* $Q_1 = Sq^1$ *and* $Q_{k+1} = Sq^{2^k}Q_k + Q_kSq^{2^k}$ *for* $k \geq 1$.

Proof If $R = (0, \ldots, 0, 1)$ in Proposition 3.5.7, then either $S = R$, $T = (0)$ or $S = (0)$, $T = R$. Since $Sq(0) = 1$, we obtain (i), and (ii) follows. To prove (iii), let $f = x$ and $g = x^a$ in (i) and use Proposition 3.5.3. The operator $D_k = \sum_{i=1}^{n} x_i^{2^k} \partial/\partial x_i$ on $P(n)$ is a derivation by the Leibniz formula and satisfies $D_k(x_i) = x_i^{2^k}$ for $1 \leq i \leq n$. Since a derivation is determined by its values on the variables x_i, (iv) follows.

For (v), $Q_1 = Sq(1) = Sq^1$ by definition. Using (iv) and the Cartan formula, $Sq^{2^k}Q_k(f) = Sq^{2^k}\sum_{i=1}^{n} x_i^{2^k}\partial f/\partial x_i = \sum_{i=1}^{n} x_i^{2^{k+1}}\partial f/\partial x_i + \sum_{i=1}^{n} x_i^{2^k}Sq^{2^k}(\partial f/\partial x_i)$. By Proposition 1.3.6, the second sum is $\sum_{i=1}^{n} x_i^{2^k}\partial(Sq^{2^k}f)/\partial x_i$. Hence using (iv), the first sum is $Q_{k+1}(f)$ and the second is $Q_kSq^{2^k}(f)$. $\qquad\square$

The Milnor basis provides an alternative description of the conjugate squares.

Proposition 3.5.11 *For* $k \geq 0$, Xq^k *is the sum of all Milnor basis elements* $Sq(R)$ *in* A_2^k.

Proof Since the total conjugate square Xq is multiplicative,

$$Xq(c(n)) = Xq(x_1)Xq(x_2)\cdots Xq(x_n) = \prod_{i=1}^{n}(x_i + x_i^2 + x_i^4 + \cdots),$$

and this is the sum of all monomials in $P(n)$ with 2-powers as exponents. Thus the terms of degree $n + k$ in $Xq(c(n))$ give the sum of all Cartan symmetric functions of degree $n + k$. This is $\sum_R c(R^+)$, where the sum is over all R such that $Sq(R) \in A_2^k$. Taking $n \geq k$, the result follows from Proposition 3.4.5. $\qquad\square$

We use the Milnor basis to calculate the dimension of A_2^d as a vector space over $\mathbb{F}_2$. It is convenient to express the result in terms of a generating function.

Definition 3.5.12 Let $V = \sum_{d \geq 0} V^d$ be a graded vector space over $\mathbb{F}_2$, such that V^d is finite-dimensional for all $d \geq 0$. The power series

$$P(V,t) = \sum_{d \geq 0} (\dim V^d) t^d$$

is called the **Poincaré series** of V. If $\dim V$ is finite, then $P(V,t)$ is a polynomial in t, the **Poincaré polynomial** of V.

For example, $\mathsf{P}^d(1)$ is the 1-dimensional vector space generated by x^d, and so $P(\mathsf{P}(1),t) = 1 + t + t^2 + \cdots + t^n + \cdots = 1/(1-t)$. For all $n \geq 1$, $P(\mathsf{P}(n),t) = 1/(1-t)^n$ for $n \geq 1$ by (1.1) and the binomial theorem.

Proposition 3.5.13 *The Poincaré series of* A_2 *is*

$$P(\mathsf{A}_2,t) = \prod_{j \geq 1} \frac{1}{1 - t^{2^j - 1}} \ .$$

Proof For $d > 0$, $Sq(R)$ has degree $d = \sum_{j \geq 1} (2^j - 1) r_j$, where $R = (r_1, r_2, \ldots)$. Thus $\dim \mathsf{A}_2^d$ is the number of solutions $R = (r_1, r_2, \ldots)$ of the equation

$$d = r_1 + 3r_2 + \cdots + (2^j - 1)r_j + \cdots , \tag{3.8}$$

where $r_j \geq 0$ for $j \geq 1$. A solution of (3.8) gives an expression for d as the sum of $|R|$ terms, of which r_j are equal to $2^j - 1$ for $j \geq 1$. Since $1/(1 - t^{2^j - 1}) = \sum_{i \geq 0} t^{i(2^j - 1)}$, this corresponds to a term of degree d in the product of power series $\prod_{j \geq 1} 1/(1 - t^{2^j - 1})$. $\qquad\square$

3.6 Remarks

The cohomology $H^*(X)$ of a topological space X over $\mathbb{F}_2$ is an A_2-algebra in a natural way. Indeed, topologists define A_2 as the algebra of universal stable operations for cohomology theory over $\mathbb{F}_2$. As discussed in Section 1.9, in the case $X = \mathbb{R}P^\infty \times \cdots \times \mathbb{R}P^\infty$ we have $H^*(X) = \mathsf{P}(n)$. The standard reference for the development of the Steenrod algebra as an algebra of stable cohomology operations is [196], which is based on lectures by Steenrod himself. Sections 3.1 and 3.2 follow Chapter I of [196] in general, but we avoid explicit use of algebraic topology, relying instead on the action of the operations Sq^k on $\mathsf{P}(n)$ to supply information about A_2.

J.-P. Serre [179] derived the Adem relations from the action of the Steenrod squares Sq^k on X, showed that they generate all stable operations in mod 2 cohomology and established the basis of admissible monomials. The work of Cartan and Serre showed that A_2 is determined by its action on $\mathsf{P}(n)$ for

$n \geq 1$. For background on symmetric functions relevant to Section 3.2, see for example Chapter 1 of [128]. For the history of algebraic topology up to 1960, see [51].

The foundations for the study of the internal structure of the algebra A_2 were laid by Adams [1] and Milnor [134]. Both gave new (and closely related) additive bases for A_2, but while [1] follows the methods of Cartan and Serre, [134] is based on the work of Milnor and J. C. Moore on Hopf algebras [135]. In Section 3.5 we have presented Milnor's basis using the methods of Cartan and Serre, and in Chapter 11 we shall describe the Hopf algebra approach. In Proposition 12.2.13, we discuss some of the bases of A_2 found by Adams.

4

Products and conjugation in A_2

4.0 Introduction

In this chapter we continue to explore the structure of the Steenrod algebra A_2. In Section 4.1 we discuss the formula for multiplying elements $Sq(R)$ of the Milnor basis. The existence of this formula is a major reason for the preference given to this basis in the literature. In Section 4.2 we prove the **Bullett–Macdonald identity**, which expresses the Adem relations in a concise form. This provides a proof that the conjugate Steenrod squares Xq^k satisfy a set of relations dual to the Adem relations. In Section 4.3, we use this to define the **conjugation** or anti-isomorphism χ of A_2, which exchanges the operations Sq^k and Xq^k. Section 4.4 contains further formulae involving χ.

4.1 The Milnor product formula

An important feature of the Milnor basis is that there is a combinatorial formula for multiplication of basis elements.

Definition 4.1.1 A **Milnor matrix** is an array

$$
X = \begin{array}{c|cccc}
 & x_{0,1} & x_{0,2} & \cdots \\
\hline
x_{1,0} & x_{1,1} & x_{1,2} & \cdots \\
x_{2,0} & x_{2,1} & x_{2,2} & \cdots \\
\vdots & \vdots & \vdots & \ddots
\end{array}
$$

of integers $x_{i,j} \geq 0$, with only finitely many nonzero entries. The row sequence $R(X) = (r_1, r_2, \ldots)$, the column sequence $S(X) = (s_1, s_2, \ldots)$ and the diagonal

54

sequence $T(X) = (t_1, t_2, \ldots)$ of X are defined by

$$r_i = \sum_{j \geq 0} 2^j x_{i,j}, \quad s_j = \sum_{i \geq 0} x_{i,j}, \quad t_k = \sum_{i+j=k} x_{i,j},$$

and the coefficient $b(X)$ is the product of mod 2 multinomial coefficients

$$b(X) = \prod_{k \geq 1} \binom{t_k}{x_{k,0} \; x_{k-1,1}, \ldots, x_{0,k}}.$$

Theorem 4.1.2 *The product of Milnor basis elements is given by*

$$Sq(R)Sq(S) = \sum_X b(X)Sq(T(X)),$$

where the sum is over all Milnor matrices X with $R(X) = R$ and $S(X) = S$.

A special case of the Milnor product formula was proved in Example 3.5.5. In this case, the Milnor matrix corresponding to the kth term in the product is

$$X = \begin{array}{c|ccc} & & b-k & 0 & \cdots \\ & a-2k & k & 0 & \cdots \\ & 0 & 0 & 0 & \cdots \\ & \vdots & \vdots & \vdots & \ddots \end{array}$$

and $b(X)$ reduces to a single binomial coefficient.

To implement the product formula, we begin by entering R in the first column, S in the first row and 0s elsewhere to get the initial Milnor matrix

$$X = \begin{array}{c|ccc} & s_1 & s_2 & \cdots \\ \hline r_1 & 0 & 0 & \cdots \\ r_2 & 0 & 0 & \cdots \\ \vdots & \vdots & \vdots & \ddots \end{array}$$

All further Milnor matrices can be generated systematically from the initial matrix as follows. We move out each r_i along its row, by subtracting powers of 2 from r_i and inserting entries $x_{i,j}$, so as to satisfy the row constraint $R(X) = R$. The column constraint $S(X) = S$ is then used to adjust the top row of the array X. Finally, by Proposition 1.4.13, the coefficient $b(X) = 1$ if and only if the entries of each diagonal of X have disjoint binary expansions.

Example 4.1.3 To evaluate the product $Sq(4,2)Sq(1,2)$, we have to consider the following Milnor matrices, where all entries not shown are 0.

$$
\begin{array}{c|cc}
 & 1 & 2 \\
\hline
4 & 0 & 0 \\
2 & 0 & 0
\end{array}
\qquad
\begin{array}{c|cc}
 & 0 & 2 \\
\hline
2 & 1 & 0 \\
2 & 0 & 0
\end{array}
\qquad
\begin{array}{c|cc}
 & 1 & 1 \\
\hline
0 & 0 & 1 \\
2 & 0 & 0
\end{array}
\qquad
\begin{array}{c|cc}
 & 0 & 2 \\
\hline
4 & 0 & 0 \\
0 & 1 & 0
\end{array}
\qquad
\begin{array}{c|cc}
 & 0 & 1 \\
\hline
0 & 0 & 1 \\
0 & 1 & 0
\end{array}
$$

As only the third and fourth matrices give $b(X) = 1 \bmod 2$, $Sq(4,2)Sq(1,2) = Sq(1,3,1) + Sq(4,2,1)$.

Example 4.1.4 As in Proposition 3.5.10, for $k \geq 1$ let $Q_k = Sq(R)$ where $R = (0,\ldots,0,1)$ with the 1 in the kth place, and let $Q_\ell = Sq(S)$ where $S = (0,\ldots,0,1)$ with the 1 in the ℓth place. Then $(Q_k)^2 = 0$ and $Q_k Q_\ell = Q_\ell Q_k$ for $k, \ell \geq 1$. In this case the initial matrix X is the only Milnor matrix, $T(X) = R + S$ and $b(X) = 1$ if $k \neq \ell$, $b(X) = 0$ if $k = \ell$.

Example 4.1.5 Let R be the sequence $(1,1,\ldots,1)$ of length k. We show by induction on k that $Sq(R) = Sq^{2^k-1} Sq^{2^{k-1}-1} \cdots Sq^3 Sq^1$. For the inductive step, the initial Milnor matrix is

$$
X = \begin{array}{c|cccc}
 & 1 & 1 & \cdots & 1 \\
\hline
2^k - 1 & 0 & 0 & \cdots & 0
\end{array}.
$$

Since both first diagonal entries are odd, $b(X) = 0 \bmod 2$. Since $y_{1,0}$ is odd for all Milnor matrices Y which arise, $b(Y) = 0 \bmod 2$ unless $y_{0,1} = 0$ and $y_{1,1} = 1$. Then by inspecting the second diagonal we see that $b(Y) = 0 \bmod 2$ unless $y_{0,2} = 0$ and $y_{1,2} = 1$. Continuing in this way, we see that the only Milnor matrix Y such that $b(Y) = 1 \bmod 2$ is

$$
Y = \begin{array}{c|cccc}
 & 0 & 0 & \cdots & 0 \\
\hline
1 & 1 & 1 & \cdots & 1
\end{array}.
$$

Since $T(Y) = (1,1,\ldots,1)$, this completes the induction.

Proof of Theorem 4.1.2 Let $R = (r_1, r_2, \ldots)$, $S = (s_1, s_2, \ldots)$. By Proposition 3.4.5, $Sq(R)Sq(S)$ is uniquely determined by $Sq(R)Sq(S)(c(d))$, where $d = \deg Sq(R) + \deg Sq(S)$. Since $d \geq \deg Sq(S)$, $Sq(S)(c(d)) = c(S^+)$ by Theorem 3.5.1, where $S^+ = (d - |S|, s_1, s_2, \ldots)$. Thus the product formula states that $Sq(R)(c(S^+))$ is the sum of Cartan symmetric functions $c(T^+)$, where $T = T(X)$ for some Milnor matrix X satisfying $R(X) = R$ and $S(X) = S$, and that $c(T^+)$ appears in this sum with coefficient $b(X)$.

We can compute $Sq(R)(c(S^+))$ by specialization of variables from the case $Sq(R)(c(n)) = c(R^+)$, where $R^+ = (n - |R|, r_1, r_2, \ldots)$ and $n \geq \deg Sq(R)$. As in Section 3.4, we work with the leading blocks of Cartan symmetric functions. Thus $Sq(R)$ acts on the block for $c(n)$ by moving r_1 1s from column 1 to column 2, r_2 1s from column 1 to column 3, and so on.

After specialization of the r_1 variables corresponding to digits moved from column 1 to column 2, we can use $Sq(R)$ to move (for example) k 1s from column 2 to column 3 and $r_1 - 2k$ 1s from column 1 to column 2, where $k < r_1/2$. More generally, specialization of the same r_1 variables allows us to use $Sq(R)$ to move $x_{1,j}$ 1s from column $j + 1$ to column $j + 2$, where $r_1 = x_{1,0} + 2x_{1,1} + 4x_{1,2} + \cdots$.

In the same way, specialization of the r_2 variables corresponding to digits moved from column 1 to column 3 allows us to move k 1s from column 2 to column 4 and $r_2 - 2k$ 1s from column 1 to column 3, where $k < r_1/2$. More generally, specialization of the same r_2 variables allows us to use $Sq(R)$ to move $x_{2,j}$ 1s from column $j + 1$ to column $j + 3$, where $r_2 = x_{2,0} + 2x_{2,1} + 4x_{2,2} + \cdots$. Thus for each $i \geq 1$ we obtain an equation $r_i = \sum_{j \geq 0} 2^j x_{i,j}$ by specializing the action of $Sq(R)$ on the r_i variables corresponding to digits in the block for $c(n)$ which are moved from column 1 to column $i + 1$. Here $x_{i,j}$ is the number of digits moved from column $j + 1$ to column $i + j + 1$. The total number of 1s ending up in column $k + 1$ is therefore $t_k = \sum_{i+j=k} x_{i,j}$, where $x_{k,0} = r_k$ and $x_{0,k} = s_k$.

Let $T = (t_1, t_2, \ldots)$, $T^+ = (d - |T|, t_1, t_2, \ldots)$. The column constraint $s_j = \sum_j x_{i,j}$ keeps track of the number of digits available in column $j + 1$ of the block for $c(d - |S|, s_1, s_2, \ldots)$ for possible moves to later columns by $Sq(R)$. The coefficient $b(X)$ keeps track of the possible choices for the ways in which these moves to later columns take place.

The same Cartan symmetric function $c(d - |T|, t_1, t_2, \ldots)$ arises regardless of how, for each k, the t_k 1s which ended up in column $k + 1$ are made up of $x_{k-j,j}$ digits 1 chosen from columns $j + 1$, $0 \leq j \leq k$. Thus we obtain $b(X)$ as a product of multinomial coefficients corresponding to the diagonals in the Milnor matrix X. $\square$

Example 4.1.6 We illustrate the combinatorics of the preceding argument using the Milnor matrix

$$
X = \begin{array}{c|cc}
 & 0 & 2 \\
\hline
2 & 1 & 0 \\
2 & 0 & 0
\end{array}
$$

in Example 4.1.3, where $R = (4, 2)$ and $S = (1, 2)$. Thus $d = \deg Sq(R) + \deg Sq(S) = 10 + 7 = 17$. Since $Sq(1, 2)(c(17)) = c(14, 1, 2)$, of degree 24, our aim is to calculate $Sq(4, 2)(c(14, 1, 2))$ as a sum $\sum_T c(T)$.

We start with the equation $Sq(4, 2)(c(24)) = c(18, 4, 2)$ and specialize variables. The polynomial $c(14, 1, 2)$ is the sum of specializations of $c(24)$, one for each monomial in $c(14, 1, 2)$, which identify 10 of the 24 variables in two sets of four and one set of two. By Proposition 1.2.5, these specializations commute with the action of $Sq(4, 2)$, and so $Sq(4, 2)(c(14, 1, 2))$ is the sum of the corresponding specializations in $c(18, 4, 2)$.

Let f be the leading term of $c(18, 4, 2)$. The Milnor matrices which arise in $Sq(R)Sq(S)$ correspond to the different ways of choosing the variables to be identified (in one set S_2 of two and two sets S_4', S_4'' of four) from the 24 variables in f, which have exponents 1 in 18 cases, 2 in 4 cases, and 4 in 2 cases. The matrix X occurs when the variables in S_2 have exponent 2 and those in S_4' and S_4'' have exponent 1 in f. This corresponds to moving the entry 1 in the initial Milnor matrix down to the second row and subtracting 2 from the entry 4 in the first column.

A monomial g in $c(T^+(X))$ arising from f by specialization has 5 variables with exponent 4, namely the two with exponent 4 in f, one arising by identification of the variables in S_2, and two arising by identification of the variables in S_4' and S_4''. The monomial g has 2 variables with exponent 2, both originally with exponent 2 in f. The remaining 10 variables in g have exponent 1, and correspond to the variables with exponent 1 in f not affected by the identifications.

The coefficient $b(X)$ is the multiplicity of $c(T^+(X))$, or equivalently of g, in $Sq(4, 2)(c(14, 1, 2))$. Consider the Cartan formula 3.5.7 applied to a monomial h in $c(14, 1, 2)$. If g is a term in $Sq(4, 2)h$, then there are 5 variables which have exponent set $\{1, 1, 1, 1, 2\}$ in h and exponent set $\{2, 2, 4, 4, 4\}$ in g, while the other variables have the same exponent in h and g.

This can only be achieved by a term in the Cartan formula for $Sq(4, 2)h$ which decomposes $(4, 2)$ as $(2, 0) + (1, 0) + (1, 0) + (0, 1) + (0, 1)$. In other words, for some variables y_i, $1 \leq i \leq 5$, the factor y_1^2 of h becomes y_1^4 in g, the factors y_2, y_3 of h become y_2^2, y_3^2 in g, and the factors y_4, y_5 of h become y_4^4, y_5^4 in g.

Thus there are $\binom{5}{2,1,2} = 30$ choices for the monomial h, according to how the 5 variables with exponent 4 in g are split as variables with exponents $1, 2$ or 4 in h. The two variables with exponent 2 in g can arise only by squaring two variables in h, so the corresponding multinomial coefficient reduces to $\binom{2}{2}$. Finally $b(X)$ is the product of these multinomial coefficients, taken over powers of 2 as exponents in g.

4.2 The Bullett–Macdonald identity

Just as the multiplicative property of the total squaring operation Sq : $P(n) \to P(n)$ expresses the Cartan formula 1.1.4 in a concise form, the Bullett–Macdonald identity expresses the Adem relations in a concise form. In order to keep track of the grading in A_2, we introduce a new variable t, and work with power series in t with coefficients in $P(n)$.

Definition 4.2.1 Let $u \in \mathbb{F}_2[[t]]$ be a power series in t. The **generalized total squaring operation** $Sq[u] : P(n) \to P(n)[[t]]$ is the linear operation

$$Sq[u] = \sum_{k \geq 0} u^k Sq^k.$$

In particular, when $u = 1$ is the identity element of $\mathbb{F}_2[[t]]$, $Sq[u] = Sq$. For a polynomial $f \in P(n)$, $Sq[t](f) = \sum_{k \geq 0} t^k Sq^k(f)$ is a polynomial in t, since $Sq^k(f) = 0$ when $k > \deg f$. More generally, the fact that $Sq^k(f)$ is nonzero for only finitely many values of k ensures that the coefficient of t^k in $Sq[u](f)$ is a polynomial in $x_1, \ldots, x_n$, whether or not the constant term of u is 0.

Proposition 4.2.2 *For all* $u \in \mathbb{F}_2[[t]]$, $Sq[u] : P(n) \to P(n)[[t]]$ *is multiplicative, i.e.* $Sq[u](fg) = Sq[u](f) \cdot Sq[u](g)$ *for all* $f, g \in P(n)$.

Proof We have $Sq[u](fg) = \sum_{k \geq 0} u^k Sq^k(fg)$, and

$$Sq[u](f) \cdot Sq[u](g) = \sum_{i \geq 0} u^i Sq^i(f) \sum_{j \geq 0} u^j Sq^j(g) = \sum_{k \geq 0} u^k \left(\sum_{i+j=k} Sq^i(f) Sq^j(g) \right).$$

The result follows from the Cartan formula 1.1.4.   □

The Bullett–Macdonald identity is a relation between the operations $Sq[u]$. We compose these by extending them to operations $Sq[u] : P(n)[[t]] \to P(n)[[t]]$, with $Sq^k(t) = t$ for all $k \geq 0$. Thus $Sq[u](\sum_{i \geq 0} f_i t^i) = \sum_{i \geq 0} Sq[u](f_i) t^i$.

Proposition 4.2.3 (Bullett–Macdonald identity) *The operations* $Sq[u]$ *on* $P(n)[[t]]$ *satisfy the relation*

$$Sq[t + t^2] Sq[1] = Sq[1 + t] Sq[t^2].$$

Proof By Proposition 4.2.2, all four operations in the relation are algebra maps of $P(n)[[t]]$. Since they all act as the identity map of $\mathbb{F}_2[[t]]$, it suffices to check that they agree on an element $x \in P^1(n)$.

The left hand side gives $Sq[t + t^2] Sq[1](x) = Sq[t + t^2](x + x^2) = (x + x^2) + (t + t^2) Sq^1(x + x^2) + (t + t^2)^2 Sq^2(x + x^2) = (x + x^2) + (t + t^2)x^2 + (t^2 + t^4)x^4$. Since $Sq[t^2](x) = x + t^2 Sq^1(x) = x + t^2 x^2$, the right hand side is $Sq[1 + t](x +$

$t^2x^2) = (x + t^2x^2) + (1 + t)Sq^1(x + t^2x^2) + (1 + t)^2Sq^2(x + t^2x^2) = (x + t^2x^2) + (1 + t)x^2 + (1 + t^2)t^2x^4$. Hence $Sq[1 + t]Sq[t^2](x) = Sq[t + t^2]Sq[1](x)$. $\qquad\square$

The Adem relations follow from the Bullett–Macdonald identity by equating coefficients of powers of t and using the grading in A_2. By Proposition 3.4.5, the resulting identities of operations on $P(n)$ for all n give corresponding identities in A_2. For example, by equating the coefficients of t and using the grading, we obtain the relations $Sq^1 Sq^{2k} = Sq^{2k+1}$ and $Sq^1 Sq^{2k+1} = 0$ for $k \geq 0$. Unfortunately the general case is not so straightforward, as we have to take linear combinations of the equations derived in this way to obtain the desired result. Recall from Proposition 3.4.6 that the Adem relation (3.1) holds in A_2 without restriction on a and b.

Proposition 4.2.4 *The identity* $Sq[t + t^2]Sq[1] = Sq[1 + t]Sq[t^2]$ *is equivalent to the set of Adem relations (3.1) with* $a, b \geq 0$.

Proof The method of proof is the same as for Proposition 2.3.2. We change the variable in the coefficient ring $\mathbb{F}_2[[t]]$ by setting $u = t + t^2$. Then $t = \sum_{i \geq 0} u^{2^i}$ by Proposition 2.1.5. The Bullett–Macdonald identity states that for $a \geq 0$, $\sum_{b \geq 0} Sq^a Sq^b$ is the coefficient of u^a in $\sum_{c,d \geq 0}(1 + t)^c t^{2d} Sq^c Sq^d$, and so, taking terms in degree $a + b$, $Sq^a Sq^b$ is the coefficient of u^a in

$$\sum_{j=0}^{a+b}(1 + t)^{a+b-j}t^{2j}Sq^{a+b-j}Sq^j,$$

when this sum is expressed in powers of u. Since $u = t + t^2$, $u^{a+1} = (1 + t)^{a+1}t^{a+1}$ and so $Sq^a Sq^b$ is the coefficient of u^{-1} in

$$\sum_{j=0}^{a+b}(1 + t)^{b-j-1}t^{2j-a-1}Sq^{a+b-j}Sq^j.$$

Thus we wish to show that for fixed a, b and j with $a \geq 0$ and $0 \leq j \leq a+b$, the coefficient of u^{-1} in $f(t) = (1 + t)^{b-j-1}t^{2j-a-1}$ is the coefficient $\binom{b-j-1}{a-2j}$ which appears in (3.1). If $2j - a - 1 \geq 0$, then $f(t) = \sum_{i \geq 0} \binom{b-j-1}{i}t^{i+2j-a-1}$ is a power series in t. Writing $t = u + u^2 + u^4 + u^8 + \cdots$, this is a power series in u, so that the coefficient of u^{-1} is 0. Hence we may assume that $j \leq a/2$, and retain only the negative powers of t in the expansion, namely

$$\sum_{i=0}^{a-2j}\binom{b-j-1}{i}t^{i+2j-a-1}. \tag{4.1}$$

The coefficient of t^{-1} here is $\binom{b-j-1}{a-2j}$, and the coefficient of u^{-1} in $t^{-1} = (u + u^2 + u^4 + u^8 + \cdots)^{-1} = u^{-1}(1 + u + u^3 + u^7 + \cdots)$ is odd. As in the proof of Proposition 2.3.2, it follows from Proposition 2.1.4 that the coefficient of u^{-1} in t^{-k} is even for $k > 1$. We conclude that the coefficient of u^{-1} in (4.1) is $\binom{b-j-1}{a-2j}$ mod 2. $\qquad\qquad\square$

4.3 Conjugation in A_2

In this section we introduce the anti-automorphism, or conjugation, χ of A_2. We begin by reworking some of the material in Section 2.3. Recall that $P[n]$ denotes the power series algebra $\mathbb{F}_2[[x_1,\ldots,x_n]]$.

Proposition 4.3.1 $Sq : P(n) \to P(n)$ and $Xq : P(n) \to P[n]$ can be extended to an inverse pair of algebra automorphisms of $P[n]$.

Proof An element $f \in P[n]$ has the form $f = \sum_I f_I x_1^{i_1} \cdots x_n^{i_n}$, with terms indexed by sequences $I = (i_1,\ldots,i_n)$ of integers > 0 and with coefficients $f_I \subset \mathbb{F}_2$. We define $Sq(f)$ and $Xq(f)$ in the natural way by $Sq(f) = \sum_I f_I Sq(x_1^{i_1} \cdots x_n^{i_n})$ and $Xq(f) = \sum_I f_I Xq(x_1^{i_1} \cdots x_n^{i_n})$. In order to see that these are well defined elements of $P[n]$, we observe that, for a given monomial h in $P(n)$, there are only finitely many monomials of lower degree. Hence there are only finitely many monomials g for which h appears in $Sq(g)$ or $Xq(g)$. Clearly Sq and Xq are algebra maps of $P[n]$.

Since Sq and Xq are multiplicative, the compositions $Sq \circ Xq$ and $Xq \circ Sq$ must be the identity map of $P[n]$ if they fix the generators. Let $x = x_i$, $1 \le i \le n$. Then $Sq \circ Xq(x) = Sq(\sum_{j \ge 0} x^{2^j}) = \sum_{j \ge 0} Sq(x^{2^j}) = \sum_{j \ge 0}(x^{2^j} + x^{2^{j+1}}) = x$, and so $Sq \circ Xq = 1$. Similarly $Xq \circ Sq(x) = Xq(x + x^2) = Xq(x) + Xq(x)^2 = \sum_{j \ge 0} x^{2^j} + \sum_{j \ge 0} x^{2^{j+1}} = x$, so $Xq \circ Sq = 1$. $\qquad\square$

We recover Proposition 2.3.3 by equating the components of $Xq \circ Sq$ and $Sq \circ Xq$ in degree k. By Proposition 3.4.5, these formulae yield the following identities in A_2.

Proposition 4.3.2 *For* $k > 0$, $\sum_{i+j=k} Sq^i Xq^j = 0$ *and* $\sum_{i+j=k} Xq^i Sq^j = 0$. $\qquad\square$

Example 4.3.3 As in Section 2.3, we can use these formulae to calculate Xq^k recursively in terms of Sq^j, $1 \le j \le k$. For example, $Xq^1 = Sq^1$, $Xq^2 = Sq^2$, $Xq^3 = Sq^2 Sq^1$, $Xq^4 = Sq^4 + Sq^3 Sq^1$ and $Xq^5 = Sq^4 Sq^1$, where we have used the Adem relations $Sq^2 Sq^2 = Sq^3 Sq^1$ and $Sq^1 Sq^1 = 0$ to express the results in the admissible basis.

Our next aim is to show that the elements $Xq^k \in A_2$ satisfy a set of conjugate Adem relations obtained by reversing the product. For example, the relations $Sq^1 Sq^2 = Sq^3$, $Sq^2 Sq^2 = Sq^3 Sq^1$ and $Sq^2 Sq^3 = Sq^5 + Sq^4 Sq^1$ have conjugate forms $Xq^2 Xq^1 = Xq^3$, $Xq^2 Xq^2 = Xq^1 Xq^3$ and $Xq^3 Xq^2 = Xq^5 + Xq^1 Xq^4$. These are easily verified using Example 4.3.3. We begin by generalizing Xq as we did for Sq in Section 4.2.

Definition 4.3.4 Let $u \in \mathbb{F}_2[[t]]$. The **generalized total conjugate square** $Xq[u] : P[n] \to P[n][[t]]$ is the linear operation

$$Xq[u] = \sum_{k \geq 0} u^k Xq^k.$$

In particular, $Xq[u] = Xq$ when $u = 1$. The same argument as for Proposition 4.2.2 shows that $Xq[u]$ is multiplicative, and as in the case of $Sq[u]$, we extend $Xq[u]$ to an operation $Xq[u] : P[n][[t]] \to P[n][[t]]$ by defining $Xq^k(t) = t$ for all $k \geq 0$. A formal calculation using Proposition 4.3.1 shows that $Sq[u]$ and $Xq[u]$ are inverse operations on $P[n][[t]]$ for all $u \in \mathbb{F}_2[[t]]$.

Proposition 4.3.5 (Conjugate Bullett–Macdonald identity) *Let* $n \geq 1$. *Then the operations* $Xq[u]$ *on* $P[n][[t]]$ *satisfy the relation*

$$Xq[t^2]Xq[1+t] = Xq[1]Xq[t+t^2].$$

Proof Since $Xq[u] = Sq[u]^{(-1)}$ is the inverse of $Sq[u]$ with respect to composition of operations on $P[n][[t]]$, this follows from Proposition 4.2.3. $\qquad\square$

The argument of Proposition 4.2.4 shows that this identity is equivalent to a set of conjugate Adem relations.

Proposition 4.3.6 *For all integers a and b, the* **conjugate Adem relations**

$$Xq^b Xq^a = \sum_{j=0}^{[a/2]} \binom{b-j-1}{a-2j} Xq^j Xq^{a+b-j}$$

hold in A_2, *where* $Xq^k = 0$ *if* $k < 0$. $\qquad\square$

Definition 4.3.7 The **conjugation** $\chi : A_2 \to A_2$ is the linear map defined on monomials Sq^A by

$$\chi(Sq^{a_1} Sq^{a_2} \cdots Sq^{a_s}) = Xq^{a_s} \cdots Xq^{a_2} Xq^{a_1}. \tag{4.2}$$

In particular, $\chi(Sq^k) = Xq^k$ for $k \geq 0$.

The map χ is well defined, since by Proposition 4.3.6 it is consistent with the Adem relations (3.1), and these, together with the relation $Sq^0 = 1$, are a set of defining relations for A_2.

Proposition 4.3.8 *The conjugation* $\chi : A_2 \to A_2$ *has the following properties.*

(i) *For all* $\theta_1, \theta_2 \in A_2$, $\chi(\theta_1\theta_2) = \chi(\theta_2)\chi(\theta_1)$,
(ii) $\chi(Xq^k) = Sq^k$ *for* $k \geq 0$,
(iii) χ^2 *is the identity map of* A_2.

Proof By definition, (i) holds when $\theta_1 = Sq^A$ and $\theta_2 = Sq^B$ are monomials in A_2. By linearity it holds for all θ_1 and θ_2. To prove (ii), we apply χ to the relations $\sum_{i+j=k} Sq^i Xq^j = 0$ and $\sum_{i+j=k} Xq^i Sq^j = 0$ of Proposition 4.3.2. This gives $\sum_{i+j=k} \chi(Xq^j)Xq^j = 0$ and $\sum_{i+j=k} Xq^i \chi(Xq^j) = 0$. It follows by induction on k that $\chi(Xq^k) = Sq^k$ for all k. Finally (ii) implies (iii), since $\chi(Sq^k) = Xq^k$ and χ^2 is an algebra map of A_2.   □

The **opposite algebra** A^{op} of an algebra A by reversing the multiplication on the same underlying vector space, so that $a * b = ba$. Thus χ is an isomorphism from A_2 to A_2^{op}.

Example 4.3.9 By the results above, A_2 can be defined using generators Xq^k for $k \geq 0$ subject to $Xq^0 = 1$ and the conjugate Adem relations. We tabulate χ in degree 9 using the admissible basis.

	Sq^9	$Sq^8 Sq^1$	$Sq^7 Sq^2$	$Sq^6 Sq^3$	$Sq^6 Sq^2 Sq^1$
Xq^9	0	1	0	0	1
$Xq^1 Xq^8$	1	0	1	0	0
$Xq^2 Xq^7$	0	0	0	0	1
$Xq^3 Xq^6$	0	0	0	1	0
$Xq^1 Xq^2 Xq^6$	0	0	1	0	0

Using Example 3.5.4, we can compute χ for the Milnor basis in degree 9. The resulting matrix is shown in Example 11.6.4, and is upper triangular for all degrees d using either the left or right order. By Proposition 3.5.11, all entries of the top row are 1s.

Proposition 4.3.10 *For* $\theta \in A_2$, $Sq^1\theta = 0$ *if and only if* $\theta = Sq^1\phi$ *for some* $\phi \in A_2$.

Proof Since $\chi(Sq^1) = Sq^1$, this follows by applying χ to Proposition 3.4.7.   □

4.4 Conjugation and the Milnor basis

In this section we collect some formulae relating the conjugation χ of A_2 to the Milnor basis. For $R = (r_1, r_2, \ldots)$, the Milnor basis element $Sq(R)$ has degree $\sum_{j \geq 1} (2^j - 1) r_j$. This should not be confused with $\deg_2 R = \sum_{j \geq 1} 2^{j-1} r_j$, to which it is related by the equation $|R| + \deg Sq(R) = 2 \deg_2 R$, where $|R| = \sum_{j \geq 1} r_j$.

Proposition 4.4.1 *Let* $a, b \geq 0$. *Then*

(i) **(Davis formula)** $Sq^a Xq^b = \sum_R \binom{\deg_2 R}{a} Sq(R)$,
(ii) **(Silverman formula)** $Xq^a Sq^b = \sum_R \binom{|R|}{b} Sq(R)$,

where the sums are over all Milnor basis elements $Sq(R)$ *of degree* $a + b$.

Proof (i) Let $Sq^a Xq^b = \sum_R m_R Sq(R)$, where $m_R \in \mathbb{F}_2$. By Proposition 3.5.11, Xq^b is the sum of all Milnor basis elements of degree b. Since $Sq^a = Sq(a)$, it follows from Theorem 4.1.2 that m_R is the sum of the coefficients $b(X)$ for Milnor matrices

$$
X = \quad
\begin{array}{c|cccc}
 & r_1 - a_1 & r_2 - a_2 & \cdots \\
\hline
a_1 & a_2 & a_3 & \cdots
\end{array}
\quad ,
$$

where $a = \sum_{j \geq 1} 2^{j-1} a_j$. Every sequence $A = (a_1, a_2, \ldots)$ of 2-degree a such that $a_j \leq r_j$ for $j \geq 1$ gives rise to one such matrix X, and $b(X) = \prod_{j \geq 1} \binom{r_j}{a_j}$.

By Proposition 1.4.11, $\binom{r_j}{a_j} = \binom{2^j r_j}{2^j a_j}$ mod 2, and $\binom{2^j r_j}{b_j} = 0$ mod 2 if b_j is not divisible by 2^j. Hence

$$
m_R = \sum_A \prod_{j \geq 1} \binom{r_j}{a_j} = \sum_A \prod_{j \geq 1} \binom{2^j r_j}{2^j a_j} = \sum_B \prod_{j \geq 1} \binom{2^j r_j}{b_j}, \qquad (4.3)
$$

where the sum is over all $B = (b_1, b_2, \ldots)$ such that $\sum_{j \geq 1} b_j = \sum_{j \geq 1} 2^j a_j = 2a$. But (4.3) is the coefficient of x^{2a} in $\prod_{j \geq 1} (1 + x)^{2^j r_j}$. Since $\deg_2 R = \sum_{j \geq 1} 2^{j-1} r_j$,

$$
\prod_{j \geq 1} (1 + x)^{2^j r_j} = (1 + x)^{2 \deg_2 R} = (1 + x^2)^{\deg_2 R}.
$$

Comparing coefficients of x^{2a}, (4.3) reduces to $m_R = \binom{\deg_2 R}{a}$.

(ii) By Proposition 3.5.11, $Xq^a Sq^b = \sum_S Sq(S) Sq(b)$, where the sum is over all Milnor basis elements $Sq(S)$ of degree a. A term $b(X) Sq(R)$ in the product

$Sq(S)Sq(b)$ arises from each Milnor matrix

$$
Y = \quad
\begin{array}{c|c}
 & b_1 \\
\hline
r_1 - b_1 & b_2 \\
r_2 - b_2 & b_3 \\
\vdots & \vdots
\end{array}
\quad .
$$

Every sequence $B = (b_1, b_2, \ldots)$ such that $|B| = b = \sum_{j \geq 1} b_j$ and $b_j \leq r_j$ for $j \geq 1$ gives rise to one such Y, with coefficient $b(Y) = \prod_{j \geq 1} \binom{r_j}{b_j}$. Comparing coefficients of x^b in the identity $\prod_{j \geq 1}(1 + x)^{r_j} = (1 + x)^{|R|}$, we have $\sum_B \prod_{j \geq 1} \binom{r_j}{b_j} = \binom{|R|}{b}$. □

Formula 4.4.1(i) leads to some useful formulae for $\chi(Sq^k) = Xq^k$.

Proposition 4.4.2 (i) $Xq^{2^n - k} = Sq^{2^{n-1}} Xq^{2^{n-1} - k} = Sq^{2^{n-1}} \cdots Sq^{2^{k-1}} Xq^{2^{k-1} - k}$ *for* $1 \leq k \leq n$. *In particular,* $Xq^{2^n - 1} = Sq^{2^{n-1}} Sq^{2^{n-2}} \cdots Sq^2 Sq^1$.
(ii) $Xq^{2^n - n - 1} = Sq^{2^{n-1}} Xq^{2^{n-1} - n - 1} + Sq^{2^{n-1} - 1} Sq^{2^{n-2} - 1} \cdots Sq^3 Sq^1$.

Proof (i) By Propositions 4.4.1(i) and 3.5.11, the first statement of (i) is equivalent to the statement that $\binom{\deg_2 R}{a}$ is odd when $a = 2^{n-1}$ and $\deg Sq(R) = 2^n - k$, where $1 \leq k \leq n$. By Proposition 2.4.6, $\mu(2^n - k) = k$, and hence $|R| \geq k$ for all R such that $\deg Sq(R) = 2^n - k$. Hence $\deg_2 R = (\deg Sq(R) + |R|)/2 \geq 2^{n-1}$. Since also $\deg_2 R \leq \deg Sq(R) < 2^n$, $\binom{\deg_2 R}{a} = 1$ by Proposition 1.4.11. The rest of (i) follows by iteration.
(ii) By Proposition 2.4.7, $2^n - n - 1 = \sum_{j=1}^{n-1}(2^j - 1)$ is the unique partition of $d = 2^n - n - 1$ as the sum of a minimum number of integers of the form $2^j - 1$. Since the number of terms has the same parity as d, all other such partitions have $\geq n + 1$ terms. Thus the minimal partition corresponds to the sequence $R = (1, 1, \ldots, 1)$ of length $n - 1$, and $|R| \geq n + 1$ for all other R with $\deg Sq(R) = d$. Applying Davis's formula 4.4.1(i), $Sq^{2^{n-1}} Xq^{2^{n-1} - n - 1}$ is the sum of all Milnor basis elements of degree $2^n - n - 1$ except $Sq(1, 1, \ldots, 1)$. As in (i), the result follows from Proposition 3.5.11 and Example 4.1.5. □

The next result gives a formula for Xq^d for all d not of the form $2^k - 1$, in terms of the last 0 in the sequence $\omega(d)$ that is followed by a 1. The case $d = 2^k - 1$ can be included in Proposition 4.4.3 by reading the second term in the formula as 0 when $k = 0$. More general formulae of this kind are discussed in Chapter 14.

Proposition 4.4.3 (Bausum's formula) *For* $1 \leq k \leq n$ *and* $2^{k-1} < j \leq 2^k$,

$$
Xq^{2^{n+1} - j} = Sq^{2^n} Xq^{2^n - j} + Sq^{2^n - 2^{k-1}} Xq^{2^n + 2^{k-1} - j}.
$$

Proof Let $d = 2^{n+1} - j$. We show that $Sq^{2^n} Xq^{2^n - j}$ is the sum of all $Sq(R)$ of degree d with $\deg_2 R \geq 2^n$, while $Sq^{2^n - 2^{k-1}} Xq^{2^n + 2^{k-1} - j}$ is the sum of all $Sq(R)$ of degree d with $\deg_2 R < 2^n$. Thus every Milnor basis element $Sq(R)$ of degree d appears in one sum and not in the other, and so the result follows from Proposition 3.5.11.

By Proposition 4.4.1(i), the first sum is $\sum_R \binom{\deg_2 R}{2^n} Sq(R)$, and the second is $\sum_R \binom{\deg_2 R}{2^n - 2^{k-1}} Sq(R)$, where the sum is over all R with $\deg Sq(R) = d$. Now $d/2 \leq (|R| + d)/2 = \deg_2 R$ and $\deg_2 R \leq d$. Since $d = 2^{n+1} - j$ and $2^{k-1} < j \leq 2^k$, we have $2^n - 2^{k-1} \leq d/2$ and $d < 2^{n+1} - 2^{k-1}$. Hence $2^n - 2^{k-1} \leq m < 2^{n+1} - 2^{k-1}$ where $m = \deg_2 R$. For numbers m in this range, $\binom{m}{2^n}$ is odd if $m \geq 2^n$ and is even if $m < 2^n$, while $\binom{m}{2^n - 2^{k-1}}$ is even if $m \geq 2^n$ and is odd if $m < 2^n$. $\qquad\square$

4.5 Remarks

Milnor [134] gave the formula for the product $Sq(R)Sq(S)$ and a formula for $\chi(Sq(R))$. In Theorem 11.6.1, we give an alternative formula for $\chi(Sq(R))$ due to Li Zaiqing [123].

The conjugation χ on A_2 was introduced in 1951 by Wu Wen-Tsun and by R. Thom [211] in the context of characteristic classes of fibre bundles. This map is the antipode for the Hopf algebra structure of A_2, which we discuss in Chapter 11. Our approach in Section 4.3 uses the formulation by S. R. Bullett and I. G. Macdonald [24] of the Adem relations as a quadratic relation in a ring of power series over A_2, which we present in Section 4.2. Our proof of the Bullett–Macdonald identity is based on the argument given in [24] (see also [191]), but avoids the direct use of residue calculus. The conjugation formulae of Section 4.4 were originally proved by Davis [49], Bausum [16] and Silverman [180].

5

Combinatorial structures

5.0 Introduction

This chapter provides a combinatorial framework for work on the Steenrod algebra and the hit problem. It concerns integer sequences, block notation for monomials, partitions of integers and order relations. Sequences and partitions have a finite number of nonzero terms, but are regarded as infinite sequences by the use of trailing 0s. We distinguish them from finite sequences with a fixed number of terms, such as the exponents of a monomial or m-tuples of scalars giving elements of a m-dimensional vector space. We consider

(1) Seq, the set of sequences $A = (a_1, a_2, a_3, \ldots)$ of integers $a_i \geq 0$ with only a finite number of nonzero terms,
(2) Dec, the set of (weakly) decreasing sequences in Seq,
(3) Part^d, the set of integer partitions of d,
(4) Bin^d, the set of binary partitions of d, with parts of the form 2^j for $j \geq 0$,
(5) Spike^d, the set of spike partitions of d, with parts of the form $2^j - 1$ for $j \geq 1$.

Our basic working tool in the hit problem for polynomials is the n-block corresponding to a monomial in $P(n)$. The block is a matrix-like array with entries 0 or 1, whose rows are the reversed binary expansions of the exponents of the monomial. Most of the constructions and processes used in the hit problem are expressed in the language of blocks, but sometimes it is simpler to describe them in terms of monomials. For example, the ω-sequence of a block is the sequence of its column sums, but the left order on blocks is the left lexicographic order on the exponent sequences of the corresponding monomials. The phrase 'adding 1 arithmetically at position (r, c) in a block' means adding 2^{c-1} to the exponent of the rth variable.

An integer partition of a positive integer d is a way of writing d as a sum of positive integers, which are called the parts of the partition. By putting the parts in decreasing order, Part^d can be identified with the set of decreasing sequences with modulus d. Alternatively, an element of Part^d can be regarded as a multiset of positive integers. For example $d = 4$ has five partitions, namely $4, 3 + 1, 2 + 2, 2 + 1 + 1$ and $1 + 1 + 1 + 1$, which can be written as $\{4\}$, $\{3, 1\}$, $\{2^2\}$, $\{2, 1^2\}$ and $\{1^4\}$, where the 'exponents' are the multiplicities of the parts. Partitions arise in the representation theory of linear groups, and we shall use them in studying $P(n)$ as a module over $\mathbb{F}_2 \text{GL}(n)$ or $\mathbb{F}_2 \text{M}(n)$. For example, descending chains of submodules of $P^d(n)$ are obtained by taking the linear span of monomials which are divisible by successively higher 2-powers of the variables.

This can be illustrated by Example 1.4.8. The submodule S of $P^3(3)$ spanned by monomials divisible by the square of some variable has dimension 9, and the quotient of S by the hit polynomials is the 6-dimensional submodule of $Q^3(3)$ spanned by the minimal spikes x^3, y^3, z^3 and the monomials $x^2 y, x^2 z, y^2 z$. The ω-sequence $\omega(f)$ is the key to describing this structure: S is spanned by the monomials with ω-sequence $(1, 1)$, while the monomial xyz, with ω-sequence (3), spans the quotient $P^3(3)/S$.

The ω-sequence of a monomial in $P^d(n)$ has all terms $\leq n$. Hence we filter Seq and Dec by the sets $\text{Seq}(n)$ and $\text{Dec}(n)$ of all such sequences. Binary partitions of d arise when we consider the blocks associated to monomials of degree d, and spike partitions of d arise when we consider spike monomials of degree d. The binary partitions which arise when working in $P(n)$ have parts with multiplicities $\leq n$, and the spike partitions have length $\leq n$.

We establish notation for the various types of sequences and partitions mentioned above in Section 5.1. Sequences are graded by 2-degree $\deg_2 A = a_1 + 2a_2 + 4a_3 + \cdots$, and we establish bijections between Seq_d and Bin^d, and between Dec_d and Spike^d, where a superscript d refers to the sum of the terms (modulus) of a sequence and a subscript d refers to its 2-degree.

Section 5.2 introduces the **dominance** and 2-**dominance** partial orders on Seq, and relates them to the left and right orders. The left order $<_l$ refines both dominance and 2-dominance on Seq, while the right order $<_r$ refines dominance for sequences of the same modulus, and refines 2-dominance for sequences of the same 2-degree.

In Section 5.3 we show that the **refinement** partial order on $\text{Bin}(d)$ gives a distributive lattice with unique maximal and minimal elements.

The bijection between Bin^d and Seq_d, ordered by 2-dominance, is a lattice isomorphism.

A positive integer d has a unique minimal spike partition, which we discuss in Section 5.4. This partition has length $\mu(d)$. In Section 5.5 we discuss Dec_d, ordered by 2-dominance, and Spike^d, ordered by reversed dominance, as partially ordered sets. In Section 5.6, we discuss maximal elements in Seq_d and Dec_d.

In Sections 5.7 and 5.8, we apply these combinatorial results to the Steenrod algebra A_2. We show that the dominance order on admissible sequences of modulus d gives a lattice Adm^d isomorphic to Spike^d, and we evaluate the excess of Milnor basis elements and of admissible monomials.

5.1 Sequences, blocks and partitions

Some of the combinatorial structures we deal with in this chapter have already been defined in Chapter 3. For convenience we repeat them here as required. We begin with sequences (cf. Definitions 3.1.4, 3.3.3).

Definition 5.1.1 The set of all sequences $R = (r_1, r_2, \ldots)$ of integers $r_i \geq 0$ with $r_i \neq 0$ for only a finite number of $i \geq 1$ is denoted by Seq. For $R \in$ Seq, its **length** $\mathrm{len}(R)$ is 0 if $r_i = 0$ for all i, and otherwise $\mathrm{len}(R)$ is the maximum i such that $r_i > 0$. By omitting trailing 0s, a sequence $R \in$ Seq may be written as a finite sequence with m terms for $m \geq \mathrm{len}(R)$. The **modulus** of $R \in$ Seq is $|R| = \sum_{i \geq 1} r_i$, and its **2-degree** is $\deg_2 R = \sum_{i \geq 1} 2^{i-1} r_i$. The set of integers $r_i > 0$ which are terms of R is denoted by $I(R)$, and the set of repeated terms of R (i.e. $r_i = r_j > 0$ where $i \neq j$) is denoted by $I_2(R)$.

We extend these definitions as follows.

Definition 5.1.2 A sequence $R \in$ Seq is **decreasing** if $r_i \geq r_{i+1}$ for $i > 0$. The set of decreasing sequences is denoted by Dec. For $d \geq 0$, we write Seq_d for the set of sequences of 2-degree d and Dec_d for the set of decreasing sequences of 2-degree d. The sequence R is n-**bounded** if $r_i \leq n$ for all i, i.e. $I(R) \subseteq Z[n] = \{1, \ldots, n\}$. The set of n-**bounded** sequences is denoted by $\mathrm{Seq}(n)$, and the set $\mathrm{Seq}(n) \cap \mathrm{Dec}$ of decreasing sequences with $r_1 \leq n$ is denoted by $\mathrm{Dec}(n)$. Thus $\mathrm{Seq}_d(n)$ is the set of n-bounded sequences of 2-degree d, and $\mathrm{Dec}_d(n)$ is the set of n-bounded decreasing sequences of 2-degree d.

Example 5.1.3

$$\mathsf{Seq}_7 = \{(7),(5,1),(3,2),(3,0,1),(1,3),(1,1,1)\},$$
$$\mathsf{Dec}_7 = \{(7),(5,1),(3,2),(1,1,1)\},$$
$$\mathsf{Seq}_7(3) = \{(3,2),(3,0,1),(1,3),(1,1,1)\},$$
$$\mathsf{Dec}_7(3) = \{(3,2),(1,1,1)\}.$$

As the above example shows, a sequence $R \in \mathsf{Seq}$ will usually be written as a finite sequence with $\mathrm{len}(R)$ terms. However, we distinguish between elements of Seq and finite sequences.

Definition 5.1.4 For $n \geq 1$, a sequence of **size** n is an n-tuple $S = (s_1,\ldots,s_n)$ of integers ≥ 0. The **modulus** $|S|$ and 2-**degree** $\deg_2 S$ are the modulus and 2-degree of the sequence obtained by adding trailing 0s to S.

One reason for distinguishing n-tuples from sequences will appear in Section 6.3, where we concatenate blocks associated to monomials in $\mathsf{P}(n)$. We can concatenate a sequence R of size m and a sequence S of size n by prefixing the entries of R to the entries of S to give a sequence of size $m+n$. We can do the same if S is a sequence, but if a sequence is split as the concatenation of R and S, then the prefix R must be regarded as a sequence of size m.

Definition 5.1.5 (cf. Definitions 2.4.1, 3.3.2, 3.3.7) The ω-**sequence** $\omega(d) = (\omega_1(d),\omega_2(d),\ldots)$ of an integer $d \geq 0$ is the sequence of its binary digits in reverse order, i.e. $\omega_j(d) = 1$ if $2^{j-1} \in \mathrm{bin}(d)$, and $\omega_j(d) = 0$ otherwise. The ω-**sequence** $\omega(f)$ of a monomial $f = x_1^{d_1}\cdots x_n^{d_n}$ in $\mathsf{P}(n)$ is the sum $\omega(f) = \sum_{i=1}^{n} \omega(d_i)$ of the ω-sequences of its exponents.

An n-**block** B is an array of integers $b_{i,j} = 0$ or 1 defined for $1 \leq i \leq n$ and $j \geq 1$. For each i there are only finitely many j's such that $b_{i,j} = 1$. The row sums $\alpha_i(B) = \sum_{j \geq 1} b_{i,j}$ form the α-**sequence** $\alpha(B) = (\alpha_1(B),\ldots,\alpha_n(B))$ of B, and the column sums $\omega_i(B) = \sum_{i=1}^{n} b_{i,j}$ form the ω-**sequence** $\omega(B) = (\omega_1(B),\omega_2(B),\ldots)$ of B. The block B has 2-**degree** $\deg_2 B = \deg_2 \omega(B)$.

Given a monomial $f = x_1^{d_1}\cdots x_n^{d_n}$ in $\mathsf{P}(n)$, we associate to f the n-block F whose ith row is $\omega(x_i^{d_i})$ for $1 \leq i \leq n$, so that $\omega(F) = \omega(f)$, and we define $\alpha(f) = \alpha(F)$.

The ω-sequence of a monomial or block is an element of Seq, but the α-sequence is a sequence of size n. The degree of the monomial f is the 2-degree of the block F or the sequence $\omega(F)$, but it is the modulus of the sequence of size n formed by its exponents.

We turn next to partitions of integers. A partition of a positive integer d is a decomposition of d as a sum of positive integers, where the order of the terms is not taken into account. We write partitions with the summands, or parts, in decreasing order. For example $d = 4$ has 5 partitions, namely (4), $(3,1)$, $(2,2)$, $(2,1,1)$ and $(1,1,1,1)$. Thus partitions of d can be identified with sequences in Dec of modulus d.

However, as a subset of Seq, the grading of Dec is by 2-degree, not by modulus, and it is often better to view a partition as an unordered sum of positive integers. For this reason we also use the alternative multiset notation for partitions. This generalizes the notation for sets by attaching 'exponents' to the elements to denote their multiplicities, for example, the partitions of 4 above are denoted by $\{4\}$, $\{3,1\}$, $\{2^2\}$, $\{2,1^2\}$ and $\{1^4\}$. By convention, we often use a Greek letter such as λ to denote a decreasing sequence, or its terms, when it is regarded as a partition, and we often use ω to denote a sequence when it is regarded as the ω-sequence of a block or a monomial. In other cases, for example to index Milnor basis elements in A_2, we use capital Roman letters for sequences.

Definition 5.1.6 A **partition** of an integer $d > 0$ is a multiset of positive integers $\lambda = \{a_1^{m_1}, \ldots, a_r^{m_r}\}$ whose sum $\sum_{i=1}^{r} m_i a_i = d$. The empty set $\emptyset$ is regarded as the unique partition of d when $d = 0$. The set of partitions of $d \geq 0$ is denoted by Part^d. The **length** of a partition is the number of its parts, so that $\text{len}(\lambda) = \sum_{i=1}^{r} m_i$ if $d > 0$ and $\text{len}(\emptyset) = 0$. A partition of d can equivalently be defined as a decreasing sequence $\lambda = (\lambda_1, \lambda_2, \ldots)$ of modulus $|\lambda| = d$, and for $1 \leq i \leq \text{len}(\lambda)$ the ith term of this sequence is called the **ith part** of λ.

A partition $\lambda = (\lambda_1, \lambda_2, \ldots, \lambda_n)$ is represented graphically by its **(Ferrers) diagram**, which we denote by the same symbol λ. This is a left justified array of 'boxes' with λ_i boxes in the ith row for $i \geq 1$. In matrix notation, the diagram of λ has a box (i,j) for $1 \leq j \leq \lambda_i$. The total number of boxes is d.

Example 5.1.7 The diagrams of $\lambda = \{4,3,2\}$ and $\mu = \{3^2, 2, 1\}$ are

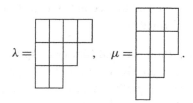

The **conjugate** (or **transpose**) $\mu = \lambda^{tr}$ of a partition is defined by reflection of its diagram in the main diagonal. Thus for $j \geq 1$, μ_j is the number of parts $\lambda_i \geq j$. The partitions in Example 5.1.7 are a conjugate pair. Thus $\mu_1 = \text{len}(\lambda)$. If the diagram of a partition λ has $< n$ equal rows, then the diagram of μ has $< n$ equal columns. Such partitions λ and μ are also called **row n-regular** and **column n-regular** respectively. In particular, for $n = 2$, the diagrams have no repeated rows and no repeated columns respectively. In Example 5.1.7, λ is row 2-regular and μ is column 2-regular.

Recall from Definition 3.3.7 or 5.1.5 that a monomial $f = x_1^{d_1} \cdots x_n^{d_n}$ in $P(n)$ is represented diagrammatically by an n-block $B = (b_{i,j})$, so that the entry $b_{i,j}$ in row i and column j is 1 if $2^{j-1} \in \text{bin}(d_i)$, and is 0 otherwise. In particular, a block corresponding to a spike with exponents in decreasing order is obtained by replacing each box in a Ferrers diagram by a 1, following our convention of omitting trailing 0s. If the spike involves $< n$ variables, we mark the zero rows by retaining at least one 0.

Definition 5.1.8 A **Ferrers** block is an n-block which corresponds to a spike monomial $f = x_1^{2^{j_1}-1} \cdots x_n^{2^{j_n}-1}$ with decreasing exponents $j_1 \geq \cdots \geq j_n \geq 0$.

Example 5.1.9 The Ferrers block corresponding to $x_1^{15} x_2^3 x_3^3$ in $P^{21}(4)$ is

$$
\begin{array}{cccc}
1 & 1 & 1 & 1 \\
1 & 1 & & \\
1 & 1 & & \\
0 & & &
\end{array}
\quad .
$$

Many of these definitions fall naturally into place if we regard the block as a basic concept. For example $\alpha(B)$ is the row sum sequence of the block B, and $\omega(B)$ is its column sum sequence. A decreasing sequence ω determines a unique Ferrers block F with ω_1 rows and $\omega = \omega(F)$. In Example 5.1.9, $\omega = (3,3,1,1)$. Then $\alpha = (\alpha_1, \ldots, \alpha_n) = \alpha(F)$ is also decreasing, and $\alpha = \omega^{tr}$ as partitions of $|\alpha(F)| = |\omega(F)|$, the number of 1s in F. The 2-degree of F, or of ω, is $d = \sum_{i>0} 2^{i-1} \omega_i$, and $d = \sum_{i>0} (2^{\alpha_i} - 1)$. The first sum gives a partition of d whose parts are 2-powers, and the second sum gives a partition of d whose parts are integers of the form $2^j - 1$.

Definition 5.1.10 A partition is called **binary** or **spike** when all its parts are of the form 2^j or $2^j - 1$ respectively. The set of binary partitions of d is denoted by Bin^d and the set of spike partitions of d by Spike^d, regarded as subsets of Part^d. We write $\text{Bin}^d(n)$ and $\text{Spike}^d(n) = \text{Spike}^d \cap \text{Part}(n)$ for the binary partitions with all multiplicities $\leq n$ and the spike partitions of length $\leq n$ respectively.

Example 5.1.11 $\mathsf{Bin}^7 = \{\{4,2,1\},\{4,1^3\},\{2^3,1\},\{2^2,1^3\},\{2,1^5\},\{1^7\}\}$ and $\mathsf{Spike}^7 = \{\{7\},\{3^2,1\},\{3,1^4\},\{1^7\}\}$. $\mathsf{Bin}^7(3)$ and $\mathsf{Spike}^7(3)$ are obtained by omitting the last two elements in each case.

The next result is illustrated by comparing Examples 5.1.3 and 5.1.11.

Proposition 5.1.12 *There are bijections between* Seq_d *and* Bin^d, *and between* Dec_d *and* Spike^d. *Under these bijections,* $\mathsf{Seq}_d(n)$ *and* $\mathsf{Dec}_d(n)$ *correspond to* $\mathsf{Bin}^d(n)$ *and* $\mathsf{Spike}^d(n)$ *respectively, and the modulus of a sequence in* Seq_d *or* Dec_d *corresponds to the length of a partition in* Bin^d, *or to* $|\alpha(F)|$, *where F is the Ferrers block of a partition in* Spike^d.

Proof Let $R = (r_1,\ldots,r_k) \in \mathsf{Seq}_d$. Then $d = \sum_{i=1}^k 2^{i-1} r_i$, and we associate to R the partition $\{1^{r_1}, 2^{r_2}, \cdots, (2^{k-1})^{r_k}\} \in \mathsf{Bin}^d$. This correspondence takes sequences with terms bounded by n to binary partitions of multiplicity bounded by n. For example, $(8,0,3,1) \subset \mathsf{Seq}_{28}(8)$ corresponds to $\{8^1, 4^3, 1^8\} \in \mathsf{Bin}^{28}(8)$.

Next let $R \in \mathsf{Dec}_d$, and let F be the unique Ferrers block with r_1 rows and $\omega(F) = R$. Then $\alpha(F) = (a_1,\ldots,a_n)$ is a decreasing sequence of size n, to which we associate the spike partition $(2^{a_1} - 1,\ldots,2^{a_n} - 1) \in \mathsf{Spike}^d$ of length $n = r_1$. This defines a bijection between $\mathsf{Dec}_d(n)$ and $\mathsf{Spike}^d(n)$. For example, $(6,3,2)$ is a decreasing sequence of 2-degree 20 and maximum term 6. The corresponding spike partition is $\{7^2, 3^1, 1^3\}$, with modulus 20 and length 6. An alternative description of the bijection is to regard the decreasing sequence R as a partition λ, take the conjugate partition $\mu = \lambda^{\mathrm{tr}}$ and form a spike partition by replacing μ_i by $2^{\mu_i} - 1$ for all i. $\qquad\square$

The following special case is important in the hit problem.

Proposition 5.1.13 *For $n,d > 0$, d has a strictly decreasing spike partition σ of length $n-1$ or n if and only if $\alpha(d+n) = n$. In this case, σ is the unique element of* $\mathsf{Spike}^d(n)$ $\omega = \sigma^{\mathrm{tr}}$ *is the unique element of* $\mathsf{Dec}_d(n)$, *and* $I(\omega) = Z[n-1]$ *or* $Z[n]$.

Proof Since $\alpha(d+n) = |\mathrm{bin}(d+n)|$, $\alpha(d+n) = n$ if and only if $d+n = 2^{a_1} + \cdots + 2^{a_n}$, or equivalently $d = (2^{a_1} - 1) + \cdots + (2^{a_n} - 1)$, where $a_1 > \cdots > a_n \geq 0$. Thus $\sigma = (2^{a_1} - 1,\ldots,2^{a_n} - 1)$ is a strictly decreasing element of $\mathsf{Spike}^d(n)$, and $\mathrm{len}(\sigma) = n$ or $n-1$ according as $a_n > 0$ or $a_n = 0$. The second statement follows from the uniqueness of the binary decomposition of $d+n$. $\qquad\square$

Example 5.1.14 Let $n = 4$ and $d = 23$. By considering the Ferrers block

$$
\begin{array}{cccc}
1 & 1 & 1 & 1 \\
1 & 1 & 1 & \\
1 & & & \\
0 & & &
\end{array}
$$

we have $\mathsf{Spike}^{23}(4) = \{\{15,7,1\}\}$ and $\mathsf{Dec}_{23}(4) = \{(3,2,2,1)\}$.

5.2 Dominance

In addition to the linear order relations $<_l$ and $<_r$ defined in Chapter 3 (see Definition 3.1.5), we need two partial orders on sequences. Recall that a partial order $\leq$ on a set A is a binary relation which is reflexive (for all $a \in A$, $a \leq a$), antisymmetric (for all $a, b \in A$, $a \leq b$ and $b \leq a$ only if $a = b$) and transitive (for all $a, b, c \in A$, $a \leq b$ and $b \leq c$ imply $a \leq c$). A set with a partial order is called a poset. A subset $B \subseteq A$ with the partial order defined by $a \leq b$ in B if and only if $a \leq b$ in A is called a subposet of A. The usual conventions for the use of the symbols $<$, $>$ and $\geq$ in relation to $\leq$ apply to partial orders.

Definition 5.2.1 For $R = (r_1, r_2, \ldots), S = (s_1, s_2, \ldots) \in \mathsf{Seq}$, R is less than or equal to S in **dominance** order, written $R \preceq S$, if for all $k \geq 1$

$$
\sum_{j=1}^{k} r_j \leq \sum_{j=1}^{k} s_j.
$$

The dominance order is important in the combinatorics of partitions and symmetric functions, and in group representation theory. In order to compare sequences arising from spike partitions and binary partitions, we introduce an exponential version of dominance.

Definition 5.2.2 For $k \geq 1$, the kth **partial 2-degree** of $R = (r_1, r_2, \ldots) \in \mathsf{Seq}$ is

$$
\widehat{r}_k = \sum_{j=1}^{k} 2^{j-1} r_j.
$$

For $S = (s_1, s_2, \ldots) \in \mathsf{Seq}$, R is less than or equal to S in **2-dominance** order, written $R \preceq_2 S$, if $\widehat{r}_k \leq \widehat{s}_k$ for all $k \geq 1$.

Thus $\widehat{R}$ is an increasing finite sequence such that $\widehat{r}_k = \deg_2 R \bmod 2^k$ for all $k \geq 1$, and $\widehat{r}_k = \deg_2 R$ for $k \geq \mathsf{len}(R)$. These conditions characterize partial

2-degree sequences of elements of Seq_d, since R can be recovered from $\widehat{R}$ using $r_k = (\widehat{r}_k - \widehat{r}_{k-1})/2^{k-1}$, where $r_0 = 0$.

Proposition 5.2.3 *The left order $<_l$ refines both dominance and 2-dominance on* Seq. *The right order $<_r$ refines dominance for sequences of the same modulus, and refines 2-dominance for sequences of the same 2-degree.*

Proof Let $R = (r_1, r_2, \ldots)$, $S = (s_1, s_2, \ldots) \in$ Seq. It is clear that if $R \prec S$ or $R \prec_2 S$ then $R <_l S$. Let $R \prec S$ where $|R| = |S|$. Then $\sum_{i=1}^{n} r_i = \sum_{i=1}^{n} s_i$, where ℓ is the maximum length of R and S. Let k be the minimum number such that $r_j = s_j$ for $j > k$ and $r_k \neq s_k$. Then $r_k > s_k$ since $\sum_{i=1}^{k-1} r_i < \sum_{i=1}^{k-1} s_i$. Hence $R <_r S$.

Similarly, if $\deg_2 R = \deg_2 S$ and $R \prec_2 S$, then, for some k, $\sum_{i=1}^{j} 2^{i-1} r_i = \sum_{i=1}^{j} 2^{i-1} s_i$ for $k \leq j \leq \ell$, and $\sum_{i=1}^{k-1} 2^{i-1} r_i < \sum_{i=1}^{k-1} 2^{i-1} s_i$. It follows that $r_j = s_j$ for $j > k$ and $r_k > s_k$. $\qquad\qquad\square$

For compatibility with the left and right orders, we use dominance to relate sequences with the same modulus, and 2-dominance to relate sequences with the same 2-degree. Regarding partitions as decreasing sequences, we shall show that conjugacy reverses dominance. This is not true for the left and right orders: for example, when $R = (4, 1, 1)$ and $S = (3, 3)$, we have $R^{\text{tr}} = (3, 1, 1, 1)$ and $S^{\text{tr}} = (2, 2, 2)$ so that for the left order $R >_l S$ and $R^{\text{tr}} >_l S^{\text{tr}}$, and for the right order $R <_r S$ and $R^{\text{tr}} <_r S^{\text{tr}}$.

Proposition 5.2.4 *Let R and S be partitions with $|R| = |S|$. Then $R \succeq S$ if and only if $R^{\text{tr}} \preceq S^{\text{tr}}$.*

Proof A pair of integers $k, \ell \geq 1$ divides the diagram of a partition into NW, NE, SW and SE quadrants of boxes (i, j), defined by $i \leq k$ and $j \leq \ell$, $i \leq k$ and $j > \ell$, $i > k$ and $j \leq \ell$, and finally $i > k$ and $j > \ell$. Let a, b, c, d be the number of boxes in the four quadrants of the diagram of R in the listed order, and let a', b', c', d' be the corresponding numbers for S. Then $a + b + c + d = |R| = |S| = a' + b' + c' + d'$, and $a + b$, $a' + b'$ depend only on k, while $a + c$, $a' + c'$ depend only on ℓ.

As it suffices to prove the implication in one direction, we may assume that $R \succeq S$. Then for all $k \geq 1$ we have $a + b \geq a' + b'$, and we must prove that $a + c \leq a' + c'$ for all $\ell \geq 1$. Given $\ell \geq 1$, let k be the largest number such that $s_k \geq \ell$. Then, with quadrants defined by (k, ℓ), the NW quadrant of the diagram of S is full and the SE quadrant is empty. Hence $a' = k\ell$ and $d' = 0$. Hence $a \leq a'$ and so $b \geq b'$. Hence $b + d \geq b' = b' + d'$, and since $|R| = |S|$, it follows that $a + c \leq a' + c'$, and so $R^{\text{tr}} \preceq S^{\text{tr}}$. $\qquad\qquad\square$

Proposition 5.2.5 *Let R and S be partitions with $|R| = |S|$ and $R \succ S$. Then* $\deg_2 R < \deg_2 S$.

Proof We have $r_1 + \cdots + r_k \geq s_1 + \cdots + s_k$ for $k \geq 1$ and $r_1 + \cdots + r_\ell = s_1 + \cdots + s_\ell$, where $\ell = \max(\mathrm{len}(R), \mathrm{len}(S))$. Hence $r_{i+1} + \cdots + r_\ell \leq s_{i+1} + \cdots + s_\ell$ for $1 \leq i < \ell$. Then $\deg_2 R = r_1 + 2r_2 + 4r_3 + \cdots + 2^{\ell-1} r_\ell = |R| + \sum_{i=1}^{\ell-1} 2^{i-1}(r_{i+1} + \cdots + r_\ell) \leq |S| + \sum_{i=1}^{\ell-1} 2^{i-1}(s_{i+1} + \cdots + s_\ell) = \deg_2 S$. This inequality is strict if any of the inequalities $r_1 + \cdots + r_k \geq s_1 + \cdots + s_k$ is strict. $\square$

The next result relates dominance and 2-dominance.

Proposition 5.2.6 *Let $R' = (r_1, r_2, r_2, r_3, r_3, r_3, r_3, \ldots)$ be the sequence derived from $R = (r_1, r_2, r_3, \ldots) \in \mathsf{Seq}$ by repeating the term r_j 2^{j-1} times for all $j \geq 1$, and let S' be the corresponding sequence derived from $S = (s_1, s_2, s_3, \ldots)$. Then $R \preceq_2 S$ if and only if $R' \preceq S'$.*

Proof By definition, $R' \preceq S'$ if and only if $\sum_{j=1}^{k} r_j' \leq \sum_{j=1}^{k} s_j'$ for $1 \leq k \leq n$. Let a be the largest integer such that $2^a - 1 \leq k$ and write $k = 2^a - 1 + b$ where $0 \leq b < 2^a$. Then for all $a \geq 1$ and $0 \leq b < 2^r$

$$\sum_{j=1}^{a} 2^{j-1} r_j + b r_{a+1} \leq \sum_{j=1}^{a} 2^{j-1} s_j + b s_{a+1}. \tag{5.1}$$

Setting $b = 0$ gives $R \preceq_2 S$. Conversely, if $R \preceq_2 S$ then $\sum_{j=1}^{a} 2^{j-1} r_j \leq \sum_{j=1}^{a} 2^{j-1} s_j$ and $\sum_{j=1}^{a+1} 2^{j-1} r_j \leq \sum_{j=1}^{a+1} 2^{j-1} s_j$. By adding $2^a - b$ times the first inequality to b times the second and dividing by 2^a, we obtain (5.1). Hence $R' \preceq S'$. $\square$

Proposition 5.2.7 *Let $d \geq 0$ and let f and g be spikes of degree d with decreasing exponent sequences A and B. Then $\omega(f) \prec_2 \omega(g)$ if and only if $A \succ B$.*

Proof Let $R = \omega(f)$ and let $S = \omega(g)$. Then $R \prec_2 S$ if and only if $R' \prec S'$, where R' and S' are derived from R and S as in Proposition 5.2.6. But R' and S' are the conjugates of the partitions A and B. For example, when $A = (7, 3, 1, 1)$, $R' = A^{\mathrm{tr}} = (4, 2, 2, 1, 1, 1, 1)$ and $R = (4, 2, 1)$. By Proposition 5.2.4, $R' \prec S'$ if and only if $B \prec A$. $\square$

5.3 Binary partitions

In this section we show that under the bijection of Proposition 5.1.12 between Seq_d and Bin^d, the 2-dominance order on Seq_d corresponds to the natural partial order on Bin^d by refinement. We begin by recalling some standard terms in relation to posets.

Definition 5.3.1 If P is a poset and $a, b \in P$, a **covers** b if $b < a$ and there is no element $c \in P$ such that $b < c < a$. A **grading** on a poset P is a function gr from P to the integers ≥ 0 such that $\text{gr}(a) = \text{gr}(b) + 1$ if a covers b, and P is then **graded** by gr.

Proposition 5.3.2 *For sequences* $R = (r_1, r_2, \ldots)$ *and* $S = (s_1, s_2, \ldots)$ *in* Seq_d,

(i) *if* $S <_l R$ *and* k *is minimal such that* $s_k < r_k$, *then* $r_k = s_k \bmod 2$,
(ii) R *covers* S *in the 2-dominance order* $\prec_2$ *if and only if for some* $k \geq 1$

$$s_j = \begin{cases} r_j - 2, & \text{if } j = k, \\ r_j + 1, & \text{if } j = k+1, \\ r_j, & \text{otherwise.} \end{cases}$$

(iii) Seq_d *is graded by the modulus* $|R|$ *of* R.

Proof (i) For some $k \geq 1$, $s_j = r_j$ for $j < k$ and $s_k < r_k$. Since $\sum_{j \geq 1} 2^{j-1} s_j = d = \sum_{j \geq 1} 2^{j-1} r_j$, by subtraction we have $\sum_{j \geq k} 2^{j-1} (r_j - s_j) = 0$. Cancelling a factor 2^{k-1}, this gives $r_k - s_k = -\sum_{j > k} 2^{j-k} (r_j - s_j) = 0 \bmod 2$.

(ii) If $R, S \in \text{Seq}_d$ are as described, then clearly $S \prec_2 R$ and $|S| = |R| - 1$. Let $T = (t_1, t_2, \ldots) \in \text{Seq}_d$ be any sequence such that R covers T. Let k be the minimal integer such that $t_k < r_k$. Let S be defined as in (ii). Then $s_j = r_j = t_j$ for $j < k$ and, by (i), $t_k \leq r_k - 2 = s_k$. Since $T \prec_2 R$ and $\sum_{j=1}^{\ell} 2^{j-1} s_j = \sum_{j=1}^{\ell} 2^{j-1} r_j$ for all $\ell > k$, $T \prec_2 S$. Since also $S \prec_2 R$ and, by hypothesis, R covers T, it follows that $T = S$.

(iii) With R and S as in (ii), $|S| = |R| - 1$. $\square$

Definition 5.3.3 Let R and S be binary partitions of d. Then R **refines** S, or $R \geq S$ in the **refinement order**, if there are binary partitions of the parts of S which can be combined to give R.

We can refine a binary partition by splitting a part 2^k such that $k \geq 1$ as $2^{k-1} + 2^{k-1}$, thus increasing the length of the partition by 1. Given binary partitions R and S of d, $S \leq R$ in the refinement order if R can be obtained from S by a finite number of steps of this type. The binary expansion of d is the unique element of minimal length $|\text{bin}(d)|$ in Bin^d, and $\{1^d\}$ is the unique element of maximal length. Every element of Bin^d can be obtained from the binary expansion $\text{bin}(d)$ of d or reduced to $\{1^d\}$ by a sequence of refinements. As the length of the partition is increased by 1 at each step, all such chains in Bin^d are of the same length, and so $\text{Bin}(d)$ is graded by length.

For a monomial f with 2-power exponents, refinement by splitting corresponds to replacement of a factor x^{2^k} by $y^{2^{k-1}} z^{2^{k-1}}$ for some variables x, y, z. Thus in terms of the bijection of Proposition 5.1.12, the splitting corresponds to replacement of a pair of consecutive entries (r_{k-1}, r_k) of $\omega(f)$ by $(r_{k-1} + 2, r_k - 1)$. By Proposition 5.3.2, this corresponds to a covering ω-sequence for the 2-dominance order on Seq_d. We summarize these results as follows.

Proposition 5.3.4 *Let* Seq_d *be ordered by 2-dominance and graded by modulus, and let* Bin^d *be ordered by refinement and graded by length. Then the bijection* $\mathsf{Seq}_d \leftrightarrow \mathsf{Bin}^d$ *of Proposition 5.1.12 is an isomorphism of graded posets. The poset* Seq_d *has unique maximal and minimal elements* $(d, 0, \ldots)$ *and* $\omega(d)$, *and* Bin^d *has unique maximal and minimal elements* $\{1^d\}$ *and* $\mathrm{bin}(d)$. $\qquad\square$

Example 5.3.5 The diagrams below show the graded posets Seq_7 and Bin^7. Note that refinement of binary partitions implies dominance, but is not equivalent to it. For example, $\{2^3, 1\} \prec \{4, 1^3\}$, but neither partition refines the other.

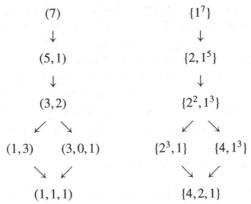

The isomorphic posets Seq_d and Bin^d are of a particularly nice type known as **distributive lattices**. We recall the definitions.

Definition 5.3.6 A **lattice** L is a poset in which every pair of elements a, b has a least upper bound, denoted by $\sup(a, b)$ or $a \vee b$, and a greatest lower bound, denoted by $\inf(a, b)$ or $a \wedge b$, where these terms have their usual meanings. The lattice L is **distributive** if for all $a, b, c \in L$

$$a \vee (b \wedge c) = (a \vee b) \wedge (a \vee c), \quad a \wedge (b \vee c) = (a \wedge b) \vee (a \wedge c),$$

the two conditions being equivalent. A subposet S of L is a **sublattice** if $a \vee b$ and $a \wedge b$ are in S when a and b are in S.

Every distributive lattice is graded, and the grading gr is related to the lattice operations by the formula $\mathrm{gr}(a) + \mathrm{gr}(b) = \mathrm{gr}(a \vee b) + \mathrm{gr}(a \wedge b)$.

Proposition 5.3.7 *For $d \geq 1$, the posets Seq_d, with 2-dominance order, and $\mathsf{Bin}(d)$, with refinement order, are isomorphic distributive lattices.*

Proof Given $R = (r_1, r_2, \ldots)$ and $S = (s_1, s_2, \ldots) \in \mathsf{Seq}_d$, let $\widehat{u}_j = \max(\widehat{r}_j, \widehat{s}_j)$ and $\widehat{v}_j = \min(\widehat{r}_j, \widehat{s}_j)$ for $j \geq 1$. These are the partial 2-degree sequences of elements $U = (u_1, u_2, \ldots)$ and $V = (v_1, v_2, \ldots) \in \mathsf{Seq}_d(n)$, and we write $U = R \vee S$ and $V = R \wedge S$. Then U is the least upper bound and V is the greatest lower bound of R and S. The distributive laws follow from the corresponding properties

$$\max(R, \min(S, T)) = \min(\max(R, S), \max(R, T)),$$
$$\min(R, \max(S, T)) = \max(\min(R, S), \min(R, T)),$$

where $\{R, S, T\}$ is a totally ordered set. $\qquad\square$

Example 5.3.8 With $R = (1, 3)$ and $S = (3, 0, 1)$, $U = (3, 2)$ and $V = (1, 1, 1)$, as shown in Example 5.3.5.

Proposition 5.3.9 *For $m, d \geq 0$, the number $b(m, d)$ of binary partitions of d of length $\leq m$ is bounded by a function of m independent of d.*

Proof Since $b(m, d) = 0$ if $\alpha(d) > m$, we may assume that $\alpha(d) \leq m$. Every binary partition of d arises from $\mathrm{bin}(d)$ by refinement. A single refinement step increases the length of the partition by 1 by replacing a part 2^j by $2^{j-1} + 2^{j-1}$ for some $j > 0$. It follows that, for $\ell = \alpha(d)$ and $k \geq \ell$, there are at most $1 + \ell + \ell(\ell+1) + \cdots + \ell(\ell+1) \cdots (k-1)$ refinements of $\mathrm{bin}(d)$ of length $\leq k$. The result follows since $\ell \leq m$. $\qquad\square$

5.4 Minimal spikes

Recall from Definition 2.4.2 that $\mu(d)$ is the minimum length of a spike partition of d. Since all its parts are odd, the length of such a partition has the same parity as d.

Proposition 5.4.1 *Let $\ell = d \bmod 2$ and let $\mu(d) \leq \ell \leq d$. Then d has a spike partition of length ℓ.*

Proof Since a spike partition of length $\mu(d)$ exists, we may assume by recursion on ℓ that d has a spike partition σ of length $\ell - 2$. Since $\ell \leq d$, σ has at least one part $\sigma_i > 1$. Then $\sigma_i = 2^{j+1} - 1 = (2^j - 1) + (2^j - 1) + 1$, where $j \geq 1$. By splitting σ_i in this way, we can convert σ to a spike partition of length ℓ. $\qquad\qquad\qquad\qquad\qquad\qquad\qquad\qquad\qquad\qquad\quad$ □

There may be more than one spike partition of d of length $\mu(d)$. For example, $17 = 15 + 1 + 1 = 7 + 7 + 3$. However, for all d there is a unique spike partition $\tau = \sigma^{\min}(d)$ such that either the parts are all distinct, for example $19 = 15 + 3 + 1$, or the parts are distinct except that the smallest part occurs twice, for example $17 = 15 + 1 + 1$. Both types cannot occur for the same d, since the largest part m is determined by $m \leq d \leq 2m$. The partition τ has minimum length $\mu(d)$, and we construct it by the 'greedy algorithm', i.e. by always maximizing the next term.

Theorem 5.4.2 *Given $d > 0$, let $d_1 = d$ and, for $i \geq 1$, let $d_{i+1} = d_i - \tau_i$, where τ_i is the largest integer of the form $2^j - 1$ such that $\tau_i \leq d_i$. Then $\tau = (\tau_1, \tau_2, \ldots)$ is a spike partition of d, and*

(i) *if $\tau_i = \tau_{i+1} > 0$, then $\tau_{i+2} = 0$;*
(ii) *if $\sigma \in \mathsf{Spike}(d)$ satisfies (i) then $\sigma = \tau$;*
(iii) *τ is the unique element of Spike^d such that $\alpha(\tau) = \sum_i \alpha(\tau_i)$ is minimal;*
(iv) *$len(\tau) = \mu(d)$.*

Proof By construction, $\tau \in \mathsf{Seq}_d$. For (i), let $\tau_i = 2^k - 1$. Then $2^k - 1 \leq d_i < 2^{k+1} - 1$, and $d_{i+1} = d_i - 2^k + 1$, so $0 \leq d_{i+1} < 2^k$. Now $\tau_{i+1} = 2^k - 1$ if and only if $2^k - 1 \leq d_{i+1} < 2^{k+1} - 1$, so we must have $d_{i+1} = 2^k - 1$ and $d_{i+2} = 0$. Hence $\tau_{i+2} = 0$.

For (ii), let $k \geq 1$ and let $1 \leq d \leq 2^{k+1} - 2$. Then the first part of any spike partition of d is $\leq 2^k - 1$. The spike partitions of numbers d in this range which satisfy (i) are either strictly decreasing, and so correspond to the $2^k - 1$ non-empty subsets of $\{1, 2, \ldots, k\}$, or they are strictly decreasing except that the last term is repeated, giving a further $2^k - 1$ choices. The construction therefore gives a bijection from the set $\{1, 2, \ldots, 2^{k+1} - 2\}$ to this set of spike partitions.

For (iii), let $\sigma \in \mathsf{Spike}^d$ be a spike partition $\neq \tau$. Then by (i) $\sigma_i = \sigma_{i+1} \geq \sigma_{i+2} > 0$ for some $i \geq 1$. Let $\sigma_i = 2^a - 1$ and $\sigma_{i+2} = 2^b - 1$, so that $a \geq b > 0$. Since $(2^a - 1) + (2^a - 1) + (2^b - 1) = (2^{a+1} - 1) + (2^{b-1} - 1) + (2^{b-1} - 1)$, we can obtain another spike partition σ' of d by replacing the parts $2^a - 1$, $2^a - 1$, $2^b - 1$ by $2^{a+1} - 1$, $2^{b-1} - 1$, $2^{b-1} - 1$ and sorting the parts into decreasing order. Then $\alpha(\sigma') < \alpha(\sigma)$ since $a + 2b - 1 < 2a + b$, proving (iii). Further $len(\sigma') \leq len(\sigma)$, and so if $\sigma \neq \tau$ we can replace σ by a spike partition σ' with smaller α and no greater length. Hence τ has minimal length $\mu(d)$, proving (iv). $\qquad\qquad\qquad\qquad\qquad\qquad\qquad\qquad\qquad\qquad\qquad$ □

Definition 5.4.3 For $d \geq 0$, the spike partition $\tau = \sigma^{\min}(d)$ given by Theorem 5.4.2 is the **minimal spike partition** of d. For $n \geq m = \mu(d)$, the spike monomial $f = x_1^{\tau_1} \cdots x_m^{\tau_m}$, or any monomial obtained from f by permuting $x_1, \ldots, x_n$, is a **minimal spike**. The sequence $\omega(f) = \omega^{\min}(d)$ is the **minimal ω-sequence** and the Ferrers block of f is the **minimal Ferrers block** for spikes in $P^d(n)$.

By Theorem 5.4.2(i), $\sigma^{\min}(d) = (2^{a_1} - 1, \ldots, 2^{a_m} - 1)$, where $m = \mu(d)$. By comparing the equations $d = \sum_{i=1}^{m}(2^{a_i} - 1)$, $2d + m = \sum_{i=1}^{m}(2^{a_i+1} - 1)$ and $(d-m)/2 = \sum_{i=1}^{m}(2^{a_i-1} - 1)$, it follows that $\sigma^{\min}(2d+m)$ is obtained by adding 1 to each part of $\sigma^{\min}(d)$, and $\sigma^{\min}((d-m)/2)$ is obtained by subtracting 1 from each part of $\sigma^{\min}(d)$. In terms of Ferrers blocks, we may think of duplication or of removal of the first column. Thus (cf. Proposition 2.4.5) we obtain the following property of the function μ.

Proposition 5.4.4 *If $\mu(d) = k$, then $\mu(2d + k) = k$, and $k - 2 \leq \mu((d - k)/2) \leq k$.*

Proof This follows by considering the number of nonzero terms when 1 is added to, or subtracted from, each term of the sequence $(a_1, \ldots, a_m)$. $\square$

Example 5.4.5 For $d = 21$, $\sigma^{\min(d)} = (15, 3, 3)$, $\alpha(\sigma^{\min}(d)) = (4, 2, 2)$ and $\omega^{\min}(d) = (3, 3, 1, 1)$. For $n = 4$, the minimal Ferrers block is shown in Example 5.1.9. The corresponding Ferrers block for $d = 45 = 31 + 7 + 7$ is obtained by duplicating the first column.

By iterating the removal of the first column from the minimal Ferrers block, we have the following algorithm for calculating $\omega^{\min}(d)$ in terms of μ.

Proposition 5.4.6 *The minimal ω-sequence $\tau = (\tau_1, \tau_2, \ldots)$ for spikes of degree d is given recursively by $d_1 = \mu(d)$ and $\tau_i = \mu(d_i)$, $d_{i+1} = (d_i - \tau_i)/2$ for $i \geq 1$.* $\square$

Example 5.4.7 Since $\sigma^{\min}(28) = (15, 7, 3, 3)$, $\mu(28) = 4$, $\omega^{\min}(28) = (4, 4, 2, 1)$, and the minimal Ferrers block is

$$
\begin{array}{cccc}
1 & 1 & 1 & 1 \\
1 & 1 & 1 & \\
1 & 1 & & \\
1 & 1 & & \\
\end{array}
$$

5.5 Spike partitions

In this section we show that the poset Dec_d of decreasing sequences of 2-degree d, ordered by 2-dominance, is a lattice. Recall from Proposition 5.1.12 that there is a bijection between Spike^d and Dec_d which associates to a spike partition σ of d the ω-sequence of a spike whose exponents are the terms of σ.

Proposition 5.5.1 *The poset* Dec_d *of decreasing sequences of 2-degree d with 2-dominance order $\preceq_2$ is isomorphic to the poset* Spike^d *of spike partitions of d with reversed dominance order $\succeq$. The minimal ω-sequence $\omega^{\min}(d)$ is the unique minimal element of* Dec_d, *and the minimal spike partition $\sigma^{\min}(d)$ is the unique minimal element of* Spike^d.

Proof By Proposition 5.2.7, the partial orders $\preceq_2$ and $\succeq$ correspond under the bijection of Proposition 5.1.12. By Theorem 5.4.2(iii), $\tau = \sigma^{\min}(d)$ is the unique element of Spike^d such that $\alpha(\tau)$ is minimal, and so it is the unique element of minimum modulus in Dec_d. Since Dec_d is a subposet of Seq_d and Seq_d is graded by the modulus, $\omega^{\min}(d)$ is the unique minimum element of Dec_d. Since the posets are isomorphic, $\sigma^{\min}(d)$ is the unique minimal element of Spike^d. $\qquad\square$

Example 5.5.2 The posets Dec_d and Spike^d are shown below for $d = 12$.

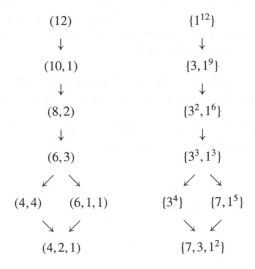

In this case, Dec_d is graded by modulus and Spike^d by α, but Example 5.5.6 shows that this is not true for all d. The bijection is given by conjugation in Dec_{12} followed by the map $k \mapsto 2^k - 1$, or equivalently by $R \mapsto (R')^{\mathrm{tr}}$, for R' as in Proposition 5.2.6.

We shall see that Dec_d is also a lattice, but the next example shows that two elements of Dec_d can have different least upper bounds in Dec_d and Seq_d. Thus in general Dec_d is not a sublattice of Seq_d.

Example 5.5.3 Let $R = (14, 1, 1, 1)$ and $S = (4, 4, 4)$ in Dec_{28}. Calculating $R \vee S$ in Seq_{28} from the partial 2-degree sequences, we obtain $(14, 1, 3)$, which is not decreasing. However $(14, 1, 3)$ is covered in Seq_{28} by $(14, 3, 2)$, which we take as $\sup(R, S)$ in Dec_{28}.

Proposition 5.5.4 *For all R and S in Dec_d, $R \wedge S$ is in Dec_d.*

Proof If $W = R \wedge S$, then the partial 2-degree sequences $\widehat{R}, \widehat{S}, \widehat{W}$ of R, S and W are related by $\widehat{w}_i = \min(\widehat{r}_i, \widehat{s}_i)$ for $i \geq 1$. Since $w_i = (\widehat{w}_i - \widehat{w}_{i-1})/2^{i-1}$, where $w_0 = 0$, W is decreasing if and only if $\widehat{w}_i - \widehat{w}_{i-1} \geq (\widehat{w}_{i+1} - \widehat{w}_i)/2$ for $i \geq 1$. The result follows by observing that if $a_1 \leq a_2 \leq a_3$ and $b_1 \leq b_2 \leq b_3$ are triples of integers such that $a_2 - a_1 \geq (a_3 - a_2)/2$ and $b_2 - b_1 \geq (b_3 - b_2)/2$, then $c_2 - c_1 \geq (c_3 - c_2)/2$ where $c_i = \min(a_i, b_i)$ for $i = 1, 2, 3$. To check this, let $c_1 = a_1$, say, and separate the four cases $(c_2, c_3) = (a_2, a_3), (a_2, b_3), (b_2, a_3)$ and (b_2, b_3). $\qquad\square$

Proposition 5.5.4 gives an alternative proof that Dec_d has a unique minimal element. We can define a lattice structure on Dec_d by $\inf(R, S) = R \wedge S$ and $\sup(R, S) = \inf\{T \in \mathrm{Dec}_d : R \vee S \preceq_2 T\}$. Since Dec_d has a unique maximal element (d), the set of such T is non-empty. We summarize as follows.

Proposition 5.5.5 *The lattices Dec_d, with 2-dominance order, and Spike^d, with reversed dominance order, are isomorphic.* $\qquad\square$

The next example shows that in general Dec_d is not graded, and so is not a distributive lattice.

Example 5.5.6 The diagram shows the interval $\{(4, 4, 2, 1) \preceq R \preceq (6, 5, 3)\}$ in the lattice Seq_{28}. Only the boxed elements are in Dec_{28}.

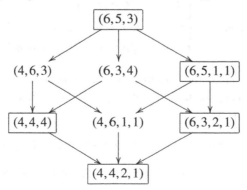

5.6 Maximal spikes

In this section we consider the subposets $\mathrm{Seq}_d(n)$ and $\mathrm{Dec}_d(n)$ of Seq_d and Dec_d whose elements are sequences with entries $\leq n$, ordered by 2-dominance. The minimal element $\omega(d)$ of Seq_d has all entries 0 or 1, and so it is the minimal element of $\mathrm{Seq}_d(n)$ for all $n \geq 1$. The poset $\mathrm{Dec}_d(n)$ is nonempty, with minimal element $\omega^{\min}(d)$, if and only if $n \geq \mu(d)$. Proposition 5.3.2 holds in $\mathrm{Seq}_d(n)$ for all n, and shows that $\mathrm{Seq}_d(n)$ is graded by modulus. In Seq_d, an element of $\mathrm{Seq}_d(n)$ can cover an element not in $\mathrm{Seq}_d(n)$. For example, $(2,2)$ covers $(0,3)$ in Seq_6. This cannot occur in $\mathrm{Dec}_d(n)$, since it is the subset of sequences $R \in \mathrm{Dec}_d$ with $r_1 \leq n$.

Proposition 5.6.1 *For all $n \geq 1$, $\mathrm{Seq}_d(n)$ is a sublattice of Seq_d.*

Proof Let $R, S \in \mathrm{Seq}_d(n)$ have partial 2-degree sequences $\widehat{R}, \widehat{S}$, so that $\widehat{r}_k = \sum_{j=1}^{k} 2^{j-1} r_j$ and $\widehat{s}_k = \sum_{j=1}^{k} 2^{j-1} s_j$ for $k \geq 1$. Since $r_{k-1}, s_{k-1} \leq n$, $\widehat{r}_k - \widehat{r}_{k-1} \leq 2^{k-1}n$ and $\widehat{s}_k - \widehat{s}_{k-1} \leq 2^{k-1}n$ for $k \geq 1$.

To show that $R \vee S$ and $R \wedge S$ are also in $\mathrm{Seq}_d(n)$, we require $a_k - a_{k-1} \leq 2^{k-1}n$ and $b_k - b_{k-1} \leq 2^{k-1}n$, where $a_k = \max(\widehat{r}_k, \widehat{s}_k)$ and $b_k = \min(\widehat{r}_k, \widehat{s}_k)$. By exchanging R and S if necessary, we may assume that $\widehat{r}_k \geq \widehat{s}_k$.

If also $\widehat{r}_{k-1} \geq \widehat{s}_{k-1}$, then the result holds since $a_k - a_{k-1} = \widehat{r}_k - \widehat{r}_{k-1}$ and $b_k - b_{k-1} = \widehat{s}_k - \widehat{s}_{k-1}$. If $\widehat{r}_{k-1} \leq \widehat{s}_{k-1}$, then $a_k - a_{k-1} = \widehat{r}_k - \widehat{s}_{k-1} \leq \widehat{r}_k - \widehat{r}_{k-1}$ and $b_k - b_{k-1} = \widehat{s}_k - \widehat{r}_{k-1} \leq \widehat{r}_k - \widehat{r}_{k-1}$, so the result holds in this case also. $\square$

By Proposition 5.3.7, the lattice $\mathrm{Seq}_d(n)$ is distributive. Since $\mathrm{Seq}_d(n)$ is a finite lattice, it has a unique maximal element. This is given by the following algorithm.

Proposition 5.6.2 *The maximal element S of $\mathrm{Seq}_d(n)$ is defined recursively for $i \geq 1$ by $s_i = \xi(n, d_i)$, where $d_1 = d$ and $d_{i+1} = (d_i - s_i)/2$, and*

$$
\xi(n, d) = \begin{cases} d, & \text{if } 0 \leq d \leq n, \\ n, & \text{if } d > n \text{ and } d = n \bmod 2, \\ n-1, & \text{if } d > n \text{ and } d \neq n \bmod 2. \end{cases}
$$

Proof It is clear from the construction that S cannot arise from a binary addition move on any sequence R with $r_i \leq n$ for all i. Hence S is the maximal element of $\mathrm{Seq}_d(n)$. $\square$

Example 5.6.3 With $n = 3$ and $d = 12$ we obtain $S = (2, 3, 1)$.

We next consider the poset $\mathrm{Dec}_d(n)$ with the partial order $\succ_2$.

Proposition 5.6.4 *For $n \geq \mu(d)$, let $M = (m_1, m_2, \ldots)$ be defined recursively by $m_i = \xi(m_{i-1}, d_i)$, where $m_0 = n$, $d_1 = d$ and $d_{i+1} = (d_i - m_i)/2$. Then*

(i) *M is the unique maximal element of $\mathrm{Dec}_d(n)$,*

(ii) *M is characterized by $m_{i-1} - m_i = 0$ or 1 for $1 \leq i < len(M)$,*

(iii) *if $R \in \mathrm{Seq}_d(n)$ and $R \succ_l M$, then $r_k > m_{k-1}$, where k is the smallest number such that $r_k > m_k$.*

Proof By definition of the function ξ, M satisfies condition (ii), and so it is decreasing. Given $R \neq M \in \mathrm{Seq}_d(n)$, let k be the smallest number such that $r_k \neq m_k$. Then $r_k = m_k \bmod 2$ by Proposition 5.3.2(i). If $r_k > m_k$ then $r_k \geq m_k + 2 > m_{k-1}$, so (iii) is proved. Since $r_k > m_{k-1} = r_{k-1}$, R is not decreasing in this case. Otherwise $r_k < m_k$, and so $R <_l M$. Hence M is the maximal element of $\mathrm{Dec}_d(n)$ in the left order, and it follows from Proposition 5.2.3 that M is also $\succ_2$-maximal in $\mathrm{Dec}_d(n)$.

To show that M is the unique maximal element of $\mathrm{Dec}_d(n)$, we shall construct $S \in \mathrm{Dec}_d(n)$ such that $S \succ_2 R$ if $R \neq M$. By considering parity as above, $r_{i-1} \geq r_i + 2$ for some i with $1 \leq i < len(R)$, where $r_0 = n$. Since $i < len(R)$, $r_{i+1} \neq 0$. Let R' be obtained from R by replacing the entries (r_i, r_{i+1}) by $(r_i + 2, r_{i+1} - 1)$. If $r_{i+1} > r_{i+2}$, then R' is decreasing and $R' \succ_2 R$. If $r_{i+1} = r_{i+2}$, let R'' be obtained from R' by replacing the entries $(r_{i+1} - 1, r_{i+2})$ by $(r_{i+1} + 1, r_{i+2} - 1)$. If $r_{i+2} > r_{i+3}$, then R'' is decreasing and $R'' \succ_2 R$. Continuing this process, we can eventually construct $S \in \mathrm{Dec}_d(n)$ such that $S \succ_2 R$ as required. $\square$

Definition 5.6.5 For $n \geq \mu(d)$, we denote the maximal element M of $\mathrm{Dec}_d(n)$ by $\omega^{\max}(n, d)$ and its conjugate by $\alpha^{\max}(n, d)$. A spike $f \in P^d(n)$ is a **maximal spike** if $\omega(f) = M$. The **maximal Ferrers block** for spikes in $P^d(n)$ is the Ferrers block corresponding to the maximal spike f whose exponents are in decreasing order.

Example 5.6.6 Let $d = 21$. Then $\omega^{\max}(4, d) = (3, 3, 3)$, whereas $\omega^{\max}(5, d) = (5, 4, 2)$.

Proposition 5.6.7 *For all $n \geq \mu(d)$, $\mathrm{Dec}_d(n)$ is a sublattice of Dec_d.*

Proof If $T = R \wedge S$ where $R, S \in \mathrm{Dec}_d(n)$, then $T \in \mathrm{Dec}_d$ by Proposition 5.5.4. Since $r_1, s_1 \leq n$, $t_1 = \min(r_1, s_1) \leq n$, and so $T \in \mathrm{Dec}_d(n)$. By Proposition 5.6.4, $\mathrm{Dec}_d(n)$ has a unique maximal element, and so the least upper bound $\inf\{T \mid R, S \preceq_2 T\}$ is well defined. Hence $\mathrm{Dec}_d(n)$ is a sublattice of Dec_d. $\square$

We next consider conditions under which $\omega^{\max}(n, d) = \omega^{\min}(d)$, so that all spikes in $P^d(n)$ have the same ω-sequence. By Proposition 5.1.13, a spike in $P^d(n)$ with strictly decreasing α-sequence is the unique decreasing spike in $P^d(n)$. It therefore satisfies the criteria for being both the maximum and minimum in Propositions 5.4.6 and 5.6.4. When $d > 1$, a necessary condition for this is that $\mu(d) = n$ or $n - 1$. For $n = 1$ or 2 and all $d \geq 0$, there is at most one element in $\mathrm{Dec}_d(n)$ and in $\mathrm{Spike}^d(n)$.

Proposition 5.6.8 *For $n \geq 5$ and $d > 2$, $\mathrm{Dec}_d(n)$ and $\mathrm{Spike}^d(n)$ have exactly one element if and only if $\alpha(d+n) = n$.*

Proof By Proposition 5.1.13, we only have to prove that if $\sigma = (\sigma_1, \ldots, \sigma_n)$ is a spike partition of d of length $\leq n$, and if σ is not strictly decreasing, then there is another such partition $\rho = (\rho_1, \ldots, \rho_n)$ except when $d = 2$ or $n \leq 4$. For this we take $\sigma = \tau$, the minimal spike partition of d, and construct ρ by modifying τ.

Since $d > 2$, τ has a part $2^a - 1$ where $a > 1$. If $\mathrm{len}(\tau) = \mu(d) \leq n - 2$, we construct ρ by splitting this as $(2^{a-1} - 1) + (2^{a-1} - 1) + 1$ and putting the resulting numbers in decreasing order. Since $n \geq 5$ we may therefore assume that $\mu(d) \geq 4$. By Theorem 5.4.2, since τ is not strictly decreasing its last four nonzero parts are $2^a - 1$, $2^b - 1$, $2^c - 1$, $2^c - 1$ where $a > b > c > 0$. We construct ρ by replacing these parts by $2^{a-1} - 1$, $2^{a-1} - 1$, $2^b - 1$, $2^{c+1} - 1$. Since $a - 1 \geq b \geq c + 1$, $\rho \in \mathrm{Spike}^d(n)$. □

Example 5.6.9 There are exceptions to the condition $\alpha(d+n) = n$ when $n = 3$ or 4. For $n = 3$ and $d > 2$, $\mathrm{Dec}_d(3)$ has exactly one element not only when $\alpha(d+3) = 3$, but also when $d = 2^k - 2$ or $2^k - 3$ and $k \geq 3$. For $n = 4$ and $d > 2$, $\mathrm{Dec}_d(4)$ has exactly one element not only when $\alpha(d+4) = 4$, but also when $d = 2^k - 3$ and $k \geq 3$. In these cases the elements of $\mathrm{Dec}_d(n)$ are the sequences $(2, \ldots, 2)$ or $(3, \ldots, 3, 1)$ of length $k - 1$, and $\alpha(d + n) = 1$ or 2. It is easy to check that $\omega = \omega^{\min}(d) = \omega^{\max}(d)$ using Theorem 5.4.2 and Proposition 5.6.4. The corresponding Ferrers blocks are

$$
\begin{array}{ccc}
 & & 1 \cdots 1 \quad 1 \\
1 \cdots 1 \quad\quad 1 \cdots 1 \quad 1 & & 1 \cdots 1 \\
1 \cdots 1 \;, \quad 1 \cdots 1 \quad\quad \text{and} & & 1 \cdots 1 \quad\quad . \\
0 \cdots 0 \quad\quad 1 \cdots 1 & & 0 \cdots 0
\end{array}
$$

5.7 Dominance in A_2

Recall from Definition 3.1.6 that a sequence $A = (a_1, \ldots, a_\ell) \in \mathrm{Seq}$ and the corresponding monomial $Sq^A = Sq^{a_1} \cdots Sq^{a_\ell}$ of degree $|A|$ in the Steenrod algebra A_2 are admissible if $a_i \geq 2a_{i+1}$ for $1 \leq i < \ell$. We denote by Adm^d

the poset of admissible sequences A with $|A| = d$ ordered by dominance. As usual, we consider the 2-dominance order on Dec_d.

Proposition 5.7.1 *For $d \geq 0$, the posets Adm^d and Dec_d are isomorphic.*

Proof The bijection will preserve the length ℓ of sequences. We define a bijection between Adm^d and Dec_d so that $A = (a_1, \ldots, a_\ell) \in \text{Adm}^d$ corresponds to $U = (u_1, \ldots, u_\ell) \in \text{Dec}_d$, where $u_\ell = a_\ell$ and for $1 \leq i < \ell$

$$u_i = a_i - (a_{i+1} + \cdots + a_\ell), \quad a_i = u_i + u_{i+1} + 2u_{i+2} + \cdots + 2^{\ell-i-1}u_\ell. \quad (5.2)$$

Then $a_i - 2a_{i+1} = (u_i + u_{i+1} + 2u_{i+2} + \cdots + 2^{\ell-i-1}u_\ell) - 2(u_{i+1} + u_{i+2} + 2u_{i+3} + \cdots + 2^{\ell-i-2}u_\ell) = u_i - u_{i+1}$, and so $a_i \geq 2a_{i+1}$ if and only if $u_i \geq u_{i+1}$. Hence $A \in \text{Adm}^d$ if and only if $U \in \text{Dec}_d$. It is easily checked that the maps $A \mapsto U$ and $U \mapsto A$ are inverses of each other, and that $|A| = \deg_2 U$.

Let $U, V \in \text{Dec}_d$ have partial 2-degree sequences $(\widehat{u}_1, \widehat{u}_2, \ldots)$ and $(\widehat{v}_1, \widehat{v}_2, \ldots)$. Then $U \succeq_2 V$ in Dec_d if and only if $\widehat{u}_i \geq \widehat{v}_i$ for $1 \leq i \leq \ell$, where $\ell = \max(\text{len}(U), \text{len}(V))$. Let $A = (a_1, \ldots, a_\ell)$ and $B = (b_1, \ldots, b_\ell)$ in Adm^d correspond to U and V respectively. Then $A \succeq B$ if and only if $\sum_{j=1}^{i} a_j \geq \sum_{j=1}^{i} b_j$ for $1 \leq i \leq \ell$.

We claim that, for all i, $\widehat{u}_i \geq \widehat{v}_i$ if and only if $\sum_{j=1}^{i} a_j \geq \sum_{j=1}^{i} b_j$. By (5.2), $\widehat{u}_i \geq \widehat{v}_i$ if and only if $\sum_{j=1}^{i} 2^{j-1}(a_j - a_{j+1} - \cdots - a_\ell) \geq \sum_{j=1}^{i} 2^{j-1}(b_j - b_{j+1} - \cdots - b_\ell)$. Adding the equation $(2^i - 1)(a_1 + \cdots + a_\ell) = (2^i - 1)d = (2^i - 1)(b_1 + \cdots + b_\ell)$ to this inequality, and dividing by 2^i, we obtain $\sum_{j=1}^{i} a_j \geq \sum_{j=1}^{i} b_j$. □

Example 5.7.2 The posets Adm^{12} and Dec_{12} are shown below.

By Proposition 5.5.1, it follows that Adm^d is also isomorphic to Spike^d. To make this explicit, recall that $U = (u_1,\ldots,u_\ell) \in \mathrm{Dec}_d$ corresponds to the spike partition of d with $u_j - u_{j+1}$ parts equal to $2^j - 1$ for $1 \leq j \leq \ell$, where $u_{\ell+1} = 0$. Now $a_i = u_i + u_{i+1} + 2u_{i+2} + \cdots + 2^{\ell-i-1}u_\ell = (u_i - u_{i+1}) + 2(u_{i+1} - u_{i+2}) + \cdots + 2^{\ell-i-1}(u_{\ell-1} - u_\ell) + 2^{\ell-i}u_\ell$. This gives a practical way to write down corresponding elements of Spike^d, Dec_d and Adm^d using Ferrers blocks.

Proposition 5.7.3 *Let* $\sigma = (2^{\alpha_1} - 1,\ldots,2^{\alpha_n} - 1) \in \mathrm{Spike}^d(n)$ *be a spike partition, and let F be the corresponding Ferrers block, with row sum sequence* $\alpha = (\alpha_1,\ldots,\alpha_n)$ *and column sum sequence* $U = \alpha^{tr} \in \mathrm{Dec}_d$. *Then the admissible sequence* $A \in \mathrm{Adm}^d$ *corresponding to* σ *and U is the column sum sequence of the array given by replacing the ith row of F by* $(2^{\alpha_i-1},\ldots,2,1)$ *if* $\alpha_i > 0$. $\quad\square$

Example 5.7.4 When $\sigma = (15,3,3,0) \in \mathrm{Spike}^{21}(4)$, F is the Ferrers block of Example 5.1.9, $\alpha(f) = (4,2,2)$, $U = (3,3,1,1)$ and $A = (12,6,2,1)$ is the column sum sequence of the array

$$
\begin{array}{cccc}
8 & 4 & 2 & 1 \\
2 & 1 & & \\
2 & 1 & & \\
0 & & &
\end{array}
$$

Since Dec_d has a unique minimal element, the same is true for Adm^d. Recall that the Milnor sequence $R = (r_1,\ldots,r_\ell)$ corresponding to the admissible sequence $A = (a_1,\ldots,a_\ell)$ is defined by $r_i = a_i - 2a_{i+1}$.

Definition 5.7.5 We denote by $A(d)$ the minimal element of Adm^d, and we denote by $R(d)$ the Milnor sequence of $A(d)$.

Thus $Sq^{A(d)}$ is the minimal admissible monomial in A_2^d in both the left and the right orders, and by Theorem 3.5.1 $Sq(R(d))$ is the minimal Milnor basis element in the right order. By applying Proposition 5.7.3 to the minimum element $\omega^{\min}(d) \in \mathrm{Dec}_d$, we obtain an algorithm for finding $A(d)$.

Proposition 5.7.6 *For* $d \geq 0$, *let* $\beta(d) = (d + \mu(d))/2$, *and define the sequence* $A = (a_1,a_2,\ldots)$ *recursively by* $d_1 = d$ *and, for* $i \geq 1$, $a_i = \beta(d_i)$ *and* $d_{i+1} = d_i - a_i$. *Then* $A = A(d)$, *the minimum element of* Adm^d.

Proof Let $\sigma^{\min}(d) = (2^{a_1} - 1, \ldots, 2^{a_m} - 1)$ be the minimal spike partition of d, where $m = \mu(d)$. Then $\beta(d) = (d + m)/2 = \sum_{i=1}^{m} 2^{a_i-1}$ is the first column sum of the array for $A(d)$ given by Proposition 5.7.3, and so its first entry $a_1 = \beta(d)$. The result follows by induction on d. The last statement follows from Proposition 5.2.3. $\qquad\qquad\qquad\qquad\qquad\qquad\qquad\qquad\qquad\qquad\qquad$ □

Example 5.7.7 Let $d = 42$. By Theorem 5.4.2, $\sigma^{\min}(d) = (31, 7, 3, 1)$, so $\mu(d) = 4$ and $\beta(d) = 23$. Then $d_2 = 19 = 15 + 3 + 1$, so $\mu(d_2) = 3$ and $\beta(d_2) = 11$. Continuing in this way, we obtain $A = (23, 11, 5, 2, 1)$. In practice it is quicker to calculate $\omega^{\min}(d) = (4, 3, 2, 1, 1)$ and $A(d)$ from the Ferrers block of the spike for $n = \mu(d)$, as in Example 5.7.4.

The isomorphism between the posets Adm^d and Dec_d is easily described using Milnor sequences. By Proposition 3.4.2, $|R| = 2a_1 - d$, where $d = |A|$, and $\deg_2 R = a_1$. By Theorem 3.5.1, the admissible basis of A_2 is triangularly related to the Milnor basis for the right order $\leq_r$, so that the admissible monomial Sq^A corresponds to the Milnor basis element $Sq(R)$, and $Sq(R) \in A_2^d$ where $d = |A| = \sum_{j\geq 1}(2^j - 1)r_j$. Given $U = (u_1, u_2, \ldots, u_\ell) \in \text{Dec}_d$ of length ℓ, let $r_i = u_i - u_{i+1}$ for $1 \leq i < \ell$ and $r_\ell = u_\ell$, so that $R = (r_1, r_2, \ldots, r_\ell)$ is the sequence of differences between successive terms of U. Since (5.2) gives $a_i - 2a_{i+1} = u_i - u_{i+1}$, R is the Milnor sequence of A.

Example 5.7.8 Let $R = (0, 2, 0, 1)$. Then R is the difference sequence of $(3, 3, 1, 1) \in \text{Dec}_{21}$, and corresponds to the minimal spike partition $(15, 3, 3) \in \text{Spike}^{21}$ and to the minimum admissible sequence $(12, 6, 2, 1) \in \text{Adm}^{21}$.

Definition 5.7.9 For $d \geq 0$, let Mil^d be the set of sequences $R \in \text{Seq}$ such that $d = \sum_{j\geq 1}(2^j - 1)r_j$. We define a partial order on Mil^d by $R \geq S$ if and only if $\rho \leq \sigma$, where the spike partitions ρ and σ of d correspond to R and S respectively.

Example 5.7.10 The posets Dec_{12} and Mil^{12} are shown below. Recall that the admissible and Milnor bases of A_2 are triangularly related in the right order $\leq_r$, but not in the left order $\leq_l$. Thus the right order, but not the left order, refines the partial order on Mil.

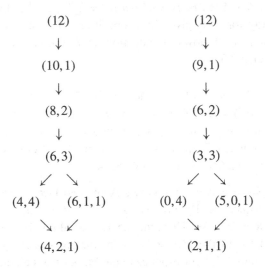

The partial order on Mil^d can be defined more directly as follows.

Proposition 5.7.11 *For $R, S \in \mathrm{Mil}^d$ let $k = \max(len(R), len(S))$. Then $R \geq S$ in Mil^d if and only if $\deg Sq(r_j, \ldots, r_k) \leq \deg Sq(s_j, \ldots, s_k)$ for $1 < j \leq k$.*

Proof Let A and $B \in \mathrm{Adm}^d$ be the admissible sequences corresponding to R and S respectively, so that by definition $R \succeq S$ if and only if $A \succeq B$, i.e. $a_1 + \cdots + a_j \geq b_1 + \cdots + b_j$ for $1 \leq j \leq k$. Since $|A| = |B| = d$, equality holds for $j = 1$, and so these inequalities are equivalent to $a_j + \cdots + a_k \leq b_j + \cdots + b_k$ for $1 < j \leq k$. The result follows, since $a_j + \cdots + a_k = \deg Sq^{a_j} \cdots Sq^{a_k} = \deg Sq(r_j, \ldots, r_k)$. $\square$

If $(P, \geq)$ is a poset, then a function ε from P to the integers ≥ 0 is **increasing** if $a \geq b$ in P implies $\varepsilon(a) \geq \varepsilon(b)$.

Proposition 5.7.12 *If the sequence $U = (u_1, u_2, \ldots) \in \mathrm{Dec}_d$, the spike partition $\sigma \in \mathrm{Spike}^d$, the admissible sequence $A = (a_1, a_2, \ldots) \in \mathrm{Adm}^d$ and the Milnor sequence $R \in \mathrm{Mil}^d$ are corresponding elements of these posets, then $u_1 = len(\sigma) = 2a_1 - d = |R|$, and this function is increasing in each case.*

Proof It is clear that these functions on the four posets correspond under the bijections defined above. To check that the functions are increasing, we need only check any one of the four cases. In the cases of Dec_d and Adm^d, this is immediate by definition of the partial order. $\square$

5.8 The excess function on A_2

In this section, we show that the function on Adm^d and Mil^d defined in Proposition 5.7.12 has a natural interpretation in terms of the action of the corresponding elements of A_2 on $P(n)$. Recall from Definition 2.4.9 that the excess $ex(\theta)$ of a nonzero element θ of A_2 is the minimum $n \geq 0$ such that $\theta(c(n)) \neq 0$, where $c(n) = x_1 x_2 \cdots x_n$ for $n \geq 1$ and $c(0) = 1$, and that $\theta(f) = 0$ for all $f \in P^d(n)$ where $d < ex(\theta)$. The following result follows immediately from Proposition 3.5.2.

Proposition 5.8.1 *The Milnor basis element* $Sq(R)$ *has excess* $|R|$. $\qquad\square$

In order to determine the excess of an admissible monomial $Sq^A \in A_2^d$, we express Sq^A in the Milnor basis. First we relate the Milnor product formula to the partial order on Mil^d.

Proposition 5.8.2 *Let* $R = (r_1, \ldots, r_k) \in Mil^d$, *let* $a \geq 2 \deg_2 R$ *and let* $R^+ = (r_0, r_1, \ldots, r_k)$ *where* $r_0 = a - \sum_{j=1}^{k} 2^j r_j$. *Then* $Sq^a Sq(R) = Sq(R^+) + \sum_S Sq(S)$ *in* A_2^{a+d}, *where all terms in the sum satisfy* $S > R^+$ *in* Mil^d.

Proof We evaluate $Sq^a Sq(R)$ using Theorem 4.1.2. Since $r \geq \sum_{j=1}^{k} 2^j r_j$, the Milnor matrix

$$X = \begin{array}{c|cccc} & 0 & 0 & \cdots & 0 \\ \hline r_0 & r_1 & r_2 & \cdots & r_k \end{array}$$

gives the term $Sq(R^+)$ in the product. A general Milnor matrix has the form

$$Y = \begin{array}{c|cccc} & x_1 & x_2 & \cdots & x_k \\ \hline y_0 & y_1 & y_2 & \cdots & y_k \end{array}$$

where $x_i + y_i = r_i$ for $1 \leq i \leq k$, and Y gives a term $b(Y)Sq(S)$, where $b(Y) \in \mathbb{F}_2$ and $S = (s_0, s_1, \ldots, s_k)$, with $s_i = x_{i+1} + y_i$ for $0 \leq i \leq k$ and $x_{k+1} = 0$. By Proposition 5.7.11, $S \geq R^+$ in Mil^d if and only if $\deg Sq(s_j, \ldots, s_k) \leq \deg Sq(r_j, \ldots, r_k)$ for $1 \leq j \leq k$. Then

$$\deg Sq(s_j, \ldots, s_k) = \deg Sq(x_{j+1}, \ldots, x_k) + \deg Sq(y_j, \ldots, y_k),$$
$$\deg Sq(r_j, \ldots, r_k) = \deg Sq(x_j, \ldots, x_k) + \deg Sq(y_j, \ldots, y_k),$$

and $\deg Sq(x_{j+1}, \ldots, x_k) \leq \deg Sq(x_j, \ldots, x_k)$, with equality if and only if $x_i = 0$ for $i \leq j \leq k$. Hence $S > R^+$ if $Y \neq X$. $\qquad\square$

This result allows us to strengthen part of Theorem 3.5.1 as follows.

Proposition 5.8.3 *Let A be an admissible sequence with corresponding Milnor sequence R. Then $Sq^A = Sq(R) + \sum_S Sq(S)$, where $S > R$ for all terms in the sum.*

Proof We argue by induction on $\ell = \text{len}(A)$. Since $Sq^a = Sq(a)$ for all $a \geq 0$, the statement is true if $\text{len}(A) = 0$ or 1. Thus let $A = (a_1, a_2, \ldots, a_\ell)$ where $a_\ell > 0$ and $\ell \geq 2$. Let $A^- = (a_2, \ldots, a_\ell)$, and let $R = (r_1, r_2, \ldots, r_\ell)$ and $R^- = (r_2, \ldots, r_\ell)$ be the Milnor sequences of A and A^-. Then $a_1 = \deg_2 R$ and $a_2 = \deg_2 R^-$, and so $a_1 \geq 2 \deg_2 R^-$ since A is admissible. By the induction hypothesis, $Sq^{A^-} = Sq(R^-) + \sum_T Sq(T)$, where $T > R^-$ for all terms in the sum. Then $Sq^A = Sq^{a_1} Sq^{A^-} = Sq^{a_1} Sq(R^-) + \sum_T Sq^{a_1} Sq(T)$.

Since $a_1 \geq 2 \deg_2 R^-$, Proposition 5.8.2 shows that $Sq^{a_1} Sq(R^-) = Sq(R) + \sum_S Sq(S)$, where $S > R$ for all terms in the sum. To deal with the remaining terms, let $T > R^-$, where $T = (t_2, \ldots, t_\ell)$ and let

$$
Y = \frac{\begin{array}{ccc} x_2 & \cdots & x_\ell \end{array}}{\begin{array}{c|ccc} y_1 & y_2 & \cdots & y_\ell \end{array}}
$$

be a Milnor matrix yielding a term Sq^S in the product $Sq^{a_1} Sq(T)$, where $S = (s_1, \ldots, s_\ell)$. Thus $x_j + y_j = t_j$ for $2 \leq j \leq \ell$, and $s_j = x_{j+1} + y_j$ for $1 \leq j \leq \ell$, where $x_{\ell+1} = 0$. To complete the inductive step, we must prove that $S > R$.

Since $T > R^-$, $\deg Sq(r_j, \ldots, r_\ell) \geq \deg Sq(t_j, \ldots, t_\ell)$ for $2 \leq j \leq \ell$, where at least one of these inequalities is strict. Then for $2 \leq j \leq \ell$,

$$
\begin{aligned}
\deg Sq(t_j, \ldots, t_\ell) &= \deg Sq(x_j, \ldots, x_\ell) + \deg Sq(y_j, \ldots, y_\ell) \\
&\geq \deg Sq(x_{j+1}, \ldots, x_\ell) + \deg Sq(y_j, \ldots, y_\ell) \\
&= \deg Sq(s_j, \ldots, s_\ell).
\end{aligned}
$$

Hence $\deg Sq(r_j, \ldots, r_\ell) \geq \deg Sq(s_j, \ldots, s_\ell)$ for $2 \leq j \leq \ell$, where at least one of these inequalities is strict. By Proposition 5.7.11, it follows that $S > R$. $\square$

Proposition 5.8.4 *The admissible monomial $Sq^A \in A_2^d$ has excess $2a_1 - d$.*

Proof By Proposition 5.7.12, the function $|R|$ is increasing on the poset Mil. By definition of excess, the excess of an element $\theta \in A_2^d$ expressed in the Milnor basis is the minimum excess of the terms $Sq(R)$ appearing in the sum. By Proposition 5.8.3, it follows that the excess of the admissible monomial Sq^A is $|R|$, where R is the Milnor sequence corresponding to A. By Proposition 5.7.12, $|R| = 2a_1 - d$. $\square$

Recall from Definition 5.7.5 that $A(d)$ is the minimal element of Adm^d and $R(d)$ is the minimal element of Mil^d.

Proposition 5.8.5 (i) *The correspondence between admissible monomials Sq^A and Milnor basis elements $Sq(R)$ preserves the excess.*

(ii) *For $d \geq 0$, $\mu(d)$ is the minimum excess of a nonzero element of A_2^d. In particular, the minimal elements $A(d) \in \text{Adm}^d$ and $R(d) \in \text{Mil}^d$ correspond to an admissible monomial $Sq^{A(d)}$ and a Milnor basis element $Sq(R(d))$ in A_2^d of minimal excess $\mu(d)$.*

Proof (i) follows from Propositions 5.7.12, 3.5.2 and 5.8.4, while (ii) follows from the poset isomorphisms of Proposition 5.7.12, since $\mu(d)$ is the minimum length of an element of Spike^d. □

Example 5.8.6 For $d = 47$, $\sigma^{\min}(d) = (31, 15, 1)$ with exponent sequence $\alpha = (5, 4, 1)$. We replace the entries in the minimal Ferrers block by 2-powers, starting from the right, and sum the rows to give

$$
\begin{array}{ccccc}
 & & 16 & 8 & 4 & 2 & 1 \\
1 \; 1 \; 1 \; 1 \; 1 & & & 8 & 4 & 2 & 1 \\
1 \; 1 \; 1 \; 1 & \longrightarrow & & & & & 1 \\
1 & & \hline \\
 & & 25 & 12 & 6 & 3 & 1
\end{array}
$$

Hence $A = A(47) = (25, 12, 6, 3, 1)$ and so $R = R(47) = (1, 0, 0, 1, 1)$. Alternatively we can first write down the $\leq_r$ minimal Milnor basis element $Sq(R) = Sq(1, 0, 0, 1, 1)$ by taking the 'frequency sequence' of $\sigma^{\min}(d)$, i.e. $r_i(d)$ is the number of times $2^i - 1$ appears as a term in $\sigma^{\min}(d)$, and then calculate A from R. Proposition 5.4.6 gives an alternative method using the function μ.

5.9 Remarks

For general information about partially ordered sets and lattices we refer to [192, Chapter 3]. There is a substantial literature on binary partitions (see e.g. [121]), in contrast to spike partitions [19, 69]. In particular, the cardinality of the set of sequences Seq_d, or equivalently of the set of ω-sequences of monomials in P^d, is the number $b(d)$ of binary partitions of d, and can be computed recursively from the relations $b(0) = 1$ and $b(2k+1) = b(2k) = b(2k-1) + b(k)$ for $k \geq 0$, or from the generating function $\sum_{d \geq 0} b(d)x^d = \prod_{j \geq 0} 1/(1 - x^{2^j})$.

The corresponding generating function for $\text{Seq}_d(n)$ or $\text{Bin}^d(n)$ is

$$\sum_{d \geq 0} b(n,d)x^d = \prod_{j \geq 0} \frac{1 - x^{2^j(n+1)}}{1 - x^{2^j}}.$$

Similarly, the generating function for the cardinality of Dec_d, or equivalently of Spike^d, is $\sum_{d \geq 0} s(d)x^d = \prod_{j \geq 1} 1/(1 - x^{2^j-1})$. The cardinality of $\text{Dec}_d(n) = \text{Spike}^d(n)$ is given by the coefficient of $x^d y^n$ in $\prod_{j \geq 0} 1/(1 - x^{2^j-1}y)$. Proposition 5.8.2 strengthens a result of K. G. Monks [146] which proves the corresponding result for the right order $<_r$.

Blocks and ω-sequences (under various names) are well known in the literature on game theory. For example, Hardy and Wright [73] explain the winning strategy in the game of Nim in essentially the following manner.

The game is between two players and starts with at least two non empty piles of objects. A move consists in taking away at least one object from a single pile of the player's choice. The players take it in turn to make a move. Nim is usually played as a misère game in which the loser is the player forced to remove the last object, but in the normal game a player wins by taking all the objects when only one pile remains.

We assume that the game is normal. The state of the game at any time may be represented by the nonzero block B whose rows are the reversed binary expansions of the numbers of objects in the piles. We call B a losing block if all entries of $\omega(B)$ are even. Otherwise B is a winning block. For example, a 1-block is a winning block and a 2-block with equal rows is a losing block.

Proposition 5.9.1 *Any move on a losing block produces a winning block.*

Proof Let $r > 0$ be a number with $\text{len}(\omega(r)) = \ell$. Then for any number $d \geq r$ we have $\omega(d-r)_\ell = 1 - \omega(d)_\ell$. Hence removing r objects from a pile of d objects changes the parity of column ℓ of the corresponding block. In particular, a losing block becomes a winning block. $\square$

Proposition 5.9.2 *There is a move on a winning block which either wins the game or produces a losing block.*

Proof Let $B = (b_{i,j})$ be a winning block. If B has only one nonzero row then removing all the objects wins the game, so we may assume that B has at least two nonzero rows. Let $\omega(B)$ have its last odd entry in column c. Then $b_{i,c} = 1$ for some row i. Let $B' = (b'_{i,j})$ be the block obtained from B by putting $b'_{i,j} = 1 - b_{i,j}$ for each j where $\omega(B)_j$ is odd, leaving the other entries of B unchanged. Then B' is a losing block. Furthermore $b'_{i,j} = b_{i,j}$ if $j > c$ and $b'_{i,c} = 0$. Hence, if row i of B represents the number d and row i of B' represents d', then $d' < d$. It follows that B' is obtained from B by removing $d - d'$ objects from the ith pile.

$\square$

From the above propositions we see that a knowledgeable player presented at some stage in the game with a winning block has a winning strategy. For example, in a game with two piles, the strategy is to equalize the numbers of objects in the piles so presenting the opposing player with a losing block. Further discussion of Nim can be found on Wikipedia.

6

The cohit module $Q(n)$

6.0 Introduction

From Chapter 3 we may interpret the hit polynomials $H(n)$ of Chapter 1 as $A_2^+ P(n)$, where $A_2^+ = \sum_{k>0} A_2^k$. A vector space basis for the cohit module $Q(n) = P(n)/H(n)$ lifts to a minimal set of polynomials which generate $P(n)$ as an A_2-module. We decompose $Q^d(n)$ as the direct sum of vector subspaces $Q^\omega(n)$, where $\omega \in \mathrm{Seq}_d(n)$. The subspaces $Q^\omega(n)$ are $\mathbb{F}_2 M(n)$-modules, and so they are filtration quotients of $Q^d(n)$. It should be emphasized that the definition of $Q^\omega(n)$ depends on the choice of a linear order relation on $\mathrm{Seq}_d(n)$. We shall always assume that the notation $Q^\omega(n)$ refers to the left order, unless otherwise specified.

In Section 6.1 we describe the actions of Steenrod squares and matrices on $P(n)$ in terms of the combinatorics of n-blocks F, which we use to represent monomials f as in Sections 3.3 and 5.1. As in Definition 3.3.7, $\alpha(f) = \alpha(F) = (\alpha_1(F), \ldots, \alpha_n(F))$ and $\omega(f) = \omega(F) = (\omega_1(F), \omega_2(F), \ldots)$ are the sequences of row and column sums of F, while the degree of f is the 2-degree $\deg_2 \omega = \sum_i 2^{i-1} \omega_i$ of $\omega = \omega(F)$.

In Section 6.2 we consider the left and right orders on the set $\mathrm{Seq}_d(n)$ of n-bounded sequences of 2-degree d, and define left and right **reducibility** of a polynomial. In each case, this leads to a filtration of $Q^d(n)$ by $\mathbb{F}_2 M(n)$-submodules, whose associated graded modules give a direct sum decomposition $Q^d(n) \cong \sum_\omega Q^\omega(n)$ as vector spaces over $\mathbb{F}_2$. This reduces the 'global' problem of finding $\dim Q^d(n)$ to the 'local' problems of finding $\dim Q^\omega(n)$ for each $\omega \in \mathrm{Seq}_d(n)$.

In Section 6.3 we introduce **concatenation** of n-blocks. This requires a treatment of **finite blocks**. In Section 6.4 we explain the **splicing** method of generating hit equations in terms of blocks, and show that $Q^\omega(n) = 0$ if ω is lower than $\omega^{\min}(d)$ in the left order. In Sections 6.5 and 6.6 we generalize the Kameko and duplication maps introduced for the 2-variable case in Sections 1.6 and 1.7.

In general, it is hard to describe $Q^\omega(n)$ as a $\mathbb{F}_2 GL(n)$-module, but there are two extreme cases for which we can do this fairly easily. We call these **tail** sequences $\omega = (1,\ldots,1)$, of 2-degree $2^\ell - 1$, and **head** sequences $\omega = (n - 1,\ldots,n - 1)$, of 2-degree $(2^\ell - 1)(n - 1)$, where ℓ is the length of ω. These are dealt with in Sections 6.7 and 6.8. Finally in Section 6.9 we find composition series for the $\mathbb{F}_2 GL(n)$-modules $Q^\omega(n)$ where ω is a tail or head sequence.

6.1 Steenrod and matrix actions on blocks

Recall from Definition 3.3.7 or 5.1.5 that monomials in $P(n)$ are represented by *n*-**blocks** whose rows are the reversed binary expansions of their exponents. The 2-**degree** of the block is the degree of the monomial. For $n = 1$ the 1-block representing x^d is the reversed binary expansion of d, and has 2-degree d.

A nonzero polynomial f is represented by a formal sum of blocks $F = B_1 + \cdots + B_k$, where no two terms are equal, so that the sum is irredundant over $\mathbb{F}_2$. The zero polynomial is represented by an empty formal sum, and not by the zero block, which represents the monomial 1. Since we normally work in $P^d(n)$ where $d > 0$, this zero block does not appear in practice, and so we write 0 for the empty formal sum. We write $Sq^k(F)$ for the sum of blocks representing $Sq^k(f)$. If $Sq^k(f) = 0$, we write $Sq^k(F) = 0$, as explained above.

Example 6.1.1 Let F and G be 1-blocks, where F has a contiguous digits 1 followed by 0, and G is obtained by interchanging the digits 0 and 1 in F, i.e. $F = 1\cdots 1\,0$, $G = 0\cdots 0\,1$. Then F represents x^{2^a-1}, where $a = \alpha(F)$, G represents x^{2^a} and $Sq^1(F) = G$ by Proposition 1.1.8. Viewed as reversed binary expansions of $2^a - 1$ and 2^a, G is obtained by adding the number 1 arithmetically to F.

For a general 1-block F of 2-degree d and $k > 0$, by Proposition 1.4.11 $Sq^k(F) = 0$ unless $\mathrm{bin}(k) \subseteq \mathrm{bin}(d)$. This leads to the following combinatorial description of how Sq^k acts on F.

Proposition 6.1.2 *Assume that $\binom{d}{k}$ is odd, so that $Sq^k(x^d) = x^{d+k}$. Let F be the 1-block representing x^d and $t = \alpha(k)$. Then $Sq^k(F)$ is the 1-block $F(t)$ obtained by taking the elements $2^{j_1},\ldots,2^{j_t}$ of $\mathrm{bin}(k)$ in decreasing order and successively forming the blocks $F(i)$, for $1 \le i \le t$, by adding 1 arithmetically at position $j_i + 1$ to $F(i - 1)$, where $F(0) = F$. Equivalently, $Sq^k(F) = Sq^{2^{j_t}} \cdots Sq^{2^{j_1}}(F)$. Also $\omega(Sq^k(F)) <_{l,r} \omega(F)$ and $\alpha(Sq^k(F)) \le \alpha(F)$.*

Proof The combinatorial process described above is simply a way of adding the numbers d and k in reversed binary arithmetic. The result is proved by induction on t using Proposition 1.1.8 and Proposition 1.4.11. Under the condition bin(k) $\subseteq$ bin(d), at each step some digit 1 in F is moved to the right, no digit 1 is moved left and the total number of digits 1 does not increase. Thus $\omega(F(i)) < \omega(F(i-1))$ in both left and right order, and $\alpha(F(i)) \le \alpha(F(i-1))$. $\qquad\qquad\qquad\qquad\qquad\qquad\qquad\qquad\qquad\qquad\qquad\qquad\square$

Example 6.1.3 Let $F = F(0) = 011111$ and $k = 10$, so that bin(k) $= \{8, 2\}$. Starting in position 4, we form $F(1) = Sq^8(F) = 0110001$. Then at position 2 we form $F(2) = Sq^2(F(1)) = 0001001$, giving $Sq^{10}(F) = Sq^2 Sq^8(F)$.

More generally, we can evaluate a Steenrod operation Sq^k on a block by using the Cartan formula 1.1.9.

Example 6.1.4 The equation $Sq^2(x^3yz) = x^5yz + x^4y^2z + x^4yz^2 + x^3y^2z^2$ is illustrated in terms of blocks by

$$
Sq^2 \begin{pmatrix} 1\,1 \\ 1\,0 \\ 1\,0 \end{pmatrix} = \begin{array}{c} 1\,0\,1 \\ 1\,0\,0 \\ 1\,0\,0 \end{array} + \begin{array}{c} 0\,0\,1 \\ 0\,1\,0 \\ 1\,0\,0 \end{array} + \begin{array}{c} 0\,0\,1 \\ 1\,0\,0 \\ 0\,1\,0 \end{array} + \begin{array}{c} 1\,1 \\ 0\,1 \\ 0\,1 \end{array} .
$$

Proposition 6.1.5 *Let* $f \in \mathsf{P}^d(n)$ *be a monomial and let* g *be a term of* $\theta(f)$, *where* $\theta \in \mathsf{A}_2^+$. *Then* $\omega(g) <_{l,r} \omega(f)$, *and* $\alpha_i(g) \le \alpha_i(f)$ *for* $1 \le i \le n$.

Proof It is enough to prove the result for $\theta = Sq^k$. Then the case $n = 1$ is treated in Proposition 6.1.2, and the general case follows from the extended Cartan formula 1.1.9. $\qquad\qquad\qquad\qquad\qquad\qquad\qquad\qquad\qquad\qquad\qquad\square$

The above observations concerning addition in reversed binary arithmetic are also involved in explaining how matrix substitution works on blocks. We first collect some elementary results concerning reversed binary arithmetic, the usual componentwise addition of sequences and the left and right order relations. Clearly $R + S \ge_l R$ and $R + S \le_r R$ for all $R, S \in$ Seq, with equality if and only if S is the zero sequence.

Proposition 6.1.6 *Let* $d = a + b$ *where* $a, b > 0$. *Then* $\omega(d) = \omega(a) + \omega(b)$ *if* bin(a) *and* bin(b) *are disjoint, and* $\omega(d) <_l \omega(a) + \omega(b)$, $\omega(d) <_r \omega(a)$ *otherwise.*

Proof Since the entries of $\omega(a)$ are the digits 0 or 1 in the reversed binary expansion of a, $\omega(d)$ is obtained from the binary addition of a and b, whereas $\omega(a) + \omega(b)$ is the term-by-term sum of integer sequences. If bin(a) and bin(b)

are disjoint, the two kinds of addition coincide. Otherwise the inequality for $\leq_l$ follows by considering the smallest 2-power in $\text{bin}(a) \cap \text{bin}(b)$, and the inequality for $\leq_r$ by considering the largest such 2-power. □

More generally, the following result holds for the right order. Recall that $\omega(r^d) = \omega(d)$ for $d \geq 0$.

Proposition 6.1.7 *For all monomials f, g we have $\omega(fg) \leq_r \omega(f)$, with equality if and only if $g = 1$.*

Proof Let $f = x_1^{a_1} \cdots x_n^{a_n}$ and $g = x_1^{b_1} \cdots x_n^{b_n}$. Then $\omega(fg) = \sum_{i=1}^{n} \omega(a_i + b_i) \leq_r \sum_{i=1}^{n} \omega(a_i) = \omega(f)$, with equality if and only if all $b_i = 0$. □

Definition 6.1.8 For $n > 1$, the **standard transvection** $U \in \text{GL}(n)$ is given by $x_1 \cdot U = x_1 + x_2$ and $x_i \cdot U = x_i$ for $i > 1$, i.e. $U = (u_{i,j})$ where $u_{i,j} = 1$ if $i = j$ or if $(i, j) = (1, 2)$, and $u_{i,j} = 0$ otherwise.

Thus $U = T_{1,2}$ in the notation of Definition 1.2.6. We next describe the expansion of $x^a y^b \cdot U = (x + y)^a y^b$ in terms of blocks. Recall that $x^r y^{a-r}$ is a term of $(x + y)^a$ if and only if $\text{bin}(r) \subseteq \text{bin}(a)$.

Proposition 6.1.9 *Let F be the 2-block F representing $x^a y^b$, and let U be the standard transvection. Then $x^a y^b \cdot U$ is the polynomial represented by $F \cdot U = \sum_r F(r)$, where $F(r)$ is the block with rows $\omega(r)$ and $\omega(a - r + b)$ and the sum is over all r such that $\text{bin}(r) \subseteq \text{bin}(a)$. For all such r, $\omega(F(r)) \leq_{l,r} \omega(F)$.*

Proof Since $F(r)$ is the block representing $x^r y^{a-r+b}$, it is clear that $F \cdot U$ is as described. For the inequalities, using Proposition 6.1.6 we have $\omega(F(r)) = \omega(r) + \omega(a - r + b) \leq_{l,r} \omega(r) + \omega(a - r) + \omega(b) = \omega(a) + \omega(b)$ since $\text{bin}(r)$ and $\text{bin}(a - r)$ are disjoint. But $\omega(a) + \omega(b) = \omega(F)$, so $\omega(F(r)) \leq_{l,r} \omega(F)$. □

Example 6.1.10 Let $F = \begin{smallmatrix} 1 & 0 & 1 \\ 1 & 1 & 0 \end{smallmatrix}$. Then $F \cdot U = F(5) + F(4) + F(1) + F(0)$,

where $F(5) = F$, $F(4) = \begin{smallmatrix} 0 & 0 & 1 \\ 0 & 0 & 1 \end{smallmatrix}$, $F(1) = \begin{smallmatrix} 1 & 0 & 0 \\ 1 & 1 & 1 \end{smallmatrix}$ and $F(0) = \begin{smallmatrix} 0 & 0 & 0 \\ 0 & 0 & 0 & 1 \end{smallmatrix}$.

The action of singular matrices on blocks is also useful. For example in the 3-variable case the linear transformation $x \mapsto y + z, y \mapsto y, z \mapsto z$ is represented by the matrix M below and acts on the monomial $x^a y^b z^c$ to give $(y + z)^a y^b z^c = \sum_r y^{r+b} z^{a-r+c}$, where the sum is again over all r such that $\text{bin}(r) \subseteq \text{bin}(a)$.

$$
\textbf{Example 6.1.11} \ \text{Let } F = \begin{matrix} 1\,1\,0 \\ 1\,0\,1 \\ 0\,1\,1 \end{matrix} \ \text{and let } M = \begin{pmatrix} 0\,1\,1 \\ 0\,1\,0 \\ 0\,0\,1 \end{pmatrix}. \text{ Then}
$$

$$
F \cdot M = \begin{matrix} 0\,0\,0 \\ 0\,0\,0\,1 \\ 0\,1\,1 \end{matrix} + \begin{matrix} 0\,0\,0 \\ 1\,1\,1 \\ 1\,1\,1 \end{matrix} + \begin{matrix} 0\,0\,0 \\ 0\,1\,1 \\ 0\,0\,0\,1 \end{matrix} + \begin{matrix} 0\,0\,0 \\ 1\,0\,1 \\ 1\,0\,0\,1 \end{matrix} \ .
$$

Proposition 6.1.12 *Let M be the singular $n \times n$ matrix obtained by replacing row i of the identity matrix I_n by the sum of the rows $j \in J$, where $i \notin J$, and let F be the n-block corresponding to the monomial $f = x_1^{a_1} \cdots x_n^{a_n}$. For each decomposition c of $\mathrm{bin}(a_i)$ as the disjoint union of $\mathrm{bin}(b_j)$, $j \in J$ let $F(c)$ be the block obtained by replacing row i of F by zero and adding b_j in binary arithmetic to row j for $j \in J$. Then $F \cdot M = \sum_c F(c)$.*

Proof As for Proposition 6.1.9, this follows by replacing x_i in f by $\sum_{j \in J} x_j$ and expanding the polynomial using Proposition 1.4.13. □

We can extend the last statement of Proposition 6.1.9 to an arbitrary matrix.

Proposition 6.1.13 *Let $M \in \mathsf{M}(n)$ be a matrix and let F be an n-block. Then $\omega(B) \leq_{l,r} \omega(F)$ for every block B in the expansion of $F \cdot M$.*

Proof The group $\mathsf{GL}(n)$ is generated by permutation matrices and the standard transvection U. Clearly, permutations of the rows of a block do not alter the ω-sequence. The result for $\mathsf{GL}(n)$ then follows from Proposition 6.1.9. Every $n \times n$ matrix of rank $k > 0$ can be written in the form $AE_k B$ where $A, B \in \mathsf{GL}(n)$ and $E_k = \left(\begin{smallmatrix} I_k & 0 \\ 0 & 0 \end{smallmatrix} \right)$, where I_k is the $k \times k$ identity matrix. Since applying E_k either sends a monomial to 0 or leaves it fixed, the result is true for E_k, and the proof is completed by composition of the actions of A, E_k and B. □

6.2 Filtrations of $\mathrm{Q}^d(n)$

The results of the previous section show that the left and right orders on ω-sequences are valuable tools for studying both the action of the Steenrod algebra and that of linear substitutions on $\mathsf{P}(n)$. We use these orders to extend the definition of ω-sequence from monomials to polynomials.

Definition 6.2.1 Let $<$ denote either the left or right order on $\mathrm{Seq}(n)$, and let $f \in \mathsf{P}^d(n)$ be nonzero. Then the ω-**sequence** $\omega(f)$ is the maximum of the sequences $\omega(g)$ for monomials g which are terms of f.

It is important to note that this definition depends on the choice of linear order. In this chapter, the notation $<$ will denote either the left order $<_l$ or the right order $<_r$ on $\mathrm{Seq}(n)$.

Example 6.2.2 Let $f = x^5yz + x^3y^2z^2 \in \mathsf{P}^7(3)$. Then $\omega(f) = \omega(x^5yz) = (3, 0, 1)$ for the left order, but $\omega(f) = \omega(x^3y^2z^2) = (1, 3)$ for the right order.

Propositions 6.1.5 and 6.1.13 give the following result.

Proposition 6.2.3 *For $f \in \mathsf{P}(n)$, $\theta \in A_2^+$ and $M \in M(n)$, $\omega(\theta(f)) < \omega(f)$ if $\theta(f) \neq 0$, and $\omega(f \cdot M) \leq \omega(f)$ if $f \cdot M \neq 0$.* $\qquad\qquad\square$

Definition 6.2.4 Given a sequence $\omega \in \mathrm{Seq}_d(n)$, $\mathsf{P}^\omega(n)$ is the vector subspace of $\mathsf{P}^d(n)$ spanned by monomials g with $\omega(g) = \omega$. If $<$ is either the left order $<_l$ or the right order $<_r$, $\mathsf{P}^{<\omega}(n)$ and $\mathsf{P}^{\leq\omega}(n)$ are the subspaces of $\mathsf{P}^d(n)$ spanned by monomials g with $\omega(g) < \omega$ and $\omega(g) \leq \omega$ respectively.

Thus for $f \in \mathsf{P}^d(n)$, $f \in \mathsf{P}^{<\omega}(n)$ if and only if $\omega(f) < \omega$, and $f \in \mathsf{P}^{\leq\omega}(n)$ if and only if $\omega(f) \leq \omega$. By Proposition 6.1.13, $\mathsf{P}^{<\omega}(n)$ and $\mathsf{P}^{\leq\omega}(n)$ are $\mathbb{F}_2M(n)$-submodules of $\mathsf{P}^d(n)$. We recall that $f \sim g$ means that f g is hit, and define a local version of the hit problem as follows.

Definition 6.2.5 A polynomial $f \in \mathsf{P}^\omega(n)$ is **left reducible** if $f = h + e$ where h is hit and $\omega(e) <_l \omega$. We define an equivalence relation on $\mathsf{P}^\omega(n)$ by $f \sim_l g$ if and only if $f - g$ is left reducible, and denote the set of equivalence classes by $Q^\omega(n)$. Similarly we write $f \sim_r g$ if $f - g$ is **right reducible**, i.e. $f - g = h + e$, where h is hit and $\omega(e) <_r \omega$.

The notation e is intended to suggest 'error terms'.

Example 6.2.6 The monomial $x^5y^5z^5$ is left reducible. To see this, consider the hit equation

$$Sq^2(x^3y^5z^5) = x^5y^5z^5 + x^4y^6z^5 + x^4y^5z^6 + x^3y^6z^6.$$

The last three monomials have ω-sequences $(1, 1, 3)$ or $(1, 3, 2)$, which are left lower than $\omega(x^5y^5z^5) = (3, 0, 3)$. Note that $(1, 3, 2)$ is not right lower than $(3, 0, 3)$, and so this hit equation does not determine whether $x^5y^5z^5$ is right reducible.

Example 6.1.10 shows that $\mathsf{P}^\omega(n)$ is not in general closed under the action of $GL(n)$. However, by Proposition 6.2.3 the vector space isomorphism $\mathsf{P}^\omega(n) \cong \mathsf{P}^{\leq\omega}(n)/\mathsf{P}^{<\omega}(n)$ may be used to define a matrix action on $\mathsf{P}^\omega(n)$ for either the left or the right order.

The spaces $Q^\omega(n)$ can be obtained formally as the graded vector spaces of a finite filtration on $Q^d(n)$. The order $<$ induces a finite filtration on $P^d(n)$ by the subspaces $P^{\leq\omega}(n)$. Intersecting with the hit polynomials gives a filtration $P^{\leq\omega}(n) \cap H^d(n)$ of $H^d(n)$, and the cohit module $Q^d(n)$ is then filtered by the quotients $Q^{\leq\omega}(n) = P^{\leq\omega}(n)/(P^{\leq\omega}(n) \cap H^d(n))$. Equivalently we have a finite cofiltration on $P^d(n)$ by the quotient spaces $P^{\geq\omega}(n) = P^d(n)/P^{\leq\omega}(n)$, and a corresponding cofiltration on $Q^d(n)$ by the quotients $Q^{\geq\omega}(n) = P^d(n)/(P^{<\omega}(n) + H^d(n))$. From either point of view, the associated graded spaces of $Q^d(n)$ are given by

$$Q^\omega(n) \cong P^\omega(n)/(P^\omega(n) \cap (P^{<\omega}(n) + H^d(n))), \qquad (6.1)$$

the set of reducibility classes of elements of $P^\omega(n)$. Taking the direct sum of the associated graded spaces of the filtration, we obtain the following result.

Proposition 6.2.7 *For the left or right order on $\mathrm{Seq}_d(n)$ there is a direct sum decomposition of vector spaces $Q^d(n) \cong \bigoplus_\omega Q^\omega(n)$, where $\omega \in \mathrm{Seq}_d(n)$.* $\square$

The modules $Q^\omega(n)$ are filtration quotients of $Q^d(n)$, and $Q^\omega(n)$ is not in general a subspace or a quotient space of $Q^d(n)$ in a natural way.

Example 6.2.8 Let $\omega = (k)$ where $1 \leq k \leq n$. Then ω is the highest element in $\mathrm{Seq}_k(n)$ in both the left and right orders, and $P^\omega(n)$ is the subspace of $P^k(n)$ spanned by the k-fold products of the variables. Clearly $Q^\omega(n) = P^\omega(n)$. Since $P^{<\omega}(n)$ is spanned by monomials divisible by x_i^2 for some i, $P^{<\omega}(n)$ is a $GL(n)$-submodule of $P^k(n)$ and $P^\omega(n)$ is the corresponding quotient module. As a $\mathbb{F}_2 GL(n)$-module, $Q^\omega(n) = P^\omega(n)$ is isomorphic to the kth exterior power $\Lambda^k(n)$ of the defining module $V(n) \cong P^1(n)$ by mapping $x_{i_1} x_{i_2} \cdots x_{i_k}$ to $x_{i_1} \wedge x_{i_2} \wedge \cdots \wedge x_{i_k}$.

The notation and result of Proposition 1.4.7 carry over immediately to $Q^\omega(n)$. Recall from Section 1.4 that $Z[n] = \{1,\ldots,n\}$. For $Y \subseteq Z[n]$, we define $Q^\omega(Y)$ to be the subspace of $Q^\omega(n)$ spanned by the monomials which are divisible by x_i if and only if $i \in Y$.

Proposition 6.2.9 *As vector spaces $Q^\omega(n) \cong \bigoplus_Y Q^\omega(Y)$, where the sum is over $Y \subseteq Z[n]$, and so $\dim Q^\omega(n) = \sum_{k=1}^n \binom{n}{k} \dim Q^\omega(Z[k])$.*

Proof The vector spaces $Q^\omega(Y)$ are independent for different choices of Y, and a suitable permutation of variables induces an isomorphism $Q^\omega(Y) \cong Q^\omega(Z[k])$, where $|Y| = k$. There are $\binom{n}{k}$ choices of Y with cardinality k. $\square$

Example 6.2.10 There are two members of $\mathsf{Seq}_3(3)$, (3) and $(1,1)$, both decreasing. By Example 1.4.8 we have $\dim Q^{(3)}(3) = 1$, and $Q^{(3)}(3)$ is generated by xyz. As a monomial in $P^{(1,1)}(3)$ cannot be divisible by all three variables, the calculation of $Q^{(1,1)}(3)$ reduces to the cases $n = 1$ or 2. Proposition 6.2.9 gives $\dim Q^{(1,1)}(3) = 3 \dim Q^3(Z[2]) + 3 \dim Q^3(Z[1])$. From Theorem 1.4.12 and Example 1.5.4, $\dim Q^{(1,1)}(3) = 6$. Hence $\dim Q^3(3) = 7$. A monomial basis for $Q^3(3)$ is given by $\{xyz, x^2y, y^2z, z^2x, x^3, y^3, z^3\}$.

When ω is a decreasing sequence, $P^\omega(n)$ contains a spike monomial, and hence $Q^\omega(n) \neq 0$. It follows from the results of Chapter 1 that the converse is true when $n = 1$ or 2, and we show in Proposition 8.3.1 that is also true when $n = 3$. Unfortunately, this fails in general for $n \geq 4$, as we show in Example 8.3.2.

6.3 Concatenation of blocks

The main purpose of introducing blocks representing monomials is to facilitate the construction of hit equations by exploiting combinatorial devices which are natural for blocks, such as **concatenation**. This requires a refinement of the definition of a block.

Definition 6.3.1 A **finite block** is a finite matrix whose entries are the integers 0 or 1. Let A be a finite block and let B be a block with the same number of rows as A. The **concatenation** of A and B is the block $A|B$ formed by the columns of A followed by the columns of B.

As for matrices, a finite block with n rows and c columns is called an (n,c)-**block**. We are not allowed to ignore trailing 0s in finite blocks or sequences. An (n,c)-block gives an n-block by adding trailing 0s. A monomial $f \in P(n)$ is represented by an (n,c)-block, and its ω-sequence $\omega(f)$ by a finite sequence of size c, if and only if all the exponents of f are $< 2^c$.

We can also concatenate two finite blocks having the same number of rows to form a finite block, but we cannot concatenate two blocks with the same number of rows. The point is that a final column of 0s is significant in the case of a finite block, and must be preserved in concatenation. We can similarly concatenate finite sequences with sequences, but we cannot concatenate two sequences. For example the concatenation of (n) with a sequence $A = (a_1, a_2, \ldots)$ is the sequence $(n, a_1, a_2, \ldots)$, and we write it as $(n)|A$.

The following statements are clear from the definitions of $\omega(F)$ as the column sum sequence of F, and of the left and right orders.

Proposition 6.3.2 *If F is an (n,c)-block and G is an n-block, then $\omega(F|G) = \omega(F)|\omega(G)$. If ρ and σ are sequences of size c and $\rho <_l \sigma$, then $\rho|\rho' <_l \sigma|\sigma'$ for any sequences $\rho',\sigma' \in$ Seq. If $\rho',\sigma' \in$ Seq and $\rho' <_r \sigma'$, then $\rho|\rho' <_r \sigma|\rho'$ for any sequences ρ and σ of size c.* ☐

If F is an (n,c)-block representing $f \in P(n)$ and G is an n-block representing $g \in P(n)$, then $F|G$ represents fg^{2^c}. Conversely, if no exponent of f exceeds $2^c - 1$, then fg^{2^c} is represented by $F|G$.

Concatenation may be extended to an (n,c)-block F and a sum of n-blocks $G = B_1 + \cdots + B_k$, representing a polynomial g, by $F|G = A|B_1 + \cdots + A|B_k$, which then represents fg^{2^c}. However, it is not useful to allow F to be a sum of finite blocks, even of the same 2-degree, unless they have the same number of columns, because the 2-degree of $F|G$ depends on the number of columns in F.

Proposition 6.3.3 *Let $f \in P^\rho(n)$, where ρ is a sequence of size c, and let $g \sim_l g' \in P^\omega(n)$. Then $fg^{2^c} \sim_l f(g')^{2^c}$ in $P^{\rho|\omega}(n)$. In particular, fg^{2^c} is left reducible if g is left reducible.*

Proof The hypothesis on f means that all exponents of all monomials which appear in f are $< 2^c$. By Definition 6.2.5, $g = g' + h + e$, where h is hit and $\omega(e) <_l \omega$. Hence $fg^{2^c} = f(g')^{2^c} + fh^{2^c} + fe^{2^c}$. There is a hit equation $h = \sum_{k>0} Sq^k(h_k)$. By Proposition 1.3.2 it follows that $h^{2^c} = \sum_{k>0} Sq^{k2^c}(h_k^{2^c})$. Then by the χ-trick 2.5.2

$$fh^{2^c} = \sum_{k>0} fSq^{k2^c}(h_k^{2^c}) \sim \sum_{k>0} Xq^{k2^c}(f)h_k^{2^c}.$$

By Proposition 6.1.5, $\omega(Xq^{k2^c}(f)) <_l \rho$. It follows that every monomial appearing in $Xq^{k2^c}(f)h_k^{2^c}$ has an ω-sequence of the form $\rho'|\omega'$, where ρ' is a sequence with c entries and $\rho' <_l \rho$. This implies that $\rho'|\omega' <_l \rho|\omega$. Hence we can write $fh^{2^c} = h' + e'$, where h' is hit and $\omega(e') <_l \rho|\omega$. Also, writing $e'' = fe^{2^c}$, we have $\omega(e'') = \omega(f)|\omega(e) <_l \rho|\omega$ by the assumption $\omega(e) <_l \omega$. Finally we obtain $fg^{2^c} = f(g')^{2^c} + h' + e' + e''$, where h' is hit and $\omega(e' + e'') <_l \rho|\omega$. Hence $fg^{2^c} \sim_l f(g')^{2^c}$. For the last statement, we put $g' = 0$. ☐

In terms of blocks we can paraphrase the above proposition as follows.

Proposition 6.3.4 *Let F, G and G' be blocks with the same number of rows and suppose F is finite. If $G \sim_l G'$ then $F|G \sim_l F|G'$. If G is left reducible, in particular if G is hit, then F|G is left reducible.* ☐

Example 6.3.5 It is not true in general that fg^{2^c} is hit if g is hit. For example, let $f = xyz$, $g = x^2$ and $c = 1$, so that $fg^{2^c} = x^5yz$. Then $\omega(f) = (3)$ and g is hit. However, x^5yz has degree 7 and so by Proposition 1.5.1 it is not hit. It is however left reducible, and an explicit equation appears in Example 6.1.4.

Proposition 6.3.3 has the following analogue for the right order.

Proposition 6.3.6 *Let $f \sim_r f'$ in $P^\rho(n)$ where ρ is a sequence of size c, and let $g \in P^\omega(n)$. Then $fg^{2^c} \sim_r f'g^{2^c}$ in $P^{\rho|\omega}(n)$.*

Proof The proof is similar to that of Proposition 6.3.3. By Definition 6.2.5 we can write $f = f' + h + e$, where h is hit and $\omega(e) <_r \rho$. Then $fg^{2^c} = f'g^{2^c} + hg^{2^c} + eg^{2^c}$ and there is a hit equation $h = \sum_{k>0} Sq^k(h_k)$. By the χ-trick 2.5.2

$$hg^{2^c} = \sum_{k>0} Sq^k(h_k)g^{2^c} \sim \sum_{k>0} h_k Xq^k(g^{2^c}).$$

By Proposition 1.3.1, $Xq^k(g^{2^c})$ is either 0 or is a sum of elements of the form $\theta(g)^{2^c}$ for some $\theta \in A_2^+$. Using Proposition 6.1.7 we obtain $\omega(hg^{2^c}) <_r \rho|\omega$. Also $\omega(eg^{2^c}) <_r \rho|\omega$ because $\omega(e) <_r \rho$. Hence $fg^{2^c} \sim_r f'g^{2^c}$, as required. □

In general $f \sim_l f'$ does not imply $fg^{2^c} \sim_l f'g^{2^c}$. In Example 6.3.10 we shall see a case where f is hit but fg^{2^c} is not hit. However, there is a useful variation of Proposition 6.3.3 obtained by imposing a restriction on the operations Sq^k used in the hit equation.

Definition 6.3.7 Let $f, f' \in P^\rho(n)$, where $\rho \in \text{Seq}(n)$ is a sequence of size c. Then $f \sim_l f'$ by a **restricted** hit equation if $f = f' + h + e$, where $e \in P^{<_l\rho}(n)$ and $h = \sum_{k=1}^{2^c-1} Sq^k(h_k)$.

Proposition 6.3.8 *Let $\rho \in \text{Seq}(n)$ be a sequence of size c and let $f, f' \in P^\rho(n)$. If $f \sim_l f'$ by a restricted hit equation, then for any sequence $\omega \in \text{Seq}(n)$ and $g \in P^\omega(n)$, $fg^{2^c} \sim_l f'g^{2^c}$ in $P^{\rho|\omega}(n)$. In particular, if f is reducible by a restricted hit equation, then fg^{2^c} is reducible.*

Proof Following the proof of Proposition 6.3.3, we see by the χ-trick 2.5.2 that

$$fg^{2^c} = f'g^{2^c} + \sum_{k=1}^{2^c-1} Sq^k(h_k)g^{2^c} + eg^{2^c} \sim f'g^{2^c} + \sum_{k=1}^{2^c-1} h_k Xq^k(g^{2^c}) + eg^{2^c}.$$

Since $k < 2^c$, Xq^k does not involve factors Sq^t for $t \geq 2^c$. Hence $Xq^k(g^{2^c}) = 0$ by Proposition 1.3.2, and $\omega(eg^{2^c}) <_l \rho|\omega$ because $\omega(e) <_l \rho$ by assumption. It follows that $fg^{2^c} \sim_l f'g^{2^r}$ as required. If f is reducible then $f' = 0$, so fg^{2^c} is reducible. □

We can paraphrase this result in terms of blocks as follows.

Proposition 6.3.9 *Let* F, F' *be sums of* (n, c)*-blocks with the same* ω*-sequence such that* $F \sim_l F'$ *by a restricted hit equation (i.e. not involving* Sq^k *for* $k \geq 2^c$*). Then* $F|G \sim_l F'|G$ *for any* n*-block* G. $\qquad\square$

The following example shows the need for the restriction on the hit equation.

Example 6.3.10 Consider the following block $B = F|G$, where $c = 2$:

$$
B = \begin{array}{cc|c}
1\,1 & 1 & \\
1\,1 & 0 & \\
0\,1 & 0 & \\
0\,1 & 0 & \\
0\,1 & 0 &
\end{array}, \ F = \begin{array}{c}
1\,1 \\
1\,1 \\
0\,1 \\
0\,1 \\
0\,1
\end{array}, \ G = \begin{array}{c}
1 \\
0 \\
0 \\
0 \\
0
\end{array}.
$$

Then F is hit by Proposition 2.5.3, since $\mu(\omega_2(F)) = \mu(5) = 3 > \omega_1(F) = 2$. However, B is not hit. This can be seen by specializing the variables in the corresponding monomial $b = x_1^7 x_2^3 x_3^2 x_4^2 x_5^2$. Setting $x_1 = x$ and $x_2 = x_3 = x_4 = x_5 = y$, we obtain the monomial $x^7 y^9 \in P^{16}(2)$, which is not hit (see the diagram following Theorem 1.8.2). By Proposition 1.4.3 it follows that b is not hit. In this case Proposition 6.3.9 does not apply, because there is no hit equation for F involving only Sq^1 and Sq^2. This can be seen using the same specialization, as $f = x_1^3 x_2^3 x_3^2 x_4^2 x_5^2$ gives $x^3 y^9 \in P^{12}(2)$, for which there is no hit equation involving only Sq^1 and Sq^2.

We can combine Propositions 6.3.4 and 6.3.9 to give a useful mechanism for manipulating concatenated blocks under restricted hit equations.

Proposition 6.3.11 *Let* F, F' *be sums of* (n, c)*-blocks with* ω*-sequence* ρ *such that* $F \sim_l F'$ *by a restricted hit equation. Let* G, G' *be* n*-blocks with* ω*-sequence* ω *such that* $G \sim_l G'$. *Then* $F|G \sim_l F'|G'$ *in* $\mathsf{P}^{\rho|\omega}(n)$. $\qquad\square$

Provided that the hit equation involving F is restricted, we may therefore manipulate the blocks F and G separately under left reducibility without altering the element of $\mathsf{Q}^{\rho|\omega}(n)$ represented by $F|G$.

We complete this section with one case where $\mathsf{Q}^\omega(n)$ can be determined. From Chapter 5 we recall that the minimum element $\omega^{\min}(d)$ of $\mathrm{Dec}_d(n)$ with the 2-dominance order is the minimum decreasing element of $\mathrm{Seq}_d(n)$ in both the left and right orders. Here we consider only the left order.

Theorem 6.3.12 *If* B *is a block in* $\mathsf{P}^d(n)$ *such that* $\omega(B) <_l \omega^{\min}(d)$, *then* B *is hit. Equivalently, for the left order,* $\mathsf{Q}^\omega(n) = 0$ *for all* $\omega <_l \omega^{\min}(d)$.

Proof Let $\omega(B) = \omega = (\omega_1, \omega_2, \ldots)$ and $\omega^{\min}(d) = \omega' = (\omega_1', \omega_2', \ldots)$, so that $\omega <_l \omega'$. Hence, for some k, $\omega_j = \omega_j'$ for $1 \leq j < k$ and $\omega_k < \omega_k'$. Let B be split as $B = A|G|H$ where G is column k of B. Then $\mu(\deg_2 H) > \deg_2 G = \omega_k$, since otherwise we could make a decreasing sequence on $\mathrm{Seq}_d(n)$ lower than $\omega^{\min}(d)$. It follows from Proposition 2.5.3 that $F = G|H$ is hit. Then by Proposition 6.3.1 $B = A|F$ is left reducible. We can therefore write $B \sim_l \sum_i B_i$ where $\omega(B_i) <_l \omega <_l \omega^{\min}(d)$ for all i. Since $\mathrm{Seq}_d(n)$ is a finite set, it follows by iteration of the procedure on the blocks B_i that B is hit. $\qquad\square$

6.4 Splicing

Splicing is a process which uses block notation to provide a visual approach to the action of Steenrod operations on monomials. It enables the construction of hit equations to which the results of Section 6.3 can be applied.

Splicing a block F consists in the construction of a new block C by the insertion of a contiguous sequence of 1s in a row r of F, starting at a 0 in column u, the 'origin' of the splicing, and finishing at the next 0 before a 1 in that row, which is then changed to 0. The block F then represents a term f in the expansion of $Sq^{2^{u-1}}(c)$, where c is the monomial represented by C.

The simplest example is 1-back splicing, which amounts to moving a digit 1 back one place if that place is originally 0. The splicing process produces a set G of blocks by attacking each row $r' \neq r$ of F'. This means adding 1 arithmetically at position (r', u), provided that $F(r', u) = 1$. Then F and G represent the same element in $Q^{\geq \omega}(n)$, where $\omega = \omega(F)$.

A common procedure is to left justify the 1s in a particular row of a block, which means that all the 1s in that row are contiguous from the first position in each block produced by the splicing process. This can be achieved in different ways: for example, we can use iterated 1-back splicing in the chosen row, or we can iterate splicing with origin at the first 0 digit in the row.

Let $F = (f_{i,j})$ denote a block such that for some row r and columns u, v we have $f_{r,t} = 0$ for $u \leq t < v$ and $f_{r,v} = 1$. Let C denote the block obtained from F by defining $c_{r,t} = 1$ for $u \leq t < v$ and $c_{r,v} = 0$, leaving other entries of F unchanged. Then

$$F = G + Sq^{2^{u-1}}(C) + E,$$

where E and G are sums of blocks such that each block in G arises by the action of $Sq^{2^{u-1}}$ on a row $r' \neq r$ of C, at column u, and each block in E arises from Steenrod operations acting at columns $< u$.

Definition 6.4.1 The process of forming C, E, G is called **splicing** the block F at (r, u). The column u is the **origin** of the splicing. The block in G corresponding to $r' \neq r$ is the block obtained by **attacking** row r' of C.

Example 6.4.2 In the following example, the block F is spliced at $(2, 1)$, and $Sq^1(C) = F + G$. Since $u = 1$, no terms E arise in this case.

$$F = \begin{matrix} 1\,1\,1\,0 \\ 0\,0\,0\,1 \end{matrix}, \ G = \begin{matrix} 0\,0\,0\,1 \\ 1\,1\,1\,0 \end{matrix}, \ C = \begin{matrix} 1\,1\,1\,0 \\ 1\,1\,1\,0 \end{matrix}.$$

Example 6.4.3 In the following example, the block F is spliced at $(2, 3)$:

$$F = \begin{matrix} 1\,1\,1\,1 \\ 1\,0\,0\,0\,1\,1 \\ 0\,1\,1 \end{matrix}, \ C = \begin{matrix} 1\,1\,1\,1 \\ 1\,0\,1\,1\,0\,1 \\ 0\,1\,1 \end{matrix}, \ Sq^4(C) = F + E + G,$$

where, from the Cartan formula and Example 6.4.2,

$$G = \begin{matrix} 1\,1\,0\,0\,1 \\ 1\,0\,1\,1\,0\,1 \\ 0\,1\,1 \end{matrix} + \begin{matrix} 1\,1\,1\,1 \\ 1\,0\,1\,1\,0\,1 \\ 0\,1\,0\,1 \end{matrix},$$

$$E = \begin{matrix} 1\,0\,0\,0\,1 \\ 1\,0\,1\,1\,0\,1 \\ 0\,0\,0\,1 \end{matrix} + \begin{matrix} 0\,0\,0\,0\,1 \\ 0\,1\,1\,1\,0\,1 \\ 0\,0\,0\,1 \end{matrix} + \begin{matrix} 0\,1\,0\,0\,1 \\ 0\,1\,1\,1\,0\,1 \\ 0\,1\,1 \end{matrix}.$$

Since the blocks in E arise from the action of Steenrod squares at columns $< u$, $\omega(E) <_l \omega(F)$. We regard these blocks as 'error' terms. On the other hand, blocks in G may have ω-sequence $>_l \omega(F)$.

We have better control in the case $v = u + 1$. The splicing process then consists of moving a digit 1 in row r of F back one place previously occupied by 0 to form C. If $f_{s,u} = 1$ where $s \neq r$, the effect of $Sq^{2^{u-1}}$ in row s of C is to add the number 1 arithmetically at (s, u). If $f_{s,v} = 1$, the resulting block G has $\omega(G) <_l \omega(F)$ and $\alpha_s(G) < \alpha_s(F)$. On the other hand, if $f_{s,v} = 0$, then $\omega(G) = \omega(F)$ and $\alpha_s(G) = \alpha_s(F)$. We shall refer to the case $v = u + k$ as k-**back splicing**.

Proposition 6.4.4 (1-**back splicing**) *Let F be a block in $\mathsf{P}^\omega(n)$ with entries $0, 1$ in row r and columns $u, u + 1$. Let C be the block formed from F by moving this digit 1 back one place, so that the entries of C in these positions are $1, 0$. Let G be the sum of blocks arising from the application of $Sq^{2^{u-1}}$ to rows of*

C other than row r. Then $F \sim_l G$, i.e. F and G represent the same element of $Q^\omega(n)$ for the left order. □

Example 6.4.5 In the example below, $F \sim_l G$ by 1-back splicing F at $(1,2)$.

$$
F = \begin{matrix} 1\,0\,1\,0\,1 \\ 1\,1 \\ 1\,1 \end{matrix} \quad , \quad G = \begin{matrix} 1\,1\,0\,0\,1 \\ 1\,0\,1 \\ 1\,1 \end{matrix} \quad + \quad \begin{matrix} 1\,1\,0\,0\,1 \\ 1\,1 \\ 1\,0\,1 \end{matrix} \quad .
$$

By iterated 1-back splicing in a nonzero row i of a block F, we can find a sum of blocks $\sum_k B_k \sim_l F$ such that all digits 1 in row i of every B_k are contiguous on the left. This gives the following result.

Proposition 6.4.6 $Q^\omega(Z[n])$ *has a spanning set consisting of monomials in which the exponent of any chosen variable x_i is an integer of the form $2^j - 1$.* □

The basis for $Q(2)$ shown in the diagram following Theorem 1.8.2 satisfies this condition for the variable y.

Proposition 6.4.7 *A block with a non trailing zero column is left reducible.*

Proof Suppose column u of a block F is zero and $f_{r,u+1} = 1$. Then 1-back splicing at (r,u) gives $G = 0$ in Proposition 6.4.4, and so F is left reducible. □

When a splicing operation is applied at (r,u) in a concatenated block $F = A|B$, where A is an (n,c)-block, then it corresponds to a restricted hit equation when $u \le c$, since the squaring operation used in splicing is $Sq^{2^{u-1}}$, and a restricted hit equation allows the use of operations Sq^k for $k < 2^c - 1$.

6.5 The Kameko maps

In the 2-variable case, the up Kameko map $\upsilon : P^d(2) \to P^{2d+2}(2)$ is the linear map defined on a monomial g by $\upsilon(g) = xyg^2$. By Proposition 1.6.1, $\upsilon(g)$ is hit if and only if g is hit. In the n-variable case, $\upsilon : P^d(n) \to P^{2d+n}(n)$ is defined by $\upsilon(g) = c(n)g^2$ where $c(n) = x_1 \cdots x_n$ is the product of the variables. Example 6.3.5, where $n = 3$ and $d = 2$, shows that this map does not always send hit polynomials to hit polynomials.

However, we show in Proposition 6.5.2 that if $c(n)g^2$ is hit then so is g. Hence the inverse map $\kappa : P^{2d+n}(n) \to P^d(n)$, defined on monomials by $\kappa(f) = g$ if $f = c(n)g^2$ and $\kappa(f) = 0$ otherwise, induces a linear map $Q^{2d+n}(n) \to Q^d(n)$. In Proposition 6.5.3 we shall prove that if $\mu(2d + n) = n$

then κ induces an isomorphism of cohit modules, whose inverse is induced by υ. In Example 6.3.5, $\mu(2d+n) = \mu(7) = 1 < n$.

We also consider a local version of the up Kameko map. The case $c = 1$ of Proposition 6.3.4 shows that if g is left reducible then $c(n)g^2$ is left reducible. Hence for any sequence $\omega \in$ Seq, υ induces an isomorphism $Q^\omega(n) \to Q^{(n)|\omega}(n)$, where $(n)|\omega$ is the sequence obtained by prefixing the sequence (n) of size 1 to ω.

Definition 6.5.1 The down Kameko map $\kappa : \mathsf{P}^{2d+n}(n) \to \mathsf{P}^d(n)$ sends a monomial $f \in \mathsf{P}^{2d+n}(n)$ to g if $f = c(n)g^2$, and to 0 otherwise. The up Kameko map $\upsilon : \mathsf{P}^d(n) \to \mathsf{P}^{2d+n}(n)$ is the linear map which sends a monomial $g \in \mathsf{P}^d(n)$ to the monomial $f = c(n)g^2 \in \mathsf{P}^{2d+n}(n)$, where $c(n) = x_1 \cdots x_n$.

The monomial $c(n)$ is represented by the n-block C consisting of a single column of 1s. If the n-block G represents g then $\upsilon(G) = C|G = F$, while $\kappa(F) = G$ if $F = C|G$, and otherwise $\kappa(F) = 0$, the empty sum of blocks. We shall frequently use the fact that if $d = n$ mod 2 then a polynomial $f \in \mathsf{P}^d(n)$ can be written uniquely in the form $f = c(n)h^2 + e$, where $\omega_1(e) < n - 1$.

Proposition 6.5.2 *The map* $\kappa : \mathsf{P}^{2d+n}(n) \to \mathsf{P}^d(n)$ *is a* $\mathbb{F}_2\mathrm{GL}(n)$-*module map, and it induces a* $\mathbb{F}_2\mathrm{GL}(n)$-*module map* $\kappa : \mathsf{Q}^{2d+n}(n) \to \mathsf{Q}^d(n)$. *Further* $\kappa \circ Sq^{2k} = Sq^k \circ \kappa$ *and* $\kappa \circ Sq^{2k+1} = 0$ *for* $k \geq 0$.

Proof We first show that κ sends hit polynomials to hit polynomials. With the notation above, it is enough to show that if $c(n)g^2$ is hit then so is g. A hit equation for $c(n)g^2$ may be written

$$c(n)g^2 = Sq^1(h_0) + \sum_{k>0} Sq^{2k}(c(n)h_k^2 + e_k), \qquad (6.2)$$

where $\omega_1(h_0) < n$ and $\omega_1(e_k) < n - 1$ for $k > 0$. By the Cartan formula, $Sq^{2k}(c(n)h_k^2) = c(n)Sq^{2k}(h_k^2) + e_k'$, where $\omega_1(e_k') < n - 1$. By Proposition 6.1.2 $\omega_1(Sq^1(h_0)) < n$ and $\omega_1(Sq^{2k}(e_k)) < n-1$. By comparing terms with $\omega_1 = n$ in (6.2) we obtain $c(n)g^2 = \sum_{k>0} c(n)(Sq^k(h_k))^2 = c(n) \sum_{k>0} (Sq^k(h_k))^2$. Hence $g = \sum_{k>0} Sq^k(h_k)$, so g is hit. The same argument proves the last statement.

Next we show that κ is a $\mathbb{F}_2\mathrm{GL}(n)$-map. Clearly $\kappa(f \cdot S) = \kappa(f) \cdot S$ if S is a permutation matrix. Let $f = c(n)g^2$ and let U be the standard transvection. Then $f \cdot U = c(n)(g \cdot U)^2 + e$, where $\omega_1(e) < n$. Hence $\kappa(e) = 0$, by definition of κ. It follows that $\kappa(f \cdot U) = \kappa(c(n)(g \cdot U)^2) = g \cdot U = \kappa(f) \cdot U$. Since $\mathrm{GL}(n)$ is generated by permutation matrices and U, this completes the proof. $\qquad\square$

In general, the up Kameko map $\upsilon : \mathsf{P}^d(n) \to \mathsf{P}^{2d+n}(n)$ does not send hit polynomials to hit polynomials. However $\upsilon : \mathsf{Q}^d(n) \to \mathsf{Q}^{2d+n}(n)$ is defined for certain degrees d, and is the inverse of the down Kameko map κ.

Proposition 6.5.3 *If $\mu(2d + n) = n$, then υ induces a map $\upsilon : Q^d(n) \to Q^{2d+n}(n)$ which is the inverse of κ. In particular, $Q^{2d+n}(n) \cong Q^d(n)$ as $\mathbb{F}_2 GL(n)$-modules.*

Proof We need to show that if $g \in P^d(n)$ is hit then $f = \upsilon(g) = c(n)g^2$ is hit. Given a hit equation $g = \sum_{k>0} Sq^k(h_k)$, the χ-trick 2.5.2 gives $c(n)g^2 = \sum_{k>0} c(n)Sq^{2k}(h_k^2) \sim \sum_{k>0} Xq^{2k}(c(n))h_k^2 = e$, say. By Proposition 6.1.2, $\omega_1(e) < n$. Since $\mu(2d+n) = n$, $\omega_1^{\min}(2d+n) = n$ and so $\omega(e) <_l \omega^{\min}(2d+n)$. It follows from Theorem 6.3.12 that e is hit. Since $f = c(n)g^2 \sim e$, f is hit. Hence $\upsilon : Q^d(n) \to Q^{2d+n}(n)$ is well defined. Clearly it is the inverse of κ. $\square$

We next consider the local version of the maps υ and κ, using the left order on ω-sequences. Given $\omega \in \mathrm{Seq}_d(n)$, υ restricts to a map $\upsilon : P^\omega(n) \to P^{(n)|\omega}(n)$, and κ to a map $\kappa : P^{(n)|\omega}(n) \to P^\omega(n)$.

Proposition 6.5.4 *For $\omega \in \mathrm{Seq}_d(n)$, $\upsilon : P^d(n) \to P^{2d+n}(n)$ induces an isomorphism $\upsilon : Q^\omega(n) \to Q^{(n)|\omega}(n)$ of $\mathbb{F}_2 GL(n)$-modules for the left order, with inverse κ.*

Proof By Proposition 6.3.3, if $g \sim_l 0$ in $P^\omega(n)$ then $c(n)g^2 \sim_l 0$ in $P^{(n)|\omega}(n)$. Hence $\upsilon : Q^\omega(n) \to Q^{(n)|\omega}(n)$ is well defined. Clearly it is surjective, and the map induced by κ is the inverse of υ. $\square$

Example 6.5.5 $Q^{(3,0,1)}(3) \cong Q^{(0,1)}(3) = 0$ for the left order.

The next proposition makes a minor simplification of the hit problem which is useful for small values of n.

Proposition 6.5.6 *For all $d' \geq 0$ and $\omega' \in \mathrm{Seq}_{d'}(n)$, $Q^{d'}(n) \cong Q^d(n)$ and $Q^{\omega'}(n) \cong Q^\omega(n)$ for the left order, for some d with $\mu(d) < n$ and some $\omega \in \mathrm{Seq}_d(n)$.*

Proof If $\mu(d') = n$ then d' and n have the same parity, and we can write $d' = 2d + n$. We can then apply Proposition 6.5.3 iteratively until $\mu(d) < n$. Using Proposition 6.5.4, a similar argument applies to the local case. $\square$

6.6 The duplication map

We next look at the situation where $\mu(d) = n - 1$ and generalize the duplication map δ of Section 1.8. This is more subtle than the up Kameko map υ. We do not have a 'down duplication map' analogous to κ, and the local version of δ is not in general an isomorphism. In terms of blocks, the duplication map δ

repeats the first column of F when this column contains exactly one 0. If f is a spike in $P^d(n)$, then $\delta(f)$ is a spike in $P^{2d+n-1}(n)$.

Definition 6.6.1 For $1 \leq i \leq n$, let c_i be the product of the variables $x_j, j \neq i$, and let $d \neq n \bmod 2$. The **duplication map** $\delta : P^d(n) \to P^{2d+n-1}(n)$ is the linear map defined on monomials by $\delta(f) = c_i f^2$ if $f = c_i g^2$ and $\delta(f) = 0$ if $\omega_1(f) < n - 1$.

The map δ matches degrees with opposite parity to n whereas, after possibly one iteration, the Kameko maps match degrees of the same parity as n. We shall frequently use the fact that, if $d \neq n \bmod 2$ then a polynomial $f \in P^d(n)$ has a unique expression of the form $f = \sum_{i=1}^n c_i f_i^2 + e$, where $\omega_1(e) < n - 1$. The next two results give criteria for $f \in P^d(n)$ to be hit or reducible when $d \neq n \bmod 2$.

Proposition 6.6.2 *Let* $f = \sum_{i=1}^n c_i f_i^2 \in P^\omega(n)$, *where* $\omega_1 = n - 1$. *Then* f *is left reducible if and only if there is a polynomial* g *such that* $f_i \sim_l x_i g$ *for* $1 \leq i \leq n$.

Proof Let $f = \sum_{i=1}^n c_i f_i^2$ be left reducible, so that $f = h + e$ where h is hit and $\omega(e) <_l \omega$. Writing $\omega = (n - 1)|\omega'$, we have $e = \sum_{i=1}^n c_i e_i^2 + e'$, where $\omega(e_i) <_l \omega'$ for $1 \leq i \leq n$ and $\omega_1(e') < n - 1$. A hit equation for h may be written

$$h = Sq^1(g_1) + \sum_{k>0} Sq^{2k}(g_{2k}) \tag{6.3}$$

where $g_j \in P^{d-j}(n)$ for all j. Since $d \neq n \bmod 2$, $\deg g_1 = n \bmod 2$ and $\deg g_{2k} \neq n \bmod 2$. Hence we may write $g_1 = c(n)g^2 + e''$, $g_{2k} = \sum_{i=1}^n c_i g_{i,k}^2 + e_k''$, where e'' and e_k'', and hence also $Sq^1(e'')$ and $Sq^{2k}(e_k'')$, all have $\omega_1 < n - 1$. Since $Sq^1(c(n)) = \sum_{i=1}^n c_i x_i^2$, $Sq^1(c(n)g^2) = \sum_{i=1}^n c_i(x_i g)^2$. Also $Sq^{2k}(c_i g_{i,k}^2) = c_i(Sq^k(g_{i,k}))^2 + e_{i,k}''$ where $\omega_1(e_{i,k}'') < n - 1$. It follows that

$$f = \sum_{i=1}^n c_i f_i^2 = \sum_{i=1}^n c_i(x_i g)^2 + \sum_{i=1}^n c_i \sum_{k>0}(Sq^k(g_{i,k}))^2 + \sum_{i=1}^n c_i(e_i)^2,$$

modulo terms with $\omega_1 < n - 1$. By equating terms in c_i, we obtain $f_i = x_i g + \sum_{k>0} Sq^k(g_{i,k}) + e_i$. Hence $f_i \sim_l x_i g$. The proof in the other direction follows by reversing the steps. $\qquad\square$

Proposition 6.6.3 *Let* $f = \sum_{i=1}^n c_i f_i^2 + e \in P^d(n)$, *where* $d \neq n \bmod 2$ *and* $\omega_1(e) < n - 1$. *If* f *is hit, then there is a polynomial* g *such that* $f_i \sim x_i g$ *for* $1 \leq i \leq n$. *Conversely, if all elements of* $P^d(n)$ *with* $\omega_1 < n - 1$ *are hit, and if* $f_i \sim x_i g$ *for* $1 \leq i \leq n$, *then* f *is hit.*

Proof We assume that $f \in \mathsf{P}^d(n)$ is hit. Then $f = \sum_{i=1}^{n} c_i f_i^2 + e$, where $\omega_1(e) < n - 1$. The preceding argument applies to f and we have

$$f = \sum_{i=1}^{n} c_i f_i^2 = \sum_{i=1}^{n} c_i (x_i g)^2 + \sum_{i=1}^{n} c_i \sum_{k>0} (Sq^k(g_{i,k}))^2$$

modulo monomials with $\omega_1 < n - 1$. Equating terms in c_i gives $f_i = x_i g + \sum_{k>0} Sq^k(g_{i,k})$. In other words $f_i \sim x_i g$, as required. The proof in the other direction again follows by reversing the steps and assuming that all elements of $\mathsf{P}^d(n)$ with $\omega_1 < n - 1$ are hit. $\square$

We can now state the global and local results for the duplication maps.

Proposition 6.6.4 *If $\mu(d) = n - 1$, then the duplication map induces a map $\delta : Q^d(n) \to Q^{2d+n-1}(n)$ of $\mathbb{F}_2 GL(n)$-modules.*

Proposition 6.6.5 *Let $\omega \in \mathrm{Seq}_d(n)$ with $\omega_1 = n - 1$. Then the duplication map induces a map $\delta : Q^\omega(n) \to Q^{(n-1)|\omega}(n)$ of $\mathbb{F}_2 GL(n)$-modules.*

Proof We prove the two propositions together. Starting with Proposition 6.6.4 we show that δ sends hit polynomials to hit polynomials. Since $\mu(d) = n - 1$, we have $\omega_1^{\min}(d) = n - 1$. It follows from Theorem 6.3.12 that any monomial in $\mathsf{P}^d(n)$ with $\omega_1 < n - 1$ is hit. The same applies to $\mathsf{P}^{2d+n-1}(n)$ because $\mu(2d + n - 1) = n - 1$, as we see from Proposition 5.4.4.

We prove that if $f \sim 0$ for some $f \in \mathsf{P}^d(n)$, then $c_i f^2 \sim 0$ in $\mathsf{P}^{2d+n-1}(n)$. Consider a hit equation $f = \sum_{k>0} Sq^k(g_k)$. Then $c_i f^2 = \sum_{k>0} c_i Sq^{2k}(g_k^2) = \sum_{k>0} Sq^{2k}(c_i g_k^2)$, modulo monomials with $\omega_1 < n - 1$, which are hit. A similar argument shows that if $f \sim_l 0 \in \mathsf{P}^\omega(n)$ then $c_i f^2 \sim_l 0 \in \mathsf{P}^{\omega'}(n)$. For $f = \sum_{i=1}^{n} c_i f_i^2 + e$ in $\mathsf{P}^d(n)$, where $\omega_1(e) < n - 1$, we have $\delta(f) = \sum_{i=1}^{n} c_i (c_i f_i^2)^2$. If f is hit, then by Proposition 6.6.3 there is a polynomial g such that $f_i \sim x_i g$ for $1 \le i \le n$.

It follows from the preceding argument that $c_i f_i^2 \sim c_i x_i^2 g^2 = x_i c(n) g^2$. The factor $c(n) g^2$ plays the role of g in Proposition 6.6.3 applied to $\mathsf{P}^{2d+n-1}(n)$. Hence $\delta(f)$ is hit. Similarly, if f is reducible in $\mathsf{P}^\omega(n)$ then $f_i \sim_l x_i g$ for $1 \le i \le n$ and some g. Then $c_i f_i^2 \sim_l x_i c(n) g^2$, which shows by Proposition 6.6.2 that $\delta(f)$ is reducible in $\mathsf{P}^{(n-1)|\omega}(n)$. This proves that δ is well defined in Proposition 6.6.5.

The proof that δ is a map of $\mathbb{F}_2 GL(n)$-modules follows the familiar pattern. It is clear that duplication commutes with the action of permutation matrices. For the standard transvection U we may assume $n > 2$, since the 2-variable case was treated in Chapter 1. For both propositions, it is enough to consider $f = \sum_{i=1}^{n} c_i f_i^2$.

We note that $c_1 \cdot U = c_1$, $c_2 \cdot U = c_1 + c_2$, $c_i \cdot U = c_i + e_i$ for $i > 2$, where $\omega_1(e_i) < n - 1$. Hence $f \cdot U = \sum_{i=1}^{n} (c_i \cdot U)(f_i \cdot U)^2 = c_1(f_1 \cdot U)^2 + (c_1 + c_2)(f_2 \cdot U)^2 + \sum_{i=3}^{n} c_i(f_i \cdot U)^2 + e$, where $\omega_1(e) < n - 1$. Then, since $\delta(e) = 0$, we have $\delta(f \cdot U) = c_1^3(f_1 \cdot U)^4 + (c_1^3 + c_2^3)(f_2 \cdot U)^4 + \sum_{i=3}^{n} c_i^3(f_i \cdot U)^4$.

On the other hand, $\delta(f) \cdot U = \sum_{i=1}^{n} c_i^3 \cdot U(f_i \cdot U)^4 = c_1^3(f_1 \cdot U)^4 + (c_1 + c_2)^3(f_2 \cdot U)^4 + \sum_{i=3}^{n} c_i^3(f_i \cdot U)^4$. The difference between $\delta(f \cdot U)$ and $\delta(f) \cdot U$ is $b = (c_1^2 c_2 + c_1 c_2^2)(f_2 \cdot U)^4$. Let $g = (x_1 x_2 x_3 \cdot x_n^3)(f_2 \cdot U)^4$. Then $Sq^1(g) = b + h$, where $\omega_2(h) < n - 1$. In particular, $h \in \mathsf{P}^{<_l (n-1)|\omega}(n)$. It follows that $\delta(f \cdot U) \sim_l \delta(f) \cdot U$, as required in Proposition 6.6.2, and $\delta(f \cdot U) \sim \delta(f) \cdot U$ in Proposition 6.6.3. $\quad\square$

6.7 Tail sequences

The cohit module $Q^\omega(n)$ is most easily determined when ω is maximal or minimal. For the left order, Theorem 6.3.12 shows that $\omega = \omega^{\min}(d)$ is the minimal sequence in the left order for which $Q^\omega(n) \neq 0$. For certain degrees d, these sequences are also maximal or minimal in $\mathrm{Seq}_d(n)$. This occurs when $\omega = (n-1, \ldots, n-1)$ and $\omega = (1, \ldots, 1)$ respectively, and we refer to these as 'head' and 'tail' sequences.

Definition 6.7.1 A **tail sequence** has the form $\tau = (1, \ldots, 1)$, with $\ell > 0$ contiguous digits 1. An n-block T is a **tail block** if $\omega(T)$ is a tail sequence.

Thus $\ell = \mathrm{len}(\tau)$ and $d = \deg_2 \tau = 2^\ell - 1$. A tail sequence τ is the minimum element of $\mathrm{Seq}_d(n)$ for both the left and right orders. Hence for $T, T' \in \mathsf{P}^\tau(n)$ the equivalence relations $T \sim T'$, $T \sim_l T'$ and $T \sim_r T'$ coincide, so that $Q^\tau(n)$ is the same for the left and right orders.

Proposition 6.7.2 *Let τ be a tail sequence, and let T, T' be blocks in $\mathsf{P}^\tau(n)$ such that the columns of T' are a permutation of the columns of T. Then $T \sim T'$.*

Proof It is enough to show that two adjacent columns C_j and C_{j+1} of T can be exchanged without altering the equivalence class of T in $Q^\tau(n)$. We may assume that the digits 1 in C_j are in rows r and s respectively, where $r \neq s$. By Proposition 6.4.4, 1-back splicing T at (r, j) produces the unique block T' which is obtained from T by exchanging C_j and C_{j+1}. Hence $T \sim T'$. $\quad\square$

We classify the blocks in $\mathsf{P}^\tau(n)$ by the set of their nonzero rows.

Definition 6.7.3 For $1 \leq k \leq \ell$, let $1 \leq i_1 \leq \cdots \leq i_k \leq n$, $Y = \{i_1, \ldots, i_k\}$ and let $\tau = (1, \ldots, 1)$ be a tail sequence of length ℓ. Then a block $T \in \mathsf{P}^\tau(n)$ is in **class** Y if, for each $i \in Y$, row i of T has at least one digit 1 and all digits in row i of T are 0 if $i \notin Y$. Row i_1 is the **preferred row** of T. The **normalized tail block** C

in class Y is defined by $c_{i_1,j} = 1$ for $1 \leq j \leq \ell - k$ and $c_{i_j, \ell-k+j} = 1$ for $1 \leq j \leq k$, and $c_{i,j} = 0$ otherwise.

Example 6.7.4 For $n = 4$, $\ell = 5$ and $Y = \{1,3,4\}$, a typical block T and the normalized block C in class Y are given by

$$
T = \begin{matrix} 0\,0\,0\,1\,0 \\ 0\,0\,0\,0\,0 \\ 1\,0\,1\,0\,0 \\ 0\,1\,0\,0\,1 \end{matrix}, \quad C = \begin{matrix} 1\,1\,1\,0\,0 \\ 0\,0\,0\,0\,0 \\ 0\,0\,0\,1\,0 \\ 0\,0\,0\,0\,1 \end{matrix}.
$$

Proposition 6.7.5 *Let τ be a tail sequence with $\mathrm{len}(\tau) \geq k$, and let $Y \subseteq Z[n] = \{1, 2, \ldots, n\}$ have cardinality $|Y| = k$. Then $Q^\tau(Y)$ has dimension 1.*

Proof By Proposition 1.5.1, a tail block cannot be hit, and so $Q^\tau(Z[k]) \neq 0$. With notation as in Definition 6.7.3, we show that if T is an arbitrary block in class $Y \subseteq Z[n]$ then $T \sim C$, the normalized block in class Y. We use induction on $k - |Y|$.

In the case $k = 1$ there is nothing to prove. By Proposition 6.7.2, $T \sim T'$ where all the digits 1 in row i_k of T' are contiguous on the left, and the first digit 1 in row i_{k-1} is in the next column after these 1s. The next step is to splice T' at $(i_{k-1}, 1)$. The result is a block $T'' \sim T'$ where T'' has a single digit 1 in row i_k. By Proposition 6.7.2 we may assume $b''_{i_k, \ell} = 1$.

We now iterate the procedure on the subblock S of T'' obtained by removing row i_k and the last column, which is of class $Y' = \{i_1, \ldots, i_{k-1}\} \subseteq Z[n-1]$. By the induction hypothesis we may assume that $S \sim C'$, the normalized block in class Y'. By construction of normalized blocks, C' is the subblock of C obtained by removing row i_k and the last column. Hence $T'' \sim C$, and it follows that $T \sim C$. $\qquad\square$

As in Section 1.4 we have the cohit space $Q^\tau(Y)$ corresponding to the variables x_i for $i \in Y$ and formed from the equivalence classes of blocks of class Y. By counting subsets Y as in Proposition 1.4.7, we have the following result.

Theorem 6.7.6 *For $n \geq 1$, let $\tau = (1, \ldots, 1)$ be a tail sequence of length ℓ. Then $\dim Q^\tau(n) = \sum_{k=1}^{\ell} \binom{n}{k}$. In particular, $\dim Q^\tau(n) = 2^n - 1$ if $\ell \geq n$.* $\qquad\square$

Since all blocks of the same class are equivalent, any choice of one block from each class Y such that $|Y| = k$, $1 \leq k \leq \ell$, will form a basis for $Q^\tau(n)$.

6.8 Head sequences

In this section we find the dimension of $Q^\omega(n)$ for a head sequence by procedures analogous to those for tail sequences in Section 6.7, in effect by switching the digits 0 and 1. This will be made more precise in Section 6.9.

Definition 6.8.1 For $n \geq 2$, a **head sequence** in Seq(n) has the form $\gamma = (n-1, \ldots, n-1)$, where $n > 1$, with $\ell > 0$ contiguous entries $n-1$, so that $\ell = \mathrm{len}(\gamma)$ and $d = \deg_2 \gamma = (n-1)(2^\ell - 1)$. A block H is a **head block** in P(n) if $\omega(H)$ is a head sequence in Seq(n).

By Proposition 5.6.2, γ is the maximum element in Seq$_d(n)$, and so it is maximal in both the left and the right orders. Thus the relations $H \sim_l H'$ and $H \sim_r H'$ coincide, so that $Q^\gamma(n)$ is the set of equivalence classes of head blocks for either relation. Note that the definition of a head sequence depends on n, whereas the definition of a tail sequence does not.

Proposition 6.8.2 *Let γ be a head sequence, and let H, H' be blocks in* P$^\gamma(n)$ *such that the columns of H' are a permutation of the columns of H. Then $H \sim_l H'$, so that H and H' represent the same element of* Q$^\gamma(n)$.

Proof We follow the proof of Proposition 6.7.2 and observe that if the digit 0 in column j of H is in row r, then by Proposition 6.4.4 splicing H at (r, j) gives $H \sim_l H'$. $\qquad\square$

Proposition 6.8.3 *Let $\gamma \in$ Seq(n) be a head sequence of length $\ell \geq n$, and let $\gamma' = (n-1)|\gamma$. Then for any $\rho \in$ Seq(n) the duplication map $\delta : Q^{\gamma|\rho}(n) \to$ Q$^{\gamma'|\rho}(n)$ is surjective.*

Proof Since $\mathrm{len}(\gamma') > n$, every block in P$^{\gamma'}(n)$ has at least one row with at least two 0s. By Proposition 6.8.2 we may assume this row has 0s in the first two columns. The result follows. $\qquad\square$

By analogy with the classification of tail sequences in Section 6.7, we classify head blocks by the rows which contain 0s.

Definition 6.8.4 For $1 \leq k \leq \ell$, let $1 \leq i_1 \leq \cdots \leq i_k \leq n$, $Y = \{i_1, \ldots, i_k\}$ and let $\gamma = (n-1, \ldots, n-1)$ be a head sequence of length ℓ. Then a block $H \in$ P$^\gamma(n)$ is in **class** Y if, for each $i \in Y$, row i of H has at least one digit 0 and all digits in row i of H are 1 if $i \notin Y$. Row i_1 is the **preferred** row of H. The **normalized head block** C in class Y is defined by $c_{i_j, k-j+1} = 0$ for $1 \leq j \leq k$ and $c_{i_1, j} = 0$ for $k < j \leq \ell$, and $c_{i,j} = 1$ otherwise.

Example 6.8.5 For $n = 4$, $\ell = 5$ and $Y = \{1,3,4\}$, a typical block H and the normalized block C in class Y are given by

$$H = \begin{matrix} 1\,1\,1\,0\,1 \\ 1\,1\,1\,1\,1 \\ 0\,1\,0\,1\,1 \\ 1\,0\,1\,1\,0 \end{matrix}, \quad C = \begin{matrix} 1\,1\,0\,0\,0 \\ 1\,1\,1\,1\,1 \\ 1\,0\,1\,1\,1 \\ 0\,1\,1\,1\,1 \end{matrix},$$

Thus a block for a head class Y is obtained from a block for a tail class Y by switching 0s and 1s, and the same is true for normalized blocks if we also reverse the order of the columns. We fix Y and use splicing operations to study head blocks of class Y. Splicing in a row in Y will generally produce terms which arise from the action of a Steenrod square on rows not in Y. Since all entries in these rows are 1, such terms have ω-sequence $<_l \gamma$, and may therefore be ignored up to reducibility. Thus these rows play no essential part in the argument.

Proposition 6.8.6 *Let γ be a head sequence of length ℓ and let $Y \subseteq Z[n]$ with $|Y| = k$ where $1 \le k \le \ell$. Then all blocks in $\mathrm{P}^Y(n)$ of class Y represent the same element in $\mathrm{Q}^Y(n)$.*

Proof We shall use splicing operations to normalize a block H in class Y. As observed above, we may ignore rows not in Y, and so we may assume that $Y = Z[k]$.

We argue by induction on k. There is nothing to prove if $k = 1$, so let $k > 1$ and assume the result for head blocks with $k - 1$ rows. By permuting columns using Proposition 6.8.2, $H \sim_l H_1$ where in H_1 all digits 0 in the last (kth) row are contiguous on the left, for example

$$H = \begin{matrix} 1\,0\,1\,0\,1 \\ 0\,1\,1\,1\,1 \\ 1\,1\,0\,1\,0 \end{matrix}, \quad H_1 = \begin{matrix} 1\,1\,1\,0\,0 \\ 1\,1\,0\,1\,1 \\ 0\,0\,0\,1\,1 \end{matrix}.$$

Next we splice H_1 at $(k, 1)$. Apart from terms with ω-sequence $<_l \gamma$, the result is a single head block H_2 with one 0 in its last row. Using Proposition 6.8.2 again, we can permute columns of H_2 to place the single 0 in this row in the first column, giving H_3. For example

$$H_2 = \begin{matrix} 1\,1\,1\,0\,0 \\ 0\,0\,1\,1\,1 \\ 1\,1\,0\,1\,1 \end{matrix}, \quad H_3 = \begin{matrix} 1\,1\,1\,0\,0 \\ 1\,0\,0\,1\,1 \\ 0\,1\,1\,1\,1 \end{matrix}.$$

Thus $H \sim_l H_3$. Let H_3' be the subblock of H_3 obtained by deleting the last row and the first column. By the induction hypothesis, H_3' can be normalized by splicing processes. When the same processes are applied to the whole block H_3, error terms E arising from Steenrod squares acting on the last row or the first column have $\omega(E) <_l \gamma$. Hence $H_3 \sim_l C$, the normalized block. This completes the inductive step.   □

Theorem 6.8.7 *For a head sequence $\gamma \in \mathrm{Seq}(n)$ of length ℓ, the normalized blocks $C(Y)$ associated with subsets $Y \subseteq Z[n]$ of cardinality $1 \le |Y| \le \ell$ form a vector space basis of $Q^\gamma(n)$. Thus $\dim Q^\gamma(n) = \sum_{i=1}^{\ell} \binom{n}{i}$. In particular, $\dim Q^\gamma(n) = 2^n - 1$ if $\ell \ge n$.*

Proof By Proposition 6.8.6, the normalized blocks $C(Y)$ span $Q^\gamma(n)$. We first prove that the corresponding monomials are not hit.

Let i_1 be the minimal element of Y and let $J = Y \setminus \{i_1\}$. We apply the singular transformation $M(Y)$ which maps x_{i_1} to $\sum_{j \in J} x_j$ and fixes x_k for $k \ne i_1$. Proposition 6.1.12 explains how to evaluate $C(Y) \cdot M(Y)$. This sum of blocks contains a spike S with all entries in row i_1 equal to 0 and all other entries equal to 1, together with blocks with ω-sequence lower than γ resulting from the superposition of 1s in binary addition. (See Example 6.1.11.) For example, when $n = 3$, $\ell = 5$ and $Y = \{1, 2, 3\}$,

$$
C(Y) = \begin{matrix} 1\,1\,0\,0\,0 \\ 1\,0\,1\,1\,1 \\ 0\,1\,1\,1\,1 \end{matrix} , \quad
S = \begin{matrix} 0\,0\,0\,0\,0 \\ 1\,1\,1\,1\,1 \\ 1\,1\,1\,1\,1 \end{matrix} .
$$

We extend this argument to prove that the normalized blocks represent linearly independent elements of $Q^\gamma(n)$. Suppose that a sum of normalized blocks $C(Y)$ is reducible, and let $C(Y)$ be a term in the sum for which $|Y|$ is minimal. The action of $M(Y)$ affects only the rows in Y, and cannot produce S from any other block $C(Y')$ in the sum, since $C(Y')$ must have a 0 entry in some row not in Y. This contradicts reducibility of the sum.   □

6.9 Tail and head modules

In this section we find composition series for the $\mathbb{F}_2 \mathrm{GL}(n)$-modules $Q^\omega(n)$, where ω is a tail or a head sequence and $n \ge 2$. We begin by identifying the reducible elements of $P^\tau(n)$ when τ is a tail sequence. Since τ is minimal in the left or right order, these are the same as the hit elements. By Propositions 6.2.9

and 6.8.6, all blocks of the same class are equivalent. This gives the following result.

Proposition 6.9.1 *If* $\tau = (1, \ldots, 1)$ *then, regarded as sums of blocks, hit polynomials in* $\mathsf{P}^\tau(n)$ *are sums with an even number of blocks in each class* $Y \subseteq Z[n]$. $\qquad\qquad\qquad\qquad\qquad\square$

Let $\tau = (1, \ldots, 1)$ be a tail sequence of length $\ell \geq 1$. We define a linear map $\epsilon : \mathsf{P}^\tau(n) \to \mathsf{P}^{(1)|\tau}(n)$ by duplicating the first column of a block in $\mathsf{P}^\tau(n)$. By Proposition 6.9.1, ϵ sends hit polynomials to hit polynomials, and so it induces a map $\overline{\epsilon} : \mathsf{Q}^\tau(n) \to \mathsf{Q}^{(1)|\tau}(n)$. Clearly $\overline{\epsilon}$ is injective for all $\ell \geq 1$, and is an isomorphism if $\ell \geq n$. The map $\overline{\epsilon}$ is a map of $\mathbb{F}_2\mathrm{GL}(n)$-modules, as it commutes with permutations of variables and also, by an argument similar to that of Proposition 1.7.4, with the standard transvection U. We use $\overline{\epsilon}$ to construct a composition series for $\mathsf{Q}^\omega(n)$, with factors isomorphic to exterior powers $\Lambda^s(n)$ of $\mathsf{V}(n) \cong \mathsf{P}^1(n)$ (see Example 6.2.8). We show in Section 10.1 that $\Lambda^s(n)$ is an irreducible $\mathbb{F}_2\mathrm{GL}(n)$-module.

From Proposition 6.7.5 and Theorem 6.7.6, any choice of tail block of class Y for each Y yields a basis of $\mathsf{Q}^\tau(n)$. In the proofs of Propositions 6.9.3 and 6.9.5 we make a different choice of normalized head and tail blocks, so as to work with the standard transvection U. The map $\overline{\epsilon}$ preserves the class of a block and sends basis elements to basis elements. If $\ell > n$, then some row of a block in $\mathsf{P}^\tau(n)$ must have at least two entries 1. By Proposition 6.8.2, we may assume that these are in the first two columns, giving the following result.

Proposition 6.9.2 *Let* $\tau = (1, \ldots, 1)$ *be a tail sequence of length* $\ell \geq 1$. *Then* $\overline{\epsilon} :$ $\mathsf{Q}^\tau(n) \to \mathsf{Q}^{(1)|\tau}(n)$ *is an injection of* $\mathbb{F}_2\mathrm{GL}(n)$-*modules, and is an isomorphism if* $\ell \geq n$. $\qquad\qquad\qquad\square$

Identifying $\mathsf{Q}^\tau(n)$ as a submodule of $\mathsf{Q}^{(1)|\tau}(n)$ via $\overline{\epsilon}$, we obtain a filtration

$$\mathsf{Q}^{(1)}(n) \subset \mathsf{Q}^{(1,1)}(n) \subset \cdots \subset \mathsf{Q}^{(1,\ldots,1)}(n), \qquad (6.4)$$

where the last sequence has length n. The first term $\mathsf{Q}^{(1)}(n)$ is the defining representation $\mathsf{P}^1(n) \cong \mathsf{V}(n)$. We shall prove that the quotient modules of the filtration (6.4) are exterior powers of $\mathsf{V}(n)$.

Proposition 6.9.3 *Let* $\tau = (1, \ldots, 1)$ *have length* $s - 1$, $2 \leq s \leq n$. *Then as* $\mathbb{F}_2\mathrm{GL}(n)$-*modules* $\mathsf{Q}^{(1)|\tau}(n)/\mathsf{Q}^\tau(n) \cong \Lambda^s(n)$, *the sth exterior power of* $\mathsf{V}(n)$.

Proof In this proof we work with the alternative normalized tail block $C'(Y)$ obtained from $C(Y)$ by reversing the order of the columns and of the rows in Y. Since $s \leq n$, the quotient $\mathsf{Q}^{(1)|\tau}(n)/\mathsf{Q}^\tau(n)$ has a vector space basis consisting of the normalized blocks $C = C'(Y)$ for $|Y| = s$ having a single digit 1 in rows in Y. Let the rows in Y be rows $i_1, \ldots, i_s$ in increasing order. Then the map $\phi : Q^{(1)|\tau}(n)/Q^\tau(n) \to \Lambda^s(n)$ which assigns to C the element $u = u(Y) = x_{i_1} \wedge \cdots \wedge x_{i_s} \in \Lambda^s(n)$ is a vector space isomorphism. It is clear that ϕ commutes with the action of permutation matrices. Since $\mathrm{GL}(n)$ is generated by permutation matrices and the standard transvection U, it suffices to check that ϕ commutes with the action of U. On $\Lambda^s(n)$, U acts by $u \cdot U = u$ unless $i_1 = 1$ and $i_2 > 2$, in which case $u \cdot U = u + v$, where v is obtained from u by replacing x_1 by x_2. We use the description of the action of U on blocks given in Proposition 6.1.9, and separate three cases.

Case 1: $1 \notin Y$. All entries of row 1 of C are 0, and so $C \cdot U = C$.

Case 2: $2 \in Y$, $1 \in Y$. The first two rows of C have the form

$$1\,0\,0\cdots 0$$

$$0\,1\,0\cdots 0$$

and $C \cdot U = C + D$, where D is the block obtained from C by exchanging the entries $c_{1,1}$ and $c_{2,1}$. Thus D is in the submodule $Q^\tau(n)$ of $Q^{(1)|\tau}(n)$.

Case 3: $2 \notin Y$, $1 \in Y$. The first two rows of C have the form

$$1\,0\,0\cdots 0$$

$$0\,0\,0\cdots 0$$

and $C \cdot U = C + D$, where D is the block obtained from C by exchanging the entries $c_{1,1}$ and $c_{2,1}$. Since $\phi(D) = v$, this completes the proof. □

We next consider head blocks. From Proposition 6.8.6 and Theorem 6.8.7, any choice of head block of class Y for each Y yields a basis of $Q^Y(n)$. The duplication map preserves the class of a block and sends basis elements to basis elements. If $\mathrm{len}(\gamma) > n$, then some row of any block must have at least two 0s. By Proposition 6.8.2, we may assume that these are in the first two columns, giving the following result.

Proposition 6.9.4 *Let* $\gamma = (n-1, \ldots, n-1)$ *be a head sequence of length* $\ell \geq 1$. *Then the duplication map* $\delta : Q^\gamma(n) \to Q^{(n-1)|\gamma}(n)$ *is an injection of* $\mathbb{F}_2 \mathrm{GL}(n)$-*modules, and is an isomorphism if* $\ell \geq n$. □

We use δ to construct a composition series for $Q^\gamma(n)$ with factors isomorphic to exterior powers of the defining representation. Identifying $Q^\gamma(n)$ as a submodule of $Q^{(n-1)|\gamma}(n)$ via δ, we obtain a filtration

$$Q^{(n-1)}(n) \subset Q^{(n-1,n-1)}(n) \subset \cdots \subset Q^{(n-1,\ldots,n-1)}(n), \qquad (6.5)$$

where the last sequence has length n. Recall from Example 6.2.8 that $Q^{(n-1)}(n) \cong \Lambda^{n-1}(n)$. We shall prove that the quotient modules of the filtration (6.5) are exterior powers of $P^1(n)$, so that (6.5) is a composition series.

Proposition 6.9.5 *Let* $\gamma = (n-1, \ldots, n-1)$ *have length* $s - 1$, $2 \le s \le n$. *Then as* $\mathbb{F}_2 GL(n)$*-modules* $Q^{(n-1)|\gamma}(n)/Q^{\gamma}(n) \cong \Lambda^{n-s}(n)$

Proof The proof is similar to that of Proposition 6.9.3. We work with the alternative normalized head block $C = C'(Y)$ obtained from $C(Y)$ by reversing the order of the rows in Y. Since $s \le n$, the quotient $Q^{(n-1)|\gamma}(n)/Q^{\gamma}(n)$ has a vector space basis consisting of the normalized blocks $C(Y)$ for $|Y| = s$ having a single 0 in rows in Y.

Let the rows *not* in Y, where all entries are 1, be rows $j_1, \ldots, j_{n-s}$ in increasing order. Then the map $\phi : Q^{(n-1)|\gamma}(n)/Q^{\gamma}(n) \to \Lambda^{n-s}(n)$ which assigns to $C = C(Y)$ the element $x_{j_1} \wedge \cdots \wedge x_{j_{n-s}} \in \Lambda^{n-s}(n)$ is a vector space isomorphism. It is clear that ϕ commutes with the action of permutation matrices. As in Proposition 6.9.3, it remains to show that ϕ commutes with the action of the standard transvection U. Recall that U fixes $y = x_{j_1} \wedge \cdots \wedge x_{j_{n-s}}$ unless $j_1 = 1$ and $j_2 > 2$, when it maps u to $y + z$ where z is obtained by replacing x_1 by x_2 in y. Again we use Proposition 6.1.9, and separate three cases.

Case 1: $2 \notin Y$. All entries of row 2 of C are 1, and so $C \cdot U = C + E$, where $E \in P^{<\gamma}(n)$.

Case 2: $2 \in Y$, $1 \in Y$. The first two rows of C have the form

$$0\ 1\ 1 \cdots 1$$
$$1\ 0\ 1 \cdots 1$$

and $C \cdot U = C + D + E$, where D is the block obtained from C by exchanging the entries $c_{1,2}$ and $c_{2,2}$, and $E \in P^{<\gamma}(n)$. Thus D is in the submodule $Q^{\gamma}(n)$ of $Q^{(n-1)|\gamma}(n)$.

Case 3: $2 \in Y$, $1 \notin Y$. The first two rows of C have the form

$$1\ 1\ 1 \cdots 1$$
$$0\ 1\ 1 \cdots 1$$

and $C \cdot U = C + D + E$, where D is the block obtained from C by exchanging the entries $c_{1,1}$ and $c_{2,1}$, and $E \in P^{<\gamma}(n)$. Since $\phi(D) = v$, this completes the proof. $\square$

6.10 Remarks

Reducibility techniques for manipulating blocks have been natural ingredients in work on the hit problem since its inception, for example in Kameko's thesis [106]. The fact that blocks left lower than the minimal spike are hit (Theorem 6.3.12) is due to Singer [187]. The Kameko maps first appeared in [106], and the duplication map in [221].

Although the map $\epsilon : \mathsf{P}^\tau(n) \to \mathsf{P}^{(1)|\tau}(n)$, which duplicates the first column of a block when $\tau = (1, \ldots, 1)$ is a tail sequence, sends hit polynomials to hit polynomials, it is not true in general that 'tail duplication' has this property. For example, tail duplication of $Sq^1(xyz) = x^2yz + xy^2z + xyz^2$ gives Singer's important example $s_8 = x^6yz + xy^6z + xyz^6$ [185, 186] of a 'strongly spike-free' polynomial which is not hit. We show in Remark 8.3.6 that s_8 is not hit.

7

Bounds for $\dim Q^d(n)$

7.0 Introduction

The dimension of the vector space $P^d(n)$ increases with d for a fixed n. By contrast, we show in Section 7.1 that there is an upper bound for $\dim Q^d(n)$ which is independent of d. The proof involves the idea of an α^+-bounded monomial. These span $Q^d(n)$, and in a given degree d the number of α^+-bounded monomials is bounded by a function of n. However, for $n > 2$, the bound is too large to be of practical use.

In Section 7.2, the 1-back splicing technique introduced in Section 6.4 is extended to multiple 1-back splicing, and is used to show that $Q^\omega(n) = 0$ if $\omega >_l \omega^{\max}(n, d)$, the maximum element of $\mathrm{Dec}_d(n)$ (compare Theorem 6.3.12). In the case where $\mathrm{Dec}_d(n)$ has only one element ω, we use the same technique to show that any block in $P^d(n)$ with ω-sequence *right* lower than $\omega^{\min}(d)$ is hit. In Chapter 14, this will be proved for every degree d.

In Section 7.3, we introduce **semi-standard** monomials and blocks. Under certain conditions on ω, $Q^\omega(n)$ is spanned by semi-standard blocks. In contrast to α^+-bounded monomials, when $n \geq 2$ the number of semi-standard monomials in a given degree is not bounded. Multiple 1-back splicing involves the interchange of entries between adjacent columns of a block, together with certain 'error' terms when applied to the action of the Steenrod algebra. By ignoring the error terms, we obtain the notion of **combinatorial splicing**. Using iterated combinatorial splicing, a block with decreasing ω-sequence is equivalent to a sum of semi-standard blocks.

In Section 7.4 we show how various notions in the previous section translate into the combinatorial theory of **Young tableaux**. The **hook-length formula** (7.1) gives the number of semi-standard blocks. In the case $d = 2^n - n - 1$ we show that the semi-standard blocks form a basis of $Q^d(n)$ and that $\dim Q^d(n) = 2^{n(n-1)/2}$. In this case we show in Chapter 22 that $Q^d(n)$ is the Steinberg representation $\mathsf{St}(n)$ of $\mathsf{GL}(n)$. For an arbitrary decreasing sequence ω of 2-degree d, we show in Proposition 7.4.5 that an ordering on $\mathsf{Dec}_d(n)$ can be chosen, dependent on ω, for which $\mathsf{Q}^\omega(n)$, defined using this order, is spanned by semi-standard monomials.

7.1 The boundedness theorem

In this section we prove the following main result.

Theorem 7.1.1 (The boundedness theorem) *For all $n \geq 1$, $\dim \mathsf{Q}^d(n)$ is bounded by a function of n independent of d.*

To prove this, we use a specially adapted spanning set of the vector space $\mathsf{Q}^d(n)$. The proof will then follow by estimating the size of this spanning set. The α-**count** $\alpha(d) = |\mathrm{bin}(d)|$ (Definition 1.4.9) is the number of 1s in the binary expansion of d, and the α-sequence of a monomial $f = x_1^{d_1} \cdots x_n^{d_n}$ is $\alpha(f) = (\alpha(d_1), \ldots, \alpha(d_n))$ (Definition 5.1.5). It is convenient to introduce a modified version of the α-sequence by first multiplying by the product $\mathsf{c}(n) = x_1 \cdots x_n$ of the variables.

Definition 7.1.2 For $d \geq 0$ let $\alpha^+(d) = \alpha(d+1)$, and for $f = x_1^{d_1} \cdots x_n^{d_n}$ in $\mathsf{P}(n)$, let $\alpha^+(f) = (\alpha(d_1 + 1), \ldots, \alpha(d_n + 1))$. The monomial f is called α^+-**bounded** if $\alpha_i^+(f) \leq i$ for $1 \leq i \leq n$.

It is clear that a spike is α^+-bounded, since in this case $\alpha(1 + d_i) = 1$ for all i. In fact this property characterizes spikes. In terms of blocks, the α^+-sequence is obtained by adding 1 arithmetically to each row of the block and then taking the α-sequence. For example, the block

$$1\ 1\ 1\ 0\ 0\ 0\ 0$$
$$1\ 1\ 0\ 1\ 0\ 0\ 0$$
$$1\ 1\ 0\ 1\ 0\ 1\ 0$$
$$1\ 0\ 1\ 1\ 0\ 0\ 1$$

is α^+-bounded, as $\alpha(1 + d_i) = i$ for $1 \leq i \leq 4$. If C is a finite (n, c)-block with all entries 1 and B is an α^+-bounded n-block, then the concatenated block $C|B$ is also α^+-bounded, as $\alpha^+(C|B) = \alpha^+(B)$. The up Kameko map υ and the duplication map δ of Chapter 6 preserve the α^+-bounded condition.

We recall from Chapter 1 that in the case $n = 1$ the spikes generate $\mathsf{P}(1)$ as a module over A_2. In this case, for $d \geq 0$ we can write $x^d = \theta(x^{2^r - 1})$ for suitable $r \geq 0$ and $\theta \in A_2$. For $n > 1$, the spikes alone are not sufficient to generate

P(n), but by Theorem 1.8.2 dim $Q^d(n) \leq 3$ for all d in the case $n = 2$. The set of α^+-bounded monomials is far from being a minimal spanning set for $Q^d(n)$, even in the case $n = 2$. For example, there are three α^+-bounded monomials in $P^5(2)$, namely $x^3 y^2$, xy^4 and y^5, but $Q^5(2) = 0$. We show in Example 7.1.5 that for all d there are at most four α^+-bounded monomials in $P^d(2)$. The monomials in the diagram at the end of Section 1.8 are α^1-bounded.

Proposition 7.1.3 *The polynomial algebra* P(n) *is generated over* A_2 *by the* α^+*-bounded monomials in* P(n). *In particular,* Q(n) *is spanned by such monomials.*

Proof The proof is by induction on the number n of rows in a block F and a subsidiary induction on the 2-degree d of F, i.e. the degree of the corresponding monomial. The result is true for $n = 1$, as previously observed. We assume that $n > 1$ and that the result is true for all numbers less than n. We may also assume that there is at least one digit 1 in each row of F, otherwise F is essentially an $(n-1)$-block. The smallest possible value of d is then $d = n$. In this case F is a spike, for which the result is true, and this starts the induction on d. The inductive hypothesis further assumes that the result is true for n and all degrees less than some given $d > n$. The proof continues in three stages, to establish the inductive step.

Stage 1. In this stage, we reduce the problem to blocks with at least one 0 in the first column. Let F have n rows and 2-degree d. If the first column of F has no 0, then F is in the image of the Kameko map υ, and we can write $F = K|L$, where K is a single column of 1s. Since $\deg L < d$, by the inductive hypothesis L is a sum of blocks of the form $\theta(B)$, where B is α^+-bounded and $\theta \neq 0 \in A_2^k$ for $k \geq 0$.

If $k = 0$ then $K|\theta(B) = K|B$ is α^+-bounded. If $k > 0$, we apply the χ-trick to K and $\theta(B)$. Clearly $\chi(\theta)(K)$ is a sum of blocks with at least one 0 in the first column, and remains so when multiplied by the square of any polynomial. Hence we can write $K|\theta(B)$ as a sum of blocks representing polynomials of the form $\phi(E)$, where $\phi \in A_2^+$, and blocks with a 0 in the first column. Since $\deg E < d$, the inductive hypothesis allows us to assume that the blocks E are α^+-bounded.

Stage 2. By Stage 1, we may start with a block $F = (f_{i,j})$ of 2-degree d with at least one 0 in its first column. Let r be the smallest number such that $f_{r,1} = 0$, and let F' denote the block obtained by replacing row r of F by a zero row. Then F' is effectively an $(n-1)$-block, to which the inductive hypothesis applies, and we may write F' as a sum of terms of the form $\theta(B')$, where $\theta \in A_2$ and B' is α^+-bounded. More precisely, $\alpha_i^+(B') \leq i$ for $1 \leq i < r$ and $\alpha_i^+(B') \leq i-1$ for $r < i \leq n$.

This time we apply the χ-trick to $\theta(B')$ and row r of F, and note that any element of A_2 acting on a row of a block with 0 in the first position produces a sum of blocks with the same property. In particular this applies to the action of $\chi(\theta)$ on row r. Hence we can express F as a sum of blocks of the form $\phi(E)$, where E is α^+-bounded as in Stage 1, and blocks $G = (g_{i,j})$ such that $\alpha^+(G) \leq i$ for $1 \leq i < r$, $\alpha^+(G) \leq i - 1$ for $r < i \leq n$ and $g_{r,1} = 0$.

Stage 3. Let G be a block of the type produced by Stage 2. The process of splicing G at $(r, 1)$ involves only Sq^1, which effectively adds 1 arithmetically to the row on which it acts, and so makes the α-count of the row equal to the original α^+-count. Hence this splicing produces an equivalence of G, modulo hits, with a sum of blocks H with a 0 in the first position of some row r' and either (i) $\alpha_{r'}(H) \leq r'$ if $r' < r$, or (ii) $\alpha_{r'}(H) \leq r' - 1$ if $r' > r$. As before, we may also assume that the hit elements have the form $\phi(E)$, where E is α^+-bounded.

In case (i), we repeat Stage 2 and use downward induction to arrive at case (ii). Thus the problem is reduced to the case where a block F has some row r with 0 in its first position and α-count $\leq r - 1$. The action of any element of A_2 preserves this 0 entry, and cannot increase the α-count. Hence a final application of Stage 2 to the block F produces blocks of the form $\phi(E)$, where E is α^+-bounded, together with blocks G satisfying the final conditions in Stage 2 and also the conditions that row r has a 0 in the first position and α-count $\leq r - 1$. Hence $\alpha_r^+(G) \leq r$ and G is α^+-bounded. This concludes Stage 3, and completes the inductive argument.   □

The following result completes the proof of Theorem 7.1.1.

Proposition 7.1.4 *For* $n \geq 1$, *the number of* α^+-*bounded monomials in* $P^d(n)$ *is bounded by a function of* n *independent of* d.

Proof Multiplication by $c(n) = x_1 \cdots x_n$ is an $\mathbb{F}_2$-isomorphism from $P^d(n)$ to $P^{d+n}(Z[n])$, the subspace of $P^{d+n}(n)$ spanned by monomials divisible by all the variables. Under this isomorphism, an α^+-bounded monomial $f = x_1^{d_1} \cdots x_n^{d_n}$ corresponds to a monomial $g = x_1^{e_1} \cdots x_n^{e_n}$ where $e_i = 1 + d_i$ and $\alpha(e_i) \leq i$ for $1 \leq i \leq n$. By writing the terms of the reversed binary expansion of each e_i in order, g corresponds to a binary composition of $n + d$ with p (nonzero) parts for some p such that $n \leq p \leq n(n+1)/2$. The binary compositions of $n + d$ with p parts are obtained by permuting a binary partition of $n + d$ with p parts. By Proposition 5.3.9, the number of such partitions is bounded by a function of n independent of d. By summing over the range $n \leq p \leq n(n+1)/2$, the same is true for the monomials g, and hence for the α^+-bounded monomials f.   □

Example 7.1.5 In the case $n = 2$, by the argument above we need to consider binary compositions of $d + 2$ with $p = 2$ or 3 nonzero parts. Hence $\mathsf{P}^d(2)$ contains no α^+-bounded monomials when $\alpha(d) > 3$. When $d + 2 = 2^a \geq 4$ the binary compositions $d + 2 = 2^{a-1} + 2^{a-1}$ and $d + 2 = 2^{a-2} + (2^{a-1} + 2^{a-2})$ give two α^+-bounded monomials in $\mathsf{P}^d(2)$, the spike $(xy)^{2^{a-1}-1}$ and the monomial $x^{2^{a-2}-1}y^{2^{a-1}+2^{a-2}-1}$. When $d + 2 = 2^a + 2^b$ with $a > b$, there are at most four α^+-bounded monomials in $\mathsf{P}^d(2)$. These correspond to the binary compositions $d + 2 = 2^a + 2^b = 2^b + 2^a = 2^{a-1} + (2^{a-1} + 2^b) = 2^{b-1} + (2^a + 2^{b-1})$. When $d + 2 = 2^a + 2^b + 2^c$ with $a > b > c$, there are three α^+-bounded monomials in $\mathsf{P}^d(2)$, corresponding to the binary compositions $d + 2 = 2^a + (2^b + 2^c) = 2^b + (2^a + 2^c) = 2^c + (2^a + 2^b)$. Thus the number of α^+-bounded monomials in $\mathsf{P}^d(2)$ is ≤ 4 for all d, the maximum value 4 being attained when $d = (2^a - 1) + (2^b - 1)$ and $a - 1 > b > 0$.

Remark 7.1.6 This situation changes dramatically when $n > 2$. For example, when $n = 3$ the number of α^+-bounded monomials rises to 236 when $d = 2340$. On the other hand, we shall see in Section 8.3 that the maximum value of $\dim \mathsf{Q}^d(3)$ is 21.

The following result is a by-product of the proof of Proposition 7.1.3.

Proposition 7.1.7 *If $d \neq n \bmod 2$, then $\mathsf{Q}^d(Z[n])$ is spanned by blocks B such that some row of B has the form $1 \cdots 1\, 0 \cdots 0$, with $\leq n - 1$ digits 1.*

Proof Since d and n have opposite parity, then there is a 0 in the first column of any block in $\mathsf{P}^d(n)$. The proof of Proposition 7.1.3 shows that $\mathsf{Q}^d(Z[n])$ is spanned by blocks such that $\alpha_i \leq n - 1$ for some i. For such a block, the digits 1 in row i can be moved left by iterated 1-back splicing (as in Proposition 6.4.6) to obtain a set of blocks B of the required form. □

Moving digits 1 to the left by 1-back splicing destroys the α^+-condition on a block. However, the argument shows that $\mathsf{Q}^d(Z[n])$ is spanned by concatenated blocks of the form $F|G$, where F has not more than $n - 1$ columns and all entries in some row of F are 1. In this row all entries of G are 0, so that G involves only $n - 1$ variables.

7.2 Combinatorial splicing

Recall from Chapter 5 that $\omega^{max}(n, d)$ is the maximal decreasing n-bounded sequence of 2-degree d in both left and right orders. The main result of this section, Proposition 7.2.4, states that if $\omega >_l \omega^{max}(n, d)$ then any block in $\mathsf{P}^\omega(n)$

is left reducible, so that $Q^\omega(n) = 0$. To prove this, we extend the technique of 1-back splicing introduced in Proposition 6.4.4, where one digit 1 of a block is pulled back to the previous column. We construct hit equations by pulling back several digits 1 in the same column simultaneously. We need a preliminary result relating the actions of Sq^{k2^t} and Xq^{k2^t}.

Proposition 7.2.1 *Let* $c(n) = x_1 \cdots x_n$ *be the product of the variables. Then* $Xq^k(c(n)) = Sq^k(c(n)) + e_k$ *for* $k \geq 0$, *where every monomial appearing in* e_k *has an exponent* ≥ 4. *More generally* $Xq^{k2^s}(c(n)^{2^s}) = Sq^{k2^s}(c(n)^{2^s}) + e_k^{2^s}$ *for all* $s \geq 0$, *where every monomial in* $e_k^{2^s}$ *has an exponent* $\geq 2^{s+2}$.

Proof By Definitions 1.1.2 and 2.2.2

$$Xq(c(n)) = \prod_{i=1}^{k} Xq(x_i) = \prod_{i=1}^{k}(x_i + x_i^2 + x_i^4 + \cdots) = \prod_{i=1}^{k}(x_i + x_i^2) + e = Sq(c(n)) + e,$$

where all monomials in e have an exponent ≥ 4. The first statement follows by comparing terms of degree k, and the second then follows from Propositions 1.3.2 and 2.2.7. $\qquad\square$

Consider two adjacent columns $u, u+1$ in a block F in $P(n)$. Let $I \subseteq Z[n]$ be a subset of the rows of F such that $f_{i,u} = 0, f_{i,u+1} = 1$ for $i \in I$, and let $k = |I|$. Let H be the block formed from F by defining $h_{i,u} = 1$ and $h_{i,u+1} = 0$ for $i \in I$, leaving the other entries of F unaltered. Thus the entries of F in column $u+1$ and rows I are moved back to column u to form H. Let J be a set of k rows of F such that $f_{j,u} = 1$ and $f_{j,u+1} = 0$ for $j \in J$. Thus I and J are disjoint. We form H^J from H by defining $h^J_{j,u} = 0$ and $h^J_{j,u+1} = 1$ for $j \in J$, leaving the other entries of H unaltered. Thus H^J is formed from H by moving the entries in column u and rows J forward to column $u+1$. For example, when $n = 4$, $u = 2$ and $I = \{2, 4\}, J = \{1, 3\}$ we might have

$$F = \begin{matrix} 1\,1\,0\,1 \\ 1\,0\,1\,1 \\ 1\,1\,0\,0 \\ 0\,0\,1\,0 \end{matrix}, \quad H = \begin{matrix} 1\,1\,0\,1 \\ 1\,1\,0\,1 \\ 1\,1\,0\,0 \\ 0\,1\,0\,0 \end{matrix}, \quad H^J = \begin{matrix} 1\,0\,1\,1 \\ 1\,1\,0\,1 \\ 1\,0\,1\,0 \\ 0\,1\,0\,0 \end{matrix}.$$

Proposition 7.2.2 *Let* F *be a block in* $P(n)$ *such that* $f_{i,u} = 0$ *and* $f_{i,u+1} = 1$ *for* $i \in I$, *where* $u \geq 1$ *and* $I \subseteq Z[n]$. *Then* $F \sim G + A + B$, *where*

(i) $G = \sum_J H^J$, *where the sum is over all* $J \subseteq Z[n]$ *such that* $|J| = |I|$ *and* $f_{j,u} = 1$ *and* $f_{j,u+1} = 0$ *for* $j \in J$, *and* H^J *is formed from* F *by exchanging the entries in columns* u *and* $u+1$ *and the rows in* I *and* J,

(ii) *A and B are sums of 'error' blocks E, E' such that $\omega(E) <_l \omega(F)$ if E is in A and $\omega(E') <_r \omega(F)$ if E' is in B,*

(iii) *if the block E appears in A and $F = F_1|F_2$, $E = E_1|E_2$ are splittings of F and E between columns $u - 1$ and u, then $\omega(E_1) <_l \omega(F_1)$.*

Proof Let $c = \prod_{i \in I} x_i$, where $|I| = k$. Let f and h be the monomials corresponding to the blocks F and H as described above. Let $f' = f/c^{2^u}$. The corresponding block F' is formed from F by changing the entries $f_{i,u+1} = 1$ to 0 for $i \in I$. Then $f = f'c^{2^u}$ and by definition of H we have $h = f'c^{2^{u-1}}$. By Proposition 7.2.1 we can write

$$ f = f'c^{2^u} = f'Sq^{k2^{u-1}}(c^{2^{u-1}}) = f'Xq^{k2^{u-1}}(c^{2^{u-1}}) + f'e^{2^{u-1}}, $$

where every monomial in $e^{2^{u-1}}$ has an exponent $\geq 2^{u+1}$. Now F and F' have the same columns from $u + 2$ onwards. It follows from Proposition 6.1.7 that $\omega(f'e^{2^{u-1}}) <_r \omega(f)$. Also, by the χ-trick, $f'Xq^{k2^{u-1}}(c^{2^{u-1}}) \sim (Sq^{k2^{u-1}}(f'))c^{2^{u-1}}$.

There are three types of block produced in the expansion of $Sq^{k2^{u-1}}(F')$ by the Cartan formula. Since F and F' have the same columns $< u$, it follows that a Steenrod square acting on such a column produces blocks E with $\omega(E) <_l \omega(F)$, and in fact satisfying (iii). Such blocks E give the sum A. The second possibility is for a Steenrod square to act on a column $> u$ or on column u in a row i where $f'_{i,u} = f'_{i,u+1} = 1$. In either case, for a block E' which arises in this way, $\omega(E') <_r \omega(F)$. Such blocks E', together with blocks representing terms of $f'e^{2^{u-1}}$, give the sum B.

Finally $Sq^{k2^{u-1}}$ can act in column u on a set of k rows J of F', where J is as in (i). Note that I and J are disjoint. This accounts for the blocks H^J. □

The process of replacing F by $G + A + B$ is referred to as **multiple 1-back splicing** at column u and rows I. We sometimes wish to view A and B as error terms, and the process of replacing F by G as **combinatorial splicing**. In Section 7.4 we shall relate combinatorial splicing to Young tableaux.

The case of 1-back splicing corresponds to $|I| = k = 1$. Since $Sq^{k2^{u-1}}$ cannot affect columns $> u$, there are no terms B in the case of 1-back splicing. Combinatorial splicing simply involves an interchange of digits 1 and 0 between adjacent columns, and is clearly an equivalence relation on sums of blocks with the same α- and ω-sequences.

Proposition 7.2.3 *Let $F \in P(n)$ be a block such that $\omega_u(F) < \omega_{u+1}(F)$ for some u. Then $F \sim A + B$, where A and B satisfy conditions (ii) and (iii) of Proposition 7.2.2.*

Proof In Proposition 7.2.2 we choose I to be maximal with the stated properties, namely $f_{i,u} = 0$, $f_{i,u+1} = 1$ for $i \in I$. Since $\omega_u(F) < \omega_{u+1}(F)$ there are no sets of type J, hence no blocks H^J, and the result follows. $\square$

The following result is analogous to that for $\omega^{\min}(d)$ given in Theorem 6.3.12. Note that although the right order is also needed in the argument, $Q^\omega(n)$ is defined using the left order. Recall from Definition 5.6.5 that $\omega^{\max}(n,d)$ is the ω-sequence of a maximal spike in $P^d(n)$.

Proposition 7.2.4 *If* $\omega \in \mathrm{Seq}_d(n)$ *and* $\omega >_l \omega^{\max}(n,d)$ *then* $Q^\omega(n) = 0$.

Proof Let $M = \omega^{\max}(n,d)$ and let F be a block in $P^\omega(n)$ where $\omega >_l M$. From Proposition 5.6.4, we can find u such that $\omega_{u+1} > m_{u+1}$, $\omega_j = m_j$ for $j \leq u$ and $\omega_{u+1} > m_u$. Hence $\omega_{u+1} > \omega_u$. By Proposition 7.2.3 $F \sim A + B$, where $\omega(B) <_r \omega(F)$ and A is a sum of blocks C such that, for splittings $C = C'|C''$, $F = F'|F''$ and $M = M'|M''$ between columns $u-1$ and u, $\omega(C') <_l \omega(F') = \omega(M')$. In particular, $\omega(A) <_l \omega(M)$. Since $\omega(B) <_r \omega(F)$, we can iterate the procedure on the blocks of B. Since there cannot be an infinite chain of blocks with strictly decreasing ω-sequence in right order, this process must halt when F becomes left reducible. Hence $Q^\omega(n) = 0$. $\square$

The following results are proved by similar methods.

Proposition 7.2.5 *If* $\mathrm{Dec}_d(n)$ *has only one element* ω, *then a block* F *in* $P^d(n)$ *such that* $\omega(F) <_r \omega$ *is hit.*

Proof We have $\omega = M = \omega^{\max}(n,d)$. We may assume that $\omega(F) >_l M$, since otherwise the result follows from Theorem 6.3.12. Then there exists u such that $\omega_i = m_i$ for $i \leq u$ and $\omega_{u+1} > m_{u+1}$. From Proposition 5.6.4, $\omega_{u+1} > m_u = \omega_u$. Hence by Proposition 7.2.3 $F \sim A + B$, where A and B are sums of blocks such that $\omega(A) <_l M$ and $\omega(B) <_r \omega(F)$. Then A is hit by Theorem 6.3.12, and the process can be iterated on the blocks of B. Since there cannot be an infinite chain of strictly decreasing ω-sequences in right order, the process must halt when F is hit. $\square$

Equivalence of blocks modulo hits can sometimes be established by combinatorial splicing.

Proposition 7.2.6 *Suppose that* $\mathrm{Dec}_d(n)$ *has only one element* ω. *For any block* B *in* $P^\omega(n)$, *let* B' *be obtained from* B *by interchanging two columns* $i < j$ *for which* $\omega_i = \omega_j$. *Then* $B' \sim B$.

Proof It is enough to prove the result for adjacent columns $j = i + 1$. In splicing at column i as in Proposition 7.2.3, we choose I such that $|I|$ is maximal. Then there is a unique J with $|J| = |I|$ and $H^J = B'$. The error terms in the splicing process are left or right lower than ω, and so they are hit by Theorem 6.3.12 and Proposition 7.2.5. Hence $B \sim B'$. $\qquad\square$

Example 7.2.7 As $\text{Dec}_{18}(3)$ has just one element $(2,2,1,1)$, the following blocks are equivalent modulo hits.

$$
\begin{array}{cccc}
1\,1\,1\,0 & 1\,1\,1\,0 & 1\,1\,0\,1 & 1\,1\,0\,1 \\
1\,0\,0\,0\,, & 0\,1\,0\,0\,, & 1\,0\,0\,0\,, & 0\,1\,0\,0\,. \\
0\,1\,0\,1 & 1\,0\,0\,1 & 0\,1\,1\,0 & 1\,0\,1\,0
\end{array}
$$

Even when $\text{Dec}_d(n)$ has more than one element, Proposition 7.2.6 is true for the minimal element $\omega^{\min}(d)$ of $\text{Dec}_d(n)$. This follows from Theorem 14.1.3, since a right lower ω-sequence cannot 2-dominate $\omega^{\min}(d)$.

The next result may be viewed as a relative version of the fact that a spike cannot appear in a linear relation $\sum_i B_i \sim 0$ on blocks in $\mathsf{P}^d(n)$. Example 7.2.9 illustrates its use in obtaining lower bounds for $\dim \mathsf{Q}^\omega(n)$.

Proposition 7.2.8 *Let $I \subseteq Z[n]$, let $\mathcal{T}$ be a set of blocks in $\mathsf{P}^d(n)$ and let $\mathcal{S}$ be the subset of $\mathcal{T}$ such that the subblock given by the rows in I form the same spike of degree e. Let $\widehat{\mathcal{S}}$ be the set of blocks formed by removing the rows I of members of $\mathcal{S}$, and suppose that $\widehat{\mathcal{S}}$ is linearly independent in $\mathsf{Q}^{d-e}(n - |I|)$. Then $\mathcal{S}$ is linearly independent in $\mathsf{Q}^d(n)$. Moreover, no linear relation on members of $\mathcal{T}$ in $\mathsf{Q}^d(n)$ can involve a member of $\mathcal{S}$.*

Proof Consider a hit equation $F = \sum_j \theta_j(H_j)$, where F is a sum of members of $\mathcal{T}$ and $\theta_j \in A_2^+$ acts on the block H_j. We can think of H_j as the superposition of the block K_j consisting of the rows I of H_j and the block L_j formed from the complementary rows. In the expansion of the right hand side of the hit equation by the Cartan formula, a typical block arises by superposing blocks in the expansions of $\phi(K_j)$ and $\psi(L_j)$, for $\phi, \psi \in A_2$. Comparison of blocks on both sides of the hit equation shows that the only contribution to a block in $\mathcal{S}$ occurs if $\phi = 1$ and K_j is the spike defining $\mathcal{S}$. Hence, by removing rows I, we obtain a hit equation for a sum of members of $\widehat{\mathcal{S}}$. This contradicts the linear independence of $\widehat{\mathcal{S}}$ in $\mathsf{Q}^{d-e}(n - |I|)$. Hence no linear relation on members of $\mathcal{T}$ can involve a member of $\mathcal{S}$. In particular, $\mathcal{S}$ is linearly independent. $\qquad\square$

Example 7.2.9 We show that $\dim Q^{(2,2)}(4) \geq 20$ by proving that the set $\mathcal{T}$ of blocks

$$
\begin{matrix}
1\,1 & 1\,1 & 1\,1 & 0\,0 & 0\,0 & 0\,0 & 1\,1 & 1\,1 & 1\,1 & 1\,0 \\
1\,1 & 0\,0 & 0\,0 & 1\,1 & 1\,1 & 0\,0 & 1\,0 & 1\,0 & 0\,0 & 1\,1 \\
0\,0 & 1\,1 & 0\,0 & 1\,1 & 0\,0 & 1\,1 & 0\,1 & 0\,0 & 1\,0 & 0\,1 \\
0\,0 & 0\,0 & 1\,1 & 0\,0 & 1\,1 & 1\,1 & 0\,0 & 0\,1 & 0\,1 & 0\,0
\end{matrix}
$$

$$
\begin{matrix}
1\,0 & 0\,0 & 1\,0 & 1\,0 & 0\,0 & 1\,0 & 1\,0 & 0\,0 & 1\,0 & 1\,0 \\
1\,1 & 1\,1 & 0\,1 & 0\,0 & 1\,0 & 0\,1 & 0\,0 & 1\,0 & 1\,0 & 0\,1 \\
0\,0 & 1\,0 & 1\,1 & 1\,1 & 1\,1 & 0\,0 & 0\,1 & 0\,1 & 0\,1 & 1\,0 \\
0\,1 & 0\,1 & 0\,0 & 0\,1 & 0\,1 & 1\,1 & 1\,1 & 1\,1 & 0\,1 & 0\,1
\end{matrix}
$$

is linearly independent in $Q^{(2,2)}(4)$. The first 6 blocks in $\mathcal{T}$ are distinct spikes, and cannot appear in any linear relation on members of $\mathcal{T}$ in $Q^{(2,2)}(4)$. Let $\mathcal{T}'$ denote the complement of these blocks in $\mathcal{T}$, and let $\mathcal{S}$ be the set of the first three blocks in $\mathcal{T}'$. The first row of each block in $\mathcal{S}$ is the same spike, and occurs nowhere else as the first row of a block of $\mathcal{T}'$. The set $\widehat{\mathcal{S}}$ of three blocks

$$
\begin{matrix}
1\,0 & 1\,0 & 0\,0 \\
0\,1 & 0\,0 & 1\,0 \\
0\,0 & 0\,1 & 0\,1
\end{matrix}
$$

forms a basis of $Q^{(1,1)}(3)$. It follows from Proposition 7.2.8 that no block in $\mathcal{S}$ can appear in a linear relation on members of $\mathcal{T}'$, nor therefore in $\mathcal{T}$. A similar argument applied to the next nine blocks of $\mathcal{T}'$, in groups of three, shows that the last two members of $\mathcal{T}$ are the only blocks which can possibly appear in a linear relation. However, specialization of variables shows that these two blocks are also linearly independent in $Q^{(2,2)}(4)$.

7.3 Semi-standard blocks

The following special type of block will be related in Section 7.4 to the theory of Young tableaux.

Definition 7.3.1 Let $B[k]$ denote the block formed by the first k rows of an n-block B. Then B is **semi-standard** if $\omega(B[k])$ is decreasing for $1 \leq k \leq n$. Equivalently, a monomial $x_1^{d_1} \cdots x_n^{d_n} \in P(n)$ is **semi-standard** if the ω-sequence of $x_1^{d_1} \cdots x_n^{d_k}$ is decreasing for $1 \leq k \leq n$.

Example 7.3.2 The set $\mathcal{T}$ of Example 7.2.9 is a complete set of semi-standard blocks in $P^{(2,2)}(4)$.

The up Kameko map υ and the duplication map δ preserve the semi-standard condition. In contrast to the α^+-bounded blocks of Definition 7.1.2, the number of semi-standard blocks in $P^d(n)$ is not bounded by a function of n. For example, in the case $n = 2$ and $d = 2^t - 1$ there are $t + 1$ semi-standard blocks of the form

$$
\begin{matrix}
1 & \cdots & 1 & 0 & \cdots & 0 \\
0 & \cdots & 0 & 1 & \cdots & 1
\end{matrix} \;,
$$

corresponding to the monomials $x^{2^a-1}y^{2^{a+b}-2^a}$ in $P^{2^t-1}(2)$, where $t = a + b$ and $a, b \geq 0$. Such a block is also α^+-bounded if and only if it is a spike or $b = 1$. For example, of the four blocks $\begin{smallmatrix}1&1\\0&0\end{smallmatrix}, \begin{smallmatrix}0&0\\1&1\end{smallmatrix}, \begin{smallmatrix}1&0\\0&1\end{smallmatrix}, \begin{smallmatrix}0&1\\1&0\end{smallmatrix}$ in $P^3(2)$, the first three are both semi-standard and α^+-bounded, while the fourth block has neither property. The table of generators for $Q^d(2)$ at the end of Chapter 1 shows the following result.

Proposition 7.3.3 *The set of blocks in* $P(2)$ *which are both semi-standard and* α^+-*bounded form a basis for* $Q(2)$. $\qquad\square$

Although the corresponding statement fails for $n > 2$, the 'special blocks' of Chapter 8 can be regarded as variations on the same theme.

Proposition 7.3.4 *A block B with decreasing ω-sequence is equivalent to a sum $\sum_i B_i$ of semi-standard blocks B_i with $\omega(B_i) = \omega(B)$ and $\alpha(B_i) = \alpha(B)$ by combinatorial splicing.*

Proof We use an order on the set of n-blocks with the same α- and ω-sequences, by writing $A \geq_\omega B$ if and only if $(\omega(A[n]), \ldots, \omega(A[1])) \geq_l (\omega(B[n]), \ldots, \omega(B[1]))$, where $A[k], B[k]$ are the subblocks given by the first k rows of A and B. Thus for some k, $\omega(A[i]) = \omega(B[i])$ for $k < i \leq n$ and $\omega(A[k]) \geq_l \omega(B[k])$. For example,

$$
\begin{matrix}
1 & & 1 & & 0\,1 \\
1 & >_\omega & 0\,1 & >_\omega & 1 \\
0\,1 & & 1 & & 1
\end{matrix} \;.
$$

Let B be an n-block which is not semi-standard. Suppose that $\omega(B[r])$ is not decreasing, and that $\omega_t(B[r]) < \omega_{t+1}(B[r])$. Let S be the set of rows $i \leq r$ in B for which $b_{i,t} = 0$ and $b_{i,t+1} = 1$, and let $k = |S|$. We show that a combinatorial splice of G in column $t+1$ and rows S produces a sum of blocks C_j such that $C_j >_\omega B$.

To obtain a typical block C_j from B, we move k digits 1 of $B[r]$ from column $t+1$ back to column t and, say, ℓ digits 1 from column t forward to column $t+1$. Then $\ell < k$, since $\omega_t(B[r]) < \omega_{t+1}(B[r])$. The other $k - \ell$ digits are moved from column t to column $t+1$ in rows $> r$. Let s be the maximum of these rows. Then $\omega(C_j[i]) = \omega(B[i])$ for $i \geq s$, since k digits 1 have been transferred in each direction between columns t and $t+1$ in the first s rows, and no other columns have been altered. Further $\omega(C_j[s-1]) >_l \omega(B[s-1])$, since k digits have been transferred from column $t+1$ to column t in the first $s-1$ rows, but only $k-1$ in the opposite direction. Hence $C_j >_\omega B$.

Thus a block B which is not semi-standard can be replaced by combinatorial splicing by a sum of blocks all of which are higher in $\geq_\omega$ order. By repeating the above procedure on each of the new blocks, we must reach a sum of semi-standard blocks with the same α- and ω-sequences. If $\omega(B)$ is not decreasing, then there are no such blocks, and so B is combinatorially equivalent to zero. $\qquad\square$

Proposition 7.3.5 *If* $\mathrm{Dec}_d(n)$ *has only one element* ω, *then* $Q^\omega(n)$ *is spanned by semi-standard blocks.*

Proof This follows immediately from Proposition 7.3.4 and the fact that the error terms in applying the combinatorial splicing process in place of Steenrod operations have ω-sequences left or right lower than $\omega^{\min}(d)$, and so are hit. $\qquad\square$

As observed at the end of the proof of Theorem 7.2.6, it follows from Theorem 14.1.3 that Proposition 7.3.5 is true more generally when $\omega = \omega^{\min}(d)$ is the minimum element of $\mathrm{Dec}_d(n)$. In general, let $\omega = \omega^{\min}(d)$ be a minimal decreasing sequence, and let B be a block in $\mathsf{P}^\omega(n)$. We denote the first row of B by R and the complementary $(n-1)$-block by $\widehat{B}$. Suppose that $R = (0,\ldots,0,1,\ldots,1)$ with all digits 1 contiguous on the right, and let $\omega' = \omega - R$. Then we can manipulate $\widehat{B}$ within its equivalence class in $Q^{\omega'}(n-1)$ without altering the first row or equivalence class of B.

To see this, consider a hit equation $\widehat{B} = \widehat{B}' + \sum_{i>0} Sq^i(H_i)$. Then, by the χ-trick 2.5.2, $B \sim B'$ modulo error terms arising from the action of the operations $\chi(Sq^i)$ on the first row. The action of any positive Steenrod operation on the first row of B is zero or moves some digit 1 to a position greater than the length of ω, and so produces blocks with ω-sequence $\leq_r \omega$. Since ω is minimal, Theorem 14.1.3 implies that such blocks are hit.

Example 7.3.6 The semi-standard blocks of Example 7.3.2 span $Q^{(2,2)}(4)$, and so $\dim Q^{(2,2)}(4) = 20$. Although there are two decreasing sequences $(4,1)$ and $(2,2)$ in $\mathrm{Dec}_6(4)$, this follows from Proposition 7.3.4, since ω-sequences of 2-degree 6 which are $<_r (2,2)$ are also $<_l (2,2)$.

Alternatively, we can argue as follows. A block in $P^{(2,2)}(4)$ has $0,1$ or 2 digits 1 in its first row. In the second case, we may assume by Proposition 7.2.6 that the first row of the block is $(0,1)$ when the block is viewed as an element of $Q^{(2,2)}(4)$. Thus, to obtain a spanning set for $Q^{(2,2)}(4)$, we consider the possible first rows $(0,0)$, $(0,1)$ and $(1,1)$ of block B, and corresponding blocks $\widehat{B}$ which span $Q^{(2,2)}(3)$, $Q^{(2,1)}(3)$ and $Q^{(1,1)}(3)$ respectively. By the results of Sections 6.7 and 6.8, $\dim Q^{(2,2)}(3) = \dim Q^{(1,1)}(3) = 6$, while it is easy to see that $\dim Q^{(2,1)}(3) = 8$ (see Proposition 8.3.4). Thus by taking $\widehat{B}$ as one of these $6+6+8 = 20$ blocks as appropriate, we obtain a spanning set for $Q^{(2,2)}(4)$.

7.4 Young tableaux

The Ferrers diagram of a partition λ was introduced in Section 5.1, as a diagram of boxes with λ_i boxes in row i. A **Young tableau** is a filling of such a box diagram, usually by entering a positive integer in each box so that the entries are strictly increasing down each column. From our point of view it is more natural to associate such a diagram or tableau with the decreasing sequence $\omega = \lambda^{\mathrm{tr}}$, so that there are ω_j boxes in column j of the diagram. As for matrices and blocks, the (i,j)th box in such a diagram is the box in row i and column j.

Definition 7.4.1 Given $\omega \in \mathrm{Dec}(n)$, a **Young tableau** T of type (ω,n) is a filling of the diagram of $\lambda = \omega^{\mathrm{tr}}$ with the (i,j)th box filled by an integer $t_{i,j} \in Z[n]$ such that $t_{i,j} < t_{i+1,j}$ for $1 \le i < \omega_j$ and $1 \le j \le \mathrm{len}(\omega)$. A Young tableau T is **semi-standard** if the entries increase weakly on rows, i.e. $t_{i,j} \le t_{i,j+1}$ for all relevant i and j.

The definition of a Young tableau requires the columns to be strictly increasing. There is a bijective correspondence between n-blocks $B = (b_{i,j})$ with decreasing positive ω-sequences ω and Young tableaux of type (ω,n), where the boxes of column j of T are filled successively from top to bottom by the row numbers i for which $b_{i,j} = 1$. The block B is constructed from the tableau T by reading the numbers in each column j of T from top to bottom and putting a digit 1 in row $t_{i,j}$ of the block for $1 \le i < \omega_j$. The shape of the tableau is formed by moving the digits in the block upwards to adjacent positions and then exchanging the 1s for boxes.

Example 7.4.2 For $\omega = (3,2,2,1,1)$, a block B in $P^\omega(4)$ and the corresponding tableau T are shown below:

$$B = \begin{matrix} 1\,1\,1 \\ 1\,0\,1\,1 \\ 0\,1\,0\,0\,1 \\ 1 \end{matrix} \;,\; T = \begin{array}{|c|c|c|c|c|} \hline 1 & 1 & 1 & 2 & 3 \\ \hline 2 & 3 & 2 \\ \cline{1-3} 4 \\ \cline{1-1} \end{array} \;.$$

Here T and B are not semi-standard.

Proposition 7.4.3 *A block B with decreasing ω-sequence is semi-standard if and only if the corresponding tableau is semi-standard.*

Proof If the tableau T corresponds to the block B, then $t_{i,j}$ is the minimal k such that $\omega_j(B[k]) \geq i$. Thus $t_{i,j} < t_{i,j+1}$ if and only if $\omega_j(B[k]) \geq i$ but $\omega_{j+1}(B[k]) < i$, where $k = t_{i,j}$. $\qquad\square$

Example 7.4.4 Let $\omega = (2,1)$. There are nine blocks in $P^\omega(3)$,

$$\begin{matrix} 1\,1 & 1\,1 & 1 & 1 & 0 & 0 & 1 & 1 & 0\,1 \\ 1 & 0 & 1\,1 & 0 & 1\,1 & 1 & 0\,1 & 1 & 1 \\ 0 & 1 & 0 & 1\,1 & 1 & 1\,1 & 1 & 0\,1 & 1 \end{matrix} \;,$$

corresponding to the tableaux

$$\begin{array}{|c|c|}\hline 1 & 1 \\\hline 2 \\ \cline{1-1}\end{array} , \begin{array}{|c|c|}\hline 1 & 1 \\\hline 3 \\ \cline{1-1}\end{array} , \begin{array}{|c|c|}\hline 1 & 2 \\\hline 2 \\ \cline{1-1}\end{array} , \begin{array}{|c|c|}\hline 1 & 3 \\\hline 3 \\ \cline{1-1}\end{array} , \begin{array}{|c|c|}\hline 2 & 2 \\\hline 3 \\ \cline{1-1}\end{array} , \begin{array}{|c|c|}\hline 2 & 3 \\\hline 3 \\ \cline{1-1}\end{array} , \begin{array}{|c|c|}\hline 1 & 2 \\\hline 3 \\ \cline{1-1}\end{array} , \begin{array}{|c|c|}\hline 1 & 3 \\\hline 2 \\ \cline{1-1}\end{array} , \begin{array}{|c|c|}\hline 2 & 1 \\\hline 3 \\ \cline{1-1}\end{array} .$$

The first eight blocks are semi-standard, and the corresponding eight monomials $x_1^3x_2$, $x_1^3x_3$, $x_1x_2^3$, $x_1x_3^3$, $x_2^3x_3$, $x_2x_3^3$, $x_1x_2^2x_3$, $x_1x_2x_3^2$ form a basis of $Q^{(2,1)}(3)$. The last block is not semi-standard, and the corresponding monomial $x_1^2x_2x_3$ is related to the others by the hit equation $Sq^1(x_1x_2x_3) = x_1^2x_2x_3 + x_1x_2^2x_3 + x_1x_2x_3^2$.

The number of semi-standard Young tableaux of type (ω, n) is given by the **hook-length formula**

$$\tau(\lambda, n) = \prod_{(i,j)} \frac{n - i + j}{h(i,j)}, \tag{7.1}$$

where $\lambda = \omega^{\mathrm{tr}}$ and the **hook-length** $h(i,j)$ of the box (i,j) is the number of boxes (i',j') such that $i' \geq i, j' = j$ or $i' = i, j' \geq j$. The product is taken over all boxes (i,j) in the diagram of λ. In Example 7.4.4, $\tau((2,1),3) = 8$.

Given $\omega \in \mathrm{Dec}_d(n)$ with conjugate partition λ, we shall show that $\tau(\lambda, n)$ is an upper bound for $\dim Q^\omega(n)$ when the ordering on $\mathrm{Seq}_d(n)$ is chosen so as to give preference to ω. We emphasize that the ordering depends on ω, and that the upper bound is not in general an upper bound for other orderings (see Example 7.4.7). In particular, it is not in general an upper bound for $\leq_l$ or $\leq_r$, and we cannot infer a global upper bound on $Q^d(n)$, even if non-decreasing ω-sequences make no contribution to $Q^d(n)$.

Proposition 7.4.5 *Let $\omega \in \mathrm{Dec}(n)$ be a decreasing sequence, and let $\leq$ be an ordering on the set of ω-sequences ω' such that $\omega' \leq \omega$ if $\deg_2 \omega' = \deg_2 \omega$ and $|\omega'| \leq |\omega|$. Then $Q^\omega(n)$, as defined by this ordering, is spanned by monomials corresponding to semi-standard tableaux with diagram $\lambda = \omega^{tr}$. In particular, $\dim Q^\omega(n) \leq \tau(\lambda, n)$.*

Proof The hypothesis on the ordering of ω-sequences implies that if $\deg_2 \omega' = \deg_2 \omega$ and either $\omega' \leq_l \omega$ or $\omega' \leq_r \omega$, then $\omega' \leq \omega$. In a multiple 1-back splicing operation $F \sim G + A + B$ as in Proposition 7.2.2, blocks H^J in G are formed from F by exchanging 0s and 1s between adjacent columns, and so the α- and ω-sequences are not changed. Thus, if $F = L|M|R$ where M is the 2-column subblock where the entries 0 and 1 are exchanged to produce M', then $F' = L|M'|R$ is a sum of blocks with the same α-sequence as F.

By Proposition 6.1.2, if E is a left or right error term arising from the action of a squaring operation as in Proposition 7.2.2, then $|\omega(E)| \leq |\omega(F)|$. It follows from Proposition 7.3.4 that any monomial $f \in P^\omega(n)$ can be written as a sum of semi-standard monomials, a hit polynomial and monomials e with $|\omega(e)| \leq |\omega(f)|$ and either $\omega(e) \leq_l \omega$ or $\omega(e) \leq_r \omega$. Thus if we define $Q^\omega(n)$ using a total ordering $\leq$ on the set of ω-sequences in degree $d = \deg_2 \omega$ which satisfies the hypothesis, then $Q^\omega(n)$ is spanned by semi-standard monomials corresponding to tableaux of shape $\lambda = \omega^{tr}$. $\qquad\square$

Example 7.4.6 The sequence $\delta_n = (n-1, n-2, \ldots, 1)$ is the unique decreasing sequence of 2-degree $d = 2^n - n - 1$. Hence $Q^d(n) = Q^\omega(n)$ where $\omega = \delta_n^{tr} = \delta_n$. Since δ_n is strictly decreasing, $\dim Q^d(n) \leq \tau(\delta_n, n)$. By induction on n, the hook-length formula gives $\tau(\delta_n, n) = 2^{n(n-1)/2}$, since

$$\frac{\tau(\delta_n, n)}{\tau(\delta_{n-1}, n-1)} = \frac{n(n+1) \cdots (2n-1)}{(2n-1)(2n-3) \cdots 3 \cdot 1} = 2^{n-1} \frac{(2n-1)!}{(2n-1)!} = 2^{n-1}.$$

We discuss this important example further in Chapters 20 and 22, and show that $Q^d(n)$ is isomorphic to the **Steinberg module** $\mathrm{St}(n)$ of dimension $2^{n(n-1)/2}$.

Example 7.4.7 The upper bound $\dim Q^\omega(n) \leq \tau(\lambda, n)$ can fail if the ordering does not satisfy the condition of Proposition 7.4.5. For example let $n = 5$ and $\omega = (3, 3)$. We have $\omega <_l (5, 0, 1)$ but $\omega >_r (5, 0, 1)$, so Proposition 7.4.5 applies only to the right order, and gives $\dim Q^\omega(5) \leq 50$.

As $Sq^2(x_1^3 x_2 x_3 x_4 x_5) = x_1^5 x_2 x_3 x_4 x_5$ mod $P^\omega(5)$, a monomial in $P^{(5,0,1)}(5)$ is left reducible, but it is not right reducible since it does not appear in any other hit equation. Hence $\dim Q^{(5,0,1)}(5) = 0$ for the left order and $\dim Q^{(5,0,1)}(5) = 5$ for the right order. The only other hit equations involving $P^\omega(5)$ are 30 independent equations in 4 variables obtained from $Sq^1(x_1^3 x_2^3 x_3 x_4)$ and 20 equations in 5 variables obtained from $Sq^1(x_1^3 x_2^3 x_3 x_4 x_5)$ by permuting the variables. Of these 20, only 15 are linearly independent. Since $\dim P^\omega(5) = 100$, we obtain $\dim Q^\omega(5) = 55$ for the left order. For the right order, $\dim Q^\omega(5) = 50$ using the 5 hit equations involving Sq^2 to reduce elements of $P^\omega(5)$ to $P^{(5,0,1)}(5)$.

7.5 Remarks

The definition of α^+-bounded 7.1.2 has been changed from that in [31] by omitting the condition 'up to permutation of rows'. Hence Theorem 7.1.3 is stronger than its original form in [31]. Unfortunately, this makes little practical difference in estimating the size of $\dim Q^d(n)$, even after dividing the old estimate by $n!$.

Proposition 7.2.4 was proved by M. F. Mothebe [149]. Proposition 7.4.5 appears in [219].

The hook-length formula (7.1) is due to Richard P. Stanley. Using the theory of Schur functions, it can be regarded as a case of the Weyl character formula. For a proof, see [175] or [60, Sections 4.3, 6.1].

8

Special blocks and a basis for Q(3)

8.0 Introduction

We consider the hit problem for degrees d where spikes $x_1^{d_1} \cdots x_n^{d_n}$ appear with sufficiently separated decreasing exponents $d_i = 2^{a_i} - 1$. In Section 8.1, we develop techniques to obtain better estimates for $\dim Q^\omega(n)$ for decreasing sequences ω with 'large heads', i.e. $a_{n-1} - a_n \geq n$. Our bases are defined by **special** n-blocks. In certain 'generic' degrees, where $a_i - a_{i+1} \geq i + 1$ for all i, Theorem 8.1.13 shows that these estimates are best possible.

The more complicated situation for sequences with 'long tails' and **tail-special** blocks is discussed in Section 8.2. In particular, we show that if there is only one decreasing ω-sequence for monomials in $P^d(n)$, then there is a reduction formula for $\dim Q^d(n)$ when $a_1 - a_2 \geq n$. Theorem 8.2.10 provides a corresponding 'generic' case $a_i - a_{i+1} \geq n - i + 1$ for all i. Neither theorem implies the other, and it remains a challenge to find the best result of this type. In particular, the conditions $a_i - a_{i+1} \geq 2$ for $1 \leq i \leq n - 1$ and $a_n \geq 0$ are sufficient for $\dim Q^d(n) = \prod_{i=1}^{n} (2^i - 1)$ when $n \leq 4$, but it is not known if this is true for $n = 5$.

In Section 8.3 we determine $\dim Q^\omega(3)$ for all $\omega \in \text{Seq}(3)$, and provide the means to obtain a monomial basis for $Q^d(3)$ for $d \geq 0$. In Section 8.4 we extend the results of Section 8.1 to n-blocks of head length $n - 1$. Following N. Sum, the special blocks of Section 8.1 are supplemented by further elements of $Q^\omega(n)$ which we call **Sum** blocks. The results of this section will not be needed until they are applied to $Q^\omega(4)$ in Chapters 29 and 30. In the case $n = 3$, they provide an alternative approach to the cases $\omega = (2, 2, 1)$ and $\omega = (2, 2, 1, 1)$.

8.1 Blocks with large head

In this section we assume $n > 1$ and consider concatenated blocks $F = H|B$, where H is a head block and the 'body' B is an arbitrary n-block. The main

139

result is the reduction theorem 8.1.11 for the dimension of the cohit module $Q^{\gamma|\rho}(n)$ in the case where $\gamma = \omega(H)$ has length $\ell \geq n$ and where $\rho = \omega(B) \in$ Seq$(n-1)$.

First we summarize techniques for replacing a block by a sum of blocks in $\mathsf{P}^\omega(n)$ representing the same element in $Q^\omega(n)$. By Definition 6.8.4, a set $Y \subseteq Z[n]$ defines a class of head blocks having a 0 in row i if and only if $i \in Y$, and the smallest i in Y is the 'preferred' row of H. Since all head blocks in the same class Y represent the same element in $Q^\gamma(n)$, we can permute the columns of H and the rows of H in Y without altering the class. In particular, we can take H to be normalized as in Definition 6.8.4, so that the preferred row of H is of the form $1 \cdots 1\, 0 \cdots 0$, with $|Y|-1$ digits 1 followed by $\ell + 1 - |Y|$ digits 0. For example, in the case $n = \ell = 5$, $i = 1$ and $Y = \{1,3,4,5\}$,

$$
H = \begin{matrix} 1\,1\,1\,0\,0 \\ 1\,1\,1\,1\,1 \\ 1\,1\,0\,1\,1 \\ 1\,0\,1\,1\,1 \\ 0\,1\,1\,1\,1 \end{matrix} \qquad\qquad (8.1)
$$

A $\sim_l$-equivalence between two blocks of the same class is established in Propositions 6.8.2 and 6.8.6 by a splicing technique which uses only restricted hit equations. Thus Proposition 6.3.11 gives the following result.

Proposition 8.1.1 *Let* $H, H' \in \mathsf{P}^\gamma(n)$ *be head blocks in the same class. Then* $H|B \sim_l H'|B'$ *for any sequence* $\rho \in$ Seq(n) *and sums of blocks* $B \sim_l B' \in \mathsf{P}^\rho(n)$, *i.e.* $H|B$ *and* $H'|B'$ *represent the same element of* $Q^{\gamma|\rho}(n)$. $\qquad\square$

Consequently, in any concatenated block $H|B$ with H a head block and $\omega(B) = \rho$, we can replace H by any block in the same class, and independently we can replace $B \in Q^\rho(n)$ by a $\sim_l$-equivalent sum of blocks, without changing the element $H|B$ in $Q^{\gamma|\rho}(n)$. In particular, we can permute the columns of H, and permute the rows of H which contain 0s, without changing B. We can also apply 1-back splicing at any position without changing the element in $Q^{\gamma|\rho}(n)$.

In general, splicing in a row can produce blocks with higher ω-sequences (see Example 6.4.3). There is a useful case where this cannot happen.

Proposition 8.1.2 *Let H be a head block in* $\mathsf{P}(n)$ *of length ℓ such that $a_{i,j} = 0$ for $k \leq j \leq \ell$, and let B be a block in* $\mathsf{P}(n)$ *such that $b_{i,1} = 1$. Then splicing $H|B$ at (i,k) does not produce terms with ω-sequences of higher left order.*

Proof With notation as in Definition 6.4.1, let $F = H|B$, so that the splicing operation gives $F \sim_l G + E$. In all cases, $\omega(E) <_l \omega(F)$. Let $G' = H'|B'$ be a block in G split in the same way as F. Then G' arises by the action of $Sq^{2^{u-1}}$ on a row r of H at column k, where $r \neq i$ and H is as in Definition 6.4.1. Since H is a head block, $a_{r,j} = 1$ for $r \neq i$ and $k \leq j \leq \ell$. Hence H' is also a head block. Since $b_{i,1} = 1$, $b'_{i,1} = 0$ and so $\omega_1(B') \leq \omega_1(B)$. In the case $\omega_1(B') = \omega_1(B)$, no entries of B after the first column are changed, and so $\omega(B') = \omega(B)$. Hence $\omega(G') \leq_l \omega(F)$. $\qquad\qquad\qquad \square$

Example 8.1.3 We illustrate this situation for $n = 3$ and $\ell = 3$. Splicing $H|B$ at $(2,2)$ gives

$$
H|B = \begin{array}{ccc|cc}
1\,1\,1 & 0\,1 \\
1\,0\,0 & 1\,0 \\
0\,1\,1 & 0\,1
\end{array}
\ \sim_l\
\begin{array}{ccc|cc}
1\,0\,0 & 1\,1 \\
1\,1\,1 & 0\,0 \\
0\,1\,1 & 0\,1
\end{array}
\ +\
\begin{array}{ccc|cc}
1\,1\,1 & 0\,1 \\
1\,1\,1 & 0\,0 \\
0\,0\,0 & 1\,1
\end{array}\ ,
$$

where both blocks have ω-sequence $(2,2,2,1,2)$.

In order to prove the reduction theorem 8.1.11 for blocks of head length $\ell \geq n$ we begin by showing that $Q^\omega(n)$ is spanned by blocks of some special type.

Definition 8.1.4 If H is a head block, a block $H|B$ in $P(n)$ is **special** if all entries of B in the preferred row of H are 0.

Some examples of special blocks are

$$
\begin{array}{cc|cc}
1\,1 & 1\,0\,1 \\
0\,1 & 0\,0\,0 \\
1\,0 & 1\,0\,1
\end{array},
\quad
\begin{array}{cc|cc}
1\,0\,1 & 0\,0 \\
1\,1\,0 & 1\,0 \\
0\,1\,1 & 0\,1
\end{array},
\quad
\begin{array}{cc|c}
1\,1\,1\,1 & 1 \\
1\,0\,0\,0 & 0 \\
0\,1\,1\,1 & 1
\end{array}\ .
$$

Reduction of a block $F = H|B$ with head H and body B to a sum of special blocks will be carried out by a sequence of splicing processes (Definition 6.4.1). We assume that $\omega(F) = \gamma|\rho$, where $\gamma = \omega(H) = (n - 1, \ldots, n - 1)$ has length n, and $\rho = \omega(B) \in \mathrm{Seq}(n)$ is an arbitrary n-bounded sequence. The results of Chapter 1 show that $Q^\omega(n)$ is spanned by special blocks in the case $n = 2$ and $\omega = (1,1)|\rho$. The next result provides a controlled way to carry out the reduction in this case.

Proposition 8.1.5 *Let $F = H|B$ be a 2-block with head H of length 2 and body B. Then F can be reduced to a sum of special blocks by splicing processes having origin ≤ 2.*

Proof The proof is by induction on the left order of the body B. There are four possibilities for the head H

$$\frac{1\,1}{0\,0}, \quad \frac{0\,0}{1\,1}, \quad \frac{0\,1}{1\,0}, \quad \frac{1\,0}{0\,1}.$$

We may assume that the block F is not special. In the first type, splicing at $(2,1)$ produces a block of the second type, but with a new body. Splicing at $(1,2)$ produces a block of the third type and restores the original body, and splicing at $(1,1)$ gives the fourth type by normalizing the head. Finally, splicing at $(1,2)$ returns to the first type, but with body lower than the original in left order on blocks. The effect is the same if we start with a different head and carry out the same sequence of four splicings. Hence by induction on left order, iteration of the routine must ultimately give rise to a special block.   □

Example 8.1.6 The first four steps of the reduction are shown below in a typical case.

$$\frac{1\,1}{0\,0}\left|\frac{1\,0\,1}{1\,1\,0}\right. \;\rightarrow\; \frac{0\,0}{1\,1}\left|\frac{0\,1\,1}{0\,1\,0}\right. \;\rightarrow\; \frac{0\,1}{1\,0}\left|\frac{1\,0\,1}{1\,1\,0}\right. \;\rightarrow\; \frac{1\,0}{0\,1}\left|\frac{1\,0\,1}{1\,1\,0}\right. \;\rightarrow\; \frac{1\,1}{0\,0}\left|\frac{0\,0\,1}{0\,0\,1}\right.$$

The effect of any similar sequence of four steps is to restore the head and to reduce the exponent sequence (a,b) of the body to $(a-1,b+1)$.

Proposition 8.1.7 *Let $F = H|B$ be an n-block with head H of length n and body B. Then F can be reduced to a sum of special blocks by splicing processes having origin $\leq n$.*

Proof The proof is by induction on n and the left order on blocks. The base case $n = 2$ is given by Proposition 8.1.5, so we assume that $n > 2$ and that the statement is true for k-blocks, $k < n$.

Let $F = H|B$ be an n-block with head H of length n and body B. If there is a 0 in the first row of H and if F is not special, then we can iteratively splice the first row of the block with origin at the first 0 in the head until all entries in the first row of the head are 1. The blocks produced this way have bodies lower in left order than B.

Thus we may start again with a block F having the property that all entries of the first row of H are 1. Now consider the block G consisting of the last $n-1$ rows of F with head of length $n-1$. By the inductive assumption, the block G is reducible to special blocks by splicing operations with origin $\leq n-1$. This is also true for F unless a splicing operation attacks the first row of F. In that case, the resulting blocks have at least two 0s in the first row of the head, because the origin of attack is lower than n.

The left order of the body may rise during this process by increasing the 2-degree of the first row of the body, but we can adopt the same device as in the case $n = 2$. After normalizing the head, we may assume that the $(1, n-1)$ and $(1, n)$ entries are 0s. Splicing at $(1, n)$ restores the first row of the body. Normalizing the head and then splicing again at $(1, n)$ reduces the 2-degree of the first row of the body, and thus makes the left order of the body lower than the original. The process ends when the block is special. This completes the inductive step. $\qquad\qquad\qquad\qquad\qquad\qquad\qquad\qquad\qquad\qquad\qquad\square$

Proposition 8.1.8 *Let* $\gamma = (n-1, \ldots, n-1)$ *be a head sequence in* $\mathrm{Seq}(n)$ *of length* $\ell \geq n$, *and let* $\rho \in \mathrm{Seq}(n-1)$. *Then* $\dim Q^{\gamma|\rho}(n) \leq (2^n - 1)$ $\dim Q^\rho(n-1)$.

Proof Proposition 8.1.7 shows that $Q^{\gamma|\rho}(n)$ is spanned by special blocks $H|B$. The only condition on $B \in P^\rho(n)$ is that it has a zero row in the preferred row of H. In order to show that $\dim Q^\omega(n) \leq (2^n - 1) \dim Q^\rho(n-1)$, we choose a set $\mathcal{B}$ of blocks representing a monomial basis of $Q^\rho(n-1)$. For each $Y \subseteq Z[n]$, let $\mathcal{B}_Y$ denote the set of blocks B_Y formed from $B \in \mathcal{B}$ by inserting a zero row at position i, the least element of Y. Let $H_Y \in Q^Y(n)$ be a block of class Y. From Proposition 8.1.1, for any element $C = H_Y|B \in Q^{\gamma|\rho}(n)$, the block B can be replaced by a sum of blocks $B_Y \in \mathcal{B}_Y$ without altering C up to left reducibility. Hence the $(2^n - 1) \dim Q^\rho(n-1)$ special blocks $H_Y|B_Y$ span $Q^\omega(n)$, and so $\dim Q^{\gamma|\rho}(n) \leq (2^n - 1) \dim Q^\rho(n-1)$. $\qquad\qquad\qquad\square$

Remark 8.1.9 The argument remains valid if ρ has some entry $\leq n$, and shows that $Q^{\gamma|\rho}(n) = 0$.

Next we consider linear independence of special blocks.

Proposition 8.1.10 *Let* $\gamma = (n-1, \ldots, n-1) \in \mathrm{Seq}(n)$ *be a head sequence of length* $\geq n$, *and let* $\rho \in \mathrm{Seq}(n-1)$. *Then* $\dim Q^{\gamma|\rho}(n) \geq (2^n - 1)$ $\dim Q^\rho(n-1)$.

Proof We show that the special blocks $H_Y|B_Y$ constructed in the proof of Proposition 8.1.8 represent linearly independent elements of $Q^\omega(n)$. In other words, no nonzero sum S of the $(2^n - 1) \dim Q^\rho(n-1)$ blocks $H_Y|B_Y$, where $B_Y \in \mathcal{B}_Y$, is left reducible. As in the proof of Theorem 6.8.7, we use the fact that if $S \sim_l 0$ then $S \cdot M \sim_l 0$ for $M \in \mathrm{M}(n)$.

We assume that $S \sim_l 0$. Let $S_{r,i}$ be the sum of the blocks in S with $|Y| = r$ and preferred row i, so that $S = \sum_{r,i} S_{r,i}$. Then the block $H_Y|B_Y$ is a term of $S_{1,i}$ if and only if row i of $H_Y|B_Y$ is zero. It follows from Proposition 6.2.9 that $S_{1,i}$ is left reducible.

Let $\widehat{H}_Y|\widehat{B}_Y$ be the subblock of $H_Y|B_Y$ obtained by deleting row i. Then $\widehat{H}_Y$ has all entries 1, and $\widehat{H}_Y|\widehat{B}_Y$ may be viewed as the result of applying the iterated up Kameko map υ to $\widehat{B}_Y$. By Proposition 6.5.4 we deduce that the elements of $\mathcal{B}$ corresponding to the subblocks $\widehat{B}_Y$ of blocks in $S_{1,i}$ are linearly dependent in $Q^\rho(n-1)$, contradicting the definition of $\mathcal{B}$ as a basis of $Q^\rho(n-1)$. Hence $S_{1,i} = 0$, and so S has no terms of class Y when $|Y| = 1$.

We use this as the base of an induction argument. Thus for $r \geq 2$ we assume that S has no terms of class Y when $|Y| < r$, and show that $S_{r,i} = 0$ for all i. We may assume that for each Y the blocks H_Y in $S_{r,i}$ are chosen as in (8.1), so that row i of H_Y has $r-1$ 1s followed by $\ell - r + 1$ 0s, and H_Y has one 0 in row $j > i$ for $j \in Y$.

Let $Y \subseteq Z[n]$ be a set of rows with $|Y| = r$ and preferred row i, and let $J = Y \setminus \{i\}$. We apply the singular transformation $M(Y)$ which maps x_i to $\sum_{j \in J} x_j$ and fixes x_k for $k \neq i$. Proposition 6.1.12 explains how to evaluate $S \cdot M(Y)$. The blocks obtained result from binary addition of integers represented by subsets of 1s in row i of a special block $H_{Y'}|B_{Y'}$ to rows $j > i$ for $j \in Y$, so that all entries of row i of $H_{Y'}$ are replaced by 0s.

First consider the case $Y' = Y$. Since row i of the block B_Y is zero, no changes take place in B_Y, other than those arising from the additions in the head subblock, where a digit 1 in row i is added to a digit 1 in row j to produce a 'carry'. All such superpositions result in blocks with left lower ω-sequence. The only remaining block is $H_i|B_Y$, where H_i is the head block which has all entries 0 in row i and all other entries 1.

This block $H_i|B_Y$ cannot be produced by the action of $M(Y)$ on any term $H_{Y'}|B_{Y'}$ in S where $Y' \neq Y$. We may assume that $H_{Y'}$ has preferred row i and is in the form (8.1). By the induction hypothesis, $|Y'| = r' \geq r$. If $r' > r$, then there are $r' - 1 \geq r$ 1s in row i of $H_{Y'}$. These cannot all be replaced by 0s in a term of $S \cdot M(Y)$ without superposition of 1s, since each row $j > i$ has only one 0. The same is true if $r' = r$ and $Y' \neq Y$, as in this case some of the rows in Y are not available to receive 1s transferred from row i.

Thus $S \cdot M(Y)$ contains a single block whose ith row is zero. Since the action of $M(Y)$ commutes with the action of A_2, $S \cdot M(Y) \sim_l 0$. The same argument as used in the case $|Y| = 1$ now shows that $S_{r,i} = 0$. This completes the inductive step. We conclude that $S = 0$, and so the blocks $H_Y|B_Y$ are linearly independent as elements of $Q^\omega(n)$. $\qquad\square$

Combining Propositions 8.1.8 and 8.1.10 gives the main result of this section.

Theorem 8.1.11 *Let* $\gamma = (n-1,\ldots,n-1)$ *have length* $\geq n$. *Then for all* $\rho \in$ Seq$(n-1)$, $\dim Q^{\gamma|\rho}(n) = (2^n - 1)\dim Q^{\rho}(n-1)$. $\qquad\square$

It is interesting to note that the above proofs use more devices on a concatenated block $H|B$ than just the independent manipulation of the blocks H and B. Sometimes it is necessary to splice across the dividing line of the concatenation. The following example illustrates a relation which cannot be obtained simply by manipulating the blocks H and B individually.

Example 8.1.12 The following block is hit.

$$H|B = \begin{array}{ccc|ccc} 1 & 1 & 0 & 1 & 0 & 0 \\ 1 & 0 & 1 & 0 & 1 & 0 \\ 0 & 1 & 1 & 0 & 0 & 1 \end{array}.$$

Neither H nor B is hit, but by splicing at $(1,3)$ we obtain $H|B \sim H_1|B_1 + H_2|B_2$, where

$$H_1|B_1 = \begin{array}{ccc|ccc} 1 & 1 & 1 & 0 & 0 & 0 \\ 1 & 0 & 0 & 1 & 1 & 0 \\ 0 & 1 & 1 & 0 & 0 & 1 \end{array}, \quad H_2|B_2 = \begin{array}{ccc|ccc} 1 & 1 & 1 & 0 & 0 & 0 \\ 1 & 0 & 1 & 0 & 1 & 0 \\ 0 & 1 & 0 & 1 & 0 & 1 \end{array}.$$

Since $H_1|B_1$ and $H_2|B_2$ have heads and tails in the same classes, they are equivalent. Hence $H|B \sim 0$.

The following global result follows from Theorem 8.1.11 by induction on n.

Theorem 8.1.13 *Let* $a_i - a_{i+1} \geq i+1$ *for* $1 \leq i < n$ *and* $a_n \geq 0$, *and let* $d = \sum_{i=1}^{n}(2^{a_i} - 1)$. *Then* $\dim Q^d(n) = \prod_{i=1}^{n}(2^i - 1)$.

Proof For $n = 1$ or 2 this is proved in Theorems 1.4.12 and 1.8.2. We use Theorem 8.1.11 to prove it by induction on n. Since $a_i > a_{i+1}$ for $1 \leq i < n$, by Proposition 5.1.13 there is only one decreasing sequence ω in $\mathrm{Dec}^d(n)$, and so $Q^d(n) = Q^{\omega}(n)$. By Proposition 6.5.3 we may assume $a_n = 0$. Then, writing $\omega = \gamma|\rho$ where γ is the head of ω, we see that $d' = \deg_2 \rho = \sum_{i=1}^{n-1}(2^{a_i} - 1)$ and d' has the same property as d with $n-1$ replacing n. Since γ has length $a_{n-1} - a_n \geq n$, Proposition 8.1.10 gives $\dim Q^d(n) = (2^n - 1)\dim Q^{\rho}(n-1) = (2^n - 1)\dim Q^{d'}(n-1)$. This establishes the inductive step. $\qquad\square$

8.2 Blocks with long tail

We recall how the χ-trick 2.5.2 works in terms of blocks. Let F be a block and let S be a subset of the entries of F. We construct the blocks G, H from F by replacing the entries 1 in S and the complement of S respectively by 0s. Then F may be viewed as the superposition of G and H. In general the superposition of two blocks is the block arising from the arithmetic addition of corresponding rows in reversed binary arithmetic. Now suppose that H satisfies the hit equation $H = \sum_{j>0} Sq^j(H_j)$. Then $F \sim \sum_{j>0} F_j$, where the blocks F_j arise from the superposition of H_j on the blocks of $Xq^j(G)$.

Frequently, as in 1-back splicing, we wish to focus on some particular blocks F_j and regard the rest of the F_i as error terms. In applications, these will have lower ω-sequences. We shall need the following elementary but useful fact about hit equations expressed in terms of blocks.

Proposition 8.2.1 *Given $m > 0$, let $F + G \in \mathsf{P}^d(n)$ be a sum of blocks, where the first row of each block in F is a tail sequence S of length m, and no block in G has this property. Let $\widehat{F}$ be the sum of blocks obtained by removing the first row of the blocks in F. If $\widehat{F} \neq 0$ in $Q(n-1)$, then $F + G \neq 0$ in $Q(n)$.*

Proof Consider a hit equation $F + G = \sum_{j>0} Sq^j(H_j)$. Since a spike cannot be a term in the image of any positive Steenrod square, the only blocks H_j which can contribute to F are those whose first row is S. Comparing blocks on both sides of the equation and removing the first row leads to a hit equation for $\widehat{F}$. This contradicts the assumption that $\widehat{F} \neq 0$ in $Q(n-1)$. $\qquad\square$

Recall from Section 6.7 that a tail block T in $\mathsf{P}^d(n)$ is in class $Y \subseteq Z[n]$ if Y is the set of rows of T which contain entries 1, and that all blocks in the same class are equivalent modulo hit polynomials. As for head blocks, the 'preferred' row of a tail block associated to Y is the smallest element in Y. If the length of the tail block exceeds n then we can arrange for the preferred row to contain two 1s. The following definition is the analogue of Definition 8.1.4 for tail blocks.

Definition 8.2.2 If T is a tail block, a block $A|T$ in $\mathsf{P}(n)$ is **tail-special** if all entries of A in the preferred row of T are 1.

As in Proposition 8.1.8, we shall prove that $Q^\omega(n)$ is spanned by tail-special blocks when ω has a sufficiently long tail. The case where the tail has length n again presents special difficulty. To resolve this, we make temporary use of a higher ω-sequence, as in the next example.

Example 8.2.3 Consider the blocks

$$F = \begin{matrix} 1\ 0\ 0 \\ 0\ 1\ 0 \\ 0\ 0\ 1 \end{matrix}\ , \quad G = \begin{matrix} 1\ 0\ 0 \\ 1\ 0\ 0 \\ 1\ 0\ 1 \end{matrix}\ .$$

Splicing $\hat{G}$ at $(3,2)$ gives

$$G \sim \begin{matrix} 0\ 1\ 0 \\ 0\ 1\ 0 \\ 1\ 1\ 0 \end{matrix}\ + \begin{matrix} 0\ 1\ 0 \\ 1\ 0\ 0 \\ 0\ 0\ 1 \end{matrix}\ + \begin{matrix} 1\ 0\ 0 \\ 0\ 1\ 0 \\ 0\ 0\ 1 \end{matrix}\ .$$

The second and third blocks are equivalent to F, and splicing the first block at $(1,1)$ also produces F. Hence $G \sim F$.

We generalize this example as follows.

Proposition 8.2.4 *For $n \geq 3$, let F denote the tail (n,n)-block corresponding to the identity matrix I_n, i.e. $f_{i,j} = 1$ if and only if $i = j$. Let G denote the block obtained from F by putting $g_{2,2} = 0$, $g_{2,1} = 1$ and $g_{r,1} = 1$ for some $r \geq 3$, leaving other entries of F unaltered. Then $F \sim G$ using only 1-back splicing.*

Proof The case $n = 3$ is explained above. For $n > 3$ columns 3 to n of G form a tail block. We may therefore interchange columns 3 and r, apply the case $n = 3$ to the first three columns and rows 1, 2 and r, and finally exchange back columns 3 and r to obtain $F \sim G$. $\qquad\square$

The following result is the analogue of Proposition 8.1.8 for sequences with long tail.

Proposition 8.2.5 *Let $\omega = \rho | \tau \in \mathrm{Seq}(n)$, where $\tau = (1, \ldots, 1)$ is a tail sequence of length $\ell \geq n$ and $\rho = (r_1, \ldots, r_m)$ with $r_i > 0$ for $1 \leq i \leq m$. Then $Q^\omega(n)$ is spanned by tail-special blocks.*

Proof As in Proposition 8.1.8, we begin with the easier case $\ell > n$.
 Case 1: $\ell > n$. Let $A | T \in Q^\omega(n)$ be a block with $\omega(A) = \rho$ and $\omega(T) = \tau$. Since $\ell > n$ we can arrange to have two 1s in the preferred row i of $T \in Q^\tau(n)$. There is nothing to prove if all entries in row i of A are 1. Otherwise we may assume, by iterated 1-back splicing in A, that $a_{i,m} = 0$, and by iterated 1-back splicing in T that $t_{i,1} = 1$. Then we 1-back splice $A | T$ at (i,m). This splicing produces an equivalence of $A | T$ with a sum of blocks $A' | T'$ such that $\alpha_i(A') = \alpha_i(A) + 1$ and T' is a tail block with $\omega(T') = \tau$. The block T' may belong to a different class from T, but its preferred row $i' \leq i$ since there are two 1s in

row i of T. The above procedure can therefore be iterated until all blocks are tail-special.

Case 2: $\ell = n$. The procedure of Case 1 works if the class of the block is not $Y = Z[n]$, because we can still arrange to have two 1s in the preferred row. Thus let $A|T$ be a block in $\mathsf{P}^\omega(n)$ where T is a tail block of length n and class $Z[n]$. Then $T \sim F$ where F is the block corresponding to the identity matrix I_n, as in Proposition 8.2.4. We may assume that $n \geq 3$, as the cases $n = 1, 2$ are covered by the results in Chapter 1.

If all entries of the first row of A are 1s, then $A|T$ is already tail-special. Otherwise, by using 1-back splicing in A and T, we may assume that $a_{1,m} = 0$ and that $t_{1,1} = 0$, $t_{1,2} = 1$. The example

$$
A|T = \begin{array}{cc|cccc}
1 & 0 & 0 & 1 & 0 & 0 \\
0 & 1 & 1 & 0 & 0 & 0 \\
0 & 0 & 0 & 0 & 1 & 0 \\
0 & 1 & 0 & 0 & 0 & 1
\end{array}
$$

illustrates the case $n = 4$, $m = 2$. We next splice at $(1, m)$. In the above example we obtain

$$
G \sim \begin{array}{cc|cccc}
1 & 1 & 1 & 0 & 0 & 0 \\
0 & 0 & 0 & 1 & 0 & 0 \\
0 & 0 & 0 & 0 & 1 & 0 \\
0 & 1 & 0 & 0 & 0 & 1
\end{array} \quad + \quad \begin{array}{cc|cccc}
1 & 1 & 1 & 0 & 0 & 0 \\
0 & 1 & 1 & 0 & 0 & 0 \\
0 & 0 & 0 & 0 & 1 & 0 \\
0 & 0 & 1 & 0 & 0 & 1
\end{array}.
$$

In general, the splicing produces an equivalence of $A|T$ with a sum of blocks $A'|T'$, where $\alpha_1(A') = \alpha_1(A) + 1$ and T' is either a tail block of class $Z[n]$ or is a block of the type G described in Proposition 8.2.4, which is equivalent to such a tail block. By iterating the process on newly formed blocks, we ultimately arrive at tail-special blocks as required, proving Case 2. $\square$

We next prove a corresponding lower bound result by the same method as for Proposition 8.1.10.

Proposition 8.2.6 *Let* $\omega = \rho|\tau \in \mathrm{Seq}(n)$, *where* $\tau = (1, \ldots, 1)$ *is a tail sequence of length* $\ell \geq n$ *and* $\rho = (r_1, \ldots, r_m)$ *with* $r_i > 0$ *for* $1 \leq i \leq m$, *and let* $\rho' = (r_1 - 1, \ldots, r_m - 1) \in \mathrm{Seq}(n - 1)$. *Then* $\dim \mathsf{Q}^\omega(n) \geq (2^n - 1) \dim \mathsf{Q}^{\rho'}(n - 1)$.

Proof We show that the set of tail-special blocks $A_Y|T_Y$ constructed above, corresponding to nonempty subsets $Y \subseteq Z[n]$ and a monomial basis $\mathcal{A}$ of $\mathsf{Q}^{\rho'}(n - 1)$, represent linearly independent elements of $\mathsf{Q}^\omega(n)$.

Let S be a sum of blocks $A_Y|T_Y$ which is left reducible, and let $S_{r,i}$ be the sum of the terms in S with $|Y| = r$ and preferred row i. Then row i of a block $A_Y|T_Y$ in $S_{1,i}$ has all entries 1, and no other term of S has this property. Further, all entries of T_Y not in row i are 0. Removing row i, Proposition 8.2.1 shows that $\mathcal{A}$ is linearly dependent modulo reducible polynomials, contradicting the definition of $\mathcal{A}$. Hence no term $A_Y|T_Y$ in S has $|Y| = 1$.

We use this as the basis of an induction argument. Thus we assume that, for some $r \geq 2$, S does not have any terms $A_Y|T_Y$ with $|Y| < r$. We may also assume that for each Y the blocks T_Y in $S_{r,i}$ are chosen so that row i of T_Y has $\ell - r + 1$ 1s followed by $r - 1$ 0s, and T_Y has one 1 in row $j > i$ for $j \in Y$.

Let $Y \subseteq Z[n]$ be a set of rows with $|Y| = r$ and preferred row i, and let $J = Y \setminus \{i\}$. We act on S by the matrix $M \in M(n)$ defined by $x_j \cdot M = x_i$ for $j \in J$ and $x_k \cdot M = x_k$ if $k \notin J$. The action of M on a polynomial is to specialize all the variables in Y to x_i.

By Proposition 6.1.9, the effect of this in terms of blocks is to add all entries 1 in rows $j \in Y$ for $j \neq i$ to row i, using binary arithmetic. Since all entries of row i of A_Y are 1s, all additions affecting A_Y result in superpositions of 1s, and so they produce blocks with ω-sequences $<_l \omega$. Thus, modulo such blocks, the effect of M on the terms of the form $A_Y|T_Y$ in S is to produce the particular block $A_Y|T_i$, where all entries of T_i are 0 except in row i where they are all 1.

Because there are no blocks in S with $|Y| < r$, the block $A_Y|T_i$ cannot be produced as a result of the action of M on any other term of S. Consider a term $A_{Y'}|T_{Y'}$ in S, where $Y' \neq Y$, $A_{Y'}$ has preferred row i and $T_{Y'}$ is chosen as above. By the induction hypothesis, $|Y'| = r' \geq r$. If $r' > r$, then there are $r' - 1 \geq r$ 0s in row i of $T_{Y'}$. All of these cannot be replaced by 1s by the action of M, since each row $j > i$ has only one digit 1.

The same is true if $r' = r$ and $Y' \neq Y$, as in this case some of the specializations of variables do not add a digit 1 to row i. Thus $S \cdot M$ contains a single block whose ith row is filled by 1s. Since the action of M commutes with the action of A_2, $S \cdot M \sim_l 0$. The same argument as in the case $|Y| = 1$ shows that $S_{r,i} = 0$. This completes the inductive step. We conclude that $S = 0$, and so the blocks $A_Y|T_Y$ represent linearly independent elements of $Q^\omega(n)$. $\square$

We can combine this result with Proposition 8.2.5 as follows.

Theorem 8.2.7 *Let $\omega = \rho|\tau \in \mathrm{Seq}(n)$, where $\tau = (1, \ldots, 1)$ is a tail sequence of length $\ell \geq n$ and where $\rho = (r_1, \ldots, r_m)$ with $r_i > 0$ for $1 \leq i \leq m$, and let $\rho' = (r_1 - 1, \ldots, r_m - 1)$. Assume that any equivalence $A \sim_l A'$ between elements $A, A' \in P^{\rho'}(n-1)$ can be achieved by a restricted hit equation using only Sq^k for $k < 2^m$. Then $\dim Q^\omega(n) = (2^n - 1) \dim Q^{\rho'}(n-1)$.*

Proof We show that, under the additional hypothesis, Proposition 8.2.5 implies that $\dim Q^\omega(n) \leq (2^n - 1) \dim Q^{\rho'}(n-1)$. The tail-special blocks $A|T$ span $Q^{\rho|\tau}(n)$, where the only condition on $A \in P^\rho(n)$ is that it has all entries 1 in the preferred row of T. Let $\mathcal{A}$ be a set of blocks representing a monomial basis of $Q^{\rho'}(n-1)$. For each $Y \subseteq Z[n]$, let $\mathcal{A}_Y$ denote the set of blocks A_Y formed from $A \in \mathcal{A}$ by inserting a new row of 1s in the preferred row i, the least element of Y. Let $T_Y \in Q^\tau(n)$ be a tail block of class Y.

From Proposition 8.1.1, the additional hypothesis implies that for any block $C = A|T_Y \in Q^\omega(n)$, the subblock A can be replaced by a sum of blocks A_Y without altering C up to left reducibility. Since the blocks T_Y span $Q^\tau(n)$, it follows that the $(2^n - 1) \dim Q^{\rho'}(n-1)$ tail-special blocks $A_Y|T_Y$ span $Q^\omega(n)$.   □

Example 8.2.8 Let $\omega = (3,1,1,1,1) = \rho|\tau \in \mathrm{Seq}(4)$. Then $\dim Q^\omega(4) = 15 \dim Q^{(2)}(3) = 45$. In this case, no extra condition is needed, as the sequence $\rho' = (2)$ corresponds to a 1-column block, representing a product of distinct variables. More generally, if $\rho = (r)$ is a sequence of length 1 and τ is a tail sequence of length $\geq n$, then $\dim Q^{\rho|\tau}(n) = (2^n - 1)\binom{n-1}{r-1}$.

Proposition 8.2.7 applies to $\omega = \gamma|\tau$ where γ is a head sequence and τ is a tail sequence, because equivalence of elements in $P^\gamma(n)$ can be achieved by splicing, as explained in Section 6.8. Example 6.3.10 shows that the hypothesis of Proposition 8.2.7 fails when $\rho' = (2,5)$.

Sometimes we can bypass the difficulty of using restricted hit equations by using Proposition 6.3.6.

Proposition 8.2.9 *Assume that there is only one element* ω *in* $\mathrm{Dec}_d(n)$. *Let* $\omega = \rho|\tau$ *where* τ *is a tail sequence of length* $\geq n$ *and* $\rho = (r_1,\ldots,r_m)$, *and let* $\rho' = (r_1 - 1,\ldots,r_m - 1)$. *Then* $\dim Q^d(n) = (2^n - 1) \dim Q^{\rho'}(n-1)$.

Proof Let $A|T$ be a block with $\omega(A) = \rho$ and $\omega(T) = \tau$. By Proposition 6.3.6, the error terms in writing $A|T$ in terms of the tail-special blocks $A_Y|T_Y$ have ω-sequences $<_r \omega$. Since $\mathrm{Dec}_d(n)$ has only one element ω, these error terms are hit by Proposition 7.2.6. Hence $\dim Q^d(n) \leq (2^n - 1) \dim Q^{\rho'}(n-1)$ by Proposition 8.2.5. The opposite inequality follows from Proposition 8.2.6.   □

Theorem 8.2.10 *Let* $d = \sum_{i=1}^n (2^{a_i} - 1)$, *where* $a_n \geq 0$ *and* $a_i - a_{i+1} \geq n-i+1$ *for* $1 \leq i < n$. *Then* $\dim Q^d(n) = \prod_{i=1}^n (2^i - 1)$.

Proof The proof follows from Proposition 8.2.9 by induction on n using a similar argument to the proof of Theorem 8.1.13, with tails instead of heads.   □

8.3 A monomial basis of $Q^d(3)$

In this section we determine the dimensions of the vector spaces $Q^\omega(3)$, using the left order. For each $d \geq 0$ a monomial basis for $Q^d(3)$ can be written down by following through the argument. The following result simplifies the problem in the 3-variable case.

Proposition 8.3.1 *If* $\omega \in \mathrm{Seq}_d(3)$ *is not decreasing, then* $Q^\omega(3) = 0$.

Proof If a block $F \in P^d(3)$ has a non-trailing zero column, then F is left reducible by Proposition 6.4.7. Otherwise if $\omega(F)$ is not decreasing then up to a row permutation F has consecutive columns $u,\, u+1$ of one of the forms

$$
\begin{array}{cccc}
1\,1 & 1\,0 & 1\,1 & 1\,1 \\
0\,1\,, & 0\,1\,, & 1\,1\,, & 0\,1\, . \\
0\,0 & 0\,1 & 0\,1 & 0\,1
\end{array}
$$

In the first case, 1-back splicing in the second row provides a left reduction of the block. The third and fourth cases are similar, splicing in the third row. In the second case, we apply Proposition 7.2.3 where $I = \{2,3\}$, $k = 2$, and note that there are no blocks of type B, because $Sq^{2^{u-1}k} = Sq^{2^u}$ acts on a zero column $u+1$ in the block F' of Proposition 7.2.2 and not on any higher column. Hence F is reduced in the left order. □

The next example shows that Proposition 8.3.1 is not true in the 4-variable case.

Example 8.3.2 There are four sequences in $\mathrm{Seq}^7(4)$ given in left order by $(3,2) >_l (3,0,1) >_l (1,3) >_l (1,1,1)$. A block with ω-sequence $(1,1,1)$ can only involve three variables. Hence, if a block in $P^{(1,3)}(4)$ is left reducible then it is hit, which is impossible because no monomial in degree $2^a - 1$ is hit for $a \geq 0$. It follows that $Q^{(1,3)}(4) \neq 0$. The situation cannot be remedied by using the right order, for which $(3,2) >_r (1,3) >_r (3,0,1) >_r (1,1,1)$, as a similar argument can be applied to a block in $P^{(3,0,1)}(4)$.

By iterated use of Proposition 6.5.4, we can further reduce the problem to sequences $\omega = (2,\ldots,2,1,\ldots,1) \in \mathrm{Dec}(3)$. Then $\omega = \gamma \,|\, \tau$, where $\gamma = (2,\ldots,2)$ is a head sequence of length $t \geq 0$ and $\tau = (1,\ldots,1)$ is a tail sequence of length $s \geq 0$.

Theorem 8.3.3 *The following table gives the dimension of* $Q^\omega(3)$ *for* $\omega = (2,\ldots,2,1,\ldots,1)$, *where* s *is the number of* 1s *and* t *is the number of* 2s.

	$s=0$	$s=1$	$s=2$	$s\geq 3$
$t=0$	1	3	6	7
$t=1$	3	8	15	14
$t=2$	6	14	21	21
$t\geq 3$	7	14	21	21

The duplication map $\delta : Q^\omega(3) \to Q^{\omega'}(3)$ *is surjective for the row* $t=3$, *and also for* $\omega = (2,2,1)$ *and* $\omega = (2,2,1,1)$.

Proof The results for $t = 0$ and $s = 0$ have already been established in Theorems 6.7.6 and 6.8.7. The results for $t \geq 3$ and $s \geq 3$ follow immediately from Propositions 8.1.10 and 8.2.9, together with results in the 2-variable case from Chapter 1. Setting $\omega' = (2)|\omega$, Proposition 6.8.3 shows that the duplication map $\delta : Q^\omega(3) \to Q^{\omega'}(3)$ is surjective for $t = 3$. The remaining cases $\omega = (2,1)$, $(2,1,1)$, $(2,2,1)$ and $(2,2,1,1)$ are treated in Propositions 8.3.4 to 8.3.9. The surjectivity of δ in the last two cases is treated in Proposition 8.3.10.   □

Proposition 8.3.4 $\dim Q^{(2,1)}(3) = 8$.

Proof There are 9 blocks in $P^\omega(3)$, of which 6 are spikes. The hit equation

$$Sq^1 \begin{pmatrix} 1 \\ 1 \\ 1 \end{pmatrix} = \begin{matrix} 1\,0 \\ 1\,0 \\ 0\,1 \end{matrix} + \begin{matrix} 1\,0 \\ 0\,1 \\ 1\,0 \end{matrix} + \begin{matrix} 0\,1 \\ 1\,0 \\ 1\,0 \end{matrix}$$

is the only relation between the other 3 monomials. Hence $\dim Q^\omega(3) = 8$.   □

Proposition 8.3.5 $\dim Q^{(2,1,1)}(3) = 15$.

Proof There are 27 blocks in $P^\omega(3)$. By the solution of the 2-variable hit problem we know that $\dim Q^\omega(2) = 3$. Hence there is a 9-dimensional subspace of $\dim Q^\omega(3)$ arising from the 12 blocks with a zero row. The other 15 blocks span $Q^\omega(Z[3])$. Each such block is a row permutation of one of the three types

$$\begin{matrix} 1\,1\,0 \\ 1\,0\,0 \\ 0\,0\,1 \end{matrix}\, , \quad \begin{matrix} 1\,0\,1 \\ 1\,0\,0 \\ 0\,1\,0 \end{matrix}\, , \quad \begin{matrix} 1\,0\,0 \\ 1\,0\,0 \\ 0\,1\,1 \end{matrix}\, .$$

By 1-back splicing the second type at $(1,2)$ and the third type at $(3,1)$, all these blocks are equivalent to a combination of the six permutations of the first type. Hence $\dim Q^\omega(Z[3]) \le 6$ and $\dim Q^\omega(3) \le 9+6 = 15$.

To prove equality, we show that the six permutations $A_1,\dots,A_6$ of the first type are linearly independent modulo hit polynomials. Consider the specializations $P(3) \to P(2)$ which map two of the variables to x and the third to y. The effect on the blocks is to add two of the three rows, so up to exchange of rows we obtain each of the three blocks

$$\begin{matrix} 1\,1\,1 \\ 1\,0\,0 \end{matrix}\, , \quad \begin{matrix} 1\,0\,1 \\ 1\,1\,0 \end{matrix}\, , \quad \begin{matrix} 0\,0\,1 \\ 0\,0\,1 \end{matrix}$$

twice. It follows that if a nonzero linear combination X of the blocks $A_1,\dots,A_6$ is hit, then A_i and A_j have the same coefficient in X if they are interchanged by exchanging rows $1\,1\,0$ and $0\,0\,1$, or by exchanging rows $1\,0\,0$ and $0\,0\,1$. Hence X is the sum of all 6 blocks. The hit equation

$$Sq^1 \left(\begin{matrix} 1\,1 \\ 1\,1 \\ 1\,0 \end{matrix} + \begin{matrix} 1\,1 \\ 1\,0 \\ 1\,1 \end{matrix} + \begin{matrix} 1\,0 \\ 1\,1 \\ 1\,1 \end{matrix} \right) = X + Y$$

shows that $X \sim Y$ where

$$Y = \begin{matrix} 1\,1 \\ 1\,1 \\ 0\,1 \end{matrix} + \begin{matrix} 1\,1 \\ 0\,1 \\ 1\,1 \end{matrix} + \begin{matrix} 0\,1 \\ 1\,1 \\ 1\,1 \end{matrix}\, .$$

However, it is clear that there is no other hit equation involving any of the blocks in Y, so Y is not hit. Hence $A_1,\dots,A_6$ are linearly independent in $Q^\omega(3)$. $\qquad\square$

Remark 8.3.6 It follows by 1-back splicing that

$$\begin{matrix} 0\,1\,1 \\ 1\,0\,0 \\ 1\,0\,0 \end{matrix} + \begin{matrix} 1\,0\,0 \\ 0\,1\,1 \\ 1\,0\,0 \end{matrix} + \begin{matrix} 1\,0\,0 \\ 1\,0\,0 \\ 0\,1\,1 \end{matrix} \sim \sum_{i=1}^{6} A_i$$

and so the polynomial $x^6yz + xy^6z + xyz^6$ is not hit.

Proposition 8.3.7 $\dim Q^{(2,2,1)}(3) = 14$.

Proof There are 27 blocks in $P^\omega(3)$, of which 6 have a zero row and are spikes. These span a 6-dimensional subspace of $Q^\omega(3)$. Then $Q^\omega(Z[3])$ is spanned by the other 21 blocks, which can be split into four types defined by row permutations of the blocks

$$
\begin{array}{cccc}
1\,1\,1 & 1\,1\,0 & 1\,1\,0 & 1\,1\,0 \\
1\,0\,0\,, & 1\,1\,0\,, & 1\,0\,1\,, & 1\,0\,0\,\cdot \\
0\,1\,0 & 0\,0\,1 & 0\,1\,0 & 0\,1\,1
\end{array}
$$

Splicing the first block at $(3,1)$ shows that there are at most three equivalence classes among the six blocks of the first type. It is clear that there are no other hit equations involving these blocks, and so the blocks of the first type give a 3-dimensional subspace of $Q^\omega(Z[3])$. Splicing the second block at $(3,2)$ reduces it to a sum of two blocks of the third type. The fourth block is equivalent to a row permutation of the third by splicing at $(3,1)$. Splicing the second block at this position produces the relation

$$
\begin{array}{ccc}
1\,1\,0 & 1\,1\,0 & 0\,0\,1 \\
1\,1\,0 \;+\; 0\,0\,1 \;+\; 1\,1\,0 & \sim_I 0. \\
0\,0\,1 & 1\,1\,0 & 1\,1\,0
\end{array}
$$

This implies that the sum of the six blocks of the third type is equivalent to zero, so that the third type spans at most a 5-dimensional subspace of $Q^\omega(Z[3])$. Hence $\dim Q^\omega(3) \leq 6 + 3 + 5 = 14$.

Finally we show that there is no other relation among the six blocks $B_1, \ldots, B_6$ of the third type. As in Case 2, we use the specializations $P(3) \to P(2)$ which map two of the variables to x and the third to y. If the blocks B_1 and B_2 are interchanged by exchanging rows $1\,0\,1$ and $0\,1\,0$, then the specialization which adds these two rows maps B_1 and B_2 to the same spike $\begin{smallmatrix}1&1&1\\1&1&0\end{smallmatrix}$ or $\begin{smallmatrix}1&1&0\\1&1&1\end{smallmatrix}$, and no other block maps to a spike. Hence B_1 and B_2 have the same coefficient in any linear combination of $B_1, \ldots, B_6$ which is hit. Since the sum of all six blocks is hit, by symmetry it suffices to prove that $B_1 + B_2$ is not hit. We may choose

$$
B_1 = \begin{array}{c} 1\,1\,0 \\ 1\,0\,1 \\ 0\,1\,0 \end{array}\,,\quad B_2 = \begin{array}{c} 1\,1\,0 \\ 0\,1\,0 \\ 1\,0\,1 \end{array}\,\cdot
$$

Let $M = \left(\begin{smallmatrix} 0 & 1 & 1 \\ 0 & 1 & 0 \\ 0 & 0 & 1 \end{smallmatrix}\right)$ be the singular matrix which maps x to $y+z$ and fixes y and z. Then

$$
(B_1 + B_2) \cdot M = \begin{matrix} 0\,0\,0 \\ 1\,1\,1 \\ 1\,1\,0 \end{matrix} + \begin{matrix} 0\,0\,0 \\ 1\,1\,0 \\ 1\,1\,1 \end{matrix} + E,
$$

where $\omega(E) <_l \omega$. Hence $(B_1 + B_2) \cdot M$ is not hit, and so $B_1 + B_2$ is not hit. It follows that $\dim Q^\omega(3) = 14$. $\square$

Proposition 8.3.8 $\dim Q^{(2,2,1,1)}(3) \leq 21$.

Proof There are 81 blocks in $P^\omega(3)$. Of these, 12 have a zero row, and from the solution of the 2-variable hit problem they span a 9-dimensional subspace of $Q^\omega(3)$. The other blocks span $Q^\omega(Z[3])$. We assemble them into types defined by row permutations of representing blocks with decreasing α-sequences. The possible decreasing α-sequences are $(4,1,1)$, $(3,2,1)$ and $(2,2,2)$.

To cut down the enumeration, we use 1-back splicing to move entries 1 in the first row of the representing block as far left as possible, and the same in the second row without disturbing the first row. This reduces the enumeration to the five types

$$
\begin{matrix} 1\,1\,1\,1 \\ 1\,0\,0\,0\,, \\ 0\,1\,0\,0 \end{matrix} \quad
\begin{matrix} 1\,1\,1\,0 \\ 1\,1\,0\,0\,, \\ 0\,0\,0\,1 \end{matrix} \quad
\begin{matrix} 1\,1\,1\,0 \\ 1\,0\,0\,1\,, \\ 0\,1\,0\,0 \end{matrix} \quad
\begin{matrix} 1\,1\,0\,0 \\ 1\,1\,0\,0\,, \\ 0\,0\,1\,1 \end{matrix} \quad
\begin{matrix} 1\,1\,0\,0 \\ 1\,0\,1\,0\,. \\ 0\,1\,0\,1 \end{matrix} \qquad (8.2)
$$

By splicing the first block at $(3,1)$, we see that the 6 blocks of the first type span at most a 3-dimensional subspace of $Q^\omega(Z[3])$. We next look at the fifth type. First note that splicing at $(3,1)$ and $(3,3)$ shows that the second and third rows of the fifth block can be switched without changing the equivalence class. Hence the 6 blocks of the fifth type span at most a 3-dimensional subspace of $Q^\omega(Z[3])$. By splicing the fifth block at $(2,2)$, we obtain the relation

$$
\begin{matrix} 1\,1\,0\,0 \\ 1\,0\,1\,0 \\ 0\,1\,0\,1 \end{matrix} + \begin{matrix} 1\,0\,1\,0 \\ 1\,1\,0\,0 \\ 0\,1\,0\,1 \end{matrix} \sim_l \begin{matrix} 1\,1\,0\,0 \\ 1\,1\,0\,0 \\ 0\,0\,1\,1 \end{matrix}.
$$

Switching rows $1,3$ and rows $2,3$ of this relation produces two further relations

$$
\begin{matrix} 0\,1\,0\,1 \\ 1\,0\,1\,0 \\ 1\,1\,0\,0 \end{matrix} + \begin{matrix} 0\,1\,0\,1 \\ 1\,1\,0\,0 \\ 1\,0\,1\,0 \end{matrix} \sim_l \begin{matrix} 0\,0\,1\,1 \\ 1\,1\,0\,0 \\ 1\,1\,0\,0 \end{matrix},
$$

$$
\begin{array}{ccc}
1\,1\,0\,0 & 1\,0\,1\,0 & 1\,1\,0\,0 \\
0\,1\,0\,1 \;+\; 0\,1\,0\,1 & \sim_l & 0\,0\,1\,1 \;. \\
1\,0\,1\,0 & 1\,1\,0\,0 & 1\,1\,0\,0
\end{array}
$$

All six blocks of type 5 appear on the left hand side of these three relations. As observed earlier, these represent only three equivalence classes, so adding the three relations produces the following equivalence for blocks of the fourth type.

$$
\begin{array}{cccc}
1\,1\,0\,0 & 1\,1\,0\,0 & 0\,0\,1\,1 \\
1\,1\,0\,0 \;+\; 0\,0\,1\,1 \;+\; 1\,1\,0\,0 & \sim_l 0. & & (8.3) \\
0\,0\,1\,1 & 1\,1\,0\,0 & 1\,1\,0\,0
\end{array}
$$

We also deduce from the relations that blocks of the fourth and fifth type together span a subspace of $Q^\omega(3)$ of dimension at most 3 with a spanning set that may be chosen from one block of the fifth type and two blocks of the fourth type.

Next we splice the fourth block at $(3,1)$ to obtain the relation

$$
\begin{array}{ccc}
1\,1\,0\,0 & 0\,0\,1\,0 & 1\,1\,0\,0 \\
1\,1\,0\,0 \;\sim_l\; 1\,1\,0\,0 \;+\; 0\,0\,1\,0 \;. & & (8.4) \\
0\,0\,1\,1 & 1\,1\,0\,1 & 1\,1\,0\,1
\end{array}
$$

This shows that each block of the fourth type is equivalent to a sum of two blocks of the second type. By permuting the rows of this relation and adding the results, as in (8.3), we see that the sum of the six blocks of the second type is 0 in $Q^\omega(3)$. Hence blocks of the second, fourth and fifth type together span a subspace of $Q^\omega(3)$ of dimension at most $5 + 1 = 6$.

Finally we consider blocks of the third type. Splicing the third block at $(2,2)$ we obtain

$$
\begin{array}{ccc}
1\,1\,1\,0 & 1\,0\,0\,1 & 1\,1\,1\,0 \\
1\,0\,0\,1 \;\sim_l\; 1\,1\,1\,0 \;+\; 1\,1\,1\,0 \;, \\
0\,1\,0\,0 & 0\,1\,0\,0 & 0\,0\,1\,0
\end{array}
$$

and further splicing the last block at $(3,1)$ gives

$$
\begin{array}{cccc}
1\,1\,1\,0 & 1\,0\,0\,1 & 1\,1\,1\,0 & 0\,0\,0\,1 \\
1\,0\,0\,1 \;\sim_l\; 1\,1\,1\,0 \;+\; 0\,0\,0\,1 \;+\; 1\,1\,1\,0 \;, & & & (8.5) \\
0\,1\,0\,0 & 0\,1\,0\,0 & 1\,1\,0\,0 & 1\,1\,0\,0
\end{array}
$$

which shows that sums of pairs of blocks of the third type are equivalent to sums of blocks of the second type. Hence blocks of the second, third, fourth and fifth types together span a subspace of $Q^\omega(3)$ of dimension at most $6+3 = 9$. Combining this with the previous results, $\dim Q^\omega(3) \leq 9+3+9 = 21$. $\square$

We next turn to the lower bound, and show how to select a basis for $Q^\omega(3)$. Our basis for the subspace $Q^\omega(Z[3])$ consists of cyclic permutations of the rows of the blocks (8.2), excluding the fourth block because of the relation (8.3).

Proposition 8.3.9 *The 6 spikes and the 15 monomials corresponding to cyclic permutation of the rows of the five blocks*

$$
\begin{array}{ccccc}
1\,1\,1\,0 & 1\,1\,1\,1 & 1\,1\,1\,0 & 1\,1\,1\,0 & 1\,1\,0\,0 \\
1\,1\,0\,1\,, & 1\,0\,0\,0\,, & 1\,1\,0\,0\,, & 1\,0\,0\,1\,, & 1\,0\,1\,0 \\
0\,0\,0\,0 & 0\,1\,0\,0 & 0\,0\,0\,1 & 0\,1\,0\,0 & 0\,1\,0\,1
\end{array}
$$

form a basis for $Q^{18}(3)$.

Proof By Proposition 1.4.7 this reduces to showing that the cyclic permutations

$$
B_1 = \begin{array}{c} 1\,1\,1\,1 \\ 1\,0\,0\,0 \\ 0\,1\,0\,0 \end{array}, \quad
B_4 = \begin{array}{c} 1\,1\,1\,0 \\ 1\,1\,0\,0 \\ 0\,0\,0\,1 \end{array}, \quad
B_7 = \begin{array}{c} 1\,1\,1\,0 \\ 1\,0\,0\,1 \\ 0\,1\,0\,0 \end{array}, \quad
B_{10} = \begin{array}{c} 0\,1\,0\,1 \\ 1\,1\,0\,0 \\ 1\,0\,1\,0 \end{array},
$$

$$
B_2 = \begin{array}{c} 1\,0\,0\,0 \\ 0\,1\,0\,0 \\ 1\,1\,1\,1 \end{array}, \quad
B_5 = \begin{array}{c} 1\,1\,0\,0 \\ 0\,0\,0\,1 \\ 1\,1\,1\,0 \end{array}, \quad
B_8 = \begin{array}{c} 1\,0\,0\,1 \\ 0\,1\,0\,0 \\ 1\,1\,1\,0 \end{array}, \quad
B_{11} = \begin{array}{c} 1\,1\,0\,0 \\ 1\,0\,1\,0 \\ 0\,1\,0\,1 \end{array},
$$

$$
B_3 = \begin{array}{c} 0\,1\,0\,0 \\ 1\,1\,1\,1 \\ 1\,0\,0\,0 \end{array}, \quad
B_6 = \begin{array}{c} 0\,0\,0\,1 \\ 1\,1\,1\,0 \\ 1\,1\,0\,0 \end{array}, \quad
B_9 = \begin{array}{c} 0\,1\,0\,0 \\ 1\,1\,1\,0 \\ 1\,0\,0\,1 \end{array}, \quad
B_{12} = \begin{array}{c} 1\,0\,1\,0 \\ 0\,1\,0\,1 \\ 1\,1\,0\,0 \end{array}
$$

of the last four blocks form a basis for $Q^\omega(Z[3])$. Using 1-back splicing and the relations (8.3), (8.4) and (8.5), we see that these 12 blocks span $Q^\omega(Z[3])$, so it remains to show that they are linearly independent in $Q^\omega(3)$.

To do this we combine the information obtained from the action of several singular matrices. We use the singular matrices M_1, M_2, M_3, M_4 corresponding to the specializations of variables $x \mapsto y$, $x \mapsto z$, $y \mapsto z$, and $x \mapsto y+z$, the other two variables being fixed in each case. Thus let $F = \sum_{i=1}^{12} a_i B_i$, where $a_i \in \mathbb{F}_2$, be a linear combination of these 12 blocks which is 0 in $Q^\omega(3)$. In applying M_i we may ignore all terms arising from superposition of entries 1, since these give blocks with ω-sequence $<_l \omega$. Using the basis

$$A_1 = \begin{matrix} 1\ 1\ 1\ 1 \\ 1\ 1\ 0\ 0 \end{matrix}, \ A_2 = \begin{matrix} 1\ 1\ 0\ 0 \\ 1\ 1\ 1\ 1 \end{matrix}, \ A_3 = \begin{matrix} 1\ 1\ 1\ 0 \\ 1\ 1\ 0\ 1 \end{matrix}$$

of $Q^{18}(2)$, we obtain $B_i \cdot M_1 \sim A_1$ for $i = 6, 12$, $B_i \cdot M_1 \sim A_2$ for $i = 2$, $B_i \cdot M_1 \sim A_3$ for $i = 5, 8$ and $B_i \cdot M_1 \sim 0$ in the other 7 cases.

It follows from the results in the 2-variable case that $a_6 + a_{12} = a_2 = a_5 + a_8 = 0$. Using M_2 and M_3 and cyclic symmetry, we obtain $a_1 = a_2 = a_3 = 0$ and $a_4 = a_7 = a_{10}$, $a_5 = a_8 = a_{11}$ and $a_6 = a_9 = a_{12}$. Finally applying M_4 to the blocks B_i for $i > 3$ gives $a_6 + a_7 + a_{12} = a_4 + a_8 + a_{10} = \sum_{i=4}^{12} a_i = 0$. It follows that $a_i = 0$ for all i, and so the 12 blocks B_i are linearly independent. $\square$

Proposition 8.3.10 *The duplication map* $\delta : Q^\omega(3) \to Q^{(2)|\omega}(3)$ *is an isomorphism for* $\omega = (2,2,1)$ *and* $\omega = (2,2,1,1)$.

Proof In both cases it suffices to prove surjectivity. For $\omega = (2,2,1)$, $Q^{(2)|\omega}(3)$ has dimension 14 and a basis consisting of special blocks. The 6 spikes for ω and $(2)|\omega$ are matched up by δ, and require no further discussion. The 6 blocks $A_Y | B_Y$ in $Q^{(2)|\omega}(3)$ for $Y \neq Z[3]$ can be chosen as duplicates of blocks in $P^\omega(3)$ by taking the two 0 entries in the preferred row of A_Y to be in the first two columns.

The two remaining basis elements of $Q^{(2)|\omega}(3)$ are represented by the blocks

$$F_1 = \begin{matrix} 0\ 1\ 1\ 0 \\ 1\ 0\ 1\ 1 \\ 1\ 1\ 0\ 0 \end{matrix}, \ F_2 = \begin{matrix} 0\ 1\ 1\ 0 \\ 1\ 0\ 1\ 0 \\ 1\ 1\ 0\ 1 \end{matrix}$$

which are not duplicates. By splicing at $(2,2)$ and $(2,3)$,

$$F_1 \sim_l \begin{matrix} 0\ 1\ 0\ 1 \\ 1\ 1\ 1\ 0 \\ 1\ 0\ 1\ 0 \end{matrix} + \begin{matrix} 0\ 1\ 1\ 0 \\ 1\ 1\ 1\ 0 \\ 1\ 0\ 0\ 1 \end{matrix},$$

and by further splicing of the first block at $(3,2)$ and of the second at $(1,1)$ and $(1,2)$, we obtain

$$F_1 \sim_l \begin{matrix} 0\,0\,1\,1 & 1\,1\,0\,0 \\ 1\,1\,1\,0 & + & 1\,1\,1\,0 \\ 1\,1\,0\,0 & 0\,0\,1\,1 \end{matrix},$$

a sum of duplicate blocks. A similar argument shows that F_2 represents an element in the image of δ. This proves the claim for $\omega = (2,2,1)$. In the case of $\omega = (2,2,1,1)$, there are three non-duplicate special blocks $A_Y|B_Y$ for $Y = Z[3]$, and these can be treated in the same way. This completes the proof of Theorem 8.3.3. $\square$

Example 8.3.11 We determine $\dim Q^7(3)$. There are two sequences in $\mathrm{Dec}^7(3)$, namely $(3,2)$ and $(1,1,1)$. Hence $Q^7(3) \cong Q^{(3,2)}(3) \oplus Q^{(1,1,1)}(3)$. The three spikes x^3y^3z, x^3yz^3 and xy^3z^3 form a basis of $Q^{(3,2)}(3)$. By Theorem 6.7.6 we have $\dim Q^{(1,1,1)}(3) = 7$. Hence $\dim Q^7(3) = 10$.

8.4 Blocks of head length $n-1$

Definition 8.4.1 An n-block $F = H|B$ with head H of length $n-1$ and nonzero body B has **range** r if the first $r-1$ rows of H have all entries 1 and the corresponding rows of B are zero, but row r of B is nonzero. Such a block is **normalized** if H is normalized (Definition 6.8.4) and row r of B is of the form $1\cdots1\,0\cdots0$.

Not all blocks have a range. Range $r = 1$ means simply that the first row of B is nonzero. We shall assume that all blocks $F = H|B$ with range r arising in a splicing process are normalized. By Proposition 8.1.1, normalizing F by splicing operations in H and B separately does not change its equivalence class in $Q^\omega(n)$, where $\omega = \omega(F)$. For example, the 5-block

$$\begin{array}{c|c}
1\,1\,1\,1 & 0\,0 \\
1\,1\,1\,1 & 0\,0 \\
1\,0\,0\,0 & 1\,0 \\
0\,1\,1\,1 & 0\,1 \\
1\,1\,1\,1 & 1\,1
\end{array}$$

is normalized, with head of length 4 and range 3.

Proposition 8.4.2 *Let* $F = H|B$ *be a normalized n-block with head H of length* $n-1$ *and range r, where* $1 \leq r \leq n-2$. *Suppose that there are at least* $r+1$ *entries* 0 *in row r of H, and a single digit* 1 *in row r of B. Then F is left reducible to special blocks.*

Proof The argument is by induction on the range r. The given conditions imply that there are 0s in positions $(r, n-r-1)$ and $(r, n-r)$ of H. In particular, for $r = 1$, there are 0s at positions $(1, n-2)$ and $(1, n-1)$. Splicing at $(1, n-1)$ moves the 1 in position $(1, n)$ into position $(1, n-1)$, making the first row of the new body zero. Since there is also a 0 at $(1, n-2)$, the new block is special.

Let $r > 1$ and assume that the result is true for blocks of range $< r$. Let F be a block of range r satisfying the given conditions. Splicing F at $(r, n-r)$ produces blocks $F' = H'|B'$ by attacking a row $r' \neq r$ of F. Row r of B' is zero and row r of H' retains a 0 at $(r, n-r-1)$. If $r' > r$, then F' is special, while if $r' < r$ then F' satisfies the given conditions and has range $< r$. This completes the induction. $\qquad\square$

Proposition 8.4.2 does not say anything about blocks with range $n-1$ or n, since the conditions cannot be realized in these cases. To obtain a more comprehensive theorem, the special blocks have to be supplemented by additional blocks. Following N. Sum, we define these additional blocks as follows.

Definition 8.4.3 An n-block $F = H|B$ of range r with head H of length $n-1$ is a **Sum** block if H is normalized and has a 0 entry in each of its last $n-r+1$ rows.

The 5-blocks

$$
\begin{array}{c|c} 1\,1\,1\,1 & 0\,0 \\ 1\,1\,1\,1 & 0\,0 \\ 1\,1\,0\,0 & 1\,0 \\ 1\,0\,1\,1 & 0\,1 \\ 0\,1\,1\,1 & 1\,1 \end{array} \, ,
\quad
\begin{array}{c|c} 1\,1\,1\,1 & 0\,0 \\ 1\,1\,1\,1 & 0\,0 \\ 1\,0\,0\,0 & 1\,0 \\ 0\,1\,1\,1 & 0\,1 \\ 1\,1\,1\,1 & 1\,1 \end{array} \, ,
\quad
\begin{array}{c|c} 1\,1\,1\,1 & 0\,0 \\ 1\,1\,1\,1 & 0\,0 \\ 1\,1\,1\,1 & 1\,0 \\ 1\,0\,0\,0 & 0\,1 \\ 0\,1\,1\,1 & 1\,1 \end{array}
$$

have range 3 and heads of length 4. The first is a Sum block, but the other two are not. An n-block of range n is automatically a Sum block. A block cannot be special and a Sum block. There are two possibilities for the normalized head of an n-block of range $n-1$, as illustrated below for $n=5$.

$$
\begin{array}{cc|cc}
1\,1\,1\,1 & 0\,0 & 1\,1\,1\,1 & 0\,0 \\
1\,1\,1\,1 & 0\,0 & 1\,1\,1\,1 & 0\,0 \\
1\,1\,1\,1 & 0\,0 \;, & 1\,1\,1\,1 & 0\,0 \; . \\
0\,0\,0\,0 & 1\,0 & 1\,0\,0\,0 & 1\,0 \\
1\,1\,1\,1 & 1\,1 & 0\,1\,1\,1 & 1\,1 \\
\end{array}
$$

Either all entries in row $n-1$ of the head are 0, or $n-2$ of them are 0 and there is one 0 in row n. In the second case, the block is a Sum block. In the first case, splicing at $(n-1,1)$ produces either a Sum block, by attacking the last row, or a block of type discussed in Proposition 8.4.2, by attacking a row $< n-1$. These observations guide the proof of the next result.

Theorem 8.4.4 *Let $F = H|B$ be an n-block with head H of length $n-1$ and nonempty body B. Then F is left reducible to a sum of special blocks and Sum blocks.*

Proof The argument is by *downward* induction on the range r. The cases $r = n$ and $r = n-1$ are already proved above. Assume that the result is true for blocks with range $> r$, where $r \le n-2$, and consider a block $F = H|B$ of range r with head H of length $n-1$. We divide the argument into two cases.

Case 1: Suppose that all entries of row r of H are 1. In normalized form the 1s in row r of F are left justified. Let G be the block consisting of the last $n-r$ rows of F. We can specify the head of G to be of length $n-r$, so that by Proposition 8.1.7 G is reducible to special blocks by splicing processes with origin $\le r$.

The same is true of F unless such a splicing operation attacks a row lower than $n-r$. In this case, either a block to which Proposition 8.4.2 applies is produced, which is therefore reducible to special blocks, or row r itself is attacked. If row r is attacked, then the new block $F' = H'|B'$ has a single 1 in its body B', because the 1s are left justified in row r. Further row r of H' contains at least r 0s. Normalizing F' and then splicing at the first 0 in row r produces either blocks to which Proposition 8.4.2 applies, or blocks with range $> r$. In the latter case, either the inductive hypothesis applies, or they are special.

Case 2: Suppose that some entry of row r of H is 0. Then there must be at least $r-1$ zeros in that row, otherwise not all columns of H can have a zero. If there are exactly $r-1$ 0s, then F is a Sum block. Hence we may assume that H has at least r 0s in row r. Splicing with origin at the first such 0 creates blocks

$F' = H'|B'$ where all the entries of row r of H' are 1, and a row $r' \neq r$ of F is attacked. If $r' < r$, then F' has the type dealt with in Proposition 8.4.2, and so it is reducible to special blocks. If $r' > r$, then we are back in Case 1. □

We use the duplication map of Section 6.6 to show that the spanning set given by special blocks and Sum blocks is best possible.

Proposition 8.4.5 *Let* $\omega = \gamma|\rho \in \mathrm{Seq}(n)$ *where* $\gamma = (n-1,\ldots,n-1)$ *is a head sequence of length* $n-1$ *and* $\rho \neq 0$. *Then* (i) *the duplication map* $\delta : Q^{\omega}(n) \to Q^{(n-1)|\omega}(n)$ *is surjective, and* (ii) $\dim Q^{\omega}(n) \geq (2^n - 1)\dim Q^{\rho}(n-1)$.

Proof Let $A|B$ be a block in $P^{(n-1)|\omega}(n)$, so that A is a head block of size $h \geq n$. If A has two equal columns, then by Proposition 8.1.1 we may permute columns so that these are columns 1 and 2. Otherwise $h = n$ and A is a permutation matrix. Since $\rho \neq 0$, we may assume by 1-back splicing in B that there is an entry 1 in the first column of B, in row j say. By permuting columns in A, we may assume that the (j,n)th entry of A is 0. Then a 1-back splicing operation in row j results in blocks $A'|B'$ where the head A' has a duplicated column, together with blocks in $P^{<(n-1)|\omega}(n)$. This proves (i), and (ii) follows by applying Proposition 8.1.11. □

Theorem 8.4.6 *Let* $\omega = \gamma|\rho \in \mathrm{Seq}(n)$, *where* $\gamma = (n-1,\ldots,n-1)$ *is a head sequence of length* $n-1$ *and* $\rho \neq 0$. *Then* $\dim Q^{\omega}(n) = (2^n - 1)\dim Q^{\rho}(n-1)$.

Proof The lower bound is given by Proposition 8.4.5. For the upper bound, we assume as *downward* inductive hypothesis on ω that the theorem is true for $\omega' >_l \omega$. All such sequences ω' have head of length $\geq n-1$. Given a block $A|B$ in $P^{\omega}(n)$ with head A of length $n-1$ and body $B \neq 0$, $A|B$ is left reducible by Theorem 8.4.4 to a sum of special blocks and Sum blocks with ω-sequences $\omega' \geq \omega$. The result follows by induction on ω, together with Proposition 8.1.1. □

8.5 Remarks

Most of the results in Sections 8.1 and 8.2 were announced by M. Kameko at the Adams conference in Manchester in 1990. The dimensions of the spaces $Q^{\omega}(3)$ were determined by Kameko [106, 107]; see also Boardman [21]. In [108] Kameko tabulated $\dim Q^{\omega}(4)$ for all $\omega \in \mathrm{Seq}(4)$, which supported his conjecture [106] $\dim Q^d(n) \leq \prod_{i=1}^{n}(2^i - 1)$. Nguyen Sum [205, 206] found counterexamples to this conjecture for all $n > 4$ (see Section 24.5) using methods similar to those of Section 8.1. It remains a challenge to find the best upper bound for $\dim Q^{\omega}(n)$ and for $\dim Q^d(n)$.

The lower bound $\dim Q^d(n) \geq \prod_{i=1}^n (2^i - 1)$ was proved by Crabb and Hubbuck [41] under the conditions $d = (2^{a_1} - 1) + (2^{a_2} - 1) + \cdots + (2^{a_n} - 1)$ where $2^{a_1 - a_2} > n, 2^{a_2 - a_3} > n - 1, \ldots, 2^{a_{n-1} - a_n} > 2$ and $a_n \geq 0$. In [154] Tran Ngoc Nam proved Theorem 8.1.13, giving a monomial basis for $Q^d(n)$ in these cases. A proof of this result is also outlined in [235].

Section 8.4 is derived from the work of Sum [209], where a spanning set for $Q^d(n)$ of the type described in Theorem 8.4.4 is obtained under conditions on d similar to those of Theorem 8.1.13.

9

The dual of the hit problem

9.0 Introduction

The dual of $P^d(n)$ is the vector space $\mathrm{Hom}(P^d(n), \mathbb{F}_2)$ of linear maps $v : P^d(n) \to \mathbb{F}_2$, and the dual of the Steenrod square $Sq^k : P^d(n) \to P^{d+k}(n)$ is the linear map $Sq_k : \mathrm{Hom}(P^{d+k}(n), \mathbb{F}_2) \to \mathrm{Hom}(P^d(n), \mathbb{F}_2)$ defined by $Sq_k(u) = v$, where $v(f) = u(Sq^k(f))$ for $f \in P^d(n)$. Since Sq_k lowers degree by k, it is often called a 'down' squaring operation. We write the evaluation $v(f)$ of $\mathrm{Hom}(P^d(n), \mathbb{F}_2)$ on $P^d(n)$ as a bilinear form $\langle v, f \rangle$. Thus Sq_k is defined by $\langle Sq_k(u), f \rangle = \langle u, Sq^k(f) \rangle$. Since $\langle Sq_a Sq_b(u), f \rangle = \langle Sq_b(u), Sq^a(f) \rangle = \langle u, Sq^b Sq^a(f) \rangle$, the algebra generated by the down squares Sq_k is isomorphic to the opposite algebra A_2^{op}, and not to A_2 itself.

The vector space $\mathrm{Hom}(P^d(n), \mathbb{F}_2)$ has a natural basis dual to the monomial basis of $P^d(n)$. We denote by $v_1, \ldots, v_n$ the basis of $DP^1(n)$ dual to the basis $x_1, \ldots, x_n$ of $P^1(n)$, so that $\langle v_i, x_j \rangle$ is 1 if $i = j$ and is 0 if $i \neq j$. We write the dual of $x_1^{d_1} \cdots x_n^{d_n}$ as $v_1^{(d_1)} \cdots v_n^{(d_n)}$, where the parenthesized exponents are called **divided powers**. In the corresponding situation over a field of characteristic 0 in place of $\mathbb{F}_2$, $v_i^{(d)} = v_i^d / d!$, which fits with the formula $v^{(r)} v^{(s)} = \binom{r+s}{r} v^{(r+s)}$ for multiplying divided powers.

This product gives a commutative graded algebra $DP(n)$ over $\mathbb{F}_2$ called a **divided power algebra**, with $DP^d(n) = \mathrm{Hom}(P^d(n), \mathbb{F}_2)$. We call an element $v_1^{(d_1)} \cdots v_n^{(d_n)}$ in $DP(n)$ corresponding to a monomial $x_1^{d_1} \cdots x_n^{d_n}$ in $P(n)$ a **d-monomial**. We use the same n-block to represent a monomial and its corresponding d-monomial.

For example, Sq_1 is evaluated on $DP^3(2)$ as follows. Since $Sq^1(x_1 x_2) = x_1^2 x_2 + x_1 x_2^2$ and $Sq^1(x_1^2) = Sq^1(x_2^2) = 0$, it follows that $\langle Sq_1(v_1^{(2)} v_2), x_1 x_2 \rangle = \langle v_1^{(2)} v_2, Sq^1(x_1 x_2) \rangle = 1$ and $\langle Sq_1(v_1^{(2)} v_2), x_1^2 \rangle = \langle Sq_1(v_1^{(2)} v_2), x_2^2 \rangle = 0$. Hence $Sq_1(v_1^{(2)} v_2) = v_1 v_2$. Similar calculations give $Sq_1(v_1 v_2^{(2)}) = v_1 v_2$, $Sq_1(v_1^{(3)}) = 0$ and $Sq_1(v_2^{(3)}) = 0$.

By dualizing the action of a matrix $A \in M(n)$ on $P(n)$, we obtain a corresponding action on $DP(n)$. In order to have a right action of $\mathbb{F}_2 M(n)$ on

164

DP(n), rather than an action of the opposite algebra, we define $\langle u \cdot A, f \rangle = \langle u, f \cdot A^{\mathrm{tr}} \rangle$ for $u \in \mathrm{DP}^d(n)$ and $f \in \mathrm{P}^d(n)$, where A^{tr} is the transpose matrix of $A \in \mathrm{M}(n)$.

Thus we dualize the hit problem using commuting actions of A_2^{op} on the left and $\mathbb{F}_2 \mathrm{M}(n)$ on the right of DP(n). The dual problem is to find the **Steenrod kernel** K(n), the subspace of elements v of DP(n) such that $Sq_k(v) = 0$ for all $k > 0$. In the example above, $\mathrm{K}^3(2)$ is spanned by $v_1^{(3)}$, $v_2^{(3)}$ and $v_1^{(2)}v_2 + v_1 v_2^{(2)}$, and $\dim \mathrm{K}^3(2) = \dim \mathrm{Q}^3(2) = 3$. The elements $v_1^{(3)}$ and $v_2^{(3)}$ are examples of 'd-spikes' or duals of spikes, and all such elements are in K(n).

In Section 9.1 we describe the structure of DP(n) and the right action of $\mathbb{F}_2 \mathrm{M}(n)$ on it. In Section 9.3 we introduce the down squares Sq_k, and in Section 9.4 we define the Steenrod kernel K(n) and the submodule J(n) of K(n) generated by d-spikes. We also define the local Steenrod kernel $\mathrm{K}^\omega(n)$ as a subspace of the dual $\mathrm{DP}^\omega(n)$ of $\mathrm{P}^\omega(n)$, where ω is an n-bounded sequence. In Section 9.5 we consider the duals of the Kameko maps of Section 6.5 and the duplication map of Section 1.7, and solve the dual hit problem for $n = 2$. In Sections 9.6, 9.7 and 9.8 we dualize the results on the tail and head modules $\mathrm{Q}^\omega(n)$ of Chapter 6. In Section 9.9 we give a basis for $\mathrm{K}^d(3)$ for degrees d of the form $2^j - 1$.

9.1 The divided power algebra DP(n)

In this section we introduce the divided power algebra DP(n). This is the graded dual of P(n). We first define DP(n) as a graded vector space over $\mathbb{F}_2$.

Definition 9.1.1 For $n \geq 1$ and $d \geq 0$, we define $\mathrm{DP}^d(n) = \mathrm{Hom}(\mathrm{P}^d(n), \mathbb{F}_2)$, the linear dual of the $\mathbb{F}_2$-vector space $\mathrm{P}^d(n)$, and we define the **divided power algebra** $\mathrm{DP}(n) = \sum_{d \geq 0} \mathrm{DP}^d(n)$, with the product and grading defined below.

A **d-monomial** in DP(n) is an element of the basis of DP(n) dual to the monomial basis of P(n). The d-monomial dual to $x_1^{r_1} \cdots x_n^{r_n}$ is denoted by $v_1^{(r_1)} \cdots v_n^{(r_n)}$, where $v_1, \ldots, v_n$ is the basis of $\mathrm{DP}^1(n)$ dual to the basis $x_1, \ldots, x_n$ of $\mathrm{P}^1(n)$. For all $v \in \mathrm{DP}^1(n)$, $v^{(1)} = v$ and $v^{(0)}$ is the identity map of $\mathbb{F}_2$. An element of DP(n), or **d-polynomial**, is a finite sum of d-monomials. The d-monomial $v_1^{(r_1)} \cdots v_n^{(r_n)}$ has **degree** $d = r_1 + \cdots + r_n$, and a d-polynomial is **homogeneous** of degree d if it is a finite sum of d-monomials of degree d.

We introduce a product on DP(n) as follows, starting with the 1-variable case. Writing $v_1 = v$, we define the product of d-monomials in DP(1) by

$$v^{(r)} v^{(s)} = \binom{r+s}{r} v^{(r+s)} \tag{9.1}$$

where the binomial coefficient is taken mod 2. Thus by Proposition 1.4.13

$$v^{(r)}v^{(s)} = \begin{cases} v^{(r+s)}, & \text{if bin}(r) \text{ and bin}(s) \text{ are disjoint,} \\ 0, & \text{otherwise.} \end{cases} \tag{9.2}$$

In particular, $v^{(0)}$ is the identity element of DP(1). A nonzero d-polynomial $u \in \text{DP}(1)$ of degree d has the form $u = a_0 + a_1 v + a_2 v^{(2)} + \cdots + a_d v^{(d)}$, where $a_i \in \mathbb{F}_2$ for $0 \le i \le d$ and $a_d \ne 0$. If $a_0 = 0$, then $u^2 = 0$ by (9.2). The element $v^{(2^j)}$ is indecomposable for all $j \ge 0$, and if $\text{bin}(d) = \{2^{j_1}, \ldots, 2^{j_s}\}$ then $v^{(d)} = v^{(2^{j_1})} \cdots v^{(2^{j_s})}$.

If we carry out the same construction with coefficients in a field of characteristic 0 instead of $\mathbb{F}_2$, then since (9.1) gives $v^{(r)}v = (r+1)v^{(r+1)}$ for $r \ge 0$, we obtain $v^{(r)} = v^r/r!$ by induction on r. This explains why $v^{(r)}$ is called the **rth divided power** of v. The product in DP(n) is defined on d-monomials by using (9.1), or equivalently (9.2), for each variable v_i, and is extended to d-polynomials by linearity.

This product is commutative and associative, and DP(n) is a graded algebra over $\mathbb{F}_2$ for all $n \ge 1$. Since $\text{P}^0(n) = \mathbb{F}_2$, the identity element of DP(n) is the identity map of $\mathbb{F}_2$. We identify DP(0) with $\mathbb{F}_2$. We write the evaluation $u(f)$ of $u \in \text{DP}^d(n)$ on $f \in \text{P}^d(n)$ as a bilinear form $\langle u, f \rangle$, so that $\langle v_i, x_j \rangle = 1$ if $i = j$ and $\langle v_i, x_j \rangle = 0$ if $i \ne j$. The bijection $x_1^{r_1} \cdots x_n^{r_n} \leftrightarrow x^{r_1} \otimes \cdots \otimes x^{r_n}$ is an isomorphism of graded algebras $\text{P}(n) \cong \text{P}(1) \otimes \cdots \otimes \text{P}(1)$, and similarly $v_1^{(r_1)} \cdots v_n^{(r_n)} \leftrightarrow v^{(r_1)} \otimes \cdots \otimes v^{(r_n)}$ is an isomorphism of graded algebras DP(n) $\cong$ DP(1) $\otimes \cdots \otimes$ DP(1). Since $u^2 = 0$ for all d-polynomials u with constant term 0, exponents in the usual sense are not needed in DP(n). The algebra DP(n) is the tensor product of the exterior subalgebras generated by $v_i^{(2^j)}$ for $1 \le i \le n$ and $j \ge 0$.

In characteristic 0, since $v^{(r)} = v^r/r!$ the binomial theorem for divided powers is simply

$$(u+v)^{(r)} = \sum_{s+t=r} u^{(s)} v^{(t)}. \tag{9.3}$$

We use this to define the rth divided power as a function $\text{DP}^1(n) \to \text{DP}^r(n)$.

Definition 9.1.2 Let $v = v_{i_1} + \cdots + v_{i_s} \in \text{DP}^1(n)$. For $r \ge 0$, the **rth divided power** $v^{(r)} \in \text{DP}^r(n)$ is defined by

$$v^{(r)} = \sum_{r_1 + \cdots + r_s = r} v_{i_1}^{(r_1)} \cdots v_{i_s}^{(r_s)}.$$

Proposition 9.1.3 gives some properties of divided powers. We call an associative, commutative graded algebra A over a field F a **divided power algebra** over F if there are functions $A^1 \to A^r$ for $r \ge 0$ satisfying these conditions. Thus $DP(n)$ is a divided power algebra over $\mathbb{F}_2$. With divided powers defined by $f^{(r)} = f^r/r!$, a polynomial algebra over a field of characteristic 0 is also a divided power algebra.

Proposition 9.1.3 (i) *For all $v \in DP^1(n)$, $v^{(0)} = 1$ and $v^{(1)} = v$.*
   (ii) *For all $u, v \in DP^1(n)$ and all $r \geq 0$, $(u+v)^{(r)} = \sum_{s+t=r} u^{(s)} v^{(t)}$.*
   (iii) *For all $v \in DP^1(n)$ and all $r, s \geq 0$, $v^{(r)} v^{(s)} = \binom{r+s}{r} v^{(r+s)}$.*

Proof (i) is clear, since by definition $v_i^{(0)} = 1$ and $v_i^{(1)} = v_i$ for $1 \leq i \leq n$. For (ii), let $u = \sum_{i=1}^n a_i v_i$ and $v = \sum_{i=1}^n b_i v_i$, where $a_i, b_i \in \mathbb{F}_2$, so that $u + v = \sum_{i=1}^n (a_i + b_i) v_i$. Then $(u+v)^{(r)} = \sum_{r_1 + \cdots + r_n = r} \prod_{i=1}^n (a_i + b_i)^{r_i} v_i^{(r_i)}$, and

$$u^{(s)} v^{(t)} = \sum_{s_1 + \cdots + s_n = s} \prod_{i=1}^n a_i^{s_i} v_i^{(s_i)} \cdot \sum_{t_1 + \cdots + t_n = t} \prod_{i=1}^n b_i^{t_i} v_i^{(t_i)}.$$

Applying (9.1) to v_i for $1 \leq i \leq n$, the coefficient of $\prod_{i=1}^n v_i^{(r_i)}$ in $\sum_{r=s+t} u^{(s)} v^{(t)}$ is $\prod_{i=1}^n \sum_{s_i + t_i = r_i} \binom{r_i}{s_i} a_i^{s_i} b_i^{t_i}$. By the binomial theorem, this is $\prod_{i=1}^n (a_i + b_i)^{r_i}$.
   (iii) is true by definition when $v = v_i$ is a basis element. We show that if it holds for u and v then it also holds for $u + v$. Expanding the factors by (ii) and using (iii) for u and v, we have

$$(u+v)^{(r)} (u+v)^{(s)} = \sum_{r_1 + r_2 = r, s_1 + s_2 = s} u^{(r_1)} v^{(r_2)} u^{(s_1)} v^{(s_2)}$$

$$= \sum_{r_1 + r_2 = r, s_1 + s_2 = s} \binom{r_1 + s_1}{r_1} \binom{r_2 + s_2}{r_2} u^{(r_1 + s_1)} v^{(r_2 + s_2)}.$$

Let $t_1 + t_2 = r + s$. Expanding the identity $(x+y)^{t_1} (x+y)^{t_2} = (x+y)^{t_1 + t_2}$ by the binomial theorem and equating coefficients, the coefficient of $u^{(t_1)} v^{(t_2)}$ is

$$\sum_{r_1 + r_2 = r} \binom{t_1}{r_1} \binom{t_2}{r_2} = \binom{t_1 + t_2}{r} = \binom{r+s}{r}.$$

Since $(u+v)^{(r+s)} = \sum_{t_1 + t_2 = r+s} u^{(t_1)} v^{(t_2)}$ by (ii), this completes the proof. □

   The product (9.1) in $DP(n)$ can be understood in terms of $P(n)$, as follows. If A is an algebra over a field F, multiplication in A gives a linear map $\mu : A \otimes A \to A$, i.e. $\mu(a \otimes b) = ab$ where $a, b \in A$, and its dual can be regarded as a linear map $\mu^* : A^* \to A^* \otimes A^*$ by identifying $(A \otimes A)^*$ with $A^* \otimes A^*$. Here $A^* = \mathrm{Hom}(A, F)$ is the linear dual of A, and the identification is defined by $(\alpha \otimes \beta)(a \otimes b) = \alpha(a)\beta(b)$ where $a, b \in A$ and $\alpha, \beta \in A^*$. The map μ^* is called the **diagonal** or **coproduct** dual to the product μ. When A is a graded algebra, the algebra $A \otimes A$ is graded by $\deg(a \otimes b) = \deg a + \deg b$ where $a, b \in A$. Taking $A = DP(n)$ and $F = \mathbb{F}_2$, the product (9.1) in $DP(n)$ is dual to a coproduct $P(n) \to P(n) \otimes P(n)$.

Proposition 9.1.4 *The dual of the product map* $\mu : DP(n) \otimes DP(n) \to DP(n)$ *is the algebra map* $\mu^* : P(n) \to P(n) \otimes P(n)$ *defined by* $\mu^*(x_i) = x_i \otimes 1 + 1 \otimes x_i$, $1 \le i \le n$.

Proof In the case $n = 1$, let $x = x_1$ and $v = v_1$. Then $DP(1) \otimes DP(1)$ has basis $v \otimes 1, 1 \otimes v$ in degree 1, and since $1v = v1 = v$ the dual map μ^* satisfies $\mu^*(x) = x \otimes 1 + 1 \otimes x$. Hence μ^* is an algebra map if and only if $\mu^*(x^m) = (x \otimes 1 + 1 \otimes x)^m = \sum_{m=r+s} \binom{m}{r} x^r \otimes x^s$. This is the statement dual to (9.1). The result for n variables follows by identifying $P(n)$ with $P(1) \otimes \cdots \otimes P(1)$ and $DP(n)$ with $DP(1) \otimes \cdots \otimes DP(1)$. $\square$

Similarly, the product μ in $P(n)$ gives a coproduct on $DP(n)$.

Proposition 9.1.5 *The dual of the multiplication map* $\mu : P(n) \otimes P(n) \to P(n)$ *is the algebra map* $\mu^* : DP(n) \to DP(n) \otimes DP(n)$ *defined on the elements* $v_i^{(m)}$, $1 \le i \le n$, *by* $\mu^*(v_i^{(m)}) = \sum_{r+s=m} v^{(r)} \otimes v^{(s)}$.

Proof From the definition, the dual of the formula $\mu(x_1^r \otimes x_1^s) = x_1^{r+s}$ in $P(1)$ is $\mu^*(v_1^{(m)}) = \sum_{r+s=m} v_1^{(r)} \otimes v_1^{(s)}$ in $DP(1)$. Using the isomorphism $DP(n) \cong DP(1) \otimes \cdots \otimes DP(1)$, the proof that μ^* is an algebra map reduces to the case $n = 1$. Writing $v_1 = v$, we wish to show that $\mu^*(v^{(r)} v^{(s)}) = \mu^*(v^{(r)}) \mu^*(v^{(s)})$.

The left hand side is $\binom{r+s}{r} \mu^*(v^{(r+s)})$ by (9.1), while the right hand side is
$$\sum_{i+j=r} (v^{(i)} \otimes v^{(j)}) \sum_{i'+j'=s} (v^{(i')} \otimes v^{(j')}) = \sum_{i+j=r} \sum_{i'+j'=s} v^{(i)} v^{(i')} \otimes v^{(j)} v^{(j')} =$$
$$\sum_{i+j=r} \sum_{i'+j'=s} \binom{i+i'}{i} \binom{j+j'}{j} v^{(i+i')} \otimes v^{(j+j')} = \sum_{a+b=r+s} (\sum_{i+j=r} \binom{a}{i} \binom{b}{j}) v^{(a)} \otimes v^{(b)}.$$
Since the identity $(1+x)^a(1+x)^b = (1+x)^{a+b}$ gives $\sum_{i+j=r} \binom{a}{i} \binom{b}{j} = \binom{a+b}{r}$ and $a + b = r + s$, this is $\binom{r+s}{r} \mu^*(v^{(r+s)})$. $\square$

These results imply that $P(n)$ and $DP(n)$ are a dual pair of graded **Hopf algebras** (see Chapter 11).

9.2 The action of matrices on $DP(n)$

Using the basis $v_1, \ldots, v_n$ of $DP^1(n)$ dual to the basis $x_1, \ldots, x_n$ of $P^1(n)$, we define a right action of $M(n)$ on $DP^d(n)$ as follows.

Definition 9.2.1 For $A = (a_{i,j}) \in M(n)$ and $1 \leq i \leq n$, we define

$$v_i \cdot A = \sum_{j=1}^{n} a_{i,j} v_j,$$

and define $f \cdot A$ for $f \in DP(n)$ by substitution in d-monomials and linearity

This definition is an example of a general duality construction for right modules over $\mathbb{F}_2 GL(n)$ or $\mathbb{F}_2 M(n)$. The usual contragredient duality for right $\mathbb{F}_2 GL(n)$-modules uses the inverse matrix (see Remark 16.5.7), and so does not extend to right $\mathbb{F}_2 M(n)$-modules.

Definition 9.2.2 Let F be a field and V a right $FGL(n)$- or $FM(n)$-module. The **transpose dual** V^{tr} of V is the right $FGL(n)$- or $FM(n)$-module $\text{Hom}(V, F)$, with the action defined by $\langle f \cdot A, v \rangle = \langle f, v \cdot A^{\text{tr}} \rangle$ for $v \in V$, $f \in \text{Hom}(V, F)$ and $A \in GL(n)$ or $M(n)$.

Some general properties of transpose duality are given in Section 10.2.

Proposition 9.2.3 *With the action defined in Definition 9.2.1,* DP$^d(n)$ *is the transpose dual of* P$^d(n)$. *Further,* DP$^1(n) \cong$ P$^1(n)$ *as* $\mathbb{F}_2 M(n)$-*modules.*

Proof It is clear from the formal similarity of Definitions 1.2.1 and 9.2.1 that $DP^1(n)$ is a right $\mathbb{F}_2 M(n)$-module, and that the map $v_i \mapsto x_i$ for $1 \leq i \leq n$ is a $\mathbb{F}_2 M(n)$-module isomorphism. For $d > 1$, substitution in d-monomials is explained by Definition 9.1.2. Since this differs from the corresponding substitution for monomials in $P^d(n)$ only by the absence of binomial coefficients, $DP^d(n)$ is a right $\mathbb{F}_2 M(n)$-module with the action given by Definition 9.2.1.

It remains to prove transpose duality. By linearity, it suffices to check this for d-monomials. Thus we wish to prove that

$$\langle v_1^{(r_1)} \cdots v_n^{(r_n)} \cdot A, x_1^{s_1} \cdots x_n^{s_n} \rangle = \langle v_1^{(r_1)} \cdots v_n^{(r_n)}, x_1^{s_1} \cdots x_n^{s_n} \cdot A^{\text{tr}} \rangle \quad (9.4)$$

where $r_1 + \cdots + r_n = s_1 + \cdots + s_n$. Since we have an action of $M(n)$ on both sides, we need only check this for a generating set of matrices in $M(n)$.

The result is clear if A is a permutation matrix or a singular matrix fixing some of the variables and mapping the rest to 0. Thus it suffices to check (9.4) for the standard transvection U and $n = 2$.

For the left hand side, $v_1^{(r_1)} v_2^{(r_2)} \cdot U = (v_1 \cdot U)^{(r_1)} (v_2 \cdot U)^{(r_2)} = (v_1 + v_2)^{(r_1)} v_2^{(r_2)} = \sum_{r_1 = a+b} v_1^{(a)} v_2^{(b)} v_2^{(r_2)} = \sum_{r_1 = a+b} \binom{b+r_2}{b} v_1^{(a)} v_2^{(b+r_2)}$, and so the left hand side of (9.4) is $\binom{b+r_2}{b}$ if $a = s_1$ and $b + r_2 = s_2$, and is 0 otherwise. Since $r_1 + r_2 = s_1 + s_2$, this coefficient is $\binom{s_2}{s_2 - r_2}$ if $r_2 \leq s_2$, and is 0 otherwise.

For the right hand side, $(x_1^{s_1} x_2^{s_2}) \cdot U^{\mathrm{tr}} = (x_1 \cdot U^{\mathrm{tr}})^{s_1} (x_2 \cdot U^{\mathrm{tr}})^{s_2} = x_1^{s_1} (x_1 + x_2)^{s_2} = x_1^{s_1} \sum_{s_2 = c+d} \binom{s_2}{c} x_1^{s_1+c} x_2^d$, the right hand side of (9.4) is $\binom{s_2}{c}$ if $s_1 + c = r_1$ and $d = r_2$, and is 0 otherwise. Since $r_1 + r_2 = s_1 + s_2$, this coefficient is $\binom{s_2}{r_1 - s_1}$ if $r_1 \geq s_1$, and is 0 otherwise. Since $s_2 - r_2 = r_1 - s_1$, this completes the proof. $\square$

Remark 9.2.4 The $\mathbb{F}_2 \mathrm{GL}(n)$-modules $\mathrm{DP}^d(n)$ and $\mathrm{P}^d(n)$ are not in general isomorphic. For example, the module $\mathrm{DP}^n(n)$ contains the product $v_1 \cdots v_n$ of the variables, which is invariant under $\mathrm{GL}(n)$, whereas the $\mathrm{GL}(n)$-invariant of lowest positive degree in $\mathrm{P}(n)$ is the Dickson invariant $\mathrm{d}_{n,n-1}$ in degree 2^{n-1} (see Chapter 15). Hence $\mathrm{DP}^n(n)$ and $\mathrm{P}^n(n)$ are not isomorphic as $\mathrm{GL}(n)$-modules for $n \geq 3$.

Since it is defined by substitution of variables, the map $f \mapsto f \cdot A$ for $A \in \mathrm{M}(n)$ is an algebra map of $\mathrm{DP}(n)$. The next result shows that it is a map of divided power algebras.

Proposition 9.2.5 *Let* $v \in \mathrm{DP}^1(n)$ *and* $A \in \mathrm{M}(n)$. *Then* $v^{(r)} \cdot A = (v \cdot A)^{(r)}$ *for all* $r \geq 0$.

Proof This follows immediately from Definition 9.2.1 when $v = v_i$, $1 \leq i \leq n$. As for Proposition 9.2.3, we need only check the statement for a generating set of matrices, and again further details are needed only for the standard transvection U and $n = 2$. Thus we may take $v = v_1 + v_2$ and $A = U$, so $v \cdot U = v_1$. Then $v^{(r)} = (v_1 \cdot U)^{(r)} = v_1^{(r)} \cdot U$, and so $v^{(r)} \cdot U = (v_1^{(r)} \cdot U) \cdot U = v_1^{(r)}$ since $U^2 = I_n$. $\square$

9.3 Down squares Sq_k

In this section we define the down Steenrod squaring operations Sq_k on $\mathrm{DP}(n)$, and show that they satisfy the **dual Cartan formula**

$$Sq_k(uv) = \sum_{i+j=k} Sq_i(u) Sq_j(v)$$

for $u, v \in \mathrm{DP}(n)$. We also show that the down squares give a left action of the opposite algebra of A_2 on $\mathrm{DP}(n)$.

Definition 9.3.1 The **down Steenrod square** $Sq_k : \mathrm{DP}^{d+k}(n) \to \mathrm{DP}^d(n)$ is the linear dual of $Sq^k : \mathrm{P}^d(n) \to \mathrm{P}^{d+k}(n)$. Thus for $u \in \mathrm{DP}^{d+k}(n)$ and $f \in \mathrm{P}^d(n)$

$$\langle Sq_k(u), f \rangle = \langle u, Sq^k(f) \rangle.$$

Definition 9.3.2 The **total down square** $Sq_* : DP(n) \to DP(n)$ is the linear map $Sq_*(u) = \sum_{k \geq 0} Sq_k(u)$, where $u \in DP(n)$.

Thus $\langle Sq_*(u), f \rangle = \langle u, Sq(f) \rangle$ for $u \in DP(n)$ and $f \in P(n)$, so that Sq_* for $DP(n)$ is the graded dual of Sq for $P(n)$. Since $Sq^k : P^{d-k}(n) \to P^d(n)$ is 0 if $k > d - k$, $Sq_k = 0$ on $DP^d(n)$ if $k > d/2$. For example, $Sq_*(v^{(2^j)}) = v^{(2^j)} + \sum_{i=0}^{j-1} v^{(2^j - 2^i)}$.

Proposition 9.3.3 $Sq_* : DP(n) \to DP(n)$ *is an algebra map, i.e.* $Sq_*(uv) = Sq_*(u)Sq_*(v)$ *for all* $u, v \in DP(n)$.

Proof The result states that the diagram

$$
\begin{array}{ccc}
DP(n) \otimes DP(n) & \xrightarrow{\ \mu\ } & DP(n) \\
{\scriptstyle Sq_* \otimes Sq_*} \big\downarrow & & \big\downarrow {\scriptstyle Sq_*} \\
DP(n) \otimes DP(n) & \xrightarrow{\ \mu\ } & DP(n)
\end{array}
$$

is commutative. Taking graded duals, this is equivalent to commutativity of the diagram

$$
\begin{array}{ccc}
P(n) \otimes P(n) & \xleftarrow{\ \mu^*\ } & P(n) \\
{\scriptstyle Sq \otimes Sq} \big\uparrow & & \big\uparrow {\scriptstyle Sq} \\
P(n) \otimes P(n) & \xleftarrow{\ \mu^*\ } & P(n) \ .
\end{array}
$$

Thus we wish to prove that $\mu^*(Sq(f)) = (Sq \otimes Sq)(\mu^*(f))$ where $f \in P(n)$ and μ^* is the coproduct in $P(n)$. Since Sq and μ^* are multiplicative, it suffices to check this on the generators $x = x_i$, $1 \leq i \leq n$. We have $\mu^*(Sq(x)) = \mu^*(x + x^2) = \mu^*(x) + (\mu^*(x))^2 = (x \otimes 1 + 1 \otimes x) + (x^2 \otimes 1 + 1 \otimes x^2)$. Since also $(Sq \otimes Sq)(\mu^*(x)) = (Sq \otimes Sq)(x \otimes 1 + 1 \otimes x) = Sq(x) \otimes Sq(1) + Sq(1) \otimes Sq(x) = (x + x^2) \otimes 1 + 1 \otimes (x + x^2)$, this completes the proof. $\square$

Proposition 9.3.4 (Cartan formula for $DP(n)$) *For $k \geq 0$ and $u, v \in DP(n)$,*

$$
Sq_k(uv) = \sum_{i+j=k} Sq_i(u)Sq_j(v).
$$

Proof This follows from Proposition 9.3.3 by equating graded parts. $\square$

The down squaring operations commute with linear substitutions.

Proposition 9.3.5 $Sq_k : DP^{d+k}(n) \to DP^d(n)$ *is a right* $\mathbb{F}_2 M(n)$-*module map.*

Proof Let $u \in DP^{d+k}(n)$ and $A \in M(n)$. Then for all $f \in P^d(n)$, $\langle Sq_k(u) \cdot A, f \rangle = \langle Sq_k(u), f \cdot A^{tr} \rangle = \langle u, Sq^k(f \cdot A^{tr}) \rangle$ and $\langle Sq_k(u \cdot A), f \rangle = \langle u \cdot A, Sq^k(f) \rangle = \langle u, Sq^k(f) \cdot A^{tr} \rangle$. Since $Sq^k(f \cdot A^{tr}) = Sq^k(f) \cdot A^{tr}$ by Proposition 1.2.5, it follows that $Sq_k(u) \cdot A = Sq_k(u \cdot A)$. $\qquad\square$

In principle, the next two results enable us to calculate $Sq_k(u)$ for any d-polynomial u.

Proposition 9.3.6 *For all* $v \in DP^1(n)$,

$$Sq_k(v^{(d)}) = \binom{d-k}{k} v^{(d-k)}.$$

Proof First let $v = v_i$, $1 \le i \le n$. Then $\langle Sq_k(v_i^{(d)}), f \rangle = \langle v_i^{(d)}, Sq^k(f) \rangle$. Since $v_i^{(d)}$ is dual to x_i^d, this is 0 unless x_i^d appears in $Sq^k(f)$. By Propositions 1.1.8 and 1.1.11, this occurs only when $f = x_i^{d-k}$, and in this case $Sq^k(f) = \binom{d-k}{k} x_i^d$. Hence $Sq_k(v_i^{(d)}) = \binom{d-k}{k} v_i^{(d-k)}$.

For the general case, let $v = v_i \cdot A$ where $A \in M(n)$. Then $v^{(d)} = v_i^{(d)} \cdot A$ by Proposition 9.2.5. Hence $Sq_k(v^{(d)}) = Sq_k(v_i^{(d)}) \cdot A$ by Proposition 9.3.5, and this is $\binom{d-k}{k} v_i^{(d-k)} \cdot A = \binom{d-k}{k} v^{(d-k)}$ by Proposition 9.2.5. $\qquad\square$

Proposition 9.3.7 *For any d-monomial* $u = v_1^{(d_1)} \cdots v_n^{(d_n)}$ *in* $DP(n)$,

$$Sq_k(u) = \sum_{k_1 + \cdots + k_n = k} Sq_{k_1}(v_1^{(d_1)}) \cdots Sq_{k_n}(v_n^{(d_n)}).$$

Proof This follows from the Cartan formula 9.3.4 by induction on n. $\qquad\square$

The **opposite algebra** A^{op} of an algebra A over a field F is defined by reversing the product in A, i.e. the product $*$ in A^{op} is related to the product in A by $a * b = ba$ for all $a, b \in A$. Thus a right A-module M can be regarded as a left A^{op}-module by defining $a \cdot x = xa$ where $a \in A$ and $x \in M$. Let $A = A_2$ and $M = DP(n)$. Then for $u \in DP(n)$ and $f \in P(n)$ we have $\langle Sq_a Sq_b(u), f \rangle = \langle Sq_b(u), Sq^a(f) \rangle = \langle u, Sq^b Sq^a(f) \rangle$. Thus the down Steenrod squares define a left action of A_2^{op} on $DP(n)$. We write Sq_k for Sq^k when it is regarded as an element of A^{op}, as a reminder that it lowers degree by k, but in general elements of A_2^{op} such as Milnor basis elements $Sq(R)$ will be given the same notation as in A_2, and likewise the graded parts of A_2^{op} will be written simply as A_2^d for $d \ge 0$. We summarize these observations as follows.

Proposition 9.3.8 *The operations* $Sq_k : DP^{d+k}(n) \to DP^d(n)$ *define a left action of the opposite algebra* A_2^{op} *on* $DP(n)$. $\qquad\square$

We extend block notation for monomials in $P(n)$, together with α- and ω-sequences of blocks, to the corresponding d-monomials in $DP(n)$. For example, the blocks

$$\begin{matrix} 1\ 1\ 0 & & 0\ 0\ 1 & & 1\ 1\ 1 \\ 0\ 0\ 1 & , & 1\ 1\ 0 & , & 1\ 1\ 1 \end{matrix}$$

represent $x_1^3 x_2^4$, $x_1^4 x_2^3$ and $x_1^7 x_2^7$ in $P(2)$, and they also represent $u = v_1^{(3)} v_2^{(4)}$, $v = v_1^{(4)} v_2^{(3)}$ and $w = v_1^{(7)} v_2^{(7)}$ in $DP(2)$.

The product uv of d-monomials u and v is easy to describe in block notation. Since $uv = 0$ unless the binary expansions of the exponents of each variable in u and v are disjoint, and otherwise exponents are added as usual, we have the **superposition** rule for multiplication in $DP(n)$: regarding the 0 entries as empty positions, we superimpose each block of one d-polynomial on each block of the other, discarding all blocks where two 1s coincide. Thus $uv = w$ in the example above, while $uw = 0$ and $vw = 0$ in $DP(2)$.

Proposition 9.3.9 *Let $\theta \in A_2^+$ and let B be a block in the expansion of $\theta(A)$, where the block A corresponds to a d-monomial in $DP(n)$. Then $\omega(B) >_{l,r} \omega(A)$ and $\alpha_i(B) \geq \alpha_i(A)$ for $1 \leq i \leq n$.*

Proof Since A_2 is generated by the elements Sq^k, $k \geq 0$, it suffices to consider $\theta = Sq^k$, $k > 0$. The result follows in this case from Proposition 6.1.5, since Sq_k is the linear dual of Sq^k. $\qquad\qquad\qquad\qquad\qquad\qquad\qquad\qquad\square$

This result is dual to Proposition 6.1.5, and Proposition 9.3.10 is dual to Proposition 6.1.13. Note that the inequalities are reversed.

Proposition 9.3.10 *Let $M \in M(n)$ and B a block in the expansion of $A \cdot M$, where the block A corresponds to a d-monomial in $DP(n)$. Then $\omega(B) \geq_{l,r} \omega(A)$.*

Proof By Proposition 9.2.3, $A \to A \cdot M$ gives an action of $M(n)$ on $DP(n)$. Hence it suffices to check the statement on generators of $M(n)$. Permutations of a block do not alter the ω-sequence, and the singular matrices which map some variables to 0 and fix the rest either fix a d-monomial or map it to 0.

Thus to complete the proof, it suffices to consider the standard transvection U and the case $n = 2$. Then $v_1^{(a)} v_2^{(b)} \cdot U = (v_1 + v_2)^{(a)} v_2^{(b)} = \sum_{r+s=a} v_1^{(r)} v_2^{(s)} v_2^{(b)} = \sum_{r+s=a} \binom{s+b}{b} v_1^{(r)} v_2^{(s+b)}$. The coefficient $\binom{s+b}{b} = 1$ mod 2 if and only if $\mathrm{bin}(s)$ and $\mathrm{bin}(b)$ are disjoint, so the corresponding ω-sequences satisfy $\omega(s) + \omega(b) = \omega(s+b)$. Since $\omega(a) \leq_{l,r} \omega(r) + \omega(s)$, for any block B in the expansion of $A \cdot U$

we have $\omega(B) = \omega(r) + \omega(s + b) = \omega(r) + \omega(s) + \omega(b) \geq_{l,r} \omega(a) + \omega(b) = \omega(A)$. □

9.4 The Steenrod kernel $\mathsf{K}(n)$

In this section we relate the action of the operations Sq_k on $DP(n)$ to the hit problem. We define the **Steenrod kernel** $\mathsf{K}^d(n)$, the subspace of $DP^d(n)$ dual to the cohit quotient $Q^d(n)$ of $P^d(n)$, and introduce dual spike monomials or **d-spikes**. Thus the hit problem is equivalent to the problem of determining $\mathsf{K}^d(n)$.

Definition 9.4.1 The **Steenrod kernel** $\mathsf{K}(n) = \sum_{d \geq 0} \mathsf{K}^d(n)$, where $\mathsf{K}^d(n)$ is the set of elements $v \in DP^d(n)$ such that $Sq_k(v) = 0$ for all $k > 0$.

The Steenrod kernel $\mathsf{K}(n)$ is dual to the cohit module $Q(n)$, and has the extra feature of being closed under the product in $DP(n)$.

Proposition 9.4.2 (i) $\mathsf{K}(n)$ *is a subalgebra of* $DP(n)$.
(ii) $\dim \mathsf{K}^d(n) = \dim Q^d(n)$ *for* $n \geq 1$ *and* $d \geq 0$.
(iii) *For* $d \geq 0$, $\mathsf{K}^d(n)$ *is the transpose dual of* $Q^d(n)$ *as a right* $\mathbb{F}_2 M(n)$*-module.*

Proof (i) follows from Proposition 9.3.3, since $\mathsf{K}(n)$ is the fixed point set of the total down square Sq_*. For (ii), we first observe that $\mathrm{Hom}(Q^d(n), \mathbb{F}_2)$ is the subspace of $\mathrm{Hom}(P^d(n), \mathbb{F}_2) = DP^d(n)$ consisting of linear maps $P^d(n) \to \mathbb{F}_2$ whose kernel contains the hit polynomials $H^d(n)$. Then for $u \in DP^d(n)$, $\langle Sq_k(u), f \rangle = \langle u, Sq^k(f) \rangle = 0$ for all $k > 0$ and $f \in P^{d-k}(n)$ if and only if $u \in \mathsf{K}^d(n)$. Hence $\mathrm{Hom}(Q^d(n), \mathbb{F}_2) = \mathsf{K}^d(n)$. Finally (iii) follows from Proposition 9.2.3. □

Definition 9.4.3 A d-monomial $u \in DP(n)$ of the form $u = v_1^{(2^{j_1}-1)} \cdots v_n^{(2^{j_n}-1)}$ is called a **d-spike**.

Proposition 9.4.4 *A d-monomial is in* $\mathsf{K}(n)$ *if and only if it is a d-spike.*

Proof By Proposition 9.4.2(ii) and the solution of the hit problem in Proposition 1.4.12, $\mathsf{K}(1)$ has $\mathbb{F}_2$-basis $1, v_1, v_1^{(3)}, v_1^{(7)}, \ldots$. Alternatively, we can calculate directly using Proposition 9.3.6. If $d = 2^j - 1$ then $\mathrm{bin}(d - k)$ and $\mathrm{bin}(k)$ are disjoint, so $\binom{d-k}{k} = 0$ and hence $Sq_k(v_1^{(d)}) = 0$ for $k > 0$. For other d, let $k = 2^j$ be the smallest power of 2 not in $\mathrm{bin}(d)$. Then $k = 2^j \in \mathrm{bin}(d - k)$, and so $\binom{d-k}{k} = 1$ and $Sq_k(v_1^{(d)}) = v_1^{(d-k)}$.

Since $K(n)$ is a subalgebra of $DP(n)$, it contains all the d-spikes. Conversely, the Cartan formula 9.3.7 and Proposition 9.3.6 show that no other d-monomials are in $K(n)$.   □

Definition 9.4.5 The **d-spike module** $J(n)$ is the $\mathbb{F}_2 GL(n)$-submodule of $K(n)$ generated by the d-spikes.

Example 9.4.6 In Section 9.0 it was shown that $v_1^{(3)}, v_2^{(3)}$ and $v_1 v_2^{(2)} + v_1^{(2)} v_2$ form a basis for $K^3(2)$. Exchanging v_1 and v_2 switches the first two elements, and if U is the standard transvection in $GL(2)$, then $v_1^{(3)} \cdot U = (v_1 + v_2)^{(3)} = v_1^{(3)} + v_2^{(3)} + v_1^{(2)} v_2 + v_1 v_2^{(2)}$. Hence $J^3(2) = K^3(2)$ is the cyclic $\mathbb{F}_2 GL(2)$-module generated by $v_1^{(3)}$. Further $v_1 v_2^{(2)} + v_1^{(2)} v_2$ is invariant under $GL(2)$, and so generates a 1-dimensional submodule.

In the case $n = 2$, the dimension of $Q^d(2)$, and hence that of $K^d(2)$, is given by Theorem 1.8.2. It follows easily from Proposition 9.4.2 that $J(2) = K(2)$. Let $d = 2^{t_1} + 2^{t_2} - 2$. If $t_1 > t_2 + 1$ and $t_2 \geq 0$, then by Proposition 9.1.3, the element $(u + v)^{(2^{t_1}-1)} v^{(2^{t_2}-1)}$ contains the d-monomial $u^{(2^{t_1}-2^{t_2}-1)} v^{(2^{t_2+1}-1)}$, which is not a d-spike, while for $t_1 = t_2$ or $t_1 = t_2 + 1$ the d-spikes themselves suffice to span $K^d(2)$. However, we shall see in Section 10.3 that $J(3) \neq K(3)$.

Proposition 9.4.7 *Let* $u = v_1^{(2^{j_1}-1)} \cdots v_n^{(2^{j_n}-1)}$ *be a d-spike and let* $A \in GL(n)$. *Then* $u \cdot A = u$ *if* A *is upper triangular and* $j_1 \leq \cdots \leq j_n$, *or if* A *is lower triangular and* $j_1 \geq \cdots \geq j_n$.

Proof First consider the case where $n = 2$ and $A = U$, the standard transvection. Then $u \cdot U = (v_1 + v_2)^{(2^{j_1}-1)} v_2^{(2^{j_2}-1)} = \sum_{i+j=2^{j_1}-1} v_1^{(i)} v_2^{(j)} v_2^{(2^{j_2}-1)}$. Since $j \leq 2^{j_1} - 1 \leq 2^{j_2} - 1$, $\text{bin}(j) \subseteq \text{bin}(2^{j_2} - 1)$ and so $v_2^{(j)} v_2^{(2^{j_2}-1)} = 0$ by (9.2) if $j \neq 0$. Hence $u \cdot U = u$. The general case follows, since an upper triangular matrix A is a product of transvections $T_{i,j}$ such that $i < j$. The case where u is decreasing and A is lower triangular follows by a similar argument, or by writing $u = u' \cdot W_0$ and $A = W_0^{-1} B W_0$, where W_0 is the permutation matrix which reverses the order of the variables and B is upper triangular.   □

Proposition 9.4.8 *The d-spike module* $J(n)$ *is a subring of* $K(n)$.

Proof The superposition rule shows that the product $u_1 u_2$ of two d-spikes is a d-spike if u_1 and u_2 involve disjoint sets of variables, and otherwise $u_1 u_2 = 0$. Given d-spikes u_1, u_2 and matrices $A_1, A_2 \in GL(n)$, it suffices to prove that a nonzero product $(u_1 \cdot A_1)(u_2 \cdot A_2) \in J(n)$ is of the form $u \cdot A$, where u is a d-spike and $A \in \mathsf{GL}(n)$. By writing $(u_1 \cdot A_1)(u_2 \cdot A_2) = (u_1(u_2 \cdot A_2 A_1^{-1})) \cdot A_1$, we may assume that A_1 is the identity element, and so we start with $u_1(u_2 \cdot A)$.

Let W_1 be a permutation matrix such that $u_1 \cdot W_1$ has exponents in decreasing order. Since $(u_1(u_2 \cdot A)) \cdot W_1 = (u_1 \cdot W_1)(u_2 \cdot A W_1)$, by renaming elements we may assume that u_1 has decreasing exponents. Similarly, let W_2 be a permutation matrix such that $u_2 \cdot W_2$ has decreasing exponents. Writing $u_2 \cdot A = (u_2 \cdot W_2) \cdot W_2^{-1} A$, and renaming elements once more, we may restrict attention to $u_1(u_2 \cdot A)$, where both u_1 and u_2 have decreasing exponents.

By the Bruhat decomposition of a matrix $A \in \mathsf{GL}(n)$ (Proposition 16.2.7), $A = B_1 W B_2$, where B_1 and B_2 are lower triangular and W is a permutation matrix. By Proposition 9.4.7, u_1 and u_2 are fixed by B_1 and B_2. Hence $u_1(u_2 \cdot A) = u_1(u_2 \cdot B_1 W B_2) = u_1(u_2 \cdot W B_2) = (u_1 \cdot B_2)(u_2 \cdot W B_2) = (u_1(u_2 \cdot W)) \cdot B_2$. Since u_1 and $u_2 \cdot W$ are d-spikes, either $u = u_1(u_2 \cdot W)$ is 0 or it is a d-spike. Hence $u_1(u_2 \cdot A)$ is of the required form. $\qquad\square$

Certain other features of the hit problem for $\mathsf{P}(n)$ can be dualized for $\mathsf{DP}(n)$. For example, $\mathsf{DP}(n)$ is the direct sum of the subspaces $\mathsf{DP}(Y)$, for $Y \subseteq Z[n]$, spanned by d-monomials $v_1^{(r_1)} \cdots v_n^{(r_n)}$ such that $r_i > 0$ if and only if $i \in Y$. This induces a corresponding decomposition of $\mathsf{K}(n)$ which dualizes Proposition 1.4.7. Arguing as in Proposition 9.4.2(ii), $\dim \mathsf{K}^d(Y) = \dim \mathsf{Q}^d(Y)$, and so $\dim \mathsf{K}^d(n) = \sum_{k=1}^n \binom{n}{k} \dim \mathsf{K}^d(Z[k])$ for $d > 0$.

The filtration of $\mathsf{P}^d(n)$ given by a linear order relation $<$ on ω-sequences of monomials (see Section 6.2) can also be dualized. We define $\mathsf{DP}^\omega(n)$, $\mathsf{DP}^{>\omega}(n)$ and $\mathsf{DP}^{\geq \omega}(n)$ as the subspaces of $\mathsf{DP}(n)$ spanned by d-monomials which have ω-sequence ω, $> \omega$ and $\geq \omega$ respectively. By Proposition 9.3.10, $\mathsf{DP}^{>\omega}(n)$ and $\mathsf{DP}^{\geq \omega}(n)$ are $\mathsf{M}(n)$-modules. Although $\mathsf{DP}^\omega(n)$ is not closed under the action of $\mathsf{M}(n)$, it can be identified with the quotient module $\mathsf{DP}^{\geq \omega}(n)/\mathsf{DP}^{>\omega}(n)$. The proof of Proposition 9.2.3 shows that $\mathsf{DP}^{>\omega}(n)$ is the transpose dual of $\mathsf{P}^{<\omega}(n)$, and $\mathsf{DP}^{\geq \omega}(n)$ is the transpose dual of $\mathsf{P}^{\leq \omega}(n)$. Thus $\mathsf{DP}^\omega(n)$ is the transpose dual of $\mathsf{P}^\omega(n)$.

This filtration on $\mathsf{DP}^d(n)$ as a $\mathsf{M}(n)$-module gives a filtration on the Steenrod kernel $\mathsf{K}^d(n)$ by the submodules $\mathsf{DP}^{\geq \omega}(n) \cap \mathsf{K}^d(n)$, and gives a filtration on the d-spike module $\mathsf{J}^d(n)$ by the submodules $\mathsf{DP}^{\geq \omega}(n) \cap \mathsf{J}^d(n)$. We shall show that the quotient module of the filtration on $\mathsf{K}^d(n)$ corresponding to ω is the transpose dual of $\mathsf{Q}^\omega(n)$.

Definition 9.4.9 For a linear order $<$ on $\mathsf{Seq}(n)$ and $\omega \in \mathsf{Seq}(n)$ of 2-degree d, $\mathsf{K}^\omega(n) = (\mathsf{DP}^{\geq \omega}(n) \cap \mathsf{K}^d(n))/(\mathsf{DP}^{>\omega}(n) \cap \mathsf{K}^d(n))$, and $\mathsf{J}^\omega(n)$ is the submodule of $\mathsf{K}^\omega(n)$ generated by the d-spikes in $\mathsf{DP}^\omega(n)$.

According to this definition, a d-polynomial f representing an element of $K^\omega(n)$ is in $K^d(n)$. However, we cannot in general choose $f \in DP^\omega(n)$, since terms with ω-sequences $> \omega$ may be required. By omitting these terms, we obtain a polynomial $\bar{f}$ which generates a submodule of $DP^\omega(n)$ isomorphic to $K^\omega(n)$, but in general $\bar{f} \notin K^d(n)$. We shall meet an example of this situation in Proposition 10.6.3. It should also be noted that $J^\omega(n)$ may be nonzero even though ω is non decreasing, so that there are no d-spikes in $DP^\omega(n)$. By Proposition 8.3.1 this situation cannot arise for $n \leq 3$, but it does arise for $n = 4$ (see Example 29.1.2).

Proposition 9.4.10 *The filtration quotient* $K^\omega(n)$ *of* $K^d(n)$ *is the transpose dual of the filtration quotient* $Q^\omega(n)$ *of* $Q^d(n)$.

Proof $K^\omega(n) = A/(A \cap B)$ where $A = DP^{\geq\omega}(n) \cap K^d(n)$ and $B = DP^{>\omega}(n)$. Since $A/(A \cap B) \cong (A + B)/B$, this quotient is isomorphic as a $M(n)$-module to $(DP^{\geq\omega}(n) \cap K^d(n) + DP^{>\omega}(n))/DP^{>\omega}(n)$. The result follows since $DP^d(n)$ is the transpose dual of $P^d(n)$ and $K^d(n)$ is the transpose dual of $Q^d(n)$. $\qquad\square$

9.5 Dual Kameko and duplication maps

In this section we consider the duals of the Kameko maps κ and υ introduced in Section 6.5, and the duplication map δ of Section 6.6.

Definition 9.5.1 The **dual up Kameko map**, $\kappa_* : DP^d(n) \to DP^{2d+n}(n)$ is the dual of the down Kameko map $\kappa : P^{2d+n}(n) \to P^d(n)$, and the **dual down Kameko map** $\upsilon_* : DP^{2d+n} \to DP^d(n)$ is the dual of the up Kameko map $\upsilon : P^d(n) \to P^{2d+n}(n)$.

Thus κ_* maps d-monomials by $\kappa_*(v_1^{(a_1)} \cdots v_n^{(a_n)}) = v_1^{(2a_1+1)} \cdots v_n^{(2a_n+1)}$, while υ_* maps d-monomials with an even exponent to 0 and d-monomials with all exponents odd by $\upsilon_*(v_1^{(2a_1+1)} \cdots v_n^{(2a_n+1)}) = v_1^{(a_1)} \cdots v_n^{(a_n)}$. The following auxiliary map will also be useful.

Definition 9.5.2 The **halving map** $c : DP(n) \to DP(n)$ is the dual of the squaring map $f \to f^2$ of $P(n)$, i.e. $\langle c(u), f \rangle = \langle u, f^2 \rangle$ for all $u \in DP(n)$ and $f \in P(n)$.

Thus $c(v_1^{(2a_1)} \cdots v_n^{(2a_n)}) = v_1^{(a_1)} \cdots v_n^{(a_n)}$, and c maps all d-monomials with an odd exponent to 0.

Proposition 9.5.3 (i) *The halving map* $c: \mathrm{DP}(n) \to \mathrm{DP}(n)$ *is a surjection such that* $c(uv) = c(u)c(v)$ *for all* $u, v \in \mathrm{DP}(n)$,

(ii) $c(u \cdot A) = c(u) \cdot A$ *for all* $u \in \mathrm{DP}(n)$ *and* $A \in \mathrm{GL}(n)$,

(iii) *for all* $k \geq 0$, $c \circ Sq_{2k} = Sq_k \circ c$ *and* $c \circ Sq_{2k+1} = 0$.

Proof $\mathrm{DP}(n)$ is the tensor product of the exterior algebras generated by $v_i^{(2^j)}$ for $1 \leq i \leq n$ and $j \geq 0$. For $j > 0$, $c(v_i^{(2^j)}) = v_i^{(2^{j-1})}$, while $c(v_i) = 0$. This implies (i), and (ii) follows from the fact that $f^2 \cdot A = (f \cdot A)^2$ for $f \in \mathrm{P}(n)$ and $A \in \mathrm{GL}(n)$. For (iii), let $u \in \mathrm{DP}(n)$ and $f \in \mathrm{P}(n)$. Then $\langle c \circ Sq_{2k}(u), f \rangle = \langle Sq_{2k}(u), f^2 \rangle = \langle u, Sq^{2k}(f^2) \rangle = \langle u, (Sq^k(f))^2 \rangle = \langle c(u), Sq^k(f) \rangle = \langle Sq_k \circ c(u), f \rangle$. Similarly $Sq^{2k+1}(f^2) = 0$ for all f implies that $c \circ Sq_{2k+1} = 0$. $\qquad\square$

Proposition 9.5.4 (i) $\kappa_*: \mathrm{DP}^d(n) \to \mathrm{DP}^{2d+n}(n)$ *is a* $\mathbb{F}_2\mathrm{GL}(n)$*-module map, and it maps* $\mathrm{K}^d(n)$ *to* $\mathrm{K}^{2d+n}(n)$ *and* $\mathrm{J}^d(n)$ *to* $\mathrm{J}^{2d+n}(n)$,

(ii) $\kappa_* c(u) = v_1 \cdots v_n u$ *for all* $u \in \mathrm{DP}(n)$,

(iii) *for all* $k \geq 0$, $Sq_{2k} \circ \kappa_* = \kappa_* \circ Sq_k$ *and* $Sq_{2k+1} \circ \kappa_* = 0$.

Proof (i) By Proposition 6.5.2, κ is a $\mathbb{F}_2\mathrm{GL}(n)$-module map, and hence by duality so is κ_*. We have $\kappa_*(v_1^{(a_1)} \cdots v_r^{(a_r)}) = (v_1^{(2a_1+1)} \cdots v_r^{(2a_r+1)})v_{r+1} \cdots v_n$ for $1 \leq r \leq n$. Taking the exponents $a_1, \ldots, a_r$ to be of the form $2^j - 1$, it follows that κ_* maps d-spikes to d-spikes. Since the $\mathbb{F}_2\mathrm{GL}(n)$-module $\mathrm{J}(n)$ is generated by the d-spikes, κ_* maps $\mathrm{J}(n)$ to $\mathrm{J}(n)$.

(ii) We may assume by linearity that u is a d-monomial. If u has an odd exponent, then $c(u) = 0$ and $v_1 \cdots v_n u = 0$, and if $u = v_1^{(2a_1)} \cdots v_n^{(2a_n)}$ then $\kappa_* c(u) = \kappa_*(v_1^{(a_1)} \cdots v_n^{(a_n)}) = v_1^{(2a_1+1)} \cdots v_n^{(2a_n+1)} = v_1 \cdots v_n u$.

(iii) Since c is surjective, we may write an arbitrary element of $\mathrm{DP}(n)$ in the form $c(u)$. Thus let $c(u) \in \mathrm{K}^d(n)$. Then $Sq_{2k}(\kappa_* c(u)) = Sq_{2k}(v_1 \cdots v_n u) = v_1 \cdots v_n Sq_{2k}(u)$, by the Cartan formula, and this is $\kappa_* c(Sq_{2k}(u))$. By Proposition 9.5.3 $c(Sq_{2k}(u)) = Sq_k c(u) = 0$, since $c(u) \in \mathrm{K}^d(n)$. Hence $Sq_{2k}\kappa_* c(u) = 0$. Similarly $Sq_{2k+1}(\kappa_* c(u)) = \kappa_* c(Sq_{2k+1}(u))$, and $c(Sq_{2k+1}(u)) = 0$ by Proposition 9.5.3. The last statement is proved similarly. $\qquad\square$

In general, the dual υ_* of the up Kameko map does not map $\mathrm{K}^{2d+n}(n)$ to $\mathrm{K}^d(n)$. For example, $(u + v + w)^{(15)} \in \mathrm{K}^{15}(3)$ since it is of the form $u^{(15)} \cdot A$, $A \in \mathrm{GL}(3)$, but $\upsilon_*(v_1 + v_2 + v_3)^{(15)}$ is the sum of all d-monomials in $\mathrm{DP}^6(3)$, which is not in $\mathrm{K}^6(3)$ since $Sq_1(v^{(6)}) = v^{(5)}$ for $v \in \mathrm{DP}^1(3)$. However, the dual of Proposition 6.5.3 is true.

Proposition 9.5.5 *If* $\mu(2d+n) = n$, *then* υ_* *induces a* $\mathbb{F}_2\mathrm{GL}(n)$*-isomorphism* $\upsilon_*: \mathrm{K}^{2d+n}(n) \to \mathrm{K}^d(n)$ *which is the inverse of* κ_*.

Proof Given $u \in K^{2d+n}(n)$, we must show that $Sq_k(v_*(u)) = 0$ for $k > 0$. For all $f \in P^{d-k}(n)$ we have $\langle Sq_k(v_*(u)), f \rangle = \langle v_*(u), Sq^k(f) \rangle = \langle u, v(Sq^k(f)) \rangle$. By Proposition 6.5.3, the hypothesis $\mu(2d+n) = n$ ensures that $v(Sq^k(f))$ is hit. Writing $v(Sq^k(f)) = \sum_{i>0} Sq^i(h_i)$, $\langle Sq_k(v_*(u)), f \rangle = \sum_{i>0} \langle u, Sq^i(h_i) \rangle = \sum_{i>0} \langle Sq_i(u), h_i \rangle = 0$ since $Sq_i(u) = 0$. Hence $Sq_k(v_*(u)) = 0$ for all $k > 0$, i.e. $v_*(u) \in K^d(n)$. Clearly v_* and κ_* are inverse maps. $\square$

A similar formal duality argument dualizes Proposition 6.5.4.

Proposition 9.5.6 *For* $\omega \in \text{Seq}^d(n)$, $v_* : DP^{2d+n}(n) \to DP^d(n)$ *restricts to an isomorphism* $v_* : K^{(n)|\omega}(n) \to K^\omega(n)$ *for the left order, with inverse* κ_*. $\square$

We can also dualize the duplication map $\delta : P^d(n) \to P^{2d+n-1}(n)$ of Section 6.6.

Definition 9.5.7 The **d-duplication map** $\delta_* : DP^{2d+n-1}(n) \to DP^d(n)$ is the linear dual of the duplication map δ. Thus $\langle \delta_*(u), f \rangle = \langle u, \delta(f) \rangle$ for $u \in DP^{2d+n-1}(n)$ and $f \in P^d(n)$.

In terms of blocks, δ_* removes the first column of a block if its first two columns are identical, and maps it to 0 otherwise. It is not a $\mathbb{F}_2 GL(n)$-module map in general: for example, $\delta_*(uv^{(2)}) = 0$, but the standard transvection U maps $uv^{(2)}$ to $uv^{(2)} + v^{(3)}$ and $\delta_*(v^{(3)}) = v$.

The next result dualizes Propositions 6.6.4 and 6.6.5.

Proposition 9.5.8 *If* $\mu(d) = n-1$, *then*

(i) δ_* *restricts to a* $\mathbb{F}_2 GL(n)$-*module map* $\delta_* : K^{2d+n-1}(n) \to K^d(n)$,

(ii) *if* $\omega \in \text{Seq}_d(n)$ *with* $\omega_1 = n-1$, *then* δ_* *restricts to a* $\mathbb{F}_2 GL(n)$-*module map* $K^{(n-1)|\omega}(n) \to K^\omega(n)$.

Proof (i) Assume that $\mu(d) = n-1$, and let $u \in K^{2d+n-1}(n)$. Then for $k > 0$ and $f \in P^d(n)$, $\langle Sq_k \delta_*(u), f \rangle = \langle \delta_*(u), Sq^k(f) \rangle = \langle u, \delta(Sq^k(f)) \rangle$. By Proposition 6.6.4, $\delta(Sq^k(f))$ is hit. Writing $\delta(Sq^k(f)) = \sum_{i>0} Sq^i(h_i)$, $\langle Sq_k \delta_*(u), f \rangle = \sum_{i>0} \langle u, Sq^i(h_i) \rangle = \sum_{i>0} \langle Sq_i(u), h_i \rangle = 0$ since $Sq_i(u) = 0$. Hence $Sq_k \delta_*(u) = 0$. It follows that δ_* maps $K^{2d+n-1}(n)$ to $K^d(n)$. Since δ is a map of $\mathbb{F}_2 GL(n)$-modules, so is δ_*. This proves (i), and (ii) is proved similarly. $\square$

For comparison with the diagram for $Q^d(2)$ in Chapter 1, we show a basis for $K^d(2)$ for $d \leq 30$ and the maps κ_* and δ_*, writing u, v for the variables in $DP(2)$.

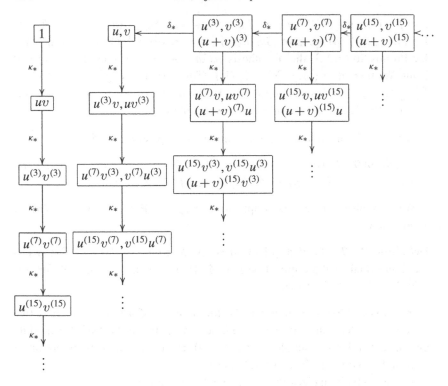

The elements listed are linearly independent, and each generates $K^d(2)$ as a cyclic $\mathbb{F}_2\mathrm{GL}(2)$-module. For all $d \geq 0$, $K^d(2)$ and $Q^d(2)$ are isomorphic $\mathbb{F}_2\mathrm{GL}(2)$-modules. For $i > j$, $(u+v)^{(2^i-1)}u^{(2^j-1)} = (u+v)^{(2^i-1)}v^{(2^j-1)}$ is the sum of all d-monomials $u^{(a)}v^{(b)}$ of degree $(2^i - 1) + (2^j - 1)$ which are divisible by $u^{(2^j-1)}v^{(2^j-1)}$. In particular, it is symmetric in u and v.

9.6 Tail and head sequences for $\mathrm{DP}(n)$

In this section we dualize the results of Section 6.9. Recall that a tail sequence $\tau = (1, \ldots, 1)$ of length $s \geq 1$ is the minimal element of $\mathrm{Seq}_d(n)$ for $d = 2^s - 1$, while a head sequence $\gamma = (n-1, \ldots, n-1)$ is the maximal element of $\mathrm{Seq}_d(n)$ for $d = (n-1)(2^s - 1)$. Since the left or right order on the spaces $\mathrm{DP}^\omega(n)$ is the reverse of that on the spaces $\mathrm{P}^\omega(n)$, $K^\tau(n)$ is a quotient space of $\mathrm{DP}^d(n)$ when τ is a tail sequence and $K^\gamma(n)$ is a subspace of $\mathrm{DP}^d(n)$ when γ is a head sequence.

The dimensions of these spaces are known from Theorems 6.7.6 and 6.8.7. By Definition 9.4.9, we have $K^\tau(n) = (K^d(n) + \mathrm{DP}^{>\tau}(n))/\mathrm{DP}^{>\tau}(n) \cong$

$K^d(n)/(K^d(n) \cap DP^{>\tau}(n))$ when τ is a tail sequence, and $K^\gamma(n) = DP^\gamma(n) \cap K^d(n)$ when γ is a head sequence.

Proposition 9.6.1 *Let $\tau = (1,\ldots,1)$ be a tail sequence of length $s \geq 1$. Then the equivalence classes mod $DP^{>\tau}(n)$ of the d-polynomials $(v_{i_1} + \cdots + v_{i_k})^{(2^s-1)}$, where $1 \leq i_1 < \cdots < i_k \leq n$ and $1 \leq k \leq s$, give a basis for $K^\tau(n)$. Thus $\dim K^\tau(n) = \sum_{k=1}^{n} \binom{n}{k}$. In particular, $\dim K^\tau(n) = 2^n - 1$ if $s \geq n$.*

Proof Let $Y = \{i_1,\ldots,i_k\}$ and $u = (v_{i_1} + \cdots + v_{i_k})^{(2^s-1)}$. Then u is obtained by the action of an element of $GL(n)$ on the d-spike $v_i^{(2^s-1)}$, where $i \in Y$, and hence $u \in K^d(n)$. By Proposition 9.1.3(ii), u is the sum of all d-monomials of degree $2^s - 1$ in $v_{i_1},\ldots,v_{i_k}$. In the notation of Section 6.7, a d-monomial in $DP^\tau(n)$ is a term of u if and only if its class $Y' \subseteq Y$. Hence the given d-polynomials u are linearly independent as elements of $K^\tau(n)$. Since $\dim K^\tau(n) = \dim Q^\tau(n)$, they span $K^\tau(n)$ by Theorem 6.7.6. □

Next we consider head sequences. In the case $s = 1$, $K^\gamma(n) = DP^\gamma(n)$ has dimension n, and the products of subsets of $n-1$ of the variables form a basis. For $s = 2$, $\dim K^\gamma(n) = \binom{n}{2} + n$ by Theorem 6.8.7, and a basis is given by the $\binom{n}{2}$ elements obtained from $(v_1 + v_2)^{(3)} v_3^{(3)} \cdots v_n^{(3)}$ by permuting the variables, together with the n products obtained similarly from $v_1^{(3)} \cdots v_{n-1}^{(3)}$. In the general case, we would like to find a basis consisting of products of $n-1$ factors of the form $v^{(2^s-1)}$ where $v \in DP^1(n)$, and which is symmetric, i.e. closed under permutations of the variables. The following observation allows us to do this.

Proposition 9.6.2 *For $s \geq k \geq 2$, the d-polynomial*

$$u(s,k) = (v_1 + v_2)^{(2^s-1)} \cdots (v_1 + v_k)^{(2^s-1)}$$

is the sum of all d-monomials in $DP(k)$ with ω-sequence $(k-1,\ldots,k-1)$ of length s. In particular, $u(s,k)$ is symmetric in $v_1,\ldots,v_k$.

Proof Since all terms in $(v_i + v_j)^{(2^s-1)}$ have ω-sequence $(1,\ldots,1)$ of length s, and the superposition rule for multiplication in $DP(n)$ implies that the ω-sequence of a product of d-monomials is the sum of the ω-sequences of the factors, every d-monomial in $u(s,k)$ has ω-sequence $(k-1,\ldots,k-1)$ of length s. Conversely, given such a d-monomial $v = v_1^{(a_1)} \cdots v_k^{(a_k)}$, a unique term of $u(s,k)$ equal to v is determined by $a_2,\ldots,a_k$. □

Proposition 9.6.3 *Let $\gamma = (n-1,\ldots,n-1)$ be a head sequence of length $s \geq 1$. Then the d-polynomials obtained by permuting the variables $v_1,\ldots,v_n$ in the d-polynomials $u(s,k)v_{k+1}^{(2^s-1)} \cdots v_n^{(2^s-1)}$ for $1 \leq k \leq s$, where $u(s,1) =$*

$v_1^{(2^s-1)}$, *form a basis for* $\mathsf{K}^\gamma(n)$. *Thus* $\dim \mathsf{K}^\gamma(n) = \sum_{k=1}^s \binom{n}{k}$, *and in particular* $\dim \mathsf{K}^\gamma(n) = 2^n - 1$ *if* $s \geq n$.

Proof The listed d-polynomials are products of factors which are given by acting on d-spikes by elements of $\mathbb{F}_2\mathrm{GL}(n)$. Hence they are in $\mathsf{K}^d(n)$. All their terms are in $\mathrm{DP}^\gamma(n)$, since all terms in $v_i^{2^s-1}$ have ω-sequence $(1,\dots,1)$, and no superposition of 1s can occur, so the ω-sequence of a product is the sum of the ω-sequences of its factors. By Proposition 9.6.2 the number of distinct d-polynomials that are obtained from $u(s,k)v_{k+1}^{(2^s-1)}\cdots v_n^{(2^s-1)}$ by permuting the variables is $\binom{n}{k}$. Hence the listed d-polynomials span $\mathsf{K}^\gamma(n)$. $\qquad\square$

In particular, $\mathsf{K}^\omega(n) = \mathsf{J}^\omega(n)$ for all $n \geq 1$ when ω is a tail or head sequence.

9.7 Dual tail modules

In this section we consider the $\mathbb{F}_2\mathrm{GL}(n)$-module structure of $\mathsf{K}^\tau(n)$, where $\tau = (1,\dots,1)$ is a tail sequence. The injective $\mathbb{F}_2\mathrm{GL}(n)$-module map $\overline{\epsilon}$: $\mathsf{Q}^\tau(n) \to \mathsf{Q}^{(1)|\tau}(n)$ of Proposition 6.9.2 can be dualized to give a surjective $\mathbb{F}_2\mathrm{GL}(n)$-module map $\mathsf{K}^{(1)|\tau}(n) \to \mathsf{K}^\tau(n)$. To describe this map, recall that $\overline{\epsilon}$ is induced by the linear map $\epsilon : \mathsf{P}^\tau(n) \to \mathsf{P}^{(1)|\tau}(n)$ which duplicates the first column of a block in $\mathsf{P}^\tau(n)$. We define $\epsilon_* : \mathrm{DP}^{(1)|\tau}(n) \to \mathrm{DP}^\tau(n)$ as the linear map which removes the first column of a block if it is the same as the second column, and otherwise maps the block to 0.

Thus ϵ_* maps the d-spike $v_i^{(2d-1)}$ to the d-spike $v_i^{(d)}$, where $d = \deg_2 \tau$. In general, the element $u = (v_{i_1} + \cdots + v_{i_k})^{(2d+1)}$ of $\mathsf{K}^{2d+1}(n)$ is the sum of all d-monomials in $v_{i_1},\dots,v_{i_k}$ of degree $2d + 1$. The equivalence class of u in $\mathsf{K}^\tau(n) = \mathsf{K}^{2d+1}(n)/(\mathsf{K}^{2d+1}(n) \cap \mathrm{DP}^{>\tau}(n))$ is the sum of all d-monomials in u with ω-sequence $(1)|\tau$, and so its image under ϵ_* is the sum of all d-monomials in $(v_{i_1} + \cdots + v_{i_k})^{(d)}$ with ω-sequence τ.

Proposition 9.7.1 *If* $\tau = (1,\dots,1)$ *is a tail sequence of length* $s \geq 1$, *then* $\epsilon_* : \mathrm{DP}^{(1)|\tau}(n) \to \mathrm{DP}^\tau(n)$ *restricts to a surjection* $\epsilon_* : \mathsf{K}^{(1)|\tau}(n) \to \mathsf{K}^\tau(n)$, *which is an isomorphism if* $s \geq n$.

Proof This follows from the preceding discussion using Proposition 9.6.1. $\quad\square$

Thus we have a sequence of surjections

$$\mathsf{K}^{(1,\dots,1)}(n) \xrightarrow{\;\epsilon_*\;} \cdots \xrightarrow{\;\epsilon_*\;} \mathsf{K}^{(1,1)}(n) \xrightarrow{\;\epsilon_*\;} \mathsf{K}^{(1)}(n) \qquad\qquad (9.5)$$

dual to (6.4), where the first sequence $(1,\ldots,1)$ has length n and the last term $K^{(1)}(n) = \mathrm{DP}^1(n)$. The next result identifies the successive kernels. This can also be deduced from Proposition 6.9.3 using Proposition 9.2.3 and the fact that the transpose dual $M^{\mathrm{tr}} \cong M$ when M is an irreducible $\mathbb{F}_2\mathrm{GL}(n)$-module (see Proposition 10.2.1(iii)).

Proposition 9.7.2 *Let* $\tau = (1,\ldots,1)$ *have length* $s - 1$, $2 \leq s \leq n$. *Then the kernel of* $\epsilon_* : K^{(1)|\tau}(n) \to K^{\tau}(n)$ *is isomorphic to* $\Lambda^s(n)$ *as an* $\mathbb{F}_2\mathrm{GL}(n)$-*module.*

Proof Given $Y = \{i_1,\ldots,i_k\} \subseteq Z[n]$, let $v_Y = v_{i_1} + \cdots + v_{i_k} \in \mathrm{DP}^1(n)$, and let $v_Y = 0$ if $Y = \emptyset$. The basis elements of $K^{(1)|\tau}(n)$ given by Proposition 9.6.1 can be written as $v_Y^{(2^s-1)}$ where $|Y| \leq s$. Given Y with $|Y| \leq s$, let $w(Y) = \sum_{X \subseteq Y} v_X^{(2^s-1)}$.

We claim that the elements $w(Y)$, where $|Y| = s$, form a basis of the kernel of $\epsilon_* : K^{(1)|\tau}(n) \to K^{\tau}(n)$. By Proposition 9.1.3(ii), $v_Y^{(2^s-1)}$ is the sum of all d-monomials of degree $2^s - 1$ in the variables v_i, $i \in Y$. If $Y' \subseteq Y$ and u is a d-monomial of degree $2^s - 1$ in the variables v_i, $i \in Y'$, then u is a term of $v_X^{(2^s-1)}$ where $X \subseteq Y$ if and only if $Y' \subseteq X$. Hence u is a term of $w(Y)$ if and only if the number of sets X such that $Y' \subseteq X \subseteq Y$ is odd.

This occurs if and only if $Y' = Y$. Thus $w(Y)$ is the sum of all the d-monomials of degree $2^s - 1$ which involve all the variables v_i for $i \in Y$. Since there is one such d-polynomial for each subset $Y \subseteq Z[n]$ with $|Y| = s$ and $\dim \mathrm{Ker}(\epsilon_*) = \binom{n}{s}$, this proves the claim.

Let $\phi : \mathrm{Ker}(\epsilon_*) \to \Lambda^s(n)$ be the linear map which assigns to $w(Y)$ the element $u(Y) = x_{i_1} \wedge \cdots \wedge x_{i_s} \in \Lambda^s(n)$, where $Y = \{i_1,\ldots,i_s\}$ with $1 \leq i_1 < \cdots < i_s \leq n$. Clearly ϕ is a vector space isomorphism which commutes with the action of permutation matrices. Thus, as in the proof of Proposition 6.9.3, it suffices to check that ϕ commutes with the action of the standard transvection U, and again we separate three cases.

Case 1: $1 \notin Y$. In this case $u(Y) \cdot U = u(Y)$. Since $w(Y)$ does not involve v_1, $w(Y) \cdot U = w(Y)$.

Case 2: $2 \in Y$, $1 \in Y$. Again $u(Y) \cdot U = u(Y)$. Since $v_1 \cdot U = v_1 + v_2$ and $v_1 + v_2 \cdot U = v_1$, U permutes the elements v_X for $X \subseteq Y$ by fixing v_X if $1 \notin X$ and exchanging v_X with $V_{X'}$ if $1 \in X$, where X' is obtained from X by inserting or removing 2. Thus $w(Y) = \sum_{X \subseteq Y} v_X^{(2^s-1)}$ is fixed by U.

Case 3: $2 \notin Y$, $1 \in Y$. In this case $u(Y) \cdot U = u(Y) + u(Y')$, where Y' is obtained from Y by replacing 1 by 2. Let $Y'' = Y \cup \{2\} = Y' \cup \{1\}$. Then $w(Y'') = A + B + C + D$ where A, B, C, D are the sums of the d-polynomials $v_X^{(2^s-1)}$ over subsets $X \subseteq Y''$ such that $1 \notin X$ and $2 \notin X$, $1 \in X$ and $2 \notin X$, $1 \notin X$ and $2 \in X$,

$1 \in X$ and $2 \in X$ respectively. Then $w(Y'') = A + B + C + D$, $w(Y) = A + B$, $w(Y') = A + C$, and $w(Y) \cdot U = A + D$, and so $w(Y) \cdot U = w(Y) + w(Y') + w(Y'')$.

However, $w(Y'') = \sum_{X \subseteq Y''} v_X^{(2^s - 1)}$ is the sum of all d-monomials of degree $2^s - 1$ involving all $s + 1$ variables v_i where $i \in Y^+$. These d-monomials must have ω-sequence $> (1)|\tau$. Hence $w(Y) \cdot U = w(Y) + w(Y')$ in $\mathsf{K}^{(1)|\tau}(n)$. $\qquad\square$

9.8 Dual head modules

In this section we consider the $\mathbb{F}_2 \mathrm{GL}(n)$-module structure of $\mathsf{K}^\gamma(n)$, where $\gamma = (n-1, \ldots, n-1)$ is a head sequence. We have a sequence of n surjections

$$\mathsf{K}^{(n-1,\ldots,n-1)}(n) \xrightarrow{\delta_*} \cdots \xrightarrow{\delta_*} \mathsf{K}^{(n-1,n-1)}(n) \xrightarrow{\delta_*} \mathsf{K}^{(n-1)}(n) \xrightarrow{\delta_*} 0 \qquad (9.6)$$

where the first sequence $(n-1, \ldots, n-1)$ has length n. The kernel of the sth d-duplication map $\delta_* : \mathsf{K}^{(n-1)|\gamma}(n) \to \mathsf{K}^\gamma(n)$ has dimension $\binom{n}{s}$, where $\mathrm{len}(\gamma) = s - 1$.

The $\mathbb{F}_2 \mathrm{GL}(n)$-module $\mathsf{K}^{(n-1)}(n) = \mathrm{DP}^{(n-1)}(n)$ has a basis consisting of products of $n-1$ of the variables $v_1, \ldots, v_n$, and is isomorphic to the $(n-1)$st exterior power $\Lambda^{n-1}(n)$ of the defining module $V(n) \cong \mathsf{P}^1(n)$. We identify $\Lambda^{n-1}(n)$ with its transpose dual $\mathrm{DP}^1(n)$ by means of the isomorphism $x_i \leftrightarrow v_i$, $1 \leq i \leq n$. We shall prove that the kernel $\mathrm{Ker}(\delta_*)$ of the sth d-duplication map is isomorphic to $\Lambda^{n-s}(n)$ for $1 \leq s \leq n$.

We begin by describing a basis for $\mathrm{Ker}(\delta_*)$. Given $Y \subseteq Z[n]$ with $|Y| = s$, let $u(Y) \in \mathrm{DP}^{(n-1)|\gamma}(n)$ be the sum of all d-monomials corresponding to blocks B in the head class Y. That is, B has exactly one 0 in the ith row if $i \in Y$, and all entries in the ith row are 1 if $i \notin Y$.

Example 9.8.1 For $n = 4$ and $Y = \{1, 2, 3\}$,

$$
u(Y) = \begin{matrix} 0\,1\,1 \\ 1\,0\,1 \\ 1\,1\,0 \\ 1\,1\,1 \end{matrix} + \begin{matrix} 0\,1\,1 \\ 1\,1\,0 \\ 1\,0\,1 \\ 1\,1\,1 \end{matrix} + \begin{matrix} 1\,0\,1 \\ 0\,1\,1 \\ 1\,1\,0 \\ 1\,1\,1 \end{matrix} + \begin{matrix} 1\,0\,1 \\ 1\,1\,0 \\ 0\,1\,1 \\ 1\,1\,1 \end{matrix} + \begin{matrix} 1\,1\,0 \\ 0\,1\,1 \\ 1\,0\,1 \\ 1\,1\,1 \end{matrix} + \begin{matrix} 1\,1\,0 \\ 1\,0\,1 \\ 0\,1\,1 \\ 1\,1\,1 \end{matrix}.
$$

Thus $u(Y)$ is the sum of $s!$ d-monomials, and is symmetric in the variables $v_i, i \in Y$. It is clear that the $\binom{n}{s}$ elements $u(Y)$ for $Y \subseteq Z[n]$ with $|Y| = s$ are linearly independent. Since none of the blocks in $u(Y)$ are duplicates, $u(Y)$ is in the kernel of $\delta_* : \mathrm{DP}^{(n-1)|\gamma}(n) \to \mathrm{DP}^\gamma(n)$.

Proposition 9.8.2 *The elements $u(Y)$ for $Y \subseteq Z[n]$ with $|Y| = s$ form a basis for* $\mathrm{Ker}(\delta_*)$.

Proof By the preceding remarks, we need only prove that $u(Y) \in K^{(n-1)|\gamma}(n)$. Since the action of Sq_k cannot decrease the number of 1s in any row, Sq_1 acts on a block in $u(Y)$ by exchanging the digit 0 in the first column with the digit 1 in the same row and the second column.

More generally, $Sq_{2^{j-1}}$ for $1 \le j \le s-1$ acts by exchanging the digit 0 in column j with the digit 1 in the same row and column $j+1$, and maps all blocks in $u(Y)$ to 0 if $j \ge s$. Thus the sum of a pair of blocks in $u(Y)$ which are exchanged by switching the first two columns is in the kernel of Sq_1, and more generally the sum of a pair of blocks in $u(Y)$ which are exchanged by switching columns j and $j+1$ is in the kernel of $Sq_{2^{j-1}}$. It follows that $u(Y)$ is in the kernel of $Sq_{2^{j-1}}$ for all $j \ge 1$. Hence $u(Y) \in K^{(n-1)|\gamma}(n)$. $\square$

Proposition 9.8.3 *Let* $\gamma = (n-1, \ldots, n-1)$ *be a head sequence of length* $s-1$, *where* $1 \le s \le n$. *Then the kernel of* $\delta_* : K^{(n-1)|\gamma}(n) \to K^{\gamma}(n)$ *is isomorphic to* $\Lambda^{n-s}(n)$ *as a* $\mathbb{F}_2 GL(n)$*-module.*

Proof Since $\text{Ker}(\delta_*)$ has a basis consisting of the elements $u(Y)$, where $Y \subseteq Z[n]$ and $|Y| = s$, a linear map $\phi : \text{Ker}(\delta_*) \longrightarrow \Lambda^{n-s}(n)$ is defined by $\phi(u(Y)) = v_{j_1} \wedge \cdots \wedge v_{j_{n-s}}$ where $j_1 < \cdots < j_{n-s}$ are the elements of $Z[n] \setminus Y$. We show that ϕ is a $\mathbb{F}_2 GL(n)$-isomorphism.

Proposition 9.8.2 shows that ϕ is a vector space isomorphism, and it is clear from the definition of $u(Y)$ that ϕ commutes with permutations of the variables. Using our standard argument on generators of $GL(n)$, it remains to check that ϕ commutes with the standard transvection U, which maps v_1 to $v_1 + v_2$ and fixes v_i for $i > 1$. Recall that U fixes $v = v_{j_1} \wedge \cdots \wedge v_{j_{n-s}}$ unless $j_1 = 1$ and $j_2 > 2$, when it maps v to $v + w$ where w is obtained by replacing v_1 by v_2 in v.

By the superposition rule for multiplication in $DP(n)$, U fixes a head block B unless it has a column with a 1 in the first row and a 0 in the second row, and in this case $B \cdot U = B + B'$, where B' is the block obtained by exchanging these two entries. Thus if the block B appears in $u(Y)$, then $B \cdot U = B$ if $2 \notin Y$, and $B \cdot U = B + B'$ if $2 \in Y$. As in Proposition 6.9.5, we separate three cases.

Case 1: $2 \notin Y$. In this case all entries in the second row of a block B in $u(Y)$ are 1, so $B \cdot U = B$ and so U fixes $u(Y)$.

Case 2: $2 \in Y, 1 \in Y$. We must show that U fixes $u(Y)$. In this case both the first and second rows of a block B in $u(Y)$ contain a 0 entry, so $B \cdot U = B + B'$ where B' has two 0s in the first row and all 1s in the second row. The same extra block B' occurs when U is applied to the block obtained by exchanging the first two rows of B. Hence the extra blocks B' cancel in pairs, so that $u(Y) \cdot U = u(Y)$.

Case 3: $2 \in Y, 1 \notin Y$. We must show that $u(Y) \cdot U = u(Y) + u(Z)$ where Z is obtained by replacing 2 by 1 in Y. In this case the first row of a block B in $u(Y)$ contains only 1s, while the second row has an entry 0. Thus $B \cdot U = B + B'$, where B' is obtained by exchanging the first two rows of B. Clearly $u(Z)$ is the sum of the blocks B' which arise in this way. $\qquad\square$

Finally we show that $\mathsf{K}^\gamma(n) = \mathsf{J}^\gamma(n)$ when γ is a head sequence.

Proposition 9.8.4 *For $1 \leq s \leq n$, the elements $u(Y)$ for $Y \subseteq Z[n]$ with $|Y| = s$ are in the d-spike module $\mathsf{J}^\gamma(n)$, where $\gamma = (n-1,\ldots,n-1)$ is a head sequence of length s.*

Proof For $s = 1$, $u(Y)$ is the product of the variables $v_i, i \notin Y$, and so it is a d-spike. In general we may assume by permuting the variables that $Y = \{1, 2, \ldots, s\}$. For $s \geq 2$, let

$$v(Y) = (v_1 + v_2)^{(2^s-1)}(v_1 + v_3)^{(2^s-1)} \cdots (v_1 + v_s)^{(2^s-1)}. \qquad (9.7)$$

We show that $v(Y)$ is the sum of all d-monomials in $DP^Y(s)$. In particular, it is symmetric in $v_1, \ldots, v_s$. To see this, note that the only term not involving v_1 is $v_2^{(2^s-1)} \cdots v_s^{(2^s-1)}$, which is represented by the $s \times s$ block with all 0 entries in the first row and all other entries 1. All other blocks representing terms in (9.7) are obtained by exchanging 0s and 1s in columns of this block, and each resulting block is obtained by a unique set of such exchanges. These blocks are exactly the blocks which represent d-monomials in $DP^Y(s)$.

Let $w(Y)$ be the product of $v(Y)$ with $v_{s+1}^{(2^s-1)} \cdots v_n^{(2^s-1)}$. Since $w(Y)$ is obtained by applying the element of $GL(n)$ which maps v_i to $v_i + v_1$ for $2 \leq i \leq s$, and fixes v_i otherwise, to the d-spike $v_2^{(2^s-1)} \cdots v_n^{(2^s-1)}$, it follows that $w(Y) \in \mathsf{J}^\gamma(n)$.

The element $w(Y)$ is the sum of all d-monomials in $P^Y(n)$ with all entries 1 in rows $> s$. This is the sum of all d-monomials in classes $C(Z)$ where $Z \subseteq Y$ and $Z \neq \emptyset$. For each such Z, let $v(Z)$ be the d-polynomial obtained by replacing $v_1, \ldots, v_s$ in (9.7) by the variables v_i for $i \in Z$ if $|Z| > 1$ and by $v(Z) = 1$ if $|Z| = 1$. We also define $w(Z)$ as the product of $v(Z)$ with $\prod_{j \notin Z} v_j^{(2^s-1)}$. Then $w(Z) \in \mathsf{J}^\gamma(n)$, and since $u(Y)$ is the sum of all d-monomials in class $C(Y)$, it follows from the inclusion-exclusion principle that $u(Y) = \sum_{Z \subseteq Y} w(Z)$. Hence $u(Y) \in \mathsf{J}^\gamma(n)$. $\qquad\square$

9.9 $\mathsf{K}^d(3)$ for $d = 2^s - 1$

To determine $\mathsf{K}^d(3)$ for all $d \geq 0$, it suffices by the Peterson conjecture (Theorem 2.5.5) and iteration of the Kameko map κ_* to consider degrees d such that $\mu(d) = 1$ or 2. Writing $d = (2^{s+t} - 1) + (2^t - 1)$ where $s, t \geq 0$, we give a basis for $\mathsf{K}^d(3)$ in the case $t = 0$ in this section, in the case $t = 1$ in

Section 10.3 and in the case $t \geq 2$ in Section 10.4. If $t > 0$, there is only one decreasing ω-sequence $\omega = (2,\ldots,2,1,\ldots,1)$ in $\mathrm{Seq}_d(3)$, with s 1s and t 2s, and $\mathsf{K}^d(3) = \mathsf{K}^\omega(3)$.

Clearly $\mathsf{K}^0(3) = \mathrm{DP}^0(3) = \mathbb{F}_2$ and $\mathsf{K}^1(3) = \mathrm{DP}^1(3)$. When $s \geq 2$ there are two decreasing ω-sequences, $\tau = (1,\ldots,1)$, of length s, and $\gamma = (3,2,\ldots,2)$, of length $s-1$, and the $\mathrm{GL}(3)$-module $\mathsf{K}^d(3)$ has a submodule $\mathsf{K}^\gamma(3)$, the Kameko image of a head module, and a quotient $\mathsf{K}^d(3)/\mathsf{K}^\gamma(3) \cong \mathsf{K}^\tau(3)$, a tail module.

We denote the basis elements v_1, v_2, v_3 of $\mathrm{DP}^1(3)$ by u, v, w. When $s = 2$, $\mathsf{K}^\gamma(3) = \mathrm{DP}^{(3)}(3)$ has dimension 1 and is generated by uvw, and $\mathsf{K}^\tau(3) = K^{(1,1)}(3)$ has dimension 6. The 6 elements $u^{(3)}, v^{(3)}, w^{(3)}, (u+v)^{(3)}, (u+w)^{(3)}, (v+w)^{(3)}$ are in $\mathsf{J}^3(3)$ and $\mathrm{DP}^\tau(3)$ and are linearly independent, so their equivalence classes mod uvw form a basis of $\mathsf{K}^\tau(3)$.

For $s = 3$, $\mathsf{K}^7(3)$ has dimension 10, with a submodule $\mathsf{K}^\gamma(3)$ of dimension 3 and a quotient $\mathsf{K}^\tau(3)$ of dimension 7. The submodule $\mathsf{K}^\gamma(3) = \kappa_*(\mathsf{K}^2(3))$. Since uv, uw, vw is a basis of $\mathsf{K}^2(3)$, $u^{(3)}v^{(3)}w, u^{(3)}vw^{(3)}, uv^{(3)}w^{(3)}$ is a basis of $\mathsf{K}^\gamma(3)$ and $u^{(7)}, v^{(7)}, w^{(7)}, (u+v)^{(7)}, (u+w)^{(7)}, (v+w)^{(7)}, (u+v+w)^{(7)}$ is a basis of $\mathsf{K}^\tau(3)$.

For $s = 4$, $\mathsf{K}^{15}(3)$ has dimension 13, with a submodule $\mathsf{K}^\gamma(3)$ of dimension 6, and a quotient $\mathsf{K}^\tau(3)$ of dimension 7. The submodule $\mathsf{K}^\gamma(3) = \kappa_*(\mathsf{K}^6(3))$. The module $\mathsf{K}^6(3) = K^{(2,2)}(3)$ is a head module with basis $u^{(3)}v^{(3)}, u^{(3)}w^{(3)}, v^{(3)}w^{(3)}, u^{(3)}(v+w)^{(3)}, v^{(3)}(u+w)^{(3)}, w^{(3)}(u+v)^{(3)}$. Hence $\mathsf{K}^\gamma(3)$ has basis $u^{(7)}v^{(7)}w, u^{(7)}vw^{(7)}, uv^{(7)}w^{(7)}, u^{(7)}v(v+w)^{(7)}, uv^{(7)}(u+w)^{(7)}, w^{(7)}u(u+v)^{(7)}$, and the quotient $\mathsf{K}^\tau(3)$ has basis $u^{(15)}, v^{(15)}, w^{(15)}, (u+v)^{(15)}, (u+w)^{(15)}, (v+w)^{(15)}, (u+v+w)^{(15)}$.

For $s \geq 5$, $\mathsf{K}^{2^s-1}(3)$ has dimension 14, with a submodule $\mathsf{K}^\gamma(3)$ of dimension 7, and a quotient $\mathsf{K}^\tau(3)$ of dimension 7. The submodule $\mathsf{K}^\gamma(3) = \kappa_*(\mathsf{K}^{2^{s-1}-2}(3))$. The module $\mathsf{K}^{2^{s-1}-2}(3) = K^{(2,\ldots,2)}(3)$ is a head module with basis consisting of the elements $w_1^{(2^{s-2}-1)}w_2^{(2^{s-2}-1)}$ where the pair of vectors (w_1, w_2) takes the values $(u,v), (u,w), (v,w), (u,v+w), (v,u+w), (w,u+v), (u+v,u+w)$. Hence $\mathsf{K}^\gamma(3)$ has basis consisting of the elements $uvw \cdot w_1^{(2^{s-1}-2)}w_2^{(2^{s-1}-2)}$ for the same seven values of (w_1, w_2). (The terms with odd exponent in the binomial expansions are annihilated by multiplication by uvw.) The quotient $\mathsf{K}^\tau(3)$ has basis $u^{(k'')}, v^{(k'')}, w^{(k'')}, (u+v)^{(k'')}, (u+w)^{(k'')}, (v+w)^{(k'')}, (u+v+w)^{(k'')}$ where $k'' = 2^s - 1$.

9.10 Remarks

The hit problem in its dual form has been studied by Alghamdi, Crabb and Hubbuck [7], [41], Boardman [21] and Repka and Selick [173]. The Steenrod kernel is called the space of primitive homology classes of $RP^\infty \times \cdots \times RP^\infty$ in [21], and the submodule generated by the dual spikes is called the 'ring of lines' in [41]. Some authors (e.g. Ault and Singer [14]) avoid working with A_2^{op} by writing Steenrod operations on $DP(n)$ on the right.

10

K(3) and Q(3) as $\mathbb{F}_2\mathrm{GL}(3)$-modules

10.0 Introduction

In this chapter we determine the $\mathbb{F}_2\mathrm{GL}(3)$-module structure of the cohit quotient $Q^d(3)$ of $P^d(3)$ and the Steenrod kernel $K^d(3)$ in $DP^d(3)$ for $d \geq 0$. As finite-dimensional modules, $P^d(3)$ and $DP^d(3)$ have composition series with factors isomorphic to one of four irreducible $\mathbb{F}_2\mathrm{GL}(3)$-modules. However, the dimension of $P^d(3)$ and $DP^d(3)$ becomes arbitrarily large as d increases, while $Q^d(3)$ and $K^d(3)$ have dimension ≤ 21 for all d. Thus we can hope to describe the structure of $Q^d(3)$ and $K^d(3)$ in fairly explicit terms.

Every irreducible $\mathbb{F}_2\mathrm{GL}(3)$-module is isomorphic to one of the four modules $I(3)$, $V(3)$, $V(3)^*$ or $St(3)$. We recall that $I(3)$ is the trivial 1-dimensional module, $V(3)$ the defining module of dimension 3 and $V(3)^*$ its contragredient dual. The fourth irreducible module $St(3)$ is the Steinberg module, of dimension 8.

For the trivial $\mathbb{F}_2\mathrm{GL}(n)$-module $I(n)$, all matrices act as the identity map. For the defining module $V(n)$, the action $v \cdot A$ of $A \in \mathrm{GL}(n)$ on a row vector v is the usual matrix multiplication. For the contragredient dual of a module V, a matrix A acts on $V^* = \mathrm{Hom}(V, \mathbb{F}_2)$ by $\langle f \cdot A, v \rangle = \langle f, v \cdot A^{-1} \rangle$, where $v \in V, f \in V^*$ and $\langle f, v \rangle = f(v)$. For the defining module $V(n)$, $V(n)^* \cong \Lambda^{n-1}(n)$, the $(n-1)$st exterior power of $V(n)$. In particular $V(3)^* \cong \Lambda^2(3)$. We give a construction for $St(3)$ in Section 10.5. Examples of all four modules appear as composition factors in $P^d(3)$ and $DP^d(3)$ for $d \leq 4$; $I(3)$ for $d = 0$, $V(3)$ for $d = 1$, $V(3)^*$ as $Q^2(3)$ or $K^2(3)$ and $St(3)$ as $Q^4(3)$ or $K^4(3)$.

It is a standard result of representation theory (see Proposition 18.4.9 and Section 18.7) that the Steinberg module $St(n)$ of $\mathrm{GL}(n)$ is a direct summand of any module in which it occurs as a composition factor. Hence an indecomposable summand of $P^d(3)$ and $DP^d(3)$ is either isomorphic to $St(3)$ or all its composition factors are isomorphic to $I(3)$, $V(3)$ or $V(3)^*$. By the

Krull–Schmidt theorem, the number of summands of each type is the same for all direct sum decompositions. For a head or tail sequence ω, we have seen in Sections 6.9, 9.7 and 9.8 that all composition factors of $Q^\omega(n)$ and $K^\omega(n)$ are exterior powers of $V(n)$ for all $n \geq 1$.

Sections 10.1 and 10.2 contain preliminary results, on representations of a finite 2-group G, and on transpose duality for $\mathbb{F}_2$GL(n)- and $\mathbb{F}_2$M(n)-modules. As $K^d(3)$ and $Q^d(3)$ are transpose duals, each determines the structure of the other by reversing composition series. The results of Section 10.1 are applied to linear independence questions in DP(3) in Sections 10.3 and 10.4.

Since the Kameko maps for P(n) and DP(n) are maps of $\mathbb{F}_2$GL(n)-modules, we may assume in studying $K^d(3)$ and $Q^d(3)$ that $\mu(d) = 1$ or 2. In the 'generic' case $d = (2^{s+t} - 1) + (2^t - 1)$, where $s \geq 2$ and $t \geq 2$, $Q^d(3)$ and $K^d(3)$ have dimension 21. We shall show that both are isomorphic to the **flag module** FL(3), the permutation representation of GL(3) on the right cosets of the subgroup L(3) of lower triangular matrices. Since $|GL(3)| = 168$ and $|L(3)| = 8$, $\dim FL(3) = 21$. In Section 10.5 we prove that FL(3) has four indecomposable summands

$$
I(3) \oplus \begin{array}{ccc} V(3) & V(3)^* & \\ \downarrow & \oplus & \downarrow \\ V(3)^* & V(3) & \end{array} \oplus St(3), \qquad (10.1)
$$

the second and third being transpose duals. When $\mu(d) = 2$, the d-spike module $J^d(3)$ is a quotient of FL(3). The results of Sections 9.9, 10.3 and 10.4 show that $K^d(3) = J^d(3)$ except in the case $d = 8$ and its iterates by the Kameko map, when $\dim(K^d(3)/J^d(3)) = 1$. We tabulate these results in Theorem 10.6.2. Sections 10.6 and 10.7 provide details of the module structure of $K^d(3)$ and $Q^d(3)$ in various cases.

Although we restrict attention to GL(3) in describing the module structure of P(3) and DP(3), it is straightforward to obtain corresponding results for M(3). The examples in degrees $d \leq 4$ discussed above also give four of the eight isomorphism types of $\mathbb{F}_2$M(3)-modules: in these cases, there are polynomials f and singular matrices A such that $f \cdot A \neq 0$. Examples of the other four are obtained by applying the up Kameko map $\upsilon(f) = x_1 x_2 x_3 f^2$ to these modules: in these cases, $f \cdot A = 0$ for all polynomials f and singular matrices A. For example, $P^1(3) \cong V(3)$ is the defining module for M(3), while its partner, the quotient module $P^{(3,1)}(3)$ of $P^5(3)$, is isomorphic to $V(3) \otimes \det$.

10.1 Some irreducible $\mathbb{F}_2\mathsf{GL}(n)$-modules

In this section we prove some general results from group representation theory which are useful in proving that certain $\mathbb{F}_2\mathsf{GL}(n)$-modules are irreducible, and that certain elements in the Steenrod kernel $\mathsf{K}(n)$ are linearly independent. The first result shows that the only irreducible representation over $\mathbb{F}_2$ of a finite group whose order is a power of 2 is the trivial 1-dimensional representation.

Proposition 10.1.1 *Let G be a finite 2-group and let V be a nonzero $\mathbb{F}_2G$-module. Then V contains a nonzero G-invariant, i.e. an element $v \in V$ such that $v \cdot g = v$ for all $g \in G$.*

Proof The equivalence relation $x \sim y$ on V given by $x \cdot g = y$ for some $g \in G$ expresses V as a finite union of G-orbits. Since the stabilizer of x is a subgroup of G, its order is a 2-power, and so the order of each orbit is also a 2-power. Thus the 2-power $|V|$ is expressed as a sum of 2-powers. Since $0 \in V$ is fixed by G, 1 is a term in this sum, and so there must be at least one other orbit of size 1. The corresponding elements are G-invariants. $\qquad\square$

Proposition 10.1.2 *Let G be a finite 2-group, let $\overline{G} = \sum_{g \in G} g \in \mathbb{F}_2G$ and let V be a right $\mathbb{F}_2G$-module. Let $x \in V$ and let X be the cyclic submodule of V generated by x. If $x \cdot \overline{G} \neq 0$, then $X \cong \mathbb{F}_2G$. In particular,*

(i) *the elements $x \cdot g$, $g \in G$, are linearly independent,*
(ii) *the only nonzero G-invariant in X is $x \cdot \overline{G}$,*
(iii) *if $y \neq 0 \in X$, then there is an element $\gamma \in \mathbb{F}_2G$ such that $y \cdot \gamma = x \cdot \overline{G}$.*

Proof Define a map $\theta : \mathbb{F}_2G \to V$ of $\mathbb{F}_2G$-modules by $\theta(1_G) = x$. Then $\mathrm{Im}(\theta) = X$ is spanned by the elements $x \cdot g$, $g \in G$, of the G-orbit of x. The map θ is injective if and only if the elements $x \cdot g$ are linearly independent. If θ is not injective, then $\mathrm{Ker}(\theta)$ contains a nonzero G-invariant, by Proposition 10.1.1. However, $\overline{G}$ is the only nonzero G-invariant in $\mathbb{F}_2G$. Hence if $\theta(\overline{G}) = x \cdot \overline{G} \neq 0$, θ is injective, and (i) and (ii) follow. Finally (iii) follows by applying Proposition 10.1.1 to the $\mathbb{F}_2G$-submodule generated by y. $\qquad\square$

Proposition 10.1.3 *Let V be a $\mathbb{F}_2\mathsf{GL}(n)$-module containing a unique nonzero element v which is invariant under the subgroup $\mathsf{L}(n)$ of lower triangular matrices and which generates V. Then V is irreducible.*

Proof Let W be a nonzero submodule of V. By restricting the action of $\mathsf{GL}(n)$ on W to that of $\mathsf{L}(n)$, we can regard W as a $\mathbb{F}_2\mathsf{L}(n)$-module. Since $\mathsf{L}(n)$ is a 2-group, W contains a nonzero $\mathsf{L}(n)$-invariant by Proposition 10.1.1. Since v is the unique such element in V, $v \in W$. Since v generates V, $W = V$. Hence V is irreducible. $\qquad\square$

Proposition 10.1.4 *For* $1 \leq s \leq n$, *the sth exterior power module* $\Lambda^s(n)$ *of the defining module* $\mathsf{V}(n)$ *is an irreducible* $\mathbb{F}_2\text{GL}(n)$-*module.*

Proof Given a basis $v_1, \ldots, v_n$ for $\mathsf{V}(n)$, the $\binom{n}{s}$ elements $v_I = v_{i_1} \wedge \cdots \wedge v_{i_s}$ indexed by sets $I = \{i_1, \ldots, i_s\}$ where $1 \leq i_1 \leq \cdots \leq i_s \leq n$ form a basis for $\Lambda^s(n)$. The element $v_1 \wedge \cdots \wedge v_s$ is invariant under $\mathsf{L}(n)$ and generates $\Lambda^s(n)$ as a $\mathbb{F}_2\text{GL}(n)$-module. By Proposition 10.1.3 it suffices to show there are no other nonzero $\mathsf{L}(n)$-invariants in $\Lambda^s(n)$.

Since $\Lambda^n(n) \cong \mathsf{I}(n)$ we may assume that $s < n$. An arbitrary element of $\Lambda^s(n)$ can be written as $f = g + h \wedge v_n$ where g and h do not involve v_n. If $i < n$ and f is fixed by the lower triangular transvection $T_{n,i}$ defined by $v_n \cdot T_{n,i} = v_n + v_i$, $v_j \cdot T_{n,i} = v_j$ for $j < n$, then $h \wedge v_i = 0$. Since $s < n$, h cannot involve all the variables $v_1, \ldots, v_{n-1}$, and so $h = 0$. Hence f does not involve v_n. By iterating the argument, it follows that f does not involve v_j for $n \geq j > s$, and so $f = v_1 \wedge \cdots \wedge v_s$. $\qquad\square$

In Propositions 10.1.5 and 10.1.6, we apply Proposition 10.1.3 to find irreducible submodules of $\mathsf{K}^4(3)$ and $\mathsf{P}^4(3)$ which are isomorphic to $\mathsf{St}(3)$. We write the variables in $\mathsf{P}(3)$ as x, y, z and the variables in $\mathsf{DP}(3)$ as u, v, w.

Proposition 10.1.5 *The Steenrod kernel* $\mathsf{K}^4(3)$ *is an irreducible* $\mathbb{F}_2\text{GL}(3)$-*module of dimension* 8.

Proof The submodule $\mathsf{DP}^{(2,1)}(3) \subset \mathsf{DP}^4(3)$ has dimension 9, with a basis consisting of the d-spikes obtained from $u^{(3)}v$ by permuting the variables and the d-monomials $u^{(2)}vw$, $uv^{(2)}w$ and $uvw^{(2)}$, each of which is mapped to uvw by Sq_1. Since $\dim \mathsf{K}^4(3) = \dim \mathsf{Q}^4(3) = 8$, the d-spikes and any two of the d-polynomials $u^{(2)}vw + uv^{(2)}w$, $u^{(2)}vw + uvw^{(2)}$ and $uv^{(2)}w + uvw^{(2)}$ form a basis for $\mathsf{K}^4(3)$.

The subgroup $\mathsf{L}(3)$ of $\text{GL}(3)$ is generated by the transvections $T_{2,1}$ and $T_{3,2}$, where $T_{2,1}$ maps v to $u+v$ and fixes u and w, and $T_{3,2}$ maps w to $v+w$ and fixes u and v. As the d-spike $u^{(3)}v$ is fixed by $T_{2,1}$ and $T_{3,2}$, it is an $\mathsf{L}(3)$-invariant. Since $u^{(2)}vw + uv^{(2)}w = (u+v)^{(3)}w + u^{(3)}w + v^{(3)}w$, it is easy to see that any d-spike generates $\mathsf{K}^4(3)$ as a $\mathbb{F}_2\text{GL}(3)$-module.

We use Proposition 10.1.2 to prove uniqueness of the invariant, with $G = \mathsf{L}(3)$ and $f = vw^{(3)}$. By applying the transvections $T_{2,1}$, $T_{3,2}$ and $T_{3,1}$, where $T_{3,1}$ maps w to $u + w$ and fixes u and v, we see that f generates $\mathsf{K}^4(3)$ as a $\mathbb{F}_2\mathsf{L}(3)$-module. Then, in the notation of Proposition 10.1.2, $\theta(\overline{G}) = vA + (u + v)A = uA$ where $A = w^{(3)} + (u+w)^{(3)} + (v+w)^{(3)} + (u+v+w)^{(3)}$, and a short calculation gives $\theta(\overline{G}) = u^{(3)}v$. The result follows from Propositions 10.1.2 and 10.1.3. $\qquad\qquad\qquad\qquad\qquad\qquad\qquad\qquad\qquad\qquad\qquad\square$

Proposition 10.1.2 also shows that the restriction of $\mathsf{St}(n)$ to $\mathsf{L}(n)$ is isomorphic to $\mathbb{F}_2\mathsf{L}(n)$ itself as a $\mathbb{F}_2\mathsf{L}(n)$-module when $n = 3$. This is a well-known property of the Steinberg module.

Proposition 10.1.6 *There is a direct sum splitting* $\mathsf{P}^4(3) = \mathsf{H}^4(3) \oplus S$, *where* $S \cong \mathsf{Q}^4(3)$ *is the* $\mathbb{F}_2\mathsf{GL}(3)$-*submodule generated by* $x^3y + x^2y^2$, *and* S *is irreducible.*

Proof Let $f = yz^3 + y^2z^2 \in S$ and let $M \subseteq S$ be the $\mathbb{F}_2G$-module generated by f, where $G = \mathsf{L}(3)$. Since $f \cdot (T_{2,1} - I_3) = xz^3 + x^2z^2$ and $f \cdot (T_{3,2} - I_3) = zy^3 + z^2y^2$, M is closed under permutations of the variables x, y, z, and hence $M = S$. Let $\theta : \mathbb{F}_2G \to M$ be the map of $\mathbb{F}_2G$-modules defined by $\theta(1) = f$. Then $\theta(\overline{G}) = yA + y^2B + (x+y)A + (x+y)^2B = xA + x^2B$ where $A = z^3 + (x+z)^3 + (y+z)^3 + (x+y+z)^3$ and $B = z^2 + (x+z)^2 + (y+z)^2 + (x+y+z)^2$. Thus $A = x^2y + xy^2$ and $B = 0$, so $\theta(\overline{G}) = x^3y + x^2y^2 \neq 0$. By Proposition 10.1.2 $\dim S = 8$ and $g = x^3y + x^2y^2$ is the unique $\mathsf{L}(3)$-invariant in S. Since f is obtained from g by permuting the variables, g generates S as a $\mathbb{F}_2\mathsf{GL}(3)$-module. Applying Proposition 10.1.3, it follows that S is irreducible. Since it contains a spike, g is not hit, so the submodule $\mathsf{H}^4(3) \cap S = 0$ since S is irreducible. Since $\dim S = 8 = \dim \mathsf{Q}^4(3)$, it follows that $\mathsf{P}^4(3) = \mathsf{H}^4(3) \oplus S$ and so $S \cong \mathsf{Q}^4(3)$. $\qquad\qquad\qquad\qquad\qquad\qquad\qquad\qquad\qquad\qquad\square$

Example 10.1.7 An isomorphism θ of $\mathsf{K}^4(3) \subset \mathsf{DP}^4(3)$ and $S \subset \mathsf{P}^4(3)$ is defined on the basis in the proof of Proposition 10.1.5 by mapping the $\mathsf{L}(3)$-invariant $u^{(3)}v$ of $\mathsf{K}^4(3)$ to the $\mathsf{L}(3)$-invariant $x^3y + x^2y^2$ of S. This determines θ on the other basis elements of $\mathsf{K}^4(3)$, for example $\theta(u^{(2)}vw + uv^{(2)}w) = x^2yz + xy^2z$. By checking that θ commutes with permutations of the variables and the standard transvection U, we can show that it is a $\mathbb{F}_2\mathsf{GL}(3)$-isomorphism. By the uniqueness of the $\mathsf{L}(3)$-invariants (or by Schur's Lemma) θ is the unique $\mathbb{F}_2\mathsf{GL}(3)$-isomorphism from $\mathsf{K}^4(3)$ to S.

10.2 Transpose duality

Recall from Definition 9.2.2 that the transpose dual V^{tr} of a finite-dimensional right $\mathbb{F}_2\mathsf{GL}(n)$- or $\mathbb{F}_2\mathsf{M}(n)$-module V is the right module V^{tr} defined by $\langle f \cdot A, v \rangle = \langle f, v \cdot A^{\mathrm{tr}} \rangle$, where $f \in V^* = \mathrm{Hom}(V, \mathbb{F}_2)$, $v \in V$ and $A \in \mathsf{GL}(n)$ or $\mathsf{M}(n)$. The $\mathbb{F}_2\mathsf{M}(n)$-modules $\mathsf{DP}^d(n)$ and $\mathsf{P}^d(n)$ are transpose duals (Proposition 9.2.3) as are $\mathsf{K}^d(n)$ and $\mathsf{Q}^d(n)$ (Proposition 9.4.2(iii)). The next result gives some general properties of this duality. Since the proof of (iii) depends on results in Chapter 18, in Example 10.2.3 we give a direct proof for the cases where $V = \Lambda^s(n)$ is an exterior power of the defining representation of $\mathrm{GL}(n)$, using the results of Section 6.9 on head and tail modules.

Proposition 10.2.1 *Let V be a finite-dimensional right $\mathbb{F}_2\mathrm{GL}(n)$- or $\mathbb{F}_2\mathrm{M}(n)$-module with transpose dual V^{tr}. Then (i) transpose duality is a bijection between submodules of V and quotient modules of V^{tr}, (ii) $(V^{tr})^{tr} \cong V$, (iii) if V is irreducible, then $V^{tr} \cong V$.*

Proof (i) Let $d = \dim V$. Then there is a bijection between subspaces of V of dimension k and subspaces of V^* of dimension $d - k$ which associates to W its annihilator $W^\perp$, defined by $f \in W^\perp$ if and only if $\langle f, w \rangle = 0$ for all $w \in W$. The dual $V^* \to W^*$ of the inclusion map $W \to V$ has kernel $W^\perp$, and so gives an isomorphism $V^*/W^\perp \cong W^*$ which we use to identify $V^*/W^\perp$ with W^*. Thus duals of subspaces of V are identified with quotient spaces of V^*. We identify the transpose dual W^{tr} of a submodule W of V with $V^{tr}/W^\perp$.

(ii) The standard isomorphism of $(V^*)^*$ with V preserves the action, since $(A^{tr})^{tr} = I_n$, the identity matrix.

(iii) Consider first the analogous situation of modules over $F(M(n, F))$, where F is a field of characteristic 0 and $M(n, F)$ is the corresponding monoid of $n \times n$ matrices over F. In this case, (iii) follows from character theory, since a matrix and its transpose have the same multiset of eigenvalues. The same argument shows that, for $F = \mathbb{F}_2$, the modules V and V^{tr} have the same Brauer character (see Section 18.5). It follows that V and V^{tr} have the same multiset of irreducible composition factors, and (iii) follows.　　□

Example 10.2.2 Proposition 9.4.2(iii) is the case where $V = \mathrm{DP}^d(n)$, $V^{tr} = \mathrm{P}^d(n)$ and $W = \mathrm{K}^d(n)$. Then $W = \mathrm{H}^d(n)^\perp$ and so $\mathrm{H}^d(n) = W^\perp$. Thus $\mathrm{Q}^d(n) = \mathrm{P}^d(n)/\mathrm{H}^d(n)$ is the transpose dual of $\mathrm{K}^d(n)$.

Example 10.2.3 In the case $n = 3$, $\mathrm{P}^2(3)$ and $\mathrm{DP}^2(3)$ are transpose duals, and the exact sequences $\mathrm{P}^{(0,1)}(3) \to \mathrm{P}^2(3) \to \mathrm{P}^{(2)}(3)$ and $\mathrm{DP}^{(2)}(3) \to \mathrm{DP}^2(3) \to \mathrm{DP}^{(0,1)}(3)$ do not split, so $V(3)^{tr} \cong V(3)$ and $(V(3)^*)^{tr} \cong V(3)^*$.

In the general case, by Proposition 6.9.3 $\mathrm{Q}^{(1)|\tau}(n)/\mathrm{Q}^\tau(n) \cong \Lambda^s(n)$ where $\tau = (1, \ldots, 1)$ is a tail sequence of length $s - 1$, and by Proposition 9.7.2 the kernel of $\epsilon_* : \mathrm{K}^{(1)|\tau}(n) \to \mathrm{K}^\tau(n)$ is isomorphic to $\Lambda^s(n)$. Since $\mathrm{Q}^\tau(n)$ and $\mathrm{K}^\tau(n)$ are transpose duals, it follows that $\Lambda^s(n)$ is isomorphic to its transpose dual.

10.3 $K^d(3)$ for $d = 2^{s+1}$

Let $\mu(d) = 2$, so that $d = (2^{s+t} - 1) + (2^t - 1)$ where $s \geq 0$ and $t > 0$. We consider the case $t = 1$ in this section, and the case $t \geq 2$ in Section 10.4. In the case $t = 1$, $K^d(3) = K^\omega(3)$ where $\omega = (2,1,\dots,1)$, of length $s + 1$. For $s = 0$, $K^2(3) = DP^{(2)}(3)$ has dimension 3 and basis uv, uw, vw. The case $s = 1$ is treated in Proposition 10.1.5. For $s = 2$, $\dim K^8(3) = 15$ and $\omega = (2,1,1)$. This is the first case where there is an element in the Steenrod kernel which cannot be obtained by linear substitution in d-spikes, since $\dim J^8(3) = 14$. There are 27 d-monomials in $DP^\omega(3)$, of which 6 are d-spikes.

Proposition 10.3.1 *The* 14 *d-polynomials*

$$(au + bv + w)^{(7)}(cu + v), \ (a'u + v)^{(7)}(b'u + w), \ u^{(7)}(a''v + w),$$

where $a,b,c,a',b',a'' \in \mathbb{F}_2$, are linearly independent elements of the d-spike module $J^8(3)$.

Proof The given d-polynomials span the $L(3)$-orbits of the d-spikes $w^{(7)}v$, $v^{(7)}w$ and $u^{(7)}w$. The orbit sum s of $w^{(7)}v$ is nonzero, because s has a term $m = u^{(5)}vw^{(2)}$. The d-monomial m is not a term of $(au + bv + w)^{(1)}(cu + v)$ when $a = 0$. When $a = 1$, the factorization $u^{(5)}w^{(2)} \cdot v$ gives one occurrence of m in the expansion, and if also $b = c = 1$ the factorization $u^{(4)}vw^{(2)} \cdot u$ gives a second occurrence of m which cancels it. Let f be a nonzero linear combination f of these 8 d-polynomials. By Proposition 10.1.2(ii), there is an element $\gamma \in \mathbb{F}_2 L(3)$ such that $f \cdot \gamma = s$. Since $s \neq 0$, it follows that $f \neq 0$, and hence the 8 d-polynomials $(au + bv + w)^{(7)}(cu + v)$, where $a,b,c \in \mathbb{F}_2$, are linearly independent.

We can extend this argument to show that any linear combination of the given 14 d-polynomials in which one of these 8 d-polynomials appears is nonzero. The point is that m is not a term of any element in the $L(3)$-orbit of $v^{(7)}w$ or $u^{(7)}w$. Thus by applying γ to a linear combination of the 14 d-polynomials which includes at least one d-polynomial of the form $(au + bv + w)^{(7)}(cu + v)$, we obtain a d-polynomial which has m as a term, and so is nonzero. Hence the problem of showing that the given set of 14 d-polynomials is linearly independent is reduced to that of showing that the remaining 6 d-polynomials are linearly independent.

To continue the reduction, we note that the d-polynomials $(a'u + v)^{(7)}(b'u + w)$ form the orbit of $v^{(7)}w$ under the action of the subgroup $G \subset L(3)$ of matrices of the form

$$\begin{pmatrix} 1 & 0 & 0 \\ a' & 1 & 0 \\ b' & 0 & 1 \end{pmatrix}.$$

Hence given a linear combination f of these 4 d-polynomials, there is an element $\gamma \in \mathbb{F}_2 G$ such that $f \cdot \gamma$ is their sum.

To see that $f \cdot \gamma \neq 0$, we observe that it has the term $u^{(5)} v^{(3)}$. This is a term in $(u+v)^{(7)} (u+w)$, but not in the other 3 elements, and not in an element of the L(3)-orbit of $u^{(7)} w$. Thus the last 6 elements are linearly independent if the last 2 are linearly independent, and this is clearly true.   □

Hence $\dim J^8(3) \geq 14$. By direct calculation, $K^8(3)$ contains the d-polynomial

$$f = u^{(3)} v^{(3)} w^{(2)} + u^{(3)} v^{(4)} w + u^{(5)} v^{(2)} w + u^{(6)} v w,$$

since $Sq_1(f) = 0$ and $Sq_2(f) = 0$. We prove in Proposition 10.3.2 that $f \notin J^8(3)$. Since $\dim K^8(3) = \dim Q^8(3) = 15$, it follows that $\dim J^8(3) = 14$, and that f together with the 14 d-polynomials of Proposition 10.3.1 form a basis for $K^8(3)$.

Proposition 10.3.2 $f \notin J^8(3)$.

Proof Let $g = x^6 yz + xy^6 z + xyz^6 \in P^8(3)$. In the notation of Case 2 of Theorem 8.3.3, g is represented by the sum of the three blocks of the third type, and the argument given there shows that g is not hit. However, g is invariant modulo hit elements, i.e. for $A \in GL(3)$, $g \cdot A = g + h$ where h is hit. The proof uses the usual argument with permutation matrices and the standard transvection U. Since g is symmetric and $g \cdot U = (x+y)^6 yz + (x+y) y^6 z + (x+y) yz^6 = g + h$ where $h = x^4 y^3 z + x^2 y^5 z + y^2 z^6$, we need only check that h is hit, and this is clear by 1-back splicing.

Since $u^{(6)} v w$ is dual to $x^6 yz$, $\langle f, g \rangle = 1$. However, if s is a d-spike, then $\langle s \cdot A, g \rangle = \langle s, g \cdot A^{\mathrm{tr}} \rangle = \langle s, g + h' \rangle$, where h' is hit. But $\langle s, g \rangle = 0$ by inspection and $\langle s, h' \rangle = 0$, since the spike dual to s cannot appear in a hit polynomial. It follows by linearity that $\langle J^8(3), g \rangle = 0$, and hence $f \notin J^8(3)$.   □

The cases $d = 2^{s+1}$ for $s \geq 3$ are similar to $d = 8$, but are simpler as $J^d(3) = K^d(3)$, with basis $(au + bv + w)^{(2^{s+1}-1)} (cu + v)$, $(a'u + v)^{(2^{s+1}-1)} (b'u + w)$, $u^{(2^{s+1}-1)} (a''v + w)$, where $a, b, c, a', b', c' \in \mathbb{F}_2$. The auxiliary d-monomials which appear in the orbit sums are $u^{(2^s+1)} v w^{(2^s-2)}$, $u^{(2^s+1)} v^{(2^s-1)}$ and $u^{(2^{s+1}-1)} v$.

10.4 $K^d(3)$ for $d = (2^{s+t} - 1) + (2^t - 1)$

Here $\omega = \gamma | \tau$, where $\gamma = (2, \dots, 2)$ is a head sequence of length t and $\tau = (1, \dots, 1)$ is a tail sequence of length s. We focus on the case $t = 2$, so that $\omega = (2, 2, 1, \dots, 1)$, of length $s + 2$. For $s = 0$, $\omega = (2, 2)$ is a head sequence, and $K^6(3)$ has dimension 6 and basis $u^{(3)} v^{(3)}$, $u^{(3)} w^{(3)}$, $v^{(3)} w^{(3)}$, $u^{(3)} (v + w)^{(3)}$,

$v^{(3)}(u+w)^{(3)}$, $w^{(3)}(u+v)^{(3)}$. For $s = 1$, $\mathsf{K}^{10}(3) = \mathsf{J}^{10}(3)$ has dimension 14 and $\omega = (2,2,1)$. The general procedure for choosing the 'test' d-monomials in the arguments of this section is discussed in Chapter 23.

Proposition 10.4.1 *A basis for* $\mathsf{K}^{10}(3)$ *is given by the* 14 *d-polynomials*

$$(au + bv + w)^{(7)}(cu \mid v)^{(3)}, \quad (u'u + b'v + w)^{(7)}u^{(3)}, \quad (a''u + v)^{(7)}u^{(3)}, \quad (10.2)$$

where $a,b,c,a',b',a'' \in \mathbb{F}_2$.

Proof As linear independence is proved in the same way as for $\mathsf{K}^8(3)$, we record only the main steps.

A sum of the given 14 d-polynomials which includes at least one d-polynomial of the form $(au + bv + w)^{(7)}(cu + v)^{(3)}$ is shown to be nonzero by using an element of $\mathbb{F}_2\mathsf{L}(3)$ to map it to a d-polynomial with the term $u^{(6)}v^{(3)}w$. We check that this is a term in the sum of the 8 d-polynomials $(au + bv + w)^{(7)}(cu + v)^{(3)}$, but is not a term of any element of the $\mathsf{L}(3)$-orbit of $w^{(7)}u^{(3)}$ or $v^{(7)}u^{(3)}$.

At the next step, we show that a linear combination of the remaining 6 d-polynomials which includes at least one of the 4 d-polynomials $(a'u + b'v + w)^{(7)}u^{(3)}$ is nonzero, by using an element of $\mathbb{F}_2 G'$ to map it to a d-polynomial with the term $u^{(7)}v^{(2)}w$. Here G' is the subgroup of $\mathsf{L}(3)$ consisting of the matrices

$$\begin{pmatrix} 1 & 0 & 0 \\ 0 & 1 & 0 \\ a' & b' & 1 \end{pmatrix}.$$

The d-monomial $u^{(7)}v^{(2)}w$ is not a term of $(a''u + v)^{(7)}u^{(3)}$. Thus the last 6 of the given d-polynomials are linearly independent if the last 2 are, and this is clearly true. Hence the set of 14 d-polynomials is linearly independent. $\qquad\square$

Proposition 10.4.2 *For* $s = 2$, $\mathsf{K}^{18}(3) = \mathsf{J}^{18}(3)$ *has dimension* 21 *and* $\omega = (2,2,1,1)$. *A basis for* $\mathsf{K}^{18}(3)$ *is given by the d-polynomials*

$$(au + bv + w)^{(15)}(cu + v)^{(3)}, \quad (a'u + b'v + w)^{(15)}u^{(3)}, \quad (c'u + v)^{(15)}u^{(3)},$$
$$(a''u + v)^{(15)}(b''u + w)^{(3)}, \quad u^{(15)}(c''v + w)^{(3)}, \quad u^{(15)}v^{(3)}$$

where $a,b,c,a',b',c',a'',b'',c'' \in \mathbb{F}_2$.

Proof The method is similar to that for $d = 8$ and $d = 10$, the six subsets of d-polynomials being removed in turn from a possible linear dependence relation. Again we give only the main steps.

A sum of the given 21 d-polynomials which includes at least one of the 8 d-polynomials $(au + bv + w)^{(15)}(cu + v)^{(3)}$ is shown to be nonzero by using an element of $\mathbb{F}_2$L(3) to map it to a d-polynomial with the term $u^{(10)}v^{(3)}w^{(5)}$. We check that this is a term in the sum of the 8 d-polynomials $(au + bv + w)^{(15)}(cu + v)^{(3)}$, but is not a term of any other element in (10.4.2).

At the next step, we show that a sum of the remaining 13 elements in the given set of d-polynomials which includes $(a'u + b'v + w)^{(15)}u^{(3)}$ for some a' and b' is nonzero, by using an element of $\mathbb{F}_2$L(3) to map it to a d-polynomial with the term $u^{(11)}v^{(2)}w^{(5)}$. This d-monomial is not a term of any element of the L(3)-orbit of a d-spike other than $v^{(15)}w^{(3)}$ or $w^{(15)}u^{(3)}$. We continue in this way to eliminate polynomials in the remaining subsets of (10.4.2) from a possible linear dependence relation, using $u^{(10)}v^{(7)}w$, $u^{(11)}v^{(7)}$, $u^{(15)}v^{(2)}w$ and $u^{(15)}v^{(3)}$ as 'test' d-monomials.   □

We conclude by discussing the case $t > 2$. In the case $s = 0$, $t = 3$, $\mathsf{K}^{14}(3)$ has basis $u^{(7)}v^{(7)}$, $u^{(7)}w^{(7)}$, $v^{(7)}w^{(7)}$, $u^{(7)}(v+w)^{(7)}$, $v^{(7)}(u+w)^{(7)}$, $w^{(7)}(u+v)^{(7)}$, $(u+v)^{(7)}(u+w)^{(7)}$. With the exponent 7 replaced by $2^t - 1$, this is also a basis for $t > 3$. For $s > 0$, the results for $t > 2$ are the same as for $t = 2$, using the duplication map. Thus $\dim \mathsf{K}^d(3) = \dim \mathsf{K}^{2^{s+2}+2}(3) = 14$ for $s = 1$ and 21 for $s \geq 2$. A basis is given by replacing the exponents in (10.2) and (10.4.2) by $(2^{s+t} - 1)$ and $(2^t - 1)$, so as to give d-polynomials in the L(3)-orbit of $u^{(2^{s+t}-1)}v^{(2^t-1)}$.

10.5 The flag module FL(3)

The defining representation V(3) of GL(3) is the $\mathbb{F}_2$GL(3)-module given by right multiplication $v \cdot A = vA$, where $v = (a_1, a_2, a_3) \in \mathbb{F}_2^3$ is a 3-dimensional row vector and $A \in$ GL(3) is a 3×3 matrix over $\mathbb{F}_2$. A **flag** in V(3) is a pair of subspaces $X = (X_1, X_2)$, where $\dim X_1 = 1$, $\dim X_2 = 2$ and $X_1 \subset X_2$. By analogy with geometry we call X_1 a 'line' and X_2 a 'plane'. The **reference flag** R is given by the line $a_2 = a_3 = 0$ and the plane $a_3 = 0$. Since there are 7 lines X_1, each contained in 3 planes, and 7 planes X_2, each containing 3 lines, there are 21 flags. Given a flag X and a matrix $A \in$ GL(3), $X \cdot A = (X_1 \cdot A, X_2 \cdot A)$ is a flag, so GL(3) permutes the flags. We define the flag module FL(3) to be the vector space of dimension 21 over $\mathbb{F}_2$ spanned by the flags.

Given $A \in$ GL(3), the flag $R \cdot A = (X_1, X_2)$ where X_1 is spanned by the first row of A and X_2 is spanned by the first two rows of A. Hence $R \cdot A = R$ if and only if $A \in$ L(3), the lower triangular subgroup. For a general matrix A, let $A = BA'$ where $B \in$ L(3) and A' is in a given set of representatives for the right cosets of L(3) in GL(3). Then $R \cdot A = (R \cdot B) \cdot A' = R \cdot A'$. Hence the flag module FL(3) can be identified with the permutation module of GL(3) given by its action on the set of right cosets $\{L(3)A : A \in GL(3)\}$. In a similar way, we consider the permutation modules $FL^1(3)$ and $FL^2(3)$ of GL(3) on the 7 lines X_1 and on the 7 planes X_2 in V(3). These may be identified with the right cosets in GL(3) of the subgroups of matrices with first row (1 0 0) and first column $\begin{pmatrix} 1 \\ 0 \\ 0 \end{pmatrix}$ respectively.

Proposition 10.5.1 *As a* $\mathbb{F}_2$GL(3)-*module,* $K^{18}(3) \cong FL(3)$.

Proof $K^{18}(3) = J^{18}(3)$ is the cyclic $\mathbb{F}_2$GL(3)-module generated by the d-spike $u^{(15)}v^{(3)}$, and so elements of the form $u^{(15)}v^{(3)} \cdot A$, $A \in GL(3)$ span $K^{18}(3)$. The flag module FL(3) is the cyclic module generated by any flag. Hence a $\mathbb{F}_2$GL(3)-module map $\phi : FL(3) \to K^{18}(3)$ is defined by $\phi(R \cdot A) = u^{(15)}v^{(3)} \cdot A$ where R is the reference flag and $A \in GL(3)$. Since $u^{(15)}v^{(3)}$ is invariant under L(3), ϕ is well defined. By (10.4.2) the set of elements $u^{(15)}v^{(3)} \cdot A'$ is a basis for $K^{18}(3)$, where A' is one of the representative matrices

$$\begin{pmatrix} a & b & 1 \\ c & 1 & 0 \\ 1 & 0 & 0 \end{pmatrix}, \begin{pmatrix} u' & b' & 1 \\ 1 & 0 & 0 \\ 0 & 1 & 0 \end{pmatrix}, \begin{pmatrix} c' & 1 & 0 \\ 1 & 0 & 0 \\ 0 & 0 & 1 \end{pmatrix}, \begin{pmatrix} a'' & 1 & 0 \\ b'' & 0 & 1 \\ 1 & 0 & 0 \end{pmatrix}, \begin{pmatrix} 1 & 0 & 0 \\ 0 & c'' & 1 \\ 0 & 1 & 0 \end{pmatrix}, \begin{pmatrix} 1 & 0 & 0 \\ 0 & 1 & 0 \\ 0 & 0 & 1 \end{pmatrix}$$

for the right cosets of L(3) in GL(3). Hence ϕ is an isomorphism. □

The map ϕ can be described more naturally by using an alternative notation for the flag $X = (X_1, X_2)$ in V(3). Given a basis w_1, w_2, w_3 for V(3) such that w_1 spans X_1 and w_1, w_2 span X_2, we denote X by (w_1, w_2). Thus w_1, w_2, w_3 are the rows of a matrix A in the coset L(3)A corresponding to X. In this notation, $(w_1, w_2) = (w_1, w_1 + w_2)$, and the map ϕ of Proposition 10.5.1 is given by $\phi(w_1, w_2) = w_1^{(15)} w_2^{(3)}$. More generally, a well defined map FL(3) $\to$ $J^d(3)$ is given by $(w_1, w_2) \mapsto w_1^{(2^{s+t}-1)} w_2^{(2^t-1)}$ for $s, t \geq 0$, where $d = (2^{s+t} - 1) + (2^t - 1)$. We use these maps to study the module structure of $J^d(3)$.

Proposition 10.5.2 *For a tail sequence τ of length ≥ 3,* $K^{\tau}(3) \cong FL^1(3)$, *the permutation representation of* GL(3) *on the 7 lines in* V(3). *For a head sequence γ of length ≥ 3,* $K^{\gamma}(3) \cong FL^2(3)$, *the permutation representation of* GL(3) *on the 7 planes in* V(3).

Proof Since the modules are independent of the length of τ or γ when it is ≥ 3, it suffices to consider length 3. An isomorphism $FL^1(3) \to K^{(1,1,1)}(3)$

is given by mapping a nonzero vector $w_1 \in V(3)$ to $w_1^{(7)}$. An isomorphism $\text{FL}^2(3) \to K^{(2,2,2)}(3)$ is given by mapping the plane spanned by w_1 and w_2 in $V(3)$ to $w_1^{(7)} w_2^{(7)}$. This map is well defined since $w_1^{(7)}(w_1 + w_2)^{(7)} = w_1^{(7)} w_2^{(7)} = (w_1 + w_2)^{(7)} w_2^{(7)}$. □

The next result gives the structure of $\text{FL}^1(3)$ and $\text{FL}^2(3)$ as $\mathbb{F}_2\text{GL}(3)$-modules.

Proposition 10.5.3 *Let A and A' be the 1-dimensional submodules of* $\text{FL}^1(3)$ *and* $\text{FL}^2(3)$ *generated by the sum of the 7 lines or planes respectively. Then* $\text{FL}^1(3) = A \oplus B$ *and* $\text{FL}^2(3) = A' \oplus B'$, *where B and B' are the 6-dimensional submodules of* $\text{FL}^1(3)$ *and* $\text{FL}^2(3)$ *given by sums of an even number of lines or planes. Then B and B' are indecomposable, and there are exact sequences*

$$V(3)^* \to B \to V(3), \ V(3) \to B' \to V(3)^*.$$

Proof We give details only for $\text{FL}^1(3)$, as those for $\text{FL}^2(3)$ are similar, using the results on head modules from Section 9.9. The map $j : \text{FL}^1(3) \to A$ which sends each line to the sum of all the lines is clearly a $\mathbb{F}_2\text{GL}(3)$-module map, and $j \circ i$ is the identity map of A, where i embeds A as a submodule. Hence we have a direct sum splitting $\text{FL}^1(3) = A \oplus B$, where $B = \ker j$ consists of sums of an even number of lines. The results of Section 9.9 allow us to identify A with the kernel of $\epsilon_* : K^{(1,1,1)} \to K^{(1,1)}(3)$ and B with $K^{(1,1)}(3)$. We have also seen that the kernel of $\epsilon_* : K^{(1,1)} \to K^{(1)}(3)$ is isomorphic to $V(3)^*$ and its image to $V(3)$.

To show that B is indecomposable, consider the $L(3)$-invariants in $K^{(1,1)}(3)$. These are spanned by $u^{(3)}$ and $u^{(3)} + v^{(3)} + (u+v)^{(3)}$. The latter corresponds to the sum of the three lines in a plane in $V(3)$, and so generates a subspace isomorphic to $V(3)^*$. Further, $u^{(3)}$ and $v^{(3)} + (u + v)^{(3)}$ generate the whole module $K^{(1,1)}(3)$, so Proposition 10.1.1 shows that $K^{(1,1)}(3)$ has no submodule isomorphic to $V(3)$. □

Proposition 10.5.4 $\text{FL}^1(3)$ *and* $\text{FL}^2(3)$ *are direct summands in* $\text{FL}(3)$ *whose intersection is the 1-dimensional direct summand generated by the sum of the 21 flags.*

Proof Consider the $\mathbb{F}_2\text{GL}(3)$-module maps $i_1 : \text{FL}^1(3) \to \text{FL}(3)$ and $j_1 : \text{FL}(3) \to \text{FL}^1(3)$ defined by $i_1(X_1) = \sum_{X_2 \supset X_1} (X_1, X_2)$ and $j_1(X_1, X_2) = X_1$. Since the composition $j_1 \circ i_1$ is the identity map of $\text{FL}^1(3)$, i_1 embeds $\text{FL}^1(3)$ as a direct summand in $\text{FL}(3)$. Similarly the composition $j_2 \circ i_2$ of the maps $i_2 : \text{FL}^2(3) \to \text{FL}(3)$ and $j_2 : \text{FL}(3) \to \text{FL}^2(3)$ defined by $i_2(X_2) = \sum_{X_1 \subset X_2} (X_1, X_2)$

and $j_2(X_1, X_2) = X_2$ is the identity map of $FL^2(3)$, so i_2 embeds $FL^2(3)$ as a direct summand in $FL(3)$.

An element of the intersection of the images of i_1 and i_2 is the sum of a set of flags which contains (X_1, X_2') and (X_1', X_2) whenever it contains (X_1, X_2). The only nonzero sum of this kind is the sum $\overline{X}$ of all the 21 flags, which is invariant under $GL(3)$. Since the map which sends each flag to this sum is a $\mathbb{F}_2 GL(3)$-map, $\overline{X}$ generates a direct summand. □

We denote the 1-dimensional summand of $FL(3)$ by $FL_0(3)$ and the indecomposable 6-dimensional summands arising from $FL^1(3)$ and $FL^2(3)$ by $FL_1(3)$ and $FL_2(3)$ respectively. To complete the determination of the structure of $FL(3)$, we show that the complementary summand of $FL^1(3) \cup FL^2(3) = FL_0(3) \oplus FL_1(3) \oplus FL_2(3)$ in $FL(3)$ is isomorphic to the irreducible module $K^4(3)$ of Proposition 10.1.5 or S of Proposition 10.1.6. This is achieved by the next two results.

Recall from Section 9.5 that the d-duplication map $\delta_* : K^{18}(3) \to K^8(3)$ is given by $\delta_*(w_1^{(15)} w_2^{(3)}) = w_1^{(7)} w_2$, where $w_1, w_2 \in DP^1(3)$.

Proposition 10.5.5 *The kernel of the surjective map $\delta_* : K^{18}(3) \to K^8(3)$ is a direct summand isomorphic to* $FL^1(3)$. *There is a $\mathbb{F}_2 GL(3)$-module map $\varepsilon_* :$ $K^{18}(3) \to K^{10}(3)$ defined by $\varepsilon_*(w_1^{(15)} w_2^{(3)}) = w_1^{(7)} w_2^{(3)}$, whose kernel is a direct summand isomorphic to* $FL^2(3)$.

Proof From the bases of $K^d(3)$ given in Sections 10.3 and 10.4, $\delta_* : K^{18}(3) \to K^8(3)$ is a surjection of $\mathbb{F}_2 GL(3)$-modules with kernel of dimension 7. For w_1, $w_2, w_3 \in DP^1(3)$, $\delta_*(w_1^{(15)} w_2^{(3)} + w_1^{(15)} w_3^{(3)} + w_1^{(15)}(w_2 + w_3)^{(3)}) = w_1^{(7)} w_2 + w_1^{(7)} w_3 + w_1^{(7)}(w_2 + w_3) = 0$. We may identify the element $w_1^{(15)} w_2^{(3)} \in K^{18}(3)$ with the flag (w_1, w_2) in $FL(3)$. By Proposition 10.5.4 the line generated by $w_1 \in FL^1(3)$ is identified with the sum $(w_1, w_2) + (w_1, w_3) + (w_1, w_2 + w_3)$ of the three flags containing w_1, where w_1, w_2, w_3 are linearly independent in $DP^1(3)$. This sum is identified with the element $w_1^{(15)} w_2^{(3)} + w_1^{(15)} w_3^{(3)} + w_1^{(15)}(w_1 + w_2)^{(3)}$ of the kernel of δ_*.

Since the element $w_1^{(7)} w_2^{(3)}$ depends only on the coset corresponding to the flag (w_1, w_2), ε_* is well defined and is a map of $\mathbb{F}_2 GL(3)$-modules. By the results of Section 10.4 it is surjective, with kernel of dimension 7. For $w_1, w_2 \in V(3)$, $\varepsilon_*(w_1^{(15)} w_2^{(3)} + w_2^{(15)} w_1^{(3)} + (w_1 + w_2)^{(15)} w_1^{(3)}) = w_1^{(7)} w_2^{(3)} + w_2^{(7)} w_1^{(3)} + (w_1 + w_2)^{(7)} w_1^{(3)} = 0$. By Proposition 10.5.4 the plane X_2 generated by linearly independent elements $w_1, w_2 \in FL^2(3)$ is identified with the sum $(w_1, w_2) + (w_2, w_1) + (w_1 + w_2, w_1)$ of the three flags containing X_2. This sum is identified with the element $w_1^{(15)} w_2^{(3)} + w_2^{(15)} w_1^{(3)} + (w_1 + w_2)^{(15)} w_1^{(3)}$ of the kernel of δ_*. □

Proposition 10.5.6 *The kernel of the surjective map* $\delta_* : K^{10}(3) \to K^4(3)$ *is a direct summand isomorphic to* FL$_1$(3). *There is a surjective* $\mathbb{F}_2$GL(3)-*module map* $\varepsilon_* : J^8(3) \to K^4(3)$ *defined by* $\varepsilon_*(w_1^{(7)}w_2) = w_1^{(3)}w_2$, *whose kernel is a direct summand isomorphic to* FL$_2$(3).

Proof The proof is similar to that of Proposition 10.5.5. We note from Section 10.4 that $\delta_* : K^{10}(3) \to K^4(3)$ is surjective with kernel of dimension 6. This kernel is spanned by elements of the form $w_1^{(7)}w_2^{(3)} + w_1^{(7)}w_3^{(3)} + w_1^{(7)}(w_1 + w_2)^{(3)}$, which can be identified as before with the sum $(w_1, w_2) + (w_1, w_3) + (w_1, w_2 + w_3)$ of the three flags containing w_1, and so with the quotient FL$_1$(3) of FL^1(3).

Since the element $w_1^{(7)}w_2$ depends only on the coset corresponding to the flag (w_1, w_2), and the relations corresponding to the sum of the three flags containing the same line are preserved, $\varepsilon_* : J^8(3) \to K^4(3)$ is well defined and is a map of $\mathbb{F}_2$GL(3)-modules. By the results of Section 10.3 it is surjective, with kernel of dimension 6. This kernel is spanned by elements of the form $w_1^{(7)}w_2 + w_2^{(7)}w_1 + (w_1 + w_2)^{(7)}w_1$ which can be identified as before with the sum $(w_1, w_2) + (w_2, w_1) + (w_1 + w_2, w_1)$ of the three flags containing X_2, and so with the quotient FL$_2$(3) of FL^2(3).   □

Proposition 10.5.7 *The direct summand of* FL(3) *complementary to* FL^1(3) $\cup$ FL^2(3) $=$ FL$_0$(3) $\oplus$ FL$_1$(3) $\oplus$ FL$_2$(3) *is isomorphic to* St(3).

Proof By Propositions 10.5.5 and 10.5.6, the kernel of the composition $\varepsilon_* \circ \delta_* : K^{18}(3) \to K^4(3)$ is isomorphic to the direct summand FL^1(3) $\oplus$ FL^2(3) of FL(3), and its image K^4(3) is isomorphic to St(3) by Example 10.1.7.   □

The results of this section may be summarized as follows. The permutation module FL(3) of GL(3) on the 21 flags in the defining module V(3) is the direct sum of four indecomposable submodules, as shown in (10.1). The summands FL$_0$(3) $\cong$ I(3) and FL$_{1,2}$(3) $\cong$ St(3) are irreducible, while FL$_1$(3) and FL$_2$(3) have two composition factors V(3) and V(3)*. The submodules FL$_0$(3) $\oplus$ FL$_1$(3) and FL$_0$(3) $\oplus$ FL$_2$(3) can be identified with the permutation modules FL^1(3) and FL^2(3) of GL(3) on the 7 lines and the 7 planes in V(3) respectively. By Proposition 10.2.1, transpose duality preserves irreducible modules and reverses composition series, so FL$_1$(3) and FL$_2$(3) are transpose duals of each other, and FL$_0$(3), FL$_{1,2}$(3) are self-dual, as is the whole module FL(3).

10.6 $K^d(3)$ and $Q^d(3)$ as $\mathbb{F}_2\text{GL}(3)$-modules

First we summarize results on $K^d(3)$ and $Q^d(3)$ obtained in Sections 8.3, 9.9, 10.3 and 10.4. By Proposition 6.5.6 we can reduce to the cases $\mu(d) = 1$ or 2 using the Kameko isomorphisms $\kappa : Q^{2d+3}(3) \to Q^d(3)$ and $\kappa_* : K^d(3) \to K^{2d+3}(3)$ when $\mu(2d+3) = 3$. When $\mu(d) = 2$, $d = (2^{s+t} - 1) + (2^t - 1)$ and there is a unique decreasing sequence $\omega = \gamma | \tau \in \text{Dec}_d(3)$, where $\gamma = (2, \ldots, 2)$ of length $t \geq 1$ and $\tau = (1, \ldots, 1)$ of length $s \geq 0$. When $\mu(d) = 1$, $d = 2^s - 1$. If $s \geq 2$ there are two decreasing sequences in $\text{Dec}_d(3)$, a tail sequence $\tau' = (1, \ldots 1)$ of length s, and $\omega' = (3) | \gamma'$ where $\gamma' = (2, \ldots, 2)$ is a head sequence of length $s - 2$. By the local Kameko isomorphisms, $Q^{\omega'}(3) \cong Q^{\gamma'}(3)$ and $K^{\omega'}(3) \cong K^{\gamma'}(3)$. We begin by reducing the case $\mu(d) = 1$ to a local problem.

Proposition 10.6.1 *Let $d = 2^s - 1$ where $s \geq 2$. Then $Q^d(3) \cong Q^{\tau'}(3) \oplus Q^{\omega'}(3)$ and $K^d(3) \cong K^{\tau'}(3) \oplus K^{\omega'}(3)$.*

Proof It is sufficient by transpose duality to prove the direct sum splitting for $K^d(3)$. Then $K^{\omega'}(3)$ is the submodule generated by the d-spike $u^{(2^{s-1}-1)}v^{(2^{s-1}-1)}w$ and $K^{\tau'}(3)$ is the quotient module with generator represented by the d-spike $u^{(2^s-1)}$. We separate the cases (i) $s = 2$ and (ii) $s > 2$.

(i) For $s = 2$, the 7 d-polynomials $f^{(3)}$ where $f \neq 0 \in \text{DP}^1(3)$ form a basis of $K^3(3)$. Their sum uvw generates $K^{\omega'}(3) = K^{(3)}(3)$, and is invariant under $\text{GL}(3)$. The map $j : f^{(3)} \to uvw$ thus satisfies $j \circ i = 1$, where i is the inclusion map for uvw. Hence $K^{\omega'}(3)$ is a direct summand of $K^3(3)$.

(ii) For $s \geq 2$, the 7 d-polynomials $f^{(2^s-1)}$, where $f \neq 0 \in \text{DP}^1(3)$, are linearly independent and span the submodule A generated by the d-spike $u^{(2^s-1)}$. In the case $s = 2$, $A = K^d(3)$, but $\dim K^d(3) > 7$ when $s > 2$. However for $s > 2$ the projection map $K^d(3) \to K^{\tau'}(3)$ restricts to an isomorphism on A. Hence A is a direct summand of $K^d(3)$, and the complementary summand is isomorphic to $K^{\omega'}(3)$. $\qquad\square$

Proposition 10.6.1 reduces the determination of the module structure of $K^d(3)$ and $Q^d(3)$ to that of $K^{\gamma|\tau}(3)$ and $Q^{\gamma|\tau}(3)$, where γ is a head sequence of length t and τ is a tail sequence of length s.

Theorem 10.6.2 *The following tables give the structure of $K^\omega(3)$ and $Q^\omega(3)$ for $\omega = (2, \ldots, 2, 1, \ldots, 1)$, where s is the number of 1s and t is the number of 2s. In all cases except $\omega = (2, 1, 1)$, $K^\omega(3) = J^\omega(3)$. In the exceptional case, $J^\omega(3)$ is the submodule $\text{St}(3) \oplus \text{FL}_2$ of $K^\omega(3)$. (In the tables, I, V denote $I(3)$, $V(3)$ etc.)*

$\mathbb{F}_2GL(3)$-module structure of $K^\omega(3)$

	$s=0$	$s=1$	$s=2$	$s \geq 3$
$t=0$	I	V	$FL_1 = \begin{matrix} V \\ \downarrow \\ V^* \end{matrix}$	$FL^1 = I \oplus FL_1$
$t=1$	V^*	St	$St \oplus \begin{matrix} I \\ \downarrow \\ FL_2 \end{matrix}$	$St \oplus FL_2$
$t=2$	$FL_2 = \begin{matrix} V^* \\ \downarrow \\ V \end{matrix}$	$St \oplus FL_1$	FL	FL
$t \geq 3$	$FL^2 = I \oplus FL_2$	$St \oplus FL_1$	FL	FL

$\mathbb{F}_2GL(3)$-module structure of $Q^\omega(3)$

	$s=0$	$s=1$	$s=2$	$s \geq 3$
$t=0$	I	V	$FL_2 = \begin{matrix} V^* \\ \downarrow \\ V \end{matrix}$	$FL^2 = I \oplus FL_2$
$t=1$	V^*	St	$St \oplus \begin{matrix} FL_1 \\ \downarrow \\ I \end{matrix}$	$St \oplus FL_1$
$t=2$	$FL_1 = \begin{matrix} V \\ \downarrow \\ V^* \end{matrix}$	$St \oplus FL_2$	FL	FL
$t \geq 3$	$FL^1 = I \oplus FL_1$	$St \oplus FL_2$	FL	FL

Proof The structure of $J^\omega(3)$ has been determined in Section 10.5. In the rest of this section, we deal with the exceptional case $\omega = (2,1,1)$, where $J^\omega(3) \neq K^\omega(3)$. The results for $Q^\omega(3)$ are obtained from those for $K^\omega(3)$ by transpose duality. $\square$

Proposition 10.6.3 *The d-polynomial*

$$f = u^{(3)}v^{(3)}w^{(2)} + u^{(3)}v^{(4)}w + u^{(5)}v^{(2)}w + u^{(6)}vw$$

generates $K^8(3)$ *as a* $\mathbb{F}_2 GL(3)$*-module. The structure of* $K^8(3)$ *is given by* $K^8(3) = A \oplus F'$, *where* $A \cong St(3)$ *and* F' *is the submodule generated by the cyclic symmetrization f' of f. The module F' is indecomposable of dimension 7, and has a submodule B of dimension 6 such that $B \cong FL_2(3)$ and $F'/B \cong I(3)$.*

Proof We have seen in Section 10.3 that $J^8(3) \subset K^8(3)$ is the submodule of dimension 14 generated by $u^{(7)}v$, and that f is in $K^8(3)$ but not in $J^8(3)$. The standard transvection U maps u to $u+v$ and fixes v and w, and a short calculation gives $f \cdot U = f + v^{(7)}w$. It follows that $v^{(7)}w$ is in the submodule of $K^d(3)$ generated by f. Hence f generates $K^d(3)$.

We have also seen in Proposition 10.5.6 that the map $\varepsilon_* : J^8(3) \to K^4(3)$ defined by $\varepsilon_*(w_1^{(7)}w_2) = w_1^{(3)}w_2$ is the projection of $J^8(3) = A \oplus B$ on a direct summand $A \cong St(3)$, and that $B \cong FL_2(3)$. A surjective map $FL^2(3) \to B$ is given by $(w_1, w_2) \mapsto w_1^{(7)}w_2 + w_2^{(7)}w_1 + (w_1+w_2)^{(7)}w_1 = w_1^{(5)}w_2^{(3)} + w_1^{(3)}w_2^{(5)}$ in $B = \text{Ker}(\varepsilon_*)$, where (w_1, w_2) denotes the plane spanned by $w_1, w_2 \in V(3)$. The map ε_* removes column 3 of a block if it duplicates column 2, and otherwise maps the block to 0.

Let f' be the cyclic symmetrization of f. Then the argument of Proposition 10.3.2 shows that $f' \notin J^8(3)$. Thus, evaluating f' on $g = x^6 yz + xy^6 z + xyz^6$, we obtain 3 nonzero terms by cyclic symmetry from $\langle u^{(6)}vw, x^6yz \rangle = 1$. Hence $\langle f', g \rangle = 1$, and it follows that $f' \notin J^8(3)$.

Next we show that f' is $GL(3)$-invariant mod B. Since f' has cyclic symmetry, it suffices to check the standard transvection U and the switch S_1 of u and v. For U, we obtain $f' \cdot U = f' + v^{(5)}w^{(3)} + v^{(5)}w^{(3)}$, and $v^{(5)}w^{(3)} + v^{(5)}w^{(3)} \in B$. For S_1, we obtain $f' \cdot S_1 = f' + s$, where s is the symmetrization of $uv^{(3)}w^{(4)} + uv^{(2)}w^{(5)}$. We obtain $s = (u+v)^{(7)}(u+w) + (v+w)^{(7)}(v+u) + (w+u)^{(7)}(w+v)$, and so $s \in B$. It follows that $\dim F'/B = 1$, so that $\dim F' = 7$ and $K^8(3) = A \oplus F'$.

Finally we prove by contradiction that F' does not split as $B \oplus C$ where $C \cong I(3)$. Since $B \cong FL_2(3)$, such a splitting implies that $F' \cong FL^2(3)$. However, f' is cyclically symmetric but not symmetric, and it is easy to check that the only cyclically symmetric element of $FL^2(3)$ is the sum of the 7 planes, which is symmetric. Hence F' is indecomposable. $\square$

Remark 10.6.4 We have $K^8(3) = K^\omega(3)$ where $\omega = (2,1,1)$, but since $f \notin DP^\omega(3)$, $K^\omega(3)$ is not a submodule of $DP^\omega(3)$. However, by omitting the term $u^{(3)}v^{(3)}w^{(2)}$, we obtain the polynomial $\bar{f} = u^{(3)}v^{(4)}w + u^{(5)}v^{(2)}w + u^{(6)}vw \in DP^\omega(3)$, which generates a submodule of $DP^\omega(3)$ isomorphic to $K^\omega(3)$. In this way we can regard $K^\omega(3)$ as a submodule of $DP^\omega(3)$.

10.7 Submodules of $Q^d(3)$

In this section we identify generators for submodules of $Q^d(3)$ in certain cases. We begin with the case where $d = 2(2^t - 1)$, so that $Q^d(3) = Q^\gamma(3)$ where $\gamma = (2,\ldots,2)$ is a head sequence of length t. In this case the results of Section 6.9 provide monomial bases for all the composition factors in the composition series (6.5) in terms of the normalized blocks $C(Y)$, $Y \subseteq Z[n]$. Since blocks in the same class Y are equivalent mod hit elements, we may choose any block from each class.

In the case $t = 3$, we have a filtration $A \subset B \subset Q^{14}(3)$, where $A \cong V(3)^*$ is spanned by the spike blocks $C_1 = \begin{smallmatrix}0&0&0\\1&1&1\\1&1&1\end{smallmatrix}, C_2 = \begin{smallmatrix}1&1&1\\0&0&0\\1&1&1\end{smallmatrix}, C_3 = \begin{smallmatrix}1&1&1\\1&1&1\\0&0&0\end{smallmatrix}$, $B/A \cong V(3)$ by $C_{1,2} = \begin{smallmatrix}0&1&1\\1&0&0\\1&1&1\end{smallmatrix}, C_{1,3} = \begin{smallmatrix}0&1&1\\1&1&1\\1&0&0\end{smallmatrix}, C_{2,3} = \begin{smallmatrix}1&1&1\\0&1&1\\1&0&0\end{smallmatrix}$, and $Q^{14}(3)/B \cong I(3)$ by $C_{1,2,3} = \begin{smallmatrix}0&1&1\\1&0&1\\1&1&0\end{smallmatrix}$. For $t > 3$, a basis of $Q^d(3)$ is given by the corresponding blocks C_Y obtained by iteration of the duplication map δ. From the table in Theorem 10.6.2, $Q^d(3) \cong FL^1(3) \cong FL_1(3) \oplus I(3)$. Thus B is a direct summand of $Q^d(3)$ isomorphic to $FL_1(3)$. We next determine the generator f of the summand $I(3)$.

Proposition 10.7.1 *The sum* $f = \sum_Y C_Y$ *over nonempty* $Y \subseteq \{1,2,3\}$ *is invariant in* $Q^d(3)$, *where* $d = 2(2^t - 1)$ *and* $t \geq 3$.

Proof Since blocks in the same class Y are equivalent modulo hit elements, f is symmetric modulo hit elements. Thus, by our usual argument, it is sufficient to show that $f \cdot U \sim f$ where U is the standard transvection. Clearly C_Y is fixed by U if $2 \notin Y$. We obtain $C_2 \cdot U \sim C_2 + C_1$, $C_{1,2} \cdot U \sim C_{1,2} + C_1$, $C_{2,3} \cdot U \sim C_{2,3} + C_{1,3}$ and $C_{1,2,3} \cdot U \sim C_{1,2,3} + C_{1,3}$. It follows that $f \cdot U \sim f$. $\qquad\square$

Next we consider the case $d = 2^s - 1$. For the submodule $Q^\tau(3)$ where $\tau = (1,\ldots,1)$ is a tail sequence of length s, the situation is similar to that for the head modules. Thus there is a filtration $A \subset B \subset Q^\tau(3)$. In the case $s = 3$, $A \cong V(3)$ is spanned by $C'_1 = \begin{smallmatrix}1&1&1\\0&0&0\\0&0&0\end{smallmatrix}, C'_2 = \begin{smallmatrix}0&0&0\\1&1&1\\0&0&0\end{smallmatrix}, C'_3 = \begin{smallmatrix}0&0&0\\0&0&0\\1&1&1\end{smallmatrix}$, $B/A \cong V(3)^*$ by $C'_{1,2} = \begin{smallmatrix}1&0&0\\0&1&1\\0&0&0\end{smallmatrix}, C'_{1,3} = \begin{smallmatrix}1&0&0\\0&0&0\\0&1&1\end{smallmatrix}, C'_{2,3} = \begin{smallmatrix}0&0&0\\1&0&0\\0&1&1\end{smallmatrix}$, and $Q^\tau(3)/B \cong I(3)$ by $C'_{1,2,3} = \begin{smallmatrix}1&0&0\\0&1&0\\0&0&1\end{smallmatrix}$. For $s > 3$, we denote by C'_Y the corresponding special blocks (Definition 8.2.2). From the table in Theorem 10.6.2, $Q^\tau(3) \cong FL^2(3) \cong FL_2(3) \oplus I(3)$. Thus B

is a direct summand of $Q^\tau(3)$ isomorphic to $FL_2(3)$. We next determine the generator f of the summand $I(3)$.

Proposition 10.7.2 *The sum* $f' = \sum_Y C'_Y$ *over nonempty* $Y \subseteq \{1,2,3\}$ *is invariant in* $Q^d(3)$, *where* $d = 2^s - 1$ *and* $s \geq 3$.

Proof The proof is similar to that of Proposition 10.7.1. Again it is easy to see that f' is symmetric modulo hit elements, and that C'_Y is fixed by U if $1 \notin Y$. We obtain $C'_1 \cdot U \sim C'_1 + C'_2$, $C'_{1,2} \cdot U \sim C'_{1,2} + C'_2$, $C'_{1,3} \cdot U \sim C'_{1,3} + C'_{2,3}$ and $C'_{1,2,3} \cdot U \sim C'_{1,2,3} + C'_{2,3}$. It follows that $f' \cdot U \sim f'$. $\qquad\square$

We next consider the contribution to $Q^d(3)$ from the ω-sequence $(3)|\gamma$, where $\gamma = (2,\ldots,2)$ is a head sequence of length $s - 2$. By Proposition 10.6.1 $Q^{(1,\ldots,1)}(3)$ is a direct summand in $Q^d(3)$. The complementary summand B is isomorphic to $Q^\gamma(3)$ via the down Kameko map κ, and so to $I(3)$, $V(3)^*$, $FL_1(3)$ and $FL^1(3)$ for $s = 2,3,4$ and $s \geq 5$ respectively. We consider generators of B and its submodules. The next result follows from Proposition 10.7.2.

Proposition 10.7.3 $Q^3(3) = Q^{(1,1)}(3) \oplus B$, *where* $B \cong I(3)$ *is generated by* $xyz + x^2y + y^2z + z^2x + x^3 + y^3 + z^3$. $\qquad\square$

Proposition 10.7.4 $Q^7(3) = Q^{(1,1,1)}(3) \oplus B$, *where* $B \cong V(3)^*$ *is generated by* $x^3y^3z + xy^2z^4$.

Proof By cyclic permutation of the variables, B contains $f_1 = x^3y^3z + xy^2z^4$, $f_2 = xy^3z^3 + x^4yz^2$ and $f_3 = x^3yz^3 + x^2y^4z$. We show that these span B modulo hit elements. We have $f_1 \sim x^3y^3z + c_{1,2,3}$, $f_2 \sim xy^3z^3 + c_{1,2,3}$, and $f_3 \sim x^3yz^3 + c_{1,2,3}$ where c_Y is a monomial corresponding to a tail block in class Y. It suffices to show that the subspace spanned by f_1, f_2 and f_3 is closed under the action of the switch S_1 which exchanges x and y and under U.

For S_1, we have $f_1 \cdot S_1 \sim f_1$, $f_2 \cdot S_1 \sim f_3$ and $f_3 \cdot S_1 \sim f_2$. For U, we first observe that since $xy^2z^4 \cdot U = (x+y)y^2z^4$, $c_{1,2,3} \cdot U \sim c_{1,2,3} + c_{2,3}$. Similarly $xy^3z^3 \cdot U \sim xy^3z^3 + c_{2,3}$. Hence $f_2 \cdot U \sim f_2$. For f_1, we have $x^3y^3z \cdot U = (x+y)^3y^3z = x^3y^3z + x^2y^4z + xy^5z + y^6z$.

To simplify this, the hit equation $xy^5z + x^2y^4z = Sq^2(xy^3z) + Sq^1(xy^3z^2)$ shows that $x^3y^3z \cdot U = x^3y^3z + c_{2,3}$ and hence $f_1 \cdot U \sim f_1$. Finally we show that $f_3 \cdot U \sim f_3 + f_2$. We have $x^3yz^3 \cdot U = (x+y)^3yz^3 = x^3yz^3 + x^2y^2z^3 + xy^3z^3 + y^4z^3$. Since $x^2y^2z^3 \sim c_{1,2,3}$ by 1-back splicing and $y^4z^3 \sim c_{2,3}$, $f_3 \cdot U \sim x^3yz^3 + xy^3z^3 \sim f_3 + f_2$, as required. $\qquad\square$

Remark 10.7.5 The proof of Theorem 8.3.3 shows that $g = x^6yz + xy^6z + xyz^6 \in P^8(3)$ is not hit, and the proof of Proposition 10.3.2 shows that g is invariant modulo hit elements. Hence g generates a trivial submodule $I(3)$ of

$Q^8(3)$. By transpose duality, Proposition 10.6.3 implies that $Q^8(3) \cong \mathrm{St}(3) \oplus C$, where the indecomposable module $C = (F')^{\mathrm{tr}}$ has a submodule $\cong \mathrm{I}(3)$ generated by g and a quotient $\cong \mathrm{FL}_1(3)$.

Remark 10.7.6 In terms of the monomial basis of $Q^{18}(3)$ of Proposition 8.3.9, a generator f for the summand $\mathrm{I}(3)$ is given by the sum of the 18 blocks given by omitting the three cyclic row permutations of

$$\begin{pmatrix} 1\,1\,1\,0 \\ 1\,1\,0\,0 \\ 0\,0\,0\,1 \end{pmatrix}.$$

Thus f is the cyclic symmetrization of the polynomial $x^{15}y^3 + x^3y^{15} + x^7y^{11} + x^{15}yz^2 + x^7y^9z^2 + x^3y^5z^{10}$. We omit the details, which as usual involve permutations of the variables and the standard transvection U.

The next result completes the identification of the submodule of $Q(3)$ generated by spikes.

Proposition 10.7.7 *Let* $\omega = (2,\dots,2,1,\dots,1)$ *with* s *1s and* t *2s, where* $s,t \geq 1$. *Then the spike* $x^{2^{s+t}-1}y^{2^t-1}$ *generates the submodule* $\mathrm{St}(3)$ *of* $Q^\omega(3)$.

Proof Let S be the submodule of $Q^\omega(3)$ generated by $x^{2^{s+t}-1}y^{2^t-1}$. We show that S has a basis given by the six spikes and any two of the monomials represented by the blocks obtained from $\begin{smallmatrix} 0 & \dots & 0 & 1 & \dots & 1 \\ 1 & \dots & 1 \\ 1 & \dots & 1 \end{smallmatrix}$ by permuting the variables.

The case $s = 2$, $t = 2$ illustrates the general argument. Consider the effect of transvections $T_{i,j}$ on the spike block

$$\begin{matrix} 1\,1\,1\,1 \\ 1\,1\,0\,0\,, \\ 0\,0\,0\,0 \end{matrix}$$

where $T_{i,j}$ adds x_i to x_j and fixes the other variables. Omitting spike blocks, $U_1 = T_{1,2}$ and $U_2 = T_{2,3}$ give sums

$$\begin{matrix} 1\,1\,1\,0 & 1\,1\,0\,1 & & 1\,1\,1\,1 & 1\,1\,1\,1 \\ 1\,1\,0\,1 + 1\,1\,1\,0\,, & 1\,0\,0\,0 + 0\,1\,0\,0 \\ 0\,0\,0\,0 & 0\,0\,0\,0 & & 0\,1\,0\,0 & 1\,0\,0\,0 \end{matrix}$$

which are hit using 1-back splicing, so these operations give no new elements of S. The transvection $T_{1,3}$ applied to the spike block above gives the sum of

16 blocks, which include two spikes and 12 blocks paired by 1-back splicing.
The sum of the remaining two blocks

$$
\begin{array}{ccc}
1\,1\,0\,0 & 0\,0\,1\,1 & 1\,1\,0\,0 \\
1\,1\,0\,0 + 1\,1\,0\,0 \sim 0\,0\,1\,1 & & \\
0\,0\,1\,1 & 1\,1\,0\,0 & 1\,1\,0\,0
\end{array}
\qquad (10.3)
$$

by 2-back splicing. Thus any two of the three blocks in (10.3), together with
the spikes, span a subspace of S of dimension 8. In the same way, it can be seen
that transvections applied to the blocks in (10.3) give no further cohits, so S is
a submodule of $Q^{18}(3)$ of dimension 8. By Theorem 10.6.2 we can identify S
with the submodule $\mathrm{St}(3)$ of $Q^{18}(3)$.

This argument can be extended to all $s,t \geq 1$ by using higher order splicing
operations to cancel the blocks produced by applying the transvection $T_{i,j}$ to a
spike block. This is effectively the same argument as for the 2-variable case,
Theorem 1.8.2, applied to rows i and j. □

For $\omega = (2,\ldots,2,1,\ldots,1)$ with s 1s and t 2s, it follows from Theorem 10.6.2
that $Q^\omega(3)$ is a cyclic $\mathbb{F}_2\mathrm{GL}(3)$-module. We next give a generator of $Q^\omega(3)$ in
each of the four special cases for which $\mathbb{F}_2$-bases of $Q^\omega(3)$ have been given
in Section 8.3. Generators in cases where $t > 2$ or $s > 2$ can be obtained from
these by duplicating the first or last columns.

Proposition 10.7.8 *The blocks*

$$
B = \begin{array}{cccc}
1\,1 & 1\,1\,0 & 1\,0\,1 & 1\,0\,1\,0 \\
1\,0 , & 1\,0\,0 , & 1\,1\,0 , & 1\,1\,0\,0 \\
0\,0 & 0\,0\,1 & 0\,1\,0 & 0\,1\,0\,1
\end{array}
$$

represent monomial generators of $Q^\omega(3)$ *for* $\omega = (2,1)$, $(2,1,1)$, $(2,2,1)$ *and*
$(2,2,1,1)$ *respectively.*

Proof Since $Q^{(2,1)}(3) \cong \mathrm{St}(3)$, the result for $\omega = (2,1)$ follows from
Proposition 10.7.7. In each case, it is sufficient to show that monomials in
the basis of $Q^\omega(3)$ given in Section 8.3 are in the $\mathbb{F}_2\mathrm{GL}(3)$-submodule M of
$Q^\omega(3)$ generated by the given block B. Since applying $T_{3,1}$ to B gives a spike,
the argument of Proposition 10.7.7 shows that M contains the spikes and the
blocks corresponding to (10.3). To see that all basis vectors of $Q^\omega(3)$ involving
only two of the three variables are in M, in the case $\omega = (2,1,1)$ we apply $T_{3,2}$
to B, and in the case $\omega = (2,2,1,1)$ we apply $T_{3,2}$ followed by $T_{3,1}$ to B. As
$Q^{(2,1,1)}(3)$ is spanned by monomials in two variables and row permutations of
B (see Proposition 8.3.5), the result for $\omega = (2,1,1)$ follows.

In the case $\omega = (2,2,1)$, by applying $T_{1,2}$ to B we find that the block $\begin{smallmatrix} 1 & 0 & 0 \\ 1 & 1 & 1 \\ 0 & 1 & 0 \end{smallmatrix}$ is in M. In terms of the proof of Proposition 8.3.7, this is a block of the first type. As $Q^{(2,2,1)}(3)$ is spanned by monomials in two variables, row permutations of B, and blocks of the first type, the result follows.

Similarly, in the case $\omega = (2,2,1,1)$ it suffices to show that M contains blocks of the five types shown in (8.2). The block B is of the fifth type. Applying the transvection $T_{2,1}$ to B gives a block $C_3 = \begin{smallmatrix} 1 & 0 & 0 & 0 \\ 1 & 1 & 1 & 0 \\ 0 & 1 & 0 & 1 \end{smallmatrix}$, and this is equivalent by 1-back splicing at $(3,1)$ to a block of the third type. Applying $T_{3,2}$ to C_3 gives a block $C_1 = \begin{smallmatrix} 1 & 0 & 0 & 0 \\ 1 & 1 & 1 & 1 \\ 0 & 1 & 0 & 0 \end{smallmatrix}$ of the first type. Since (8.4) shows that each block of the fourth type is equivalent to a sum of two blocks of the second type, it remains to show that M contains a block of the second type. Applying $T_{3,1}$ to C_3 shows that $C_2 + C_3' \in M$, where $C_2 = \begin{smallmatrix} 1 & 1 & 0 & 0 \\ 1 & 1 & 1 & 0 \\ 0 & 0 & 0 & 1 \end{smallmatrix}$ is of the second type and $C_3' = \begin{smallmatrix} 1 & 0 & 0 & 1 \\ 1 & 1 & 1 & 0 \\ 0 & 1 & 0 & 0 \end{smallmatrix}$ is of the third type. Hence $C_2 \in M$. $\qquad\square$

10.8 Remarks

The module structure of $Q^d(3)$ and of $K^d(3)$ was determined by J. M. Boardman [21] (see also [7] and [173]). Submodules of dimension 1 of $Q^d(n)$ are of particular interest in algebraic topology because of Singer's transfer map $\phi_n(\mathbb{F}_2) : \text{Tor}_n^{\mathcal{A}_2}(\mathbb{F}_2, \mathbb{F}_2) \to Q(n)^{\text{GL}(n)}$ [186]. We do not know whether the $\mathbb{F}_2\text{GL}(4)$-module $Q^\omega(4)$ is cyclic for all $\omega \in \text{Seq}(4)$.

11

The dual of the Steenrod algebra

11.0 Introduction

In this chapter we show that the Steenrod algebra A_2 is a **graded Hopf algebra**. This means that it is a graded algebra with additional structure, the main element of which is a **coproduct**. Informally, a product takes two objects and combines them into one larger object, but a coproduct takes one object and describes the possible ways of splitting it into two parts of the same kind. Coproducts are dual to products, in the sense that diagrams involving products can be turned into diagrams involving coproducts by 'reversing the arrows'. This process was illustrated in Chapter 9, where it was shown that the product for the polynomial algebra $P(n)$ gives a coproduct for the divided polynomial algebra $DP(n)$, and vice versa. The algebras $P(n)$ and $DP(n)$ provide a basic example of a pair of dual Hopf algebras.

Section 11.1 provides an introduction to Hopf algebras. The algebraic formalism is expressed in terms of tensor products: the product in A_2 can be regarded as a bilinear map $\mu : A_2 \otimes A_2 \to A_2$ of graded algebras. The coproduct is a map $\phi : A_2 \to A_2 \otimes A_2$ of graded algebras such that $\phi(Sq^k) = \sum_{i+j=k} Sq^i \otimes Sq^j$. We shall see that this is essentially the Cartan formula.

If this property is to be used to define ϕ, it is necessary to show that it is compatible with the Adem relations. For example, since $Sq^2Sq^2 = Sq^3Sq^1$ is an Adem relation, we need to show that $\phi(Sq^2)\phi(Sq^2) = \phi(Sq^3)\phi(Sq^1)$. When this equation is expanded by using the formula $\phi(Sq^k) = \sum_{i+j=k} Sq^i \otimes Sq^j$, we find that the Adem relations $Sq^1Sq^1 = 0$ and $Sq^1Sq^2 = Sq^3$ are also involved in the verification. We avoid this difficulty by using the action of A_2 on $P(n)$, $n \geq 1$, to define the coproduct in Section 11.2. In particular, we rely on the fact (Proposition 3.4.5) that an element $\theta \in A_2^d$ is determined by the polynomial $\theta(x_1 x_2 \cdots x_n)$ for $n \geq d$.

211

As well as a coproduct, or diagonal, a Hopf algebra has two further structure maps, the counit, or **augmentation**, and the **antipode**, or conjugation. For A_2, the counit $\varepsilon : A_2 \to \mathbb{F}_2$ maps an element to its component in grading 0, i.e. it is the linear map defined by $\varepsilon(Sq^0) = 1$ and $\varepsilon(Sq^k) = 0$ for $k \geq 1$. The antipode is the map $\chi : A_2 \to A_2$. In Section 11.3 we show that, with these definitions, A_2 is a Hopf algebra.

When A is a finite dimensional algebra over a field F, the dual space $A^* = \mathrm{Hom}(A, F)$ is a coalgebra, with structure maps obtained by dualizing those of A. The same is true if $A = \sum_{d \geq 0} A^d$ is a graded algebra such that A^d is finite dimensional for all degrees d. In Section 11.4 we explain how, with the same restrictions, the dual space H^* of a Hopf algebra H is also a Hopf algebra, and $(H^*)^* \cong H$.

In Section 11.5 we introduce the dual A_2^* of the Steenrod algebra A_2, and show that it is a polynomial algebra $\mathbb{F}_2[\xi_1, \xi_2, \ldots]$ with generators ξ_i of degree $2^i - 1$ for $i \geq 1$. We give formulae for evaluating the coproduct and the conjugate of the generators ξ_i. In Section 11.6 we apply these results to express the conjugate $\chi(Sq(R))$ of a Milnor basis element in a manner similar to Milnor's product formula.

11.1 Hopf algebras

In this section we review background material on Hopf algebras. We begin by defining algebras and coalgebras over an arbitrary field.

Definition 11.1.1 An (associative) **algebra** A is a vector space over a field F, together with a linear map $\mu : A \otimes A \to A$, the **product**, and an injective linear map $\eta : F \to A$, the **unit**, such that the following diagrams are commutative.

$$
\begin{array}{ccc}
A \otimes A \otimes A & \xrightarrow{\mu \otimes \mathrm{id}} & A \otimes A \\
{\scriptstyle \mathrm{id} \otimes \mu} \downarrow & & \downarrow {\scriptstyle \mu} \\
A \otimes A & \xrightarrow{\mu} & A
\end{array}
\qquad
\begin{array}{ccc}
F \otimes A \xrightarrow{\eta \otimes \mathrm{id}} & A \otimes A & \xleftarrow{\mathrm{id} \otimes \eta} A \otimes F \\
& {\scriptstyle \cong} \searrow \; \downarrow {\scriptstyle \mu} \; \swarrow {\scriptstyle \cong} & \\
& A &
\end{array}
$$

Here the isomorphisms $F \otimes A \to A$ and $A \otimes F \to A$ are given by scalar multiplication $\alpha \otimes x \mapsto \alpha x$ and $x \otimes \alpha \mapsto \alpha x$, for $\alpha \in F$ and $x \in A$. Writing $\mu(x_1 \otimes x_2) = x_1 x_2$, these diagrams encode the associative law $(x_1 x_2) x_3 = x_1 (x_2 x_3)$ and the identity law $1_A x = x = x 1_A$, where $1_A = \eta(1) \in A$ is the identity element. The remaining standard axioms for an algebra are encoded in the properties of tensor products. The algebra A is **commutative** if the diagram

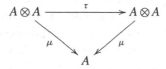

commutes, where $\tau(x_1 \otimes x_2) = x_2 \otimes x_1$ for $x_1, x_2 \in A$.

Definition 11.1.2 A **subalgebra** of an algebra A is a subspace S such that $\mu(S \otimes S) \subseteq S$ and $1_A = \eta(1) \in S$. An **ideal** in A is a subspace $I \neq A$ such that $\mu(A \otimes I) \subseteq I$ and $\mu(I \otimes A) \subseteq I$.

If I is an ideal in the algebra A, the maps μ and η induce an algebra structure on the quotient space A/I with identity $1_A + I \neq I$. With this algebra structure, A/I is a **quotient algebra** of A.

Definition 11.1.3 A (coassociative) **coalgebra** C is a vector space over F together with a linear map $\phi : C \to C \otimes C$, the **coproduct** or **diagonal**, and a surjective linear map $\varepsilon : C \to F$, the **counit** or **augmentation**, such that the following diagrams are commutative.

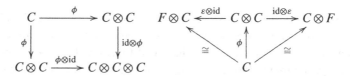

These diagrams encode the **coassociative law** and the **counit law** for C. The isomorphisms $C \to F \otimes C$, $C \to C \otimes F$ are given by $x \mapsto 1 \otimes x$, $x \mapsto x \otimes 1$. The coalgebra C is **cocommutative** if the diagram

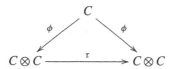

commutes, where τ switches the factors. A **primitive** element is an element $x \in C$ such that $\phi(x) = x \otimes 1 + 1 \otimes x$.

Example 11.1.4 The polynomial algebra $\mathsf{P}(n)$ has a coproduct $\phi : \mathsf{P}(n) \to \mathsf{P}(n) \otimes \mathsf{P}(n)$, given by $\phi(x_i) = x_i \otimes 1 + 1 \otimes x_i$ for each generator x_i, $1 \leq i \leq n$, $\phi(1) = 1 \otimes 1$ and $\phi(fg) = \phi(f)\phi(g)$ for $f, g \in \mathsf{P}(n)$. In this case, there are no relations to check other than commutativity, so ϕ is well defined. For example, if $x \in \mathsf{P}^1(n)$ then $\phi(x) = x \otimes 1 + 1 \otimes x$ and $\phi(x^k) = \sum_{i+j=k} \binom{k}{i} x^i \otimes x^j$ for $k \geq 0$. In particular, $\phi(x_i^{2^j}) = x_i^{2^j} \otimes 1 + 1 \otimes x_i^{2^j}$ for $j \geq 1$, i.e. $x_i^{2^j}$ is primitive.

We remark that $\mathsf{P}(n) \otimes \mathsf{P}(n)$ is isomorphic as a graded algebra to $\mathsf{P}(2n)$, with generators $x_i \otimes 1$ and $1 \otimes x_i$ for $1 \leq i \leq n$. The coproduct on $\mathsf{P}(n)$ is coassociative and cocommutative. This is again easy to check: since all the maps are algebra homomorphisms, it suffices to evaluate them on the generators $x_i \in \mathsf{P}(n)$. The counit $\varepsilon : \mathsf{P}(n) \to \mathbb{F}_2$ maps a polynomial to its constant term.

Example 11.1.5 The divided polynomial algebra $\mathsf{DP}(n)$ has a coproduct $\phi : \mathsf{DP}(n) \to \mathsf{DP}(n) \otimes \mathsf{DP}(n)$, given by $\phi(v_i^{(m)}) = \sum_{r+s=m} v_i^{(r)} \otimes v_i^{(s)}$ for each generator v_i, $1 \leq i \leq n$, $\phi(1) = 1 \otimes 1$ and $\phi(fg) = \phi(f)\phi(g)$ for $f, g \in \mathsf{DP}(n)$. The generators v_i are primitive. To show that ϕ is well defined, we need to check that it is compatible with the relations $v_i^{(r)} v_i^{(s)} = \binom{r+s}{r} v_i^{(r+s)}$. This is the same calculation as in the proof of Proposition 9.1.5: applying ϕ to both sides of this relation and collecting terms in $v_i^{(a)} v_i^{(b)}$ where $a + b = r + s$, the statement reduces to the identity $\binom{a+b}{r} = \sum_{r_1+r_2=r} \binom{a}{r_1} \binom{b}{r_2}$.

As for $\mathsf{P}(n)$, $\mathsf{DP}(n) \otimes \mathsf{DP}(n)$ is isomorphic as a graded algebra to $\mathsf{DP}(2n)$, with generators $v_i \otimes 1$ and $1 \otimes v_i$ for $1 \leq i \leq n$. The coproduct on $\mathsf{DP}(n)$ is coassociative and cocommutative. As for $\mathsf{P}(n)$, the counit map on $\mathsf{DP}(n)$ maps a d-polynomial to its constant term.

Example 11.1.6 Given a set X and a field F, the F-vector space V with basis X is a coalgebra with coproduct $\phi(x) = x \otimes x$ and counit $\varepsilon(x) = 1$ for all $x \in X$. The dual space $V^* = \mathrm{Hom}(V, F)$ is the set of functions $f : X \to F$, and V^* is an algebra with product $(fg)(x) = f(x)g(x)$ for $f, g \in V^*$ and $x \in X$, and identity element e defined by $e(x) = 1$ for all $x \in X$. This is an example of a general construction: a coalgebra structure on a F-vector space V induces an algebra structure on V^*. The converse is true in the finite-dimensional case.

The **Sweedler notation**

$$\phi(x) = \sum_x x' \otimes x'' \tag{11.1}$$

is helpful in writing formulae involving coproducts. The coassociative property $(\mathrm{id} \otimes \phi) \circ \phi(x) = (\phi \otimes \mathrm{id}) \circ \phi(x)$ becomes $\sum_x x' \otimes x'' \otimes x'''$, the counit property becomes $\sum_x \varepsilon(x') x'' = x = \sum_x x' \varepsilon(x'')$ and the cocommutative property becomes $\sum_x x' \otimes x'' = \sum_x x'' \otimes x'$.

Definition 11.1.7 A **subcoalgebra** of a coalgebra C is a vector subspace S such that $\phi(S) \subseteq S \otimes S$ and the restriction of ε to S is surjective. A **coideal** in C is a vector subspace I such that $\phi(I) \subseteq C \otimes I + I \otimes C$ and $\varepsilon(I) = 0$.

If I is a coideal of the coalgebra C, the maps ϕ and ε induce a coalgebra structure on the quotient space C/I. With this structure, C/I is a **quotient coalgebra** of C.

Example 11.1.8 For $k \geq 1$, the polynomial subalgebra $S(k) = \mathbb{F}_2[x_1^{2^k}, \ldots, x_n^{2^k}]$ is a subcoalgebra of $P(n)$, and the ideal $I(k) = \langle x_1^{2^k}, \ldots, x_n^{2^k} \rangle$ is a coideal. The quotient $P(n)/I(k)$ has a $\mathbb{F}_2$-basis given by cosets of monomials with all exponents $< 2^k$. In particular, the quotient algebra $P(n)/I(1)$ has dimension 2^n and a basis given by the cosets of square-free monomials in $x_1, \ldots, x_n$. It is isomorphic to the exterior algebra $\Lambda(n)$ over $\mathbb{F}_2$, with a coproduct defined on the generators by $\phi(x_i + I) = (x_i + I) \otimes (1 + I) + (1 + I) \otimes (x_i + I)$.

For $k \geq 1$, the finite subalgebra $T(k)$ spanned by d-monomials $v_1^{(i_1)} \cdots v_n^{(i_n)}$ with all exponents $i_1, \ldots, i_n < 2^k$ is a subalgebra of $DP(n)$, and the ideal $J(k)$ generated by the elements $v_i^{(2^j)}$ for $1 \leq i \leq n$ and $0 \leq j < k$ is a coideal. The quotient $DP(n)/J(k)$ has a $\mathbb{F}_2$-basis given by cosets of d-monomials with all exponents divisible by 2^k. In particular, the subalgebra $T(1)$ of $DP(n)$ has dimension 2^n and a basis given by d-monomials which are products $v_{i_1} \cdots v_{i_r}$ of distinct generators. Since $v_i^2 = 0$ for $1 \leq i \leq n$, $T(1)$ is also isomorphic to $\Lambda(n)$, and the coproduct is defined on the generators by $\phi(v_i) = v_1 \otimes 1 + 1 \otimes v_i$.

Thus $P(n)$ and $DP(n)$ are both algebras and coalgebras. By Propositions 9.1.4 and 9.1.5, these structures are compatible in the following sense.

Proposition 11.1.9 *Let B be a vector space over a field F with an algebra structure given by the product $\mu : B \otimes B \to B$ and the unit map $\eta : F \to B$, and a coalgebra structure given by the coproduct $\phi : B \to B \otimes B$ and the counit map $\varepsilon : B \to F$. Then μ and η are coalgebra maps if and only if ϕ and ε are algebra maps.*

Proof Both statements can be expressed by the same four commutative diagrams. Thus μ is a coalgebra map if the diagrams

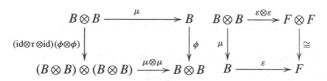

commute, while η is a coalgebra map if the diagrams

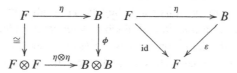

commute. However, the two left hand diagrams above commute if ϕ is a map of algebras, and the two right hand diagrams above commute if ε is a map of algebras. □

Definition 11.1.10 A **bialgebra** over a field F is a vector space B over F with an algebra structure given by maps $\mu : B \otimes B \to F$ and $\eta : F \to B$ and a coalgebra structure given by maps $B \to B \otimes B$ and $\varepsilon : B \to F$ such that the four diagrams of Proposition 11.1.9 commute.

Writing $\phi(x) = \sum_x x' \otimes x''$ and $\mu(x, y) = xy$ where $x, y \in B$ and $\eta(1) = 1_B$ for the identity element of B, the four conditions for a bialgebra can be expressed by $\sum_{xy}(xy)' \otimes (xy)'' = \sum_x \sum_y x'y' \otimes x''y''$, $\varepsilon(xy) = \varepsilon(x)\varepsilon(y)$ for all $x, y \in B$ and $\phi(1_B) = 1_B \otimes 1_B$, $\varepsilon(1_B) = 1$.

Example 11.1.11 The group algebra FG of a group G over a field F has a coalgebra structure defined by $\phi(g) = g \otimes g$ and $\varepsilon(g) = 1$ for all $g \in G$. The product in G is linked to the coproduct by the equation $\phi(gh) = \phi(g)\phi(h)$ for all $g, h \in G$. This equation can be read in two ways. Defining the product in $FG \otimes FG$ by $(g_1 \otimes g_2)(h_1 \otimes h_2) = g_1 h_1 \otimes g_2 h_2$, it states that $\phi : FG \to FG \otimes FG$ preserves the product. On the other hand, defining the coproduct in $FG \otimes FG$ by $\phi(g \otimes h) = (g \otimes h) \otimes (g \otimes h)$, it states that the product map $\mu : FG \otimes FG \to FG$ preserves the coproduct.

Definition 11.1.12 An **antipode** for a bialgebra B over a field F is a F-linear map $\chi : B \to B$ such that the diagrams

$$
\begin{array}{ccc}
B \otimes B & \xrightarrow{\mathrm{id} \otimes \chi} & B \otimes B \\
\phi \uparrow & & \downarrow \mu \\
B \xrightarrow{\varepsilon} F & \xrightarrow{\eta} & B
\end{array}
\qquad
\begin{array}{ccc}
B \otimes B & \xrightarrow{\chi \otimes \mathrm{id}} & B \otimes B \\
\phi \uparrow & & \downarrow \mu \\
B \xrightarrow{\varepsilon} F & \xrightarrow{\eta} & B
\end{array}
$$

commute. A **Hopf algebra** is a bialgebra with an antipode.

We shall see that the antipode is unique if it exists. The conditions on χ can be written as $\sum_x x' \chi(x'') = \varepsilon(x)1_B = \sum_x \chi(x')x''$, where $x \in B$.

Example 11.1.13 We have seen that $P(n)$ and $DP(n)$ are bialgebras. With the antipode defined as the identity map, they are Hopf algebras. (Over a general field, the antipode of a polynomial algebra is *minus* the identity map.) For a group algebra FG, the definition reduces to $\chi(g)g = e = g\chi(g)$ for $g \in G$, where $e \in G$ is the identity element. Thus the antipode is the inverse map $\chi(g) = g^{-1}$.

If M is a monoid (such as $M(n, \mathbb{F}_2)$) and if some element of M has no inverse, then FM is a bialgebra, but not a Hopf algebra.

The following result is helpful in understanding the role of the antipode. We use the Sweedler notation $\phi(x) = \sum_x x' \otimes x''$ introduced in (11.1).

Proposition 11.1.14 *Given a bialgebra B over a field F, let $A = \text{Hom}(B, B)$ be the F-space of linear maps $f : B \to B$. Then*

(i) *with the 'convolution' product $f * g$ of $f, g \in A$ defined by $(f * g)(x) = \sum_x f(x')g(x'')$, where $x \in B$, and the identity element $1_A = \eta \circ \varepsilon$, A is an algebra over F;*

(ii) *an antipode for B is an inverse in the algebra A for the identity map id_B of B;*

(iii) *if B has an antipode, then it is unique.*

Proof (i) Since $f, g \in A$ are linear, the convolution product $f * g$ is bilinear. Hence it defines a map $A \otimes A \to A$, which is associative since $((f * g) * h)(x) = \sum_x f(x') \otimes g(x'') \otimes h(x''') = (f * (g * h))(x)$ for $f, g, h \in A$ and $x \in B$. The counit property $\sum_x \varepsilon(x')x'' = x = \sum_x x'\varepsilon(x'')$ implies that for $f \in A$ and $x \in B$

$$(1_A * f)(x) = ((\eta \circ \varepsilon) * f)(x) = \sum_x \varepsilon(x')f(x'') = f(\sum_x \varepsilon(x')x'') = f(x),$$

and so $1_A * f = f$, and similarly $f * 1_A = f$. Hence 1_A is an identity element for the convolution product.

(ii) In terms of the convolution product, the definition of an antipode χ for B may be written as $\chi * \text{id}_B = \eta \circ \varepsilon = \text{id}_B * \chi$. Hence χ is an inverse for id_B in A.

(iii) Let χ and χ' be antipodes for B. Then $\chi = \chi * 1_A = \chi * (\text{id}_B * \chi') = (\chi * \text{id}_B) * \chi' = 1_A * \chi' = \chi'$. $\qquad\square$

Definition 11.1.15 Let H be a Hopf algebra over a field F. A **Hopf subalgebra** of H is a F-subspace B which is a subalgebra and a subcoalgebra of H and which is closed under the antipode χ of H, i.e. $\chi(B) \subseteq B$. A **Hopf ideal** of H is a F-subspace I which is an ideal and a coideal of H and which is closed under χ. If I is a Hopf ideal in A, the structure maps μ, ϕ, η, ε and χ of A induce a Hopf algebra structure on the quotient space A/I. With this structure, A/I is a **quotient Hopf algebra** of A.

Example 11.1.16 Since the antipode in $P(n)$ and $DP(n)$ is the identity map, the ideals $I(k)$ in $P(n)$ and $J(k)$ in $DP(n)$ of Example 11.1.8 are Hopf ideals, and so $P(n)/I(k)$ and $DP(n)/J(k)$ are finite dimensional Hopf algebras.

11.2 The coproduct $A_2 \to A_2 \otimes A_2$

In this section and the next we shall show that the Steenrod algebra A_2 is a **graded Hopf algebra**, as also are $P(n)$ and $DP(n)$. In other words, it is a graded algebra with a Hopf algebra structure such that the coproduct, counit and antipode are graded maps.

If $A = \sum_{i \geq 0} A^i$ and $B = \sum_{j \geq 0} B^j$ are graded vector spaces over a field F, then their graded tensor product $A \otimes B = \sum_{k \geq 0} (A \otimes B)^k$, where $(A \otimes B)^k = \sum_{i+j=k} A^i \otimes B^j$. If A and B are graded algebras, then $A \otimes B$ is a graded algebra, with product defined on elements of the form $a \otimes b$ by $(a' \otimes b')(a'' \otimes b'') = a'a'' \otimes b'b''$ and extended by bilinearity.

Recall from Section 3.3 that for a sequence $R = (r_1, \ldots, r_\ell) \in$ Seq with $|R| = n$, the Cartan symmetric function $c(R)$ is the sum of all monomials $f = x_1^{2^{a_1}} \cdots x_n^{2^{a_n}}$ such that $\omega(f) = R$. Thus $c(R)$ is a symmetric polynomial in $P^{n+d}(n)$ where $n + d = \deg_2 R = \sum_{i=1}^{\ell} 2^{i-1} r_i$, and its terms are represented by n-blocks with one entry 1 in each row. We denote by $C^{n+d}(n)$ the subspace spanned by the Cartan symmetric functions $c(R)$ in $P^{n+d}(n)$.

By splitting n-blocks into their first r and last $n - r$ rows, we can write

$$c(R) = \sum c(R')c(R''), \qquad (11.2)$$

where the sum is over ordered pairs (R', R'') such that $R = R' + R''$, and where $c(R')$ and $c(R'')$ are Cartan symmetric functions in the first r and the last $n - r$ variables, so that $|R'| = r$ and $|R''| = n - r$. We include the cases $r = 0$ and $r = n$. For example, when $n = 4$ and $r = 2$, $c(1,2,1) = c(1,1)c(0,1,1) + c(0,1,1)c(1,1) + c(0,2)c(1,0,1) + c(1,0,1)c(0,2)$.

An isomorphism $P(n) \to P(r) \otimes P(n - r)$ of graded algebras is given by $x_i \mapsto x_i \otimes 1$ for $i \leq r$ and $x_i \mapsto 1 \otimes x_{i-r}$ for $i > r$. Thus (11.2) gives an injective linear map

$$\phi_{n,r} : C^{n+d}(n) \longrightarrow \sum_{d'+d''=d} C^{r+d'}(r) \otimes C^{n-r+d''}(n-r) \qquad (11.3)$$

defined by $\phi_{n,r}(c(R)) = \sum c(R') \otimes c(R'')$, where $\deg_2 R' = r + d'$, $\deg_2 R'' = n - r + d''$. We remark that there may be several terms with the same d' and d''.

By Proposition 3.4.5, an element $\theta \in A_2^d$ is determined uniquely by the symmetric function $\theta(c(n)) \in C^{n+d}(n)$ when $n \geq d$. The basis elements $c(R)$ of $C^{n+d}(n)$ are indexed by binary partitions of $n + d$ of length n, or equivalently by spike partitions of d of length $\leq n$. Thus for $n \geq d$, $\dim C^{n+d}(n) = \dim A_2^d$, the number of spike partitions of d (Proposition 3.5.13).

Proposition 11.2.1 *There is a commutative diagram of* $\mathbb{F}_2$-*isomorphisms*

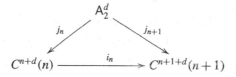

where $n \geq d$, $j_n(\theta) = \theta(\mathsf{c}(n))$ *and* $i_n(\mathsf{c}(r_1, r_2, \ldots, r_\ell)) = \mathsf{c}(r_1 + 1, r_2, \ldots, r_\ell)$.

Proof The maps j_n and j_{n+1} are isomorphisms by Proposition 3.4.5. The diagram commutes because the action of $\theta \in A_2^d$ on the product $\mathsf{c}(n)$ cannot affect more than d of the variables. $\qquad\qquad\square$

The maps (11.3) are compatible with the isomorphisms i_n, in the sense that the diagram below commutes for $r \geq d$, and the similar diagram relating $\phi_{n,r}$ with $\phi_{n+1,r}$ commutes for $n - r \geq d$.

$$
\begin{array}{ccc}
C^{n+d}(n) & \xrightarrow{\;\phi_{n,r}\;} & \sum_{d'+d''=d} C^{r+d'}(r) \otimes C^{n-r+d''}(n-r) \\
{\scriptstyle i_n}\big\downarrow{\scriptstyle\cong} & & {\scriptstyle\cong}\big\downarrow{\scriptstyle i_r\otimes\mathrm{id}} \\
C^{n+1+d}(n+1) & \xrightarrow{\;\phi_{n+1,r+1}\;} & \sum_{d'+d''=d} C^{r+1+d'}(r+1) \otimes C^{n-r+d''}(n-r)
\end{array}
$$

Thus we define $\phi : A_2 \to A_2 \otimes A_2$ so that the following diagram commutes when r and $n - r$ are $\geq d$. By Proposition 11.2.1, j_n and $j_r \otimes j_{n-r}$ are isomorphisms.

$$
\begin{array}{ccc}
A_2^d & \xrightarrow{\;\phi\;} & \sum_{d'+d''=d} A_2^{d'} \otimes A_2^{d''} \\
{\scriptstyle j_n}\big\downarrow{\scriptstyle\cong} & & {\scriptstyle\cong}\big\downarrow{\scriptstyle j_r\otimes j_{n-r}} \\
C^{n+d}(n) & \xrightarrow{\;\phi\;} & \sum_{d'+d''=d} C^{r+d'}(r) \otimes C^{n-r+d''}(n-r)
\end{array}
$$

Definition 11.2.2 The **coproduct** (or diagonal map) $\phi : A_2 \to A_2 \otimes A_2$ is defined by $\phi(\theta) = \sum \theta' \otimes \theta''$, where for $\theta \in A_2^d$ and $r, n - r \geq d$

$$
\theta(x_1 x_2 \cdots x_n) = \sum \theta'(x_1 \cdots x_r)\theta''(x_{r+1} \cdots x_n). \tag{11.4}
$$

The map ϕ is well defined, since this sum is independent of r when r and $n - r$ are $\geq d$. In general, the sum will contain several terms with the same d' and d''.

Example 11.2.3 Let $\theta = Sq(1,1)$. Then $\theta(c(n)) = c(n-2,1,1)$ is the monomial symmetric function $\sum x_1^4 x_2^2 x_3 \cdots x_n$, and splits into four sums

$$c(r-2,1,1)c(n-r) = Sq(1,1)(x_1\cdots x_r)Sq(0)(x_{r+1}\cdots x_n),$$
$$c(r-1,0,1)c(n-r-1,1) = Sq(0,1)(x_1\cdots x_r)Sq(1)(x_{r+1}\cdots x_n),$$
$$c(r-1,1)c(n-r-1,0,1) = Sq(1)(x_1\cdots x_r)Sq(0,1)(x_{r+1}\cdots x_n),$$
$$c(r)c(n-r-2,1,1) = Sq(0)(x_1\cdots x_r)Sq(1,1)(x_{r+1}\cdots x_n),$$

and so $\phi(Sq(1,1)) = Sq(1,1)\otimes 1 + Sq(0,1)\otimes Sq(1) + Sq(1)\otimes Sq(0,1) + 1\otimes Sq(1,1)$.

Proposition 11.2.4 *The coproduct* $\phi : A_2 \to A_2 \otimes A_2$ *satisfies*

(i) $\phi(Sq^k) = \sum_{i+j=k} Sq^i \otimes Sq^j$, *for all* $k \geq 0$,
(ii) $\phi(Sq(R)) = \sum_{S+T=R} Sq(S) \otimes Sq(T)$, *for all* $R \in$ Seq.

Proof Both formulae follow directly from the definition of ϕ, using the Cartan formula 1.1.4 and Proposition 3.5.7. □

It follows from (ii) (or from Proposition 3.5.10(i)) that $Q_k = Sq(0,\ldots,0,1)$ is primitive, i.e. $\phi(Q_k) = Q_k \otimes 1 + 1 \otimes Q_k$ for $k \geq 1$.

Definition 11.2.5 The **counit** (or **augmentation**) $\varepsilon : A_2 \to \mathbb{F}_2$ is the linear map which associates to an element of A_2 its component in grading 0.

Proposition 11.2.6 *With ϕ and ε as structure maps, the Steenrod algebra A_2 is a cocommutative coalgebra, i.e. the following diagrams commute:*

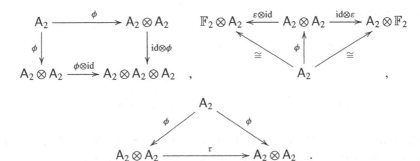

Proof Let $\theta \in A_2^d$ and $r,s,t \geq d$. Coassociativity of ϕ follows by rewriting the expansion

$$\theta(x_1\cdots x_r y_1\cdots y_s z_1\cdots z_t) = \sum \theta'(x_1\cdots x_r)\theta''(y_1\cdots y_s)\theta'''(z_1\cdots z_t)$$

in two ways. For cocommutativity of ϕ, we observe that $\theta(x_1\cdots x_r y_1\cdots y_r)$ is symmetric in all the variables, and in particular it is invariant under exchanging the x's with the y's. The counit properties state that $\phi(Sq^0) = Sq^0 \otimes Sq^0$ and that for $\theta \in A_2^+$, $\phi(\theta)$ has the form $\theta \otimes 1 + \sum \theta' \otimes \theta'' + 1 \otimes \theta$, where all elements θ' and θ'' are in A_2^+. These statements are clear by considering terms in $\theta(x_1 \cdots x_r y_1 \cdots y_s)$ of the form $f(x_1, \ldots, x_r) y_1 \cdots y_s$ or $x_1 \cdots x_r g(y_1, \ldots, y_s)$. □

11.3 The Hopf algebra structure of A_2

In this section we show that the Steenrod algebra A_2 is a Hopf algebra. We begin by showing that A_2 is a bialgebra with the coproduct ϕ and counit ε of Section 11.2.

Proposition 11.3.1 *The coproduct* $\phi : A_2 \to A_2 \otimes A_2$ *is a homomorphism of algebras, i.e. if* $\theta_1, \theta_2 \in A_2$ *and* $\phi(\theta_1) = \sum_{\theta_1} \theta_1' \otimes \theta_1''$, $\phi(\theta_2) = \sum_{\theta_2} \theta_2' \otimes \theta_2''$, *then* $\phi(\theta_1 \theta_2) = \sum_{\theta_1} \sum_{\theta_2} \theta_1' \theta_2' \otimes \theta_1'' \theta_2''$.

Proof By Proposition 3.4.5, $\theta \in A_2^d$ is determined by $\theta(c(n))$ when $n \geq d$. Hence θ and all θ', θ'' are determined by equation (11.4). It follows that for $n \geq 1$ and $f, g \in P(n)$, the coproduct $\phi : A_2 \to A_2 \otimes A_2$ satisfies the equation

$$\theta(fg) = \sum \theta'(f)\theta''(g), \text{ where } \phi(\theta) = \sum \theta' \otimes \theta''. \qquad (11.5)$$

Thus for $\theta_1, \theta_2 \in A_2$, $\theta_1 \theta_2(fg) = \theta_1(\sum_{\theta_2} \theta_2'(f) \theta_2''(g)) = \sum_{\theta_2} \theta_1(\theta_2'(f) \theta_2''(g)) = \sum_{\theta_1} \sum_{\theta_2} \theta_1'(\theta_2'(f)) \theta_1''(\theta_2''(g)) = \sum_{\theta_1} \sum_{\theta_2} \theta_1' \theta_2'(f) \theta_1'' \theta_2''(g)$. By (11.4) $\phi(\theta_1 \theta_2) = \sum_{\theta_1} \sum_{\theta_2} \theta_1' \theta_2' \otimes \theta_1'' \theta_2'' = \left(\sum_{\theta_1} \theta_1' \otimes \theta_1''\right)\left(\sum_{\theta_2} \theta_2' \otimes \theta_2''\right) = \phi(\theta_1)\phi(\theta_2)$. □

Proposition 11.3.2 *For a monomial* Sq^A *in* A_2, $\phi(Sq^A) = \sum_{A=B+C} Sq^B \otimes Sq^C$.

Proof This follows immediately from Propositions 11.2.4 and 11.3.1. □

The next result shows that the conjugation $\chi : A_2 \to A_2$ is an antipode.

Proposition 11.3.3 *For all* $\theta \in A_2^+$,

$$\sum \theta' \chi(\theta'') = 0 = \sum \chi(\theta')\theta'',$$

where $\phi(\theta) = \sum \theta' \otimes \theta''$.

Proof By linearity we may assume that $\theta = Sq^A$ is a monomial. Let $\theta = \theta_1 \theta_2$, where $\theta_2 = Sq^k$, $k > 0$. Then $\sum_\theta \theta' \chi(\theta'') = \sum_{\theta_1 \theta_2} (\theta_1 \theta_2)' \chi((\theta_1 \theta_2)'') = \sum_{\theta_1} \sum_{\theta_2} \theta_1' \theta_2' \chi(\theta_1'' \theta_2'') = \sum_{\theta_1} \sum_{\theta_2} \theta_1' \theta_2' \chi(\theta_2'') \chi(\theta_1'') = \sum_{\theta_1} \theta_1' \left(\sum_{\theta_2} \theta_2' \chi(\theta_2'')\right) \chi(\theta_1'')$, by Propositions 11.3.1 and 4.3.8. The inner sum $\sum_{\theta_2} \theta_2' \chi(\theta_2'') = \sum_{i=0}^k Sq^i Xq^{k-i} = 0$ by Proposition 4.3.2. Hence $\sum \theta' \chi(\theta'') = 0$, and similarly $\sum \chi(\theta')\theta'' = 0$. □

Example 11.3.4 As in Proposition 3.5.10, let $Q_k = Sq(0, \ldots, 0, 1)$, where the 1 is in the kth place. Since Q_k is a primitive, it follows from Proposition 11.3.3 that $\chi(Q_k) = Q_k$.

Proposition 11.3.5 *The conjugation map χ is a map of coalgebras, i.e. if $\theta \in A_2$ and $\phi(\theta) = \sum \theta' \otimes \theta''$, then $\phi(\chi(\theta)) = \sum \chi(\theta') \otimes \chi(\theta'')$ and $\varepsilon(\chi(\theta)) = \varepsilon(\theta)$.*

Proof The second statement is clear since $\chi(Sq^0) = Sq^0$ and χ maps A_2^+ to A_2^+ by Proposition 11.3.3. By Propositions 11.3.1 and 4.3.8, it suffices to check the first statement for $\theta = Sq^k$, $k \geq 1$. Since $\phi(Sq^k) = \sum_{i+j=k} Sq^i \otimes Sq^j$, we have to prove that

$$\phi(Xq^k) = \sum_{i+j=k} Xq^i \otimes Xq^j.$$

By Proposition 3.5.11, Xq^k is the sum of all Milnor basis elements $Sq(R)$ of grading k. By Proposition 11.2.4(ii), $\phi(Sq(R)) = \sum_{S+T=R} Sq(S) \otimes Sq(T)$. Hence $\phi(Xq^k) = \sum Sq(S) \otimes Sq(T)$, where the sum is over all $Sq(S)$ and $Sq(T)$ whose degrees i and j satisfy $i + j = k$. But $Xq^i \otimes Xq^j$ is the sum of all $Sq(S) \otimes Sq(T)$ where $Sq(S)$ has degree i and $Sq(T)$ has degree j. $\qquad\square$

To summarize these results, we make the following definitions.

Definition 11.3.6 A graded algebra $A = \sum_{d \geq 0} A^d$ over a field F is **connected** if $A^0 \cong F$. If a connected, graded algebra A is a Hopf algebra over F, then it is a **connected, graded Hopf algebra** if the coproduct ϕ, the counit ε and the antipode χ are graded maps, i.e. ϕ maps A^d to $(A \otimes A)^d = \sum_{d=d'+d''} A^{d'} \otimes A^{d''}$, ε maps A^d to 0 for $d > 0$ and $\varepsilon : A^0 \cong F$, and χ maps A^d to A^d for $d \geq 0$. A Hopf algebra is **cocommutative** if the coproduct is commutative.

Theorem 11.3.7 *The Steenrod algebra A_2 is a connected, graded, cocommutative Hopf algebra, with coproduct ϕ, counit ε and conjugation χ.* $\qquad\square$

Proof The required properties are stated in Section 11.1. These are verified by Propositions 11.2.6, 11.3.1, 11.3.3, and easily seen properties of ε. $\qquad\square$

We conclude this section with an important example of a quotient Hopf algebra of A_2.

Example 11.3.8 Let $I = A_2 Sq^1 A_2$ be the 2-sided ideal in A_2 generated by Sq^1. It follows from the Adem relation $Sq^1 Sq^{2k} = Sq^{2k+1}$ that the admissible monomials Sq^A, where at least one term of A is odd, form a $\mathbb{F}_2$-basis for I.

The Milnor basis elements $Sq(R)$, where R has at least one odd term, also give a $\mathbb{F}_2$-basis for I. To see this, we use the formula $Q_{k+1} = Sq^{2^k} Q_k + Q_k Sq^{2^k}$

of Proposition 3.5.10(v) to show that $Q_k \in I$ by induction on k, and then apply Milnor's product formula. Since Sq^1 is primitive, I is a coideal, and since Sq^1 is invariant under χ, so is I. The quotient algebra A_2/I is generated by the cosets of Sq^{2^j} for $j \geq 1$. We shall see in Section 13.3 that A_2/I is isomorphic to the Hopf algebra obtained from A_2 by doubling the grading.

11.4 Duality for Hopf algebras

Given vector spaces V and W over a field F, we denote by $\text{Hom}(V,W)$ the vector space of linear maps $V \to W$. The dual of V is $V^* = \text{Hom}(V,F)$. Given $h \in \text{Hom}(V,W)$, the dual (or transpose) $h^* \in \text{Hom}(W^*,V^*)$ of h is defined by $h^*(g) = g \circ h$, where $g \in W^*$.

There is a linear map $i : V^* \otimes W^* \to (V \otimes W)^*$ defined on elements of the form $f \otimes g \in V^* \otimes W^*$ by $i(f \otimes g)(v \otimes w) = f(v)g(w)$ for all $v \in V$ and $w \in W$. The map i is an isomorphism if V and W are finite-dimensional; in general, it is injective but not surjective. Thus if A is a finite-dimensional algebra over F, we can dualize the product map $\mu : A \otimes A \to A$ to obtain a map $\phi : A^* \to A^* \otimes A^*$ given by $\phi = i^{-1} \circ \mu^*$. For $x, y \in A$ and $f \in A^*$, the coproduct ϕ is given by

$$f(xy) = f(\mu(x \otimes y)) = \phi(f)(x \otimes y) = \sum_f f'(x)f''(y). \quad (11.6)$$

We also define the counit $\varepsilon : A^* \to F$ to be the dual of the unit map $\eta : F \to A$, using the canonical identification of $\text{Hom}(F,F)$ with F. Thus $\varepsilon(f) = f(1_A)$ for $f \in A^*$, where $1_A = \eta(1)$ is the identity element of A. Then ϕ and ε define a coalgebra structure on A^*, which is cocommutative if and only if A is commutative.

In the same way, if C is a coalgebra over F with coproduct ϕ and counit ε, we can define an algebra structure on C^* by $\mu = \phi^* \circ i$ and $\eta = \varepsilon^*$. Since i and not i^{-1} enters here, we need not assume that C is finite-dimensional. For $f, g \in C^*$ and $x \in C$, the product μ is given by

$$(fg)(x) = \mu(f \otimes g)(x) = (f \otimes g)\phi(x) = (f \otimes g)\sum_x x' \otimes x'' = \sum_x f(x')g(x'').$$
$$(11.7)$$

The identity element of C^* is $\eta(1) = \varepsilon : C \to F$. The algebra C^* is commutative if and only if the coalgebra C is cocommutative.

Thus when B is a finite-dimensional bialgebra, B^* is also a bialgebra. Finally, if $\chi : B \to B$ is an antipode for B, it follows by reversing the arrows in the diagrams of Definition 11.1.12, and identifying $(B \otimes B)^*$ with $B^* \otimes B^*$, that

$\chi^* : B^* \to B^*$ is an antipode for B^*. Thus $(\chi^*(f))(x) = f(\chi(x))$ for $x \in B$ and $f \in B^*$.

If V is a finite-dimensional vector space over F, then as in Chapters 9 and 10 we write $\xi(v) = \langle \xi, v \rangle$, where $\xi \in V^*$ and $v \in V$. If $v_1, \ldots, v_n$ is a basis for V, then the dual basis $\xi_i, \ldots, \xi_n$ for V^* is defined by $\langle \xi_i, v_j \rangle = 1$ if $i = j$, and $\langle \xi_i, v_j \rangle = 0$ otherwise. Recall that if W is a finite-dimensional vector space over F, then a bilinear map $V \otimes W \to F$ is nonsingular if and only if the corresponding linear map $V \to W^*$ is an isomorphism. Writing $\phi(x) = \sum_x x' \otimes x''$ as in (11.1), the structure maps on the dual of a Hopf algebra can be defined as follows.

Proposition 11.4.1 *Let H be a finite-dimensional Hopf algebra over a field F, and let $H^* = \mathrm{Hom}(H, F)$. Then H^* is a Hopf algebra where the product is defined by $\langle fg, x \rangle = \sum_x \langle f, x' \rangle \langle g, x'' \rangle$, the identity element is the counit of H, the coproduct is defined by $\langle f, xy \rangle = \sum_f \langle f', x \rangle \langle f'', y \rangle$, the counit is the evaluation map $f \mapsto f(1_H) = \langle f, 1_H \rangle$ and the antipode χ^* is defined by $\langle \chi^*(f), x \rangle = \langle f, \chi(x) \rangle$, where $f, g \in H^*$ and $x, y \in H$. The dual of H^* is H, and H^* is commutative if and only if H is cocommutative.*

Proof The proof is a formal matter of dualizing diagrams. The duals of the two diagrams of Definition 11.1.1 for H are the two diagrams of Definition 11.1.3 for H^*, and the duals of the two diagrams of Definition 11.1.3 for H are the two diagrams of Definition 11.1.1 for H^*. The duals of the four diagrams of Proposition 11.1.9 for H are the corresponding four diagrams for H^*, after exchanging the diagrams for the unit and counit. The duals of the two diagrams of Definition 11.1.12 for H are the corresponding diagrams for H^*.   □

The Steenrod algebra A_2 and the algebras $P(n)$ and $DP(n)$ are not finite dimensional, but they are graded Hopf algebras which are finite dimensional in each degree. Proposition 11.4.1 is easily adapted to give a graded dual for such a Hopf algebra.

Example 11.4.2 It follows from Examples 11.1.4 and 11.1.5 that the polynomial algebra $P(n)$ and the divided polynomial algebra $DP(n)$ are a pair of dual graded Hopf algebras. The exterior algebra $\Lambda(n)$ is a self-dual graded Hopf algebra.

The monomials $x_i^{2^k}$ of $P(n)$, and the generators v_i of $DP(n)$, are primitive. In each case these elements span the vector subspace of primitive elements. The primitive monomial $x_i^{2^k}$ is dual to the indecomposable d-monomial $v_i^{(2^k)}$ in $DP(n)$, and the primitive element v_i is dual to the indecomposable element x_i in $P(n)$. The primitive and indecomposable elements of dual Hopf algebras are related in this manner in general.

The following result summarizes the relation of duality as it applies to Hopf subalgebras and quotients. It can also be extended to graded Hopf algebras which are finite dimensional in each degree.

Proposition 11.4.3 (i) *Let A be a finite-dimensional Hopf algebra over a field F, I a Hopf ideal in A, and $B = A/I$, the quotient Hopf algebra. We regard $B^* - Hom(B,F)$ as a subspace of $A^* = Hom(A,F)$ by composition with the quotient map $A \to B$, i.e. if $f \in A^*$, then $f \in B^*$ if and only if $\langle f,x \rangle = 0$ for all $x \in I$. Then B^* is a Hopf subalgebra of A^*.*

(ii) *Conversely, given a Hopf subalgebra C of A^*, let I be the ideal in A such that $x \in I$ if and only if $\langle f,x \rangle = 0$ for all $f \in C$. Then I is a Hopf ideal, and $C^* \cong A/I$ as Hopf algebras. Thus duality defines a $1-1$ correspondence between the quotient Hopf algebras of A and the Hopf subalgebras of A^*.*

Proof (i) Since $B = A/I$ is a quotient Hopf algebra of A, we have the commutative diagrams

$$
\begin{array}{ccc}
A \otimes A & \xleftarrow{\phi_A} & A \\
{\scriptstyle j \otimes j}\downarrow & & \downarrow{\scriptstyle j} \\
B \otimes B & \xleftarrow{\phi_B} & B ,
\end{array}
\qquad
\begin{array}{ccc}
A & \xleftarrow{\mu_A} & A \otimes A \\
{\scriptstyle j}\downarrow & & \downarrow{\scriptstyle j \otimes j} \\
B & \xleftarrow{\mu_B} & B \otimes B ,
\end{array}
\qquad
\begin{array}{ccc}
A & \xleftarrow{\chi_A} & A \\
{\scriptstyle j}\downarrow & & \downarrow{\scriptstyle j} \\
B & \xleftarrow{\chi_B} & B ,
\end{array}
$$

where $j : A \to B$ is the quotient map. Dualizing all the maps and identifying $(A \otimes A)^*$ with $A^* \otimes A^*$ and $(B \otimes B)^*$ with $B^* \otimes B^*$, we obtain the commutative diagrams

$$
\begin{array}{ccc}
A^* \otimes A^* & \xrightarrow{\mu_{A^*}} & A^* \\
{\scriptstyle j^* \otimes j^*}\uparrow & & \uparrow{\scriptstyle j^*} \\
B^* \otimes B^* & \xrightarrow{\mu_{B^*}} & B^* ,
\end{array}
\qquad
\begin{array}{ccc}
A^* & \xrightarrow{\phi_{A^*}} & A^* \otimes A^* \\
{\scriptstyle j^*}\uparrow & & \uparrow{\scriptstyle j^* \otimes j^*} \\
B^* & \xrightarrow{\phi_{B^*}} & B^* \otimes B^* ,
\end{array}
\qquad
\begin{array}{ccc}
A^* & \xrightarrow{\chi_{A^*}} & A^* \\
{\scriptstyle j^*}\uparrow & & \uparrow{\scriptstyle j^*} \\
B^* & \xrightarrow{\chi_{B^*}} & B^* .
\end{array}
$$

Hence B^* is a Hopf subalgebra of A^*.

(ii) Since C is a Hopf subalgebra of A^*, we have the commutative diagrams

$$
\begin{array}{ccc}
A^* \otimes A^* & \xrightarrow{\mu_{A^*}} & A^* \\
{\scriptstyle i \otimes i}\uparrow & & \uparrow{\scriptstyle i} \\
C \otimes C & \xrightarrow{\mu_C} & C ,
\end{array}
\qquad
\begin{array}{ccc}
A^* & \xrightarrow{\phi_{A^*}} & A^* \otimes A^* \\
{\scriptstyle i}\uparrow & & \uparrow{\scriptstyle i \otimes i} \\
C & \xrightarrow{\phi_C} & C \otimes C ,
\end{array}
\qquad
\begin{array}{ccc}
A^* & \xrightarrow{\chi_{A^*}} & A^* \\
{\scriptstyle i}\uparrow & & \uparrow{\scriptstyle i} \\
C & \xrightarrow{\chi_C} & C ,
\end{array}
$$

where $i : C \to A^*$ is the inclusion map. Dualizing as in (i), we obtain the commutative diagrams

$$
\begin{array}{ccc}
A \otimes A \xleftarrow{\ \phi_A\ } A & A \xleftarrow{\ \mu_A\ } A \otimes A & A \xleftarrow{\ \chi_A\ } A \\
{\scriptstyle i^* \otimes i^*}\Big\downarrow \qquad \Big\downarrow {\scriptstyle i^*} & {\scriptstyle i^*}\Big\downarrow \qquad \Big\downarrow {\scriptstyle i^* \otimes i^*} & {\scriptstyle i^*}\Big\downarrow \qquad \Big\downarrow {\scriptstyle i^*} \\
C^* \otimes C^* \xleftarrow{\ \phi_{C^*}\ } C^*, & C^* \xleftarrow{\ \mu_{C^*}\ } C^* \otimes C^*, & C^* \xleftarrow{\ \chi_{C^*}\ } C^*.
\end{array}
$$

Hence C^* is the quotient subalgebra A^*/I, where $I = \mathrm{Ker}(i^*)$. Regarding $x \in A$ as a linear map $A^* \to F$, $x \in I$ if and only if $x(f) = \langle f, x \rangle = 0$ for all $f \in C$. $\quad\square$

Example 11.4.4 For $A = P(n)$, the ideal $I(k)$ generated by $x_i^{2^k}$, $1 \le i \le n$, is a Hopf ideal. The quotient $B = P(n)/I(k)$ is spanned by (cosets of) monomials with all exponents $< 2^k$, represented by (n,k)-blocks. The dual B^* is the subalgebra $T(k)$ of $DP(n)$ spanned by d-monomials with all exponents $< 2^k$.

For $A = DP(n)$, the ideal $J(k)$ generated by $v_i^{(2^j)}$, where $0 \le j < k$ and $1 \le i \le n$, is a Hopf ideal. The quotient $B = DP(n)/J(k)$ is spanned by (cosets of) d-monomials with all exponents divisible by 2^k, represented by blocks whose first k columns are zero. The dual B^* is the subalgebra $S(k)$ of $P(n)$ generated by $x_i^{2^k}$ for $1 \le i \le n$, which is spanned by monomials with all exponents divisible by 2^k.

11.5 The dual algebra A_2^*

In this section we introduce the dual of the Steenrod algebra A_2. Since A_2 is a graded Hopf algebra over $\mathbb{F}_2$ and A_2^d is finite-dimensional for $d \ge 0$, its graded dual $A_2^* = \sum_{d \ge 0} (A_2^d)^*$ is also a Hopf algebra over $\mathbb{F}_2$. Since A_2 is cocommutative and non-commutative, A_2^* is commutative and non-cocommutative.

By Proposition 11.4.1, the coproduct in A_2 and the product in A_2^* are related by

$$\langle \xi\eta, \theta \rangle = \sum_\theta \langle \xi, \theta' \rangle \langle \eta, \theta'' \rangle \tag{11.8}$$

where $\theta \in A_2$ and $\xi, \eta \in A_2^*$. To calculate with (11.8), we combine dual bases for A_2^d and $(A_2^*)^d$ for each $d \ge 0$ into dual bases for A_2 and A_2^*. In particular, we use the Milnor basis $\{Sq(R)\}$ of A_2, indexed by sequences $R = (r_1, r_2, \ldots) \in$ Seq, and denote the dual basis for A_2^* by $\{\xi^R\}$, so that $\langle \xi^R, Sq(S) \rangle = 1$ if $S = R$ and $\langle \xi^R, Sq(S) \rangle = 0$ otherwise, where $R, S \in$ Seq.

If $R = (0, 0, \ldots)$, then $\{\xi^R\} = 1$, the identity element of A_2^*. For $r > 0$, since $Sq(r, 0, 0, \ldots) = Sq^r$, $\xi^{(r)}$ is the element of A_2^* dual to Sq^r, and we write the dual element $\xi^{(1)}$ to Sq^1 as ξ_1. Since $\phi(Sq^2) = Sq^2 \otimes 1 + Sq^1 \otimes Sq^1 + 1 \otimes Sq^2$, (11.8) gives $\langle \xi_1^2, Sq^2 \rangle = \langle \xi_1, Sq^2 \rangle \langle \xi_1, 1 \rangle + \langle \xi_1, Sq^1 \rangle \langle \xi_1, Sq^1 \rangle + \langle \xi_1, 1 \rangle \langle \xi_1, Sq^2 \rangle$, and so $\xi_1^2 = \xi^{(2)}$.

In degree 3, we have basis elements $\xi^{(3)}$ and $\xi^{(0,1)}$ dual to $Sq(3) = Sq^3$ and $Sq(0,1)$ respectively. Since $\xi_1^3 = \xi_1 \xi^{(2)}$ and $Sq^1 \otimes Sq^2$ is a term in $\phi(Sq^3)$ but not in $\phi(Sq(0,1))$, ξ_1^3 is dual to Sq^3 and so $\xi_1^3 = \xi^{(3)}$. In general, using the factorization $\xi_1^r = \xi_1 \xi_1^{r-1}$ for $r > 1$ and applying 11.2.4(ii) to (11.8), we obtain

$$\langle \xi_1^r, Sq(r_1, r_2, \ldots) \rangle = \begin{cases} \langle \xi_1^{r-1}, Sq(r_1 - 1, r_2, \ldots) \rangle, & \text{if } r_1 > 0, \\ 0, & \text{if } r_1 = 0. \end{cases}$$

It follows by induction that ξ_1^r is dual to Sq^r for all $r > 0$, i.e. $\xi_1^r = \xi^{(r)}$. Thus the polynomial algebra $\mathbb{F}_2[\xi_1]$ is a subalgebra of A_2^*. In Theorem 11.5.1, we show that A_2^* is a polynomial algebra on generators $\xi_k = \xi^{R_k}$, where $R_k = (0, \ldots, 0, 1)$, and $Sq(R_k) = Q_k$. From the discussion for degree 3 above, it follows that $\xi_2 = \xi^{(0,1)}$ is indecomposable, and, using (11.8), Proposition 11.2.4(ii) shows that ξ_k is indecomposable for $k \geq 1$.

By dualizing the product in A_2 using (11.8), we obtain a coproduct in A_2^*, which we again denote by ϕ. Clearly $\phi(\xi_1) = \xi_1 \otimes 1 + 1 \otimes \xi_1$, and since ϕ is multiplicative

$$\phi(\xi_1^k) = \sum_{k=i+j} \binom{k}{i} \xi_1^i \otimes \xi_1^j.$$

In particular, $\phi(\xi_1^{2^a}) = \xi_1^{2^a} \otimes 1 + 1 \otimes \xi_1^{2^a}$ for $a \geq 0$. To compute $\phi(\xi_k)$, we use Theorem 4.1.2 to find all products of Milnor basis elements which contain the term $Sq(R_k)$. For $k = 2$, $Sq(1)Sq(2) = Sq(3)$ and $Sq(2)Sq(1) = Sq(3) + Sq(0,1)$, so by (11.8) we find $\phi(\xi_2) = \xi_2 \otimes 1 + \xi_1^2 \otimes \xi_1 + 1 \otimes \xi_2$. This is an example of the non-cocommutativity of A_2^*.

Theorem 11.5.1 *The algebra A_2^* is isomorphic as a graded algebra to the polynomial algebra $\mathbb{F}_2[\xi_1, \xi_2, \ldots,]$, where ξ_k has degree $2^k - 1$ for $k \geq 1$.*

Proof As above, we consider the basis of A_2^* which is dual to the Milnor basis of A_2, writing ξ^R for the basis element dual to $Sq(R)$ for each sequence $R = (r_1, r_2, \ldots) \in \text{Seq}$. Thus

$$\langle \xi^{(R)}, Sq(S) \rangle = \begin{cases} 1, & \text{if } R = S, \\ 0, & \text{if } R \neq S. \end{cases} \tag{11.9}$$

The product map in A_2^* is the graded dual ϕ^* of the coproduct ϕ in A_2. Hence $\langle \xi^{(S)} \xi^{(T)}, Sq(R) \rangle = \langle \phi^*(\xi^{(S)} \otimes \xi^{(T)}), Sq(R) \rangle = \langle \xi^{(S)} \otimes \xi^{(T)}, \phi(Sq(R)) \rangle = \langle \xi^{(S)} \otimes \xi^{(T)}, \sum_{R=S'+T'} Sq(S') \otimes Sq(T') \rangle = \sum_{R=S'+T'} \langle \xi^{(S)}, Sq(S') \rangle \langle \xi^{(T)}, Sq(T') \rangle$.

This evaluates to 1 if and only if $S = S'$, $T = T'$ for some term in the sum, i.e. if and only if $R = S + T$. It follows from (11.9) that $\xi^S \xi^{(T)} = \xi^{S+T}$.

The sequence $R = (r_1, r_2, \ldots r_k, 0, \ldots)$ can be written in the form $R = \sum_{i=1}^{k} r_i R_i$ where, as above, $R_i = (0, \ldots, 0, 1)$ with the ith term equal to 1. Thus writing $\xi_i = \xi^{R_i}$ for $i \geq 1$, we obtain $\xi^{(R)} = \xi_1^{r_1} \xi_2^{r_2} \cdots \xi_k^{r_k}$, a monomial in the ξ_i, $i \geq 1$. In this notation, the product in A_2^* is the usual product of monomials.   □

To complete the description of A_2^*, we need to calculate the coproduct ϕ in A_2^* and the antipode, which we again denote by χ. Since ϕ and χ are algebra maps, it suffices to compute them on the generators ξ_k, $k \geq 1$. Since $\chi^2 = 1$ for the antipode χ of A_2, the antipode χ of A_2^* also satisfies $\chi^2 = 1$.

A positive integer k can be written as the sum of positive integers in 2^{k-1} ways, called **compositions** of k. For example, 4 has the eight compositions 4, $3+1$, $1+3$, $2+2$, $2+1+1$, $1+2+1$, $1+1+2$ and $1+1+1+1$. We write each composition $a_1 + \cdots + a_r$ as a sequence $A = (a_1, \ldots, a_r)$ of size r, so that $1 \leq r \leq k$ and $|A| = k$.

Proposition 11.5.2 *The coproduct* $\phi : A_2^* \longrightarrow A_2^* \otimes A_2^*$ *and the conjugation* $\chi : A_2^* \longrightarrow A_2^*$ *are given on the generators* ξ_k *of* $A_2^* = \mathbb{F}_2[\xi_1, \xi_2, \ldots]$ *by*

(i) $\phi(\xi_k) = \sum_{i+j=k} \xi_i^{2^j} \otimes \xi_j$, *where* $\xi_0 = 1$,
(ii) $\chi(\xi_k) = \sum_A \xi(A)$, *where the sum is over all compositions* $A = (a_1, \ldots, a_r)$ *of* k, *and*

$$\xi(A) = \xi_{a_1} \xi_{a_2}^{2^{a_1}} \xi_{a_3}^{2^{a_1 + a_2}} \cdots \xi_{a_r}^{2^{a_1 + \cdots + a_{r-1}}}.$$

Note that $\xi(A)$ is not written in standard form as a monomial in commuting variables. If $a_i = a_j = a$, then the exponent of ξ_a in ξ^A is not given explicitly. All 2^{k-1} terms are distinct, so that there is no cancellation in the sum. This can be seen as follows: if $A = (a_1, \ldots, a_r)$ and $B = (b_1, \ldots, b_s)$ are compositions of k such that $\xi(A) = \xi(B)$, then $a_1 = b_1$ since exactly one exponent is odd, and dividing by ξ_{a_1} and taking 2^{a_1}th roots gives $A = B$ by induction.

Example 11.5.3 We give some examples of these formulae. For ϕ we have

$$\phi(\xi_1) = \xi_1 \otimes 1 + 1 \otimes \xi_1,$$
$$\phi(\xi_2) = \xi_2 \otimes 1 + \xi_1^2 \otimes \xi_1 + 1 \otimes \xi_2,$$
$$\phi(\xi_3) = \xi_3 \otimes 1 + \xi_2^2 \otimes \xi_1 + \xi_1^4 \otimes \xi_2 + 1 \otimes \xi_3,$$
$$\phi(\xi_4) = \xi_4 \otimes 1 + \xi_3^2 \otimes \xi_1 + \xi_2^4 \otimes \xi_2 + \xi_1^8 \otimes \xi_3 + 1 \otimes \xi_4,$$

and for χ, after writing the terms in standard form,

$$\chi(\xi_1) = \xi_1,$$
$$\chi(\xi_2) = \xi_2 + \xi_1^3,$$
$$\chi(\xi_3) = \xi_3 + \xi_1\xi_2^2 + \xi_1^4\xi_2 + \xi_1^7,$$
$$\chi(\xi_4) = \xi_4 + \xi_1\xi_3^2 + \xi_2^5 + \xi_1^8\xi_3 + \xi_1^3\xi_2^4 + \xi_1^9\xi_2^2 + \xi_1^{12}\xi_2 + \xi_1^{15}.$$

Proof of Proposition 11.5.2 (i) Recall that ξ_k is dual to $Sq(R_k)$ where $R_k \in$ Seq has kth term 1 and all other terms 0. Hence $\xi^R\xi^S$ is a term in $\phi(\xi_k)$ if and only if $Sq(R_k)$ is a term in $Sq(R)Sq(S)$. If $X = (x_{i,j})$ is a Milnor matrix giving rise to the term $Sq(R_k)$ in the product $Sq(R)Sq(S)$, then all entries of X must be 0 except for a single 1 in the kth diagonal, i.e. $x_{i,j} = 1$ for some (i,j) with $i+j=k$.

For each such (i,j) with $i,j > 0$, the row condition $r_i = \sum_{j \geq 0} 2^j x_{i,j}$ implies that $R = (0, \ldots, 2^j, 0, \ldots) = 2^j R_i$, and the column condition $s_j = \sum_{i \geq 0} x_{i,j}$ implies that $S = S_j$. This gives the term $\xi^{(R)}\xi^{(S)} = \xi_i^{2^j}\xi_j$ in $\phi(\xi_k)$. In the same way, the case $(i,j) = (k,0)$ gives the term $\xi_k \otimes 1$ and the case $(i,j) = (0,k)$ gives the term $1 \otimes \xi_k$.

(ii) It follows from (i) that $\chi(\xi_1) = \xi_1$ and that for $k > 1$

$$\chi(\xi_k) = \xi_k + \sum_{i=1}^{k-1} \xi_{k-i}^{2^i} \chi(\xi_i). \tag{11.10}$$

Using induction on k, we assume that (ii) holds for $1 \leq i \leq k-1$. Then for such i, $\chi(\xi_i) = \sum_A \xi(A)$, where the sum is over all 2^{i-1} sequences $A = (a_1, \ldots, a_\ell)$ of positive integers with $|A| = i$ and length ℓ with $1 \leq \ell \leq i$. For each sequence A of length ℓ, let B be the corresponding sequence $B = A|(k-i)$ of length $\ell+1$. Then $|B| = k$ and $\xi(B) = \xi(A)\xi_{k-i}^{2^i}$. Substituting in (11.10) and including the case $B = (k)$ to give the term $\xi(B) = \xi_k$, this completes the induction. $\square$

Proposition 11.5.4 *The elements $\xi_1^{2^j}, j \geq 0$, are primitive in* A_2^*.

Proof Since $\phi(\xi_1) = 1 \otimes \xi_1 + \xi_1 \otimes 1$, ξ_1 is primitive. For $k \geq 1$ it follows that $\phi(\xi_1^k) = (1 \otimes \xi_1 + \xi_1 \otimes 1)^k$. In the case $k = 2^j$, this reduces to $1 \otimes \xi_1^{2^j} + \xi_1^{2^j} \otimes 1$, i.e. $\xi_1^{2^j}$ is primitive. $\square$

11.6 Conjugation in the Milnor basis

In this section we give a formula for $\chi(Sq(R))$ in A_2 resembling Milnor's formula for $Sq(R)Sq(S)$, Theorem 4.1.2.

Theorem 11.6.1 *The conjugate of the Milnor basis element Sq(R) is*

$$\chi(Sq(R)) = \sum_X b(X) Sq(T(X) - S(X))$$

where the sum is taken over all Milnor matrices X such that $x_{0j} = 0$ for all $j \geq 1$ and $R(X) = R$, and $Sq(T(X) - S(X)) = 0$ if $s_i > t_i$ for some i.

Proof Since $\chi : A_2 \to A_2$ and $\chi : A_2^* \to A_2^*$ are dual maps and the monomial basis of A_2^* is dual to the Milnor basis of A_2, $Sq(S)$ appears in the Milnor basis expansion of $\chi(Sq(R))$ if and only if ξ^R is a term in $\chi(\xi^S)$.

Let $\eta_k = \chi(\xi_k)$ for $k \geq 1$, so that by (11.10)

$$\eta_k = \xi_k + \xi_{k-1}^2 \eta_1 + \cdots + \xi_1^{2^{k-1}} \eta_{k-1}.$$

Given $S = (s_1, \ldots, s_n, 0, \ldots) \in Seq$, let $\chi(\xi^S) = \eta^S = \eta_1^{s_1} \cdots \eta_n^{s_n}$ be a monomial in the conjugate generators $\eta_1, \eta_2, \ldots$. We wish to express η^S as a polynomial in the generators $\xi_1, \xi_2, \ldots$ of A_2^*. To complete the proof, we shall show by induction on n that

$$\eta^S = \sum_{T(X) - S(X) = S} b(X) \xi^{R(X)}, \tag{11.11}$$

where X runs through Milnor matrices of the form

$$X = \begin{array}{c|ccccc} & 0 & 0 & \cdots & 0 & 0 \\ \hline x_{1,0} & x_{1,1} & x_{1,2} & \cdots & x_{1,n-1} & 0 \\ x_{2,0} & x_{2,1} & x_{2,2} & \cdots & 0 & 0 \\ \vdots & \vdots & \vdots & \ddots & \vdots & \vdots \\ x_{n-1,0} & x_{n-1,1} & 0 & \cdots & 0 & 0 \\ x_{n,0} & 0 & 0 & \cdots & 0 & 0 \end{array},$$

with $\leq n$ nonzero diagonals, so that the diagonal sum $t_k = 0$ when $k > n$.

For $n = 1$, $\eta_i = \xi_1$ and there is only one array X to consider, with $x_{1,0} = s_1$ and $x_{i,j} = 0$ otherwise. For the inductive step, we write $S = U + V$, where $U = (s_1, \ldots, s_{n-1}, 0, \ldots)$ and $V = (0, \ldots, 0, s_n, 0, \ldots)$, so that $\eta^S = \eta^U \eta^V$. Expanding $\eta^V = \eta_n^{s_n}$ by the binomial theorem, we obtain

$$\eta^V = \sum_{a_1 + \cdots + a_n = s_n} \binom{s_n}{a_1, \ldots, a_n} \xi_n^{a_1} (\xi_{n-1}^2 \eta_1)^{a_2} \cdots (\xi_1^{2^{n-1}} \eta_{n-1})^{a_{n-1}}.$$

The terms in this sum are parameterized by sequences $A = (a_1,\ldots,a_n) \in \mathrm{Seq}$ with $|A| = s_n$. We identify these sequences with Milnor matrices

$$
Z = \begin{array}{c|ccccc}
 & 0 & 0 & \cdots & 0 & 0 \\
\hline
0 & 0 & 0 & \cdots & z_{1,n-1} & 0 \\
0 & 0 & 0 & & 0 & 0 \\
\vdots & \vdots & & & \vdots & \vdots \\
0 & z_{n-1,1} & 0 & \cdots & 0 & 0 \\
z_{n,0} & 0 & 0 & \cdots & 0 & 0
\end{array}
$$

with only one nonzero diagonal, by writing $a_i = z_{i,n-i}$ for $1 \le i \le n$. Then $R(Z) = (2^{n-1}a_1,\ldots,2a_{n-1},a_n,0,\ldots)$, $S(Z) = (a_{n-1},\ldots,a_1,0,\ldots)$, $T(Z) = V$ and $b(Z) = \binom{s_n}{a_1,\ldots,a_n}$. In this notation $\eta^V = \sum_{T(Z)=V} b(Z)\xi^{R(Z)}\eta^{S(Z)}$. Hence $\eta^S = \eta^U\eta^V = \sum_{T(Z)=V} b(Z)\xi^{R(Z)}\eta^{S(Z)+U}$. Since the sequence $S(Z) + U$ has length $< n$, the induction hypothesis gives

$$
\eta^{S(Z)+U} = \sum_{T(Y)-S(Y)=S(Z)+U} b(Y)\xi^{R(Y)}, \tag{11.12}
$$

where Y runs through Milnor matrices

$$
Y = \begin{array}{c|ccccc}
 & 0 & 0 & \cdots & 0 & 0 \\
\hline
y_{1,0} & y_{1,1} & y_{1,2} & \cdots & y_{1,n-2} & 0 \\
y_{2,0} & y_{2,1} & y_{2,2} & \cdots & 0 & 0 \\
\vdots & \vdots & \vdots & \ddots & \vdots & \vdots \\
y_{n-2,0} & y_{n-2,1} & 0 & \cdots & 0 & 0 \\
y_{n-1,0} & 0 & 0 & \cdots & 0 & 0
\end{array},
$$

with at most $n-1$ nonzero diagonals. Thus

$$
\eta^S = \sum_{T(Z)=V} b(Z)\xi^{R(Z)} \sum_{T(Y)-S(Y)=S(Z)+U} b(Y)\xi^{R(Y)}.
$$

For each pair of arrays Y,Z, let $X = Y + Z$, so that $x_{i,j} = y_{i,j}$ for $i+j < n$, $x_{i,j} = z_{i,j}$ for $i+j = n$ and $x_{i,j} = 0$ for $i+j > n$. Then $R(X) = R(Y) + R(Z)$, $S(X) = S(Y)+S(Z)$, $T(X) = T(Y)+T(Z)$, $b(X) = b(Y)b(Z)$ and $T(X) - S(X) = (T(Y) - S(Y)) + (T(Z) - S(Z)) = (S(Z) + U) + (V - S(Z)) = S$. Conversely, it is clear that any Milnor matrix X of the given form gives arrays Y and Z as above, and so gives a term in the sum. This completes the induction. $\qquad\square$

Example 11.6.2 To evaluate $\chi(Sq(4,2))$, we consider the eight arrays

0	0	0	0 0	0	0	0	0 0
4 0 ,	2 1 ,	0 2 ,	0 0 1 ,	4 0 ,	2 1 ,	0 2 ,	0 0 1 .
2 0	2 0	2 0	2 0 0	0 1	0 1	0 1	0 1 0

The third array and the last two give zero terms, and the fourth and sixth arrays give the same term $Sq(0,1,1)$, which therefore cancels. Hence $\chi(Sq(4,2)) = Sq(4,2) + Sq(1,3) + Sq(3,0,1)$.

It follows from Theorem 11.6.1 that the matrix representing $\chi : A_2^d \to A_2^d$ for the Milnor basis in left or right order is triangular.

Proposition 11.6.3 *For any $R \in$ Seq, $\chi(Sq(R)) = Sq(R) + \sum_S Sq(S)$, where $S \leq_{l,r} R$ for all terms in the sum.*

Proof The initial Milnor matrix produces a summand $Sq(R)$. Thus it suffices to show that $S < R$ in both the left and right orders for the summand $Sq(T(X) - S(X))$ produced by any other Milnor matrix $X = (x_{i,j})$.

Since $t_1 \leq r_1$ for all X, $t_1 - s_1 \leq r_1$, so $Sq(T(X) - S(X)) <_l Sq(R)$ if $s_1 > 0$. Hence we may assume that $x_{i,1} > 0$ for all i. Then $t_2 \leq r_2$, so $t_2 - s_2 \leq r_2$ and $Sq(T(X) - S(X)) <_l Sq(R)$ if $s_2 > 0$. Hence we may assume that $x_{i,2} = 0$ for all i. Iterating this argument gives the result for the left order.

For the right order, let the last nonzero diagonal of X (excluding column 0) be the kth, so that $x_{i,j} = 0$ if $i + j > k$ and $x_{i,j} \neq 0$ for some i,j with $i + j = k$. In particular, rows $\geq k$ if X are the same as for the initial Milnor matrix. Thus $t_j = r_j$ for $j > k$ and $s_j = 0$ for $j \geq k$. This gives $t_k - s_k = t_k = \sum_{i+j=k} x_{i,j} > r_k$, since $x_{k,0} = r_k$ and $x_{i,j} > 0$ for some other entry on the kth diagonal of X. Hence $Sq(T(X) - S(X)) <_r Sq(R)$. $\qquad\square$

Example 11.6.4 We tabulate χ in degree 9 using the Milnor basis and the right order. (Compare Example 4.3.9.) For the left order, the last two rows and columns are exchanged, so that the matrix is the same in this case. By Proposition 3.5.11, every entry in the first row is 1.

	$Sq(9)$	$Sq(6,1)$	$Sq(3,2)$	$Sq(0,3)$	$Sq(2,0,1)$
$\chi(Sq(9))$	1	1	1	1	1
$\chi(Sq(6,1))$	0	1	0	1	1
$\chi(Sq(3,2))$	0	0	1	1	1
$\chi(Sq(0,3))$	0	0	0	1	0
$\chi(Sq(2,0,1))$	0	0	0	0	1

Note that it is easier to calculate this matrix by working in A_2^* and transposing, rather than by using Theorem 11.6.1. For example, the last column follows from the calculation $\chi(\xi_2^3) = (\chi(\xi_2))^3 = (\xi_1^3 + \xi_2)^3 = \xi_1^9 + \xi_1^6\xi_2 + \xi_1^3\xi_2^2 + \xi_2^3$.

11.7 Remarks

The theory of Hopf algebras, in relation to their applications to algebraic topology, was developed in the early 1960s by Milnor and Moore [135]. This remained for many years the standard reference in the subject for topologists. The 1980s saw a vast expansion of the subject of Hopf algebras, much of it centred on quantum groups and related to applications in mathematical physics. Details regarding Hopf algebras can be found in many sources; in Section 11.1 we follow [112, Chapter III], which is recommended to the reader interested in the applications of Hopf algebra methods to Lie groups and algebras and to the low-dimensional topology of knots and braids.

Many of the results of Sections 11.2, 11.3 and 11.5 can be found in [196, Chapter 2], following Milnor and Moore. In particular, note that a connected graded bialgebra over a field always has an antipode χ, which can be defined recursively on degree by the formula $\chi(x) = -x - \sum_i \chi(x_i')x_i''$, where the coproduct $\phi(x) = x \otimes 1 + \sum_i x_i' \otimes x_i'' + 1 \otimes x$.

The Hopf algebra structure of A_2 and the structure of A_2^* were elucidated by Milnor in his influential paper [134]. Theorem 11.6.1 is due to Li Zaiqing [123]. The triangularity of χ for various orders on the Milnor basis, including the right and left orders, is proved more simply in [47] using the dual algebra. This paper also contains a wealth of information about conjugation invariants in A_2^* and related algebras.

12

Further structure of A_2

12.0 Introduction

In this chapter we study the internal structure of the Steenrod algebra A_2 in more detail. In Section 12.1 we discuss formulae involving the conjugation χ extending the results of Section 4.4 which express products of the form $Xq^a Sq^b$ and $Sq^a Xq^b$ in the Milnor basis. These formulae will be applied to nilpotence questions in A_2 in Chapter 13.

In Section 12.2 we focus on the Milnor basis elements $Sq_t^k = Sq(0,\ldots,0,k)$, where k is in the tth place. These elements play an important part in the structure of the Hopf subalgebras of A_2. Two special families of Hopf subalgebras are introduced in Sections 12.3 and 12.4.

The subalgebra A_q of Section 12.3 can be regarded as an algebra of 'Steenrod qth power operations', where $q = 2^t$, and can be defined by taking the elements Sq_t^k as generators and imposing relations corresponding to the Adem relations. In the case $t = 1$, $Sq_1^k = Sq^k$ and $A_q = A_2$.

The finite subalgebras $A_2(\lambda)$ of Section 12.4 are indexed by column 2-regular partitions λ. They too have bases formed by suitable subsets of the Milnor basis. In the case $\lambda = (n+1, n, \ldots, 1)$, $A_2(\lambda)$ is the subalgebra generated by $Sq^1, Sq^2, Sq^4, \cdots, Sq^{2^n}$, and we denote it by $A_2(n)$. The importance of the family of subalgebras $A_2(n)$ has been recognized since the earliest studies of A_2.

In Section 12.5 we discuss a number of bases of A_2 which, like the admissible basis, consist of monomials in the Steenrod squares Sq^k, and also restrict to a basis of $A_2(n)$ for all n. This fails for the admissible basis: for example, $Sq^2 Sq^1 Sq^2$ is the sum of two admissibles Sq^5 and $Sq^4 Sq^1$, neither of which is in $A_2(1)$. The combinatorial structure which underlies the construction of these bases is the partially ordered set of '2-atomic numbers' $a(s,t) = 2^s(2^t - 1)$, where $s \geq 0$ and $t \geq 1$, where $a(s,t) \leq a(s',t')$ if $s+t \leq s'+t'$

and $t \leq t'$. In Section 12.6 we introduce a further monomial basis, due to D. Arnon, for which the excess of each basis element is the exponent of its last factor.

12.1 Some conjugation formulae

In this section we collect some formulae in A_2 involving the conjugation χ. We begin by deriving formulae involving products of the form $Sq^a Xq^b$ and $Xq^a Sq^b$ from the Bullett–Macdonald identity, Proposition 4.2.3.

Proposition 12.1.1 *For all $a, b \geq 0$,*

$$\text{(i) } Sq^a Xq^b = \sum_{|j-a| \leq \min(a,b)} \binom{a+b}{2a-j} Xq^{a+b-j} Sq^j,$$

$$\text{(ii) } Xq^a Sq^b = \sum_{j=0}^{[a/2]} \binom{(a+b)+(a-2j)-1}{a-2j} Sq^{a+b-j} Xq^j.$$

Proof (i) It follows from Propositions 4.2.3 and 4.3.1 that

$$Xq[1+t]Sq[t+t^2] = Sq[t^2]Xq[1].$$

On the right hand side, $Sq^a Xq^b$ is the term of degree $a + b$ in the coefficient of t^{2a}. From the left hand side, the coefficient of $Xq^{a+b-j} Sq^j$ in this sum is the coefficient of t^{2a} in $(1+t)^{a+b-j}(t+t^2)^j = (1+t)^{a+b}t^j$, which is $\binom{a+b}{2a-j}$. The conditions $0 \leq 2a-j \leq a+b$, $j \geq 0$ and $a+b-j \geq 0$ are equivalent to $a-c \leq j \leq a+c$, where $c = \min(a,b)$.

(ii) We use a different specialization of the Bullett–Macdonald identity

$$Sq[st+t^2]Sq[s^2] = Sq[st+s^2]Sq[t^2],$$

identifying the variables s and t with elements of the power series ring $\mathbb{F}_2[[u]]$ by setting $s = 1/(1+u) = 1+u+u^2+\cdots$ and $t = u^2/(1+u) = u^2 + u^3 + u^4 + \cdots$. This gives $Sq(u^2)Sq(1/(1+u^2)) = Sq(1)Sq(u^4/(1+u^2))$, which we may rewrite in the form $Xq(v)Sq(1) = Sq(1/(1+v))Xq(v^2/(1+v))$ by conjugating and writing $v = u^2$. On the left hand side, $Xq^a Sq^b$ is the term of degree $a+b$ in the coefficient of v^a. From the right hand side, the coefficient of $Sq^{a+b-j} Xq^j$ in this sum is the coefficient of v^a in $v^{2j}(1+v)^{-a-b}$. Expressing this binomial coefficient in standard form, we obtain $\binom{-a-b}{a-2j} = \binom{(a+b)+(a-2j)-1}{a-2j}$. $\square$

Part (i) of the next result is useful in obtaining certain nilpotence results in A_2 (see Section 13.5), while iterating (ii) gives an expression for Xq^{2^n} as a sum of monomials. For example, $Xq^{16} = Sq^{16} + Sq^8 Sq^8 + Sq^8 Sq^4 Sq^4 + Sq^8 Sq^4 Sq^2 Sq^2$.

Proposition 12.1.2 (i) *For* $r, s > 0$, $Xq^{2^r} Sq^{2^r(2^s-1)} = Sq^{2^{r-1}(2^{s+1}-1)} Xq^{2^{r-1}}$.
  (ii) **(Straffin's formula)** *For* $r > 0$, $Xq^{2^r} = Sq^{2^r} + Sq^{2^{r-1}} Xq^{2^{r-1}}$.

Proof Let $a = 2^{r-1}(2^{s+1} - 1)$ and $b = 2^{r-1}$ in Proposition 12.1.1(i). Since $a + b = 2^{r+s}$, $\binom{a+b}{a-2j} = 0 \bmod 2$ except when $j = 2a$ or $j = a - b$. For (i), we have $a > b$, and since $j \le a + \min(a, b)$, only the term $j = 2a$ is nonzero. For (ii), we have $s = 0$, $a = b$, and both values of j give nonzero terms. $\square$

Example 12.1.3 Note that the composition $Xq^{2^r} Sq^{2^r}$ is fixed by χ, and by setting $s = 1$ in (i) we obtain $Sq^{3 \cdot 2^{r-1}} Xq^{2^{r-1}} = Xq^{2^r} Sq^{2^r} = Sq^{2^{r-1}} Xq^{3 \cdot 2^{r-1}}$. For example, $Sq^6 Xq^2 = Xq^4 Sq^4 = Sq^2 Xq^6$. The other composition $Sq^{2^r} Xq^{2^r} = Xq^{2^{r+1}} + Sq^{2^{r+1}}$, by (ii).

Proposition 12.1.4 (i) $Sq^{2^s} Xq^{2^r} = Xq^{2^r} Sq^{2^s} + Xq^{2^r - 2^s} Sq^{2^{s+1}}$ *for* $0 \le s \le r - 2$.
  (ii) $Sq^{2^{r-1}} Xq^{2^r} = Xq^{2^r} Sq^{2^{r-1}} + Xq^{2^{r-1}} Sq^{2^r} + Xq^{2^r + 2^{r-1}}$ *for* $r \ge 1$.
  (iii) $Sq^{2^r} Sq^{2^r} = Sq^{2^{r-1}} Xq^{2^r} Sq^{2^{r-1}} + Sq^{2^{r-1}} Sq^{2^{r-1}} Xq^{2^r}$ *for* $r \ge 1$.

Proof For (i), let $a = 2^s$, $b = 2^r$ in Proposition 12.1.1(i). By Proposition 1.4.11, the only values of j such that $\binom{2^s + 2^r}{2^{s+1} - j}$ is odd are $j = 2^s$ and $j = 2^{s+1}$. In the case $s = r - 1$ the term for $j = 0$ also contributes, giving (ii). Finally (iii) follows from (ii) and Example 12.1.3, since $Sq^{2^{r-1}}(Sq^{2^{r-1}} Xq^{2^r} + Xq^{2^r} Sq^{2^{r-1}}) = Sq^{2^{r-1}}(Xq^{2^{r-1}} Sq^{2^r} + Xq^{3 \cdot 2^{r-1}}) = (Xq^{2^r} + Sq^{2^r}) Sq^{2^r} + Xq^{2^r} Sq^{2^r} = Sq^{2^r} Sq^{2^r}$. $\square$

Example 12.1.5 In the case $r = 2$, the relations of Proposition 12.1.4 are

$$Sq^1 Xq^4 = Xq^4 Sq^1 + Xq^3 Sq^2$$
$$Sq^2 Xq^4 = Xq^4 Sq^2 + Xq^2 Sq^4 + Xq^6$$
$$Sq^4 Xq^4 = Sq^2 Xq^4 Sq^2 + Sq^2 Sq^2 Xq^4.$$

These relations are preserved under doubling of exponents.

Using the filtration of A_2 by the sequence of finite subalgebras $A_2(n)$ (Section 12.4), we can express relations (i) and (iii) of Proposition 12.1.4 in the following form, due to C. T. C. Wall.

Proposition 12.1.6 (i) *For* $0 \le s \le r - 2$, $Sq^{2^s} Sq^{2^r} + Sq^{2^r} Sq^{2^s} \in A_2(r-1)$,
  (ii) *for* $r \ge 1$, $Sq^{2^r} Sq^{2^r} + Sq^{2^{r-1}} Sq^{2^r} Sq^{2^{r-1}} + Sq^{2^{r-1}} Sq^{2^{r-1}} Sq^{2^r} \in A_2(r-1)$.

Proof Both results follow from Proposition 12.1.4 by substituting for Xq^{2^r} using Straffin's formula, Proposition 12.1.2(i). $\square$

The following family of relations between elements of the form $Sq^a Xq^b$ was found by M. G. Barratt and H. R. Miller. The function $\beta(d) = (d + \mu(d))/2$ was introduced in Proposition 5.7.6 (see also Section 14.3).

Proposition 12.1.7 *For $0 \leq j < \beta(d)$ and $m \geq 0$,*

$$\sum_{i=0}^{d} \binom{m+i}{j} Sq^i Xq^{d-i} = 0.$$

Since the coefficients $\binom{m+i}{j}$ are unchanged by adding a 2-power $> j$ to m, there is only a finite number of distinct relations for each j.

Example 12.1.8 Let $d = 4$ and $j = 2$. Then for $m \equiv 0, 1, 2, 3 \bmod 4$ respectively we obtain $Sq^1 Xq^3 + Sq^2 Xq^2 = 0$, $Sq^2 Xq^2 + Sq^3 Xq^1 = 0$, $Xq^4 + Sq^3 Xq^1 + Sq^4 = 0$, $Xq^4 + Sq^1 Xq^3 + Sq^4 = 0$.

Proof of Proposition 12.1.7 Expanding each term of the relation in the Milnor basis using Davis's formula 4.4.1(i), the relation is equivalent to

$$\sum_{i=0}^{d} \binom{m+i}{j} \binom{\deg_2 R}{i} = 0 \bmod 2 \tag{12.1}$$

for $m, d \geq 0$, $0 \leq j < \beta(d)$ and all Milnor basis elements $Sq(R) \in A_2^d$. Recall that $\deg_2 R = (|R| + d)/2$. Since $\mu(d) \leq |R| \leq d$, $\beta(d) \leq \deg_2 R \leq d$. Writing $\deg_2 R = k$, (12.1) follows from the identity

$$\sum_{i=j-m}^{k} (-1)^i \binom{m+i}{j} \binom{k}{i} = 0, \text{ for } m \geq 0, \ 0 \leq j < k, \tag{12.2}$$

where the binomial coefficients have their usual meaning as integers. We prove (12.2) by induction on m. For the base case $m = 0$, we take the jth derivative of the identity $(1-x)^k = \sum_{r=0}^{k} (-1)^r \binom{r}{k} x^r$, divide by $j!$ and set $x = 1$. The inductive step follows from the 'Pascal triangle' recursion $\binom{m+1+i}{j} = \binom{m+i}{j} + \binom{m+i}{j-1}$ by taking the alternating sum $\sum_i (-1)^i \binom{k}{i}$ and applying the inductive hypothesis to both terms on the right hand side. $\square$

12.2 The Milnor elements Sq_t^k

In this section we introduce a linear operation Sq_t^k on $P(n)$ for $t \geq 1$ and $k \geq 0$ by generalizing the definition of $Sq^k = Sq_1^k$ in Chapter 1. In Proposition 12.2.9 we identify Sq_t^k with the Milnor basis element $Sq(0, \ldots, 0, k)$ of length t. In Section 12.3 we consider the subalgebra of A_2 generated by the elements Sq_t^k, $k \geq 1$ as the algebra of Steenrod qth powers, where $q = 2^t$.

Definition 12.2.1 Let $q = 2^t$ where $t \geq 1$. The **total Steenrod qth power** Sq_t : $P(n) \to P(n)$ is the algebra map defined by $\mathrm{Sq}_t(1) = 1$ and $\mathrm{Sq}_t(x_i) = x_i + x_i^q$ for $1 \leq i \leq n$. For $k, d \geq 0$, the **kth Steenrod qth power** $\mathrm{Sq}_t^k : P^d(n) \to P^{d+k(q-1)}(n)$ is the linear map given by restricting Sq_t to $P^d(n)$ and projecting on to $P^{d+k(q-1)}(n)$.

Thus $\mathrm{Sq}_t = \sum_{k \geq 0} \mathrm{Sq}_t^k$ is the formal sum of its graded parts. The following properties of the elements Sq_t^k are straightforward generalizations of the corresponding results in Section 1.1.

Proposition 12.2.2 *For all $x \in P^1(n)$, $\mathrm{Sq}_t(x) = x + x^q$. Thus $\mathrm{Sq}_t^0(x) = x$, $\mathrm{Sq}_t^1(x) = x^q$ and $\mathrm{Sq}_t^k(x) = 0$ for all $k > 1$.* □

Proposition 12.2.3 (**Cartan formula**) *For polynomials $f, g \in P(n)$ and $t \geq 1$, $k \geq 0$,*

$$\mathrm{Sq}_t^k(fg) = \sum_{i+j=k} \mathrm{Sq}_t^i(f)\mathrm{Sq}_t^j(g). \quad \square$$

Proposition 12.2.4 *For $t \geq 1$, Sq_t^0 is the identity map of $P(n)$.* □

Example 12.2.5 In $P(2) = \mathbb{F}_2[x, y]$ we have

$$\mathrm{Sq}_t^1(xy) = \mathrm{Sq}_t^1(x)\mathrm{Sq}_t^0(y) + \mathrm{Sq}_t^0(x)\mathrm{Sq}_t^1(y) = x^q y + x y^q.$$

Proposition 12.2.6 *For all $x \in P^1(n)$,*

$$\mathrm{Sq}_t^k(x^d) = \binom{d}{k} x^{d+k(2^t-1)},$$

where the binomial coefficient is reduced mod 2. In particular $\mathrm{Sq}_t^k(x^{2^s}) = x^{2^{s+t}}$ if $k = 2^s$, and $\mathrm{Sq}_t^k(x^{2^s}) = 0$ otherwise for $k > 0$. □

Proposition 12.2.7 *Let $f = x_1^{d_1} \cdots x_n^{d_n}$ be a monomial in $P(n)$. Then*

$$\mathrm{Sq}_t^k(f) = \sum_{k_1+\cdots+k_n=k} \mathrm{Sq}_t^{k_1}(x_1^{d_1}) \cdots \mathrm{Sq}_t^{k_n}(x_n^{d_n}). \quad \square$$

The following result explains why Sq_t^k is called a qth power, where $q = 2^t$.

Proposition 12.2.8 *For $f \in P^d(n)$, $\mathrm{Sq}_t^k(f) = 0$ if $k > d$ and $\mathrm{Sq}_t^d(f) = f^q$.* □

We next identify the operation Sq_t^k on $P(n)$ with an element in A_2.

Proposition 12.2.9 *For $k \geq 0$ and $t \geq 1$, $\mathrm{Sq}_t^k = Sq(R)$, where $R = (0, \cdots, 0, k)$ has length t.*

Proof The operations Sq_t^k and $Sq(R)$ on $P(n)$ both have degree $k(q-1)$. The element $Sq(R)$ is characterized by the property that $Sq(R)(c(n)) = c(R^+)$, for all $n \geq k$, where $c(n) = x_1 \cdots x_n$ and $c(R^+)$ is the Cartan symmetric function in $P(n)$ associated with the sequence $R^+ = (n-k, 0, \ldots, 0, k)$. The leading term of $c(R^+)$ is $f = (x_1 \cdots x_k)^{2^t} x_{k+1} \cdots x_n$, and Propositions 12.2.3 and 12.2.7 imply that $Sq_t^k(c(n))$ is the symmetrization of f. $\qquad\square$

The following results are straightforward generalizations of Propositions 1.3.2 and 1.3.3.

Proposition 12.2.10 *For* $f \in P(n)$ *and* $s, k \geq 0$, $Sq_t^k(f^{2^s}) = (Sq_t^j(f))^{2^s}$ *if* $k = 2^s j$, *and* $Sq_t^k(f^{2^s}) = 0$ *otherwise.* $\qquad\square$

Proposition 12.2.11 *For* $f, g \in P(n)$ *and* $s, k \geq 0$,

$$Sq_t^k(gf^{2^s}) = \sum_{i+2^s j = k} Sq_t^i(g)(Sq_t^j(f))^{2^s}. \qquad\square$$

In particular, $Sq_t^k(gf^{2^s}) = Sq_t^k(g)f^{2^s}$ if $k < 2^s$, and $Sq_t^{2^s}(gf^{2^s}) = Sq_t^{2^s}(g)f^{2^s} + g(Sq_t^1(f))^{2^s}$. If $\deg g < 2^s$, then $Sq_t^{2^s}(gf^{2^s}) = g(Sq_t^1(f))^{2^s}$.

In Proposition 12.2.13 we construct a number of bases for A_2 by using compositions of the elements Sq_t^k. These are closely related to the Milnor basis. The table below shows one such basis in degrees ≤ 9.

0	1	2	3	4	5	6	7	8	9
1	Sq_1^1	Sq_1^2	Sq_1^3	Sq_1^4	Sq_1^5	Sq_1^6	Sq_1^7	Sq_1^8	Sq_1^9
		Sq_2^1	$Sq_1^1 Sq_2^1$	$Sq_1^2 Sq_2^1$	$Sq_1^3 Sq_2^1$	$Sq_1^4 Sq_2^1$	$Sq_1^5 Sq_2^1$	$Sq_1^6 Sq_2^1$
				Sq_2^2	$Sq_1^1 Sq_2^2$	$Sq_1^2 Sq_2^2$	$Sq_1^3 Sq_2^2$
						Sq_3^1	$Sq_1^1 Sq_3^1$	$Sq_1^2 Sq_3^1$
									Sq_2^3

The proof of Theorem 3.4.4 used the Cartan symmetric function $c(R^+)$ for the $<_r$-minimal sequence R^+ in the expansion of $Sq^A(c(d))$, where A is admissible. However, in Proposition 12.2.13 we use the $<_r$-maximal summand in the expansion of $Sq_t^k(c(n))$ as a sum of Cartan symmetric functions. We use the following result to determine this summand.

Proposition 12.2.12 *For* $r \geq 1$ *and* $k \geq 0$, *let* $R_t^k = (-k, \ldots, 0, k)$ *where* k *is in the* $(t+1)$*th place. Then for any sequence* $R = (r_1, r_2, \ldots)$ *with* $r_1 \geq k$,

$$Sq_t^k(c(R)) = \binom{r_{t+1} + k}{k} c(R') + \sum_S c(S)$$

where $R' = R + R_t^k = (r_1 - k, r_2, \ldots, r_t, r_{t+1} + k, r_{t+2}, \ldots)$, *and* $S <_r R'$ *for each term* $c(S)$ *in the sum.*

Proof Let the block B represent a monomial in $c(R)$, so that each row of B contains exactly one 1. The effect of Sq_t^k on B is to produce a number of blocks of the same kind by moving certain 1s t columns to the right. The entries which are moved are in columns $s+1$ corresponding to summands 2^s in partitions of k as sums of 2-powers. Thus the $<_r$-maximal block B' which can be produced corresponds to the partition $k = 1 + 1 + \cdots + 1$, and is obtained by moving k entries 1 from the first column of B to the $(t+1)$th column.

In the symmetric polynomial $Sq_t^k(c(R))$, the same block B' with $r_{t+1} + k$ entries 1 in column $t+1$ occurs when Sq_t^k is applied to any of the blocks corresponding to the different choices of k rows where entries 1 are moved. Thus $c(R')$ appears in $Sq_t^k(c(R))$ with coefficient $\binom{r_{t+1}+k}{k}$. $\qquad\square$

Proposition 12.2.13 (Adams bases) *The set of monomials* $Sq_1^{k_1} Sq_2^{k_2} \cdots Sq_r^{k_r}$ *where* $r \geq 1$ *and* $k_1, k_2, \ldots, k_r \geq 1$, *together with the identity element* $Sq^0 = 1$, *form a vector space basis for* A_2. *More generally, the set of all similar monomials in the elements* Sq_t^k, *where the suffixes are taken a fixed but arbitrary order, together with 1, form a basis for* A_2.

Proof In each degree $k > 0$ we show that the monomials $Sq_1^{k_1} Sq_2^{k_2} \cdots Sq_r^{k_r}$ of degree $k = \sum_{i=1}^r k_i(2^i - 1)$ are a basis for A_2^k, by applying them to $c(n) = x_1 \cdots x_n$ for $n \geq \sum_{i=1}^r k_i$. By Proposition 12.2.12 the $<_r$-maximal term in $Sq_r^{k_r}(c(n))$ is $c(n - k_r, 0, \ldots, 0, k_r)$, where k_r is in the $(r+1)$th place. Applying $Sq_{r-1}^{k_{r-1}}$ to $c(n - k_r, 0, \ldots, 0, k_r)$, the $<_r$-maximal term is $c(n - k_r - k_{r-1}, 0, \ldots, 0, k_{r-1}, k_r)$, and this is also the $<_r$-maximal term of $Sq_{r-1}^{k_{r-1}} Sq_r^{k_r}(c(n))$. Continuing in this way, the $<_r$-maximal term of $Sq_1^{k_1} \cdots Sq_r^{k_r}(c(n))$ is $c(n - \sum_{i=1}^r k_i, k_1, \ldots, k_r)$. It follows that the elements $Sq_1^{k_1} \cdots Sq_r^{k_r}$ are linearly independent in A_2. By counting the number of monomials of degree k and using Proposition 3.5.13, it follows that these monomials also span A_2.

The final $<_r$-maximal term $c(n - \sum_{i=1}^r k_i, k_1, \ldots, k_r)$ is independent of the order in which the operations $Sq_i^{k_i}$ are applied. For example, we may apply $Sq_1^{k_1}$ first to obtain the $<_r$-maximal term $c(k - k_1, k_1)$, then $Sq_2^{k_2}$ to obtain the $<_r$-maximal term $c(k - k_1 - k_2, k_1, k_2)$, and so on. Thus a basis is obtained when the suffixes are taken in any fixed but arbitrary order. $\qquad\square$

12.3 The subalgebras A_q, $q = 2^t$

For $q = 2^t$, $t \geq 1$, the Milnor basis elements Sq_t^k generate a Hopf subalgebra A_q of A_2, which we refer to as the 'mod q Steenrod algebra'. This subalgebra has an additive basis given by Milnor basis elements $Sq(R)$ where the jth term of the sequence R is 0 if j is not divisible by t. This is clear from the product formula, Theorem 4.1.2, since a Milnor matrix X which arises when two such elements are multiplied must have entries $x_{ij} = 0$ unless i and j are both divisible by t, and hence $i + j$ is divisible by t. Hence we make the following definition.

Definition 12.3.1 For $t \geq 1$ and $R = (r_1, r_2, \ldots) \in$ Seq, let $S = R_t$ be the sequence defined by $s_j = r_k$ if $j = kt$ for $k \geq 1$, and $s_j = 0$ otherwise. Let $Sq_t(R) = Sq(R_t)$ be the corresponding Milnor basis element of A_2. The **mod** q **Steenrod algebra** A_q is the subalgebra of A_2 with $\mathbb{F}_2$-basis $\{Sq_t(R), R \in$ Seq$\}$.

Proposition 3.5.13 can be generalized as follows.

Proposition 12.3.2 *The Poincaré series* $P(A_q, \tau) = \prod_{j \geq 1} 1/(1 - \tau^{q^j - 1})$. $\square$

The algebra A_q can alternatively be defined by generators $Sq_t^k = Sq_t(k)$ and Adem relations, as for A_2 itself. Note that the coefficients in these relations are in $\mathbb{F}_2$.

Proposition 12.3.3 *Let $a, b \geq 0$ and let $q = 2^t$ where $t \geq 1$. Then*

$$Sq_t^a Sq_t^b = \sum_{j=0}^{[a/q]} \binom{(q-1)(b-j)-1}{a-qj} Sq_t^{a+b-j} Sq_t^j,$$

where the binomial coefficient is taken mod 2.

Proof By the Milnor product formula 4.1.2

$$Sq_t^a Sq_t^b = \sum_{k=0}^{b} \binom{a+b-(q+1)k}{b-k} Sq_t(a+b-(q+1)k, k).$$

(Compare Example 3.5.5.) Using this to expand both sides of the relation in the Milnor basis, and equating coefficients of $Sq_t(a+b-(q+1)k, k)$, we see that the relation is equivalent to the identity

$$\binom{a+b-(q+1)k}{b-k} = \sum_{j=k}^{[a/q]} \binom{(q-1)(b-j)-1}{a-qj} \binom{a+b-(q+1)k}{j-k},$$

of mod 2 binomial coefficients, where $0 \leq k \leq b$. The change of variables $a' = a - qk$, $b' = b - k$ and $j' = j - k$ reduces this to the case $k = 0$, namely

$$\binom{a'+b'}{b'} = \sum_{j=0}^{[a'/q]} \binom{(q-1)(b'-j')-1}{a'-qj'} \binom{a'+b'}{j'}. \tag{12.3}$$

Replacing a', b' by a, b, we prove (12.3) by induction on $a + b$. The base case $(a, b) = (0, 0)$ holds since $\binom{c}{0} = 1$ for all integers c. We assume that the cases $(a - 1, b)$ and $(a, b - 1)$ hold, and prove the case (a, b). For this we use the identity

$$\binom{c-1}{d} + \binom{c-1}{d-1} = \binom{c}{d} = \binom{c-q}{d} + \binom{c-q}{d-q} \tag{12.4}$$

for mod 2 binomial coefficients, which follows from $(1 + x)^c = (1 + x)^q (1 + x)^{c-q}$ since $(1 + x)^q = 1 + x^q$ mod 2. Writing $c = (q - 1)(b - j)$ and $d = a - qj$, we can expand each term on the right of (12.3) in the form

$$
\binom{c-1}{d}\binom{a+b}{j} = \binom{c-1}{d}\binom{a+b-1}{j} + \binom{c-1}{d}\binom{a+b-1}{j-1}
$$
$$
= \binom{c-1}{d-1}\binom{a+b-1}{j} + \binom{c-q}{d}\binom{a+b-1}{j}
$$
$$
+ \binom{c-q}{d-q}\binom{a+b-1}{j} + \binom{c-1}{d}\binom{a+b-1}{j-1}.
$$

The first term sums over j to $\binom{(a-1)+b}{b}$ and the second to $\binom{a+(b-1)}{b-1}$ by the inductive hypothesis. Since $c - q = (q - 1)(b - j - 1) - 1$ and $d - q = a - q(j + 1)$, the third and fourth terms cancel on summing over j. Since $\binom{a+b-1}{b} + \binom{a+b-1}{b-1} = \binom{a+b}{b}$, this completes the inductive step. □

As in Chapter 3, the Adem relations lead to a basis of admissible monomials for A_q.

Definition 12.3.4 For $A = (a_1, a_2, \ldots, a_\ell) \in$ Seq, $Sq_t^A = Sq_t^{a_1} Sq_t^{a_2} \cdots Sq_t^{a_\ell}$ is a **monomial** in A_q. The monomial Sq_t^A is **admissible** if $a_i \geq q a_{i+1}$ for $1 \leq i < \ell$.

Proposition 12.3.5 (i) *The set of admissible monomials is a* $\mathbb{F}_2$-*basis for* A_q.
(ii) *The elements* $Sq_t^{2^j}$, $j \geq 0$, *are a minimal generating set for* A_q *as an algebra over* $\mathbb{F}_2$.

Proof As in Proposition 3.1.7, we can use the Adem relations 12.3.3 to raise a monomial in the left order if it is not admissible, and thus show that every element of A_q is a sum of admissible monomials. Proposition 12.3.2 then implies (i), while (ii) is proved by generalizing the argument of Proposition 3.2.4. □

Since $\phi(Sq_t^k) = \sum_{i+j=k} Sq_t^i \otimes Sq_t^j$ by Proposition 11.2.4(ii), A_q is a Hopf subalgebra of A_2. Proposition 11.2.4 can be generalized as follows.

Proposition 12.3.6 *The diagonal map* $\phi : A_q \to A_q \otimes A_q$ *satisfies*

(i) $\phi(Sq_t^k) = \sum_{i+j=k} Sq_t^i \otimes Sq_t^j$, *for all* $k \geq 0$.

(ii) $\phi(Sq_t(R)) = \sum_{S+T=R} Sq_t(S) \otimes Sq_t(T)$, *for all* $R \in$ Seq. $\qquad\square$

We can also describe the dual algebra A_q^* as a Hopf quotient algebra of A_2^*. The required formulae follow from the corresponding results in Section 11.5.

Proposition 12.3.7 *Let* $t \geq 1$ *and* $q = 2^t$. *The dual algebra* A_q^* *is the quotient* A_2^*/I_q, *where* I_q *is the Hopf ideal in* A_2^* *generated by the elements* ξ_j *such that* t *does not divide* j. *As an algebra, it is isomorphic to the polynomial algebra over* $\mathbb{F}_2$ *with generators* ξ_{kt}, $k \geq 1$ *of degree* $q^k - 1$. *The coproduct in* A_q^* *is defined by* $\phi(\xi_{kt}) = \sum_{i=0}^{k} \xi_{(k-i)t}^{q^i} \otimes \xi_{it}$, *and the conjugation by* $\chi(\xi_{kt}) = \sum_A \xi(A)$, *where the sum is over compositions* $A = (a_1, \ldots, a_s)$ *of* k, *and* $\xi(A) = \xi_{a_1 t} \xi_{a_2 t}^{q^{a_1}} \xi_{a_3 t}^{q^{a_1+a_2}} \cdots \xi_{a_s t}^{q^{a_1+\cdots+a_{s-1}}}$. $\qquad\square$

12.4 The subalgebras $A_2(\lambda)$

In this section we define a family of Hopf subalgebras $A_2(\lambda)$ of A_2, indexed by partitions $\lambda = (\lambda_1, \ldots, \lambda_\ell)$ which are column 2-regular, i.e. $\lambda_i - \lambda_{i+1} = 0$ or 1 for $1 \leq i \leq \ell$, where $\lambda_{\ell+1} = 0$. When $\lambda = (n+1, n, \ldots, 1)$, this is the important subalgebra generated by $Sq^1, Sq^2, Sq^4, \cdots, Sq^{2^n}$. In deference to traditional notation, we denote this subalgebra by $A_2(n)$ and *not* by $A_2(\lambda)$. For example, $A_2(0)$ has dimension 2 and is generated by Sq^1, while $A_2(1)$ has dimension 8 and is generated by Sq^1 and Sq^2. When $\lambda = (1, \ldots, 1)$ with $|\lambda| = k > 1$, $A_2(\lambda)$ is the exterior algebra generated by the primitive elements $Q_1, \ldots, Q_k$.

Definition 12.4.1 (i) For $n \geq 0$, we denote by $A_2(n)$ the vector subspace of A_2 spanned by the Milnor basis elements $Sq(R)$, where $R = (r_1, r_2, \ldots)$ and $0 \leq r_i \leq 2^{n+2-i} - 1$ for $i \geq 1$.

(ii) For a column 2-regular partition λ of length $\ell > 1$, we denote by $A_2(\lambda)$ the vector subspace of A_2 spanned by the Milnor elements $Sq(R)$, where $R = (r_1, \ldots, r_\ell)$ and $0 \leq r_i \leq 2^{\lambda_i} - 1$ for $1 \leq i \leq \ell$.

Thus $A_2(\lambda)$ has dimension $2^{|\lambda|}$, with a unique element $Sq(2^{\lambda_1} - 1, \ldots, 2^{\lambda_\ell} - 1)$ of highest degree $\sum_{i=1}^{\ell} (2^i - 1)(2^{\lambda_i} - 1)$. In particular $A_2(n)$ has dimension $2^{(n+1)(n+2)/2}$, with a unique element $Sq(2^{n+1} - 1, 2^n - 1, \ldots, 1)$ of highest degree $(n-1)2^{n+2} + n + 5$. The Poincaré series of $A_2(\lambda)$ is

$$P(A_2(\lambda), t) = \prod_{j=1}^{\ell} \frac{1 - t^{2^{\lambda_j}(2^j - 1)}}{1 - t^{2^j - 1}}.$$

As the factors of this product have the form $(1 - T^{2^{\lambda_j}})/(1 - T) = \prod_{i=0}^{\lambda_j - 1} (1 + T^{2^i})$, where $T = t^{2^j - 1}$, we can write

$$P(A_2(\lambda), t) = \prod_{j=1}^{\ell} \prod_{i=0}^{\lambda_j - 1} (1 + t^{2^i(2^j - 1)}). \tag{12.5}$$

Example 12.4.2 $A_2(1)$ has Poincaré series $(1+t)(1+t^2)(1+t^3)$ and top degree element $Sq^5 Sq^1 = Sq(3, 1)$, while $A_2(2)$ has Poincaré series $(1+t)(1+t^2)(1+t^4)(1+t^3)(1+t^6)(1+t^7)$ and top degree element $Sq(7, 3, 1) = Sq^{17} Sq^5 Sq^1$. The admissible basis of A_2 does not provide a basis for $A_2(n)$ for $n \geq 1$. However, for all λ the unique element of highest degree in $A_2(\lambda)$ is an admissible monomial, which can be found using the correspondence 3.4.1 between Milnor basis elements and admissible monomials.

Proposition 12.4.3 $A_2(\lambda)$ *is a Hopf subalgebra of* A_2.

Proof By Proposition 11.2.4, $A_2(\lambda)$ is closed under the coproduct ϕ. We shall use Theorem 4.1.2 to prove that it is closed under multiplication. Let $R = (r_1, \ldots, r_{n+1})$ and $S = (s_1, \ldots, s_{n+1})$ where $r_i, s_i < 2^{n+2-i}$ for $1 \leq i \leq n+1$. Let $X = (x_{i,j})$ be a Milnor matrix for the product $Sq(R)Sq(S)$ with coefficient $b(X) = 1 \bmod 2$, and let $Sq(T) = Sq(t_1, t_2, \ldots)$ be the summand of the product produced by X, so that $t_k = \sum_{i+j=k} x_{i,j}$ for $k \geq 1$.

Since $r_i = \sum_{j \geq 0} 2^j x_{i,j}$ and $r_i < 2^{\lambda_i - 1}$, $x_{i,j} \leq 2^{\lambda_i - j}$ for $i \geq 1$. Since λ is column 2-regular, $\lambda_i - j \leq \lambda_{i+j} = \lambda_k$. Also $x_{0,k} \leq 2^{\lambda_k}$ since $s_j = \sum_{i \geq 0} x_{i,j}$ and $s_k < 2^{\lambda_k}$. Thus all the terms $x_{i,j}$ on the kth diagonal of X are $\leq 2^{\lambda_k}$. Since $b(X) = 1 \bmod 2$, each multinomial coefficient $\binom{t_k}{x_{k,0}\ x_{k-1,1}, \ldots, x_{0,k}}$ is odd, and so by Proposition 1.4.13 $\mathrm{bin}(t_k)$ is the disjoint union of the sets $\mathrm{bin}(x_{i,j})$, $i+j=k$. Hence $t_k < 2^{\lambda_k}$. □

Proposition 12.4.4 *The elements* $Sq_i(2^j)$, $1 \leq i \leq \ell$, $0 \leq j < \lambda_i$, *generate* $A_2(\lambda)$.

Proof Recall that $Sq_i(k) = Sq(0, \ldots, 0, k)$, with k in the ith place. The basis elements $Sq(R)$ of $A_2(\lambda)$ are indexed by $R = (r_1, \ldots, r_\ell)$, $0 \leq r_i \leq 2^{\lambda_i} - 1$. We use the right order $\leq_r$ on the sequences R in each degree. If we expand the product $Sq(R)Sq(S)$ by the product formula, where R and S are such that $\mathrm{bin}(r_i)$ and $\mathrm{bin}(s_i)$ are disjoint for $1 \leq i \leq n+1$, then the initial Milnor matrix X gives a nonzero term $Sq(R+S)$ in the expansion, and this is $\leq_r$-maximal.

Thus if R has at least two nonzero terms r_i and r_j, $i < j$, we can factor $Sq(R)$ nontrivially as $Sq(r_1, \ldots, r_i)Sq(0, \ldots, 0, r_{i+1}, \ldots, r_{n+1}) \bmod <_r$-lower terms. It follows by induction on the $\leq_r$ order that the elements $Sq_i(k)$, $1 \leq i \leq \ell$, $1 \leq k < 2^{\lambda_i}$, span $A_2(\lambda)$. If k is not a 2-power, then we can factor $Sq_t(k)$ nontrivially as $Sq_i(j)Sq_i(k-j) \bmod <_r$-lower terms, where $1 \leq j < k$ and j and $k - j$ have disjoint binary expansions. The result follows by induction on the number of terms in the binary expansions and the $\leq_r$ order. $\qquad\square$

Proposition 12.4.5 *The elements* Sq^{2^j}, $0 \leq j \leq n$, *are a minimal generating set for* $A_2(n)$.

Proof Since $Sq^k = Sq(k)$, the element $Sq^k \in A_2(n)$ for $k < 2^{n+1}$, and so $Sq^{2^j} \in A_2(n)$ for $0 \leq j \leq n$. As $Sq_i^{2^j}$ has degree $2^{i+j} - 2^j$, all the generators $Sq_i(2^j)$, $1 \leq i \leq i+j \leq n+1$, of $A_2(n)$ given by Proposition 12.4.4 have grading $< 2^{n+1}$. However, by Proposition 3.2.4, all elements of degree $< 2^{n+1}$ in A_2 are in the subalgebra of A_2 generated by the elements Sq^{2^j}, $0 \leq j \leq n$. $\qquad\square$

Combining this argument with Proposition 12.2.13, we obtain further bases for $A_2(n)$.

Proposition 12.4.6 *The monomials* $Sq_1(k_1)Sq_2(k_2) \cdots Sq_t(k_r)$, *where* $r \geq 1$ *and* $1 \leq k_i < 2^{n+2-i}$, *together with* $Sq^0 = 1$, *form a vector space basis for* $A_2(n)$. *More generally, the set of similar monomials in the elements* $Sq_t(k)$, *with suffixes* r *taken in a fixed but arbitrary order, together with* 1, *form a basis for* $A_2(n)$. $\qquad\square$

Proposition 12.4.7 *Every finitely generated subalgebra of* A_2 *is finite.*

Proof Every finite set of elements of A_2 can be expressed in terms of finitely many elements Sq^{2^j}, and hence it is a subset of $A_2(n)$ for some n. $\qquad\square$

We next consider the duals of the Hopf subalgebras $A_2(\lambda)$. The dual of a Hopf subalgebra H of A_2 is a Hopf quotient algebra H^* of A_2^*. Recall from Section 11.1 that $H^* = A_2^*/I$, where I is a Hopf ideal in A_2^*, i.e. an ideal such that $\phi(I) \subseteq I \otimes A_2^* + A_2^* \otimes I$ and $\chi(I) = I$. By taking $j_k = \lambda_k$ for $1 \leq k \leq \ell$ and $j_k = 0$ for $k > \ell$ in Proposition 12.4.8, we obtain the Hopf quotient algebra of A_2^* dual to $A_2(\lambda)$.

Proposition 12.4.8 *Let* $n_k = 2^{j_k}$ *for* $k \geq 1$, *where* $j_{k-1} \leq j_k + 1$ *for* $k \geq 1$. *Then the ideal* I *in* A_2^* *generated by the elements* $\xi_k^{n_k}$ *is a Hopf ideal.*

Proof Since $\phi(\xi_k^{n_k}) = (\phi(\xi_k))^{n_k}$ and n_k is a 2-power, it follows from Proposition 11.5.2(i) that $\phi(\xi_k^{n_k}) = \sum_{i=0}^k \xi_{k-i}^{2^i n_k} \otimes \xi_i^{n_k}$. By induction on i, $j_{k-i} \leq j_k + i$. Hence $\xi_{k-i}^{2^i n_k} \in I$ for $i < k$, so that the ith term in the sum is in $I \otimes A_2^*$, while the kth term $1 \otimes \xi_k^{n_k} \in A_2^* \otimes I$.

By Proposition 11.5.2(ii) $\chi(\xi_k) = \sum_A \xi(A)$, where the sum is over compositions $A = (a_1, \ldots, a_r)$ of k, and $\xi(A)$ is divisible by $\xi_s^{2^{k-s}}$ where $s = a_r$, so that $1 \leq s \leq k$ and $k - s = a_1 + \cdots + a_{r-1}$. Then $\xi(A)$ is divisible by

$\xi_s^{2^{k-s}}$. Since n_k is a 2-power, $\chi(\xi_k^{n_k}) = \sum_A \xi(A)^{n_k}$, and $\xi(A)^{n_k}$ is divisible by $\xi_s^{2^{k-s}n_k} = \xi_s^{2^{k-s+j_k}}$, and hence also by $\xi_s^{n_s}$, since $k-s+j_k \geq j_s$. Hence $\xi(A)^{n_k} \in I$, and it follows that $\chi(\xi_k^{n_k}) \in I$. $\qquad\square$

12.5 2-atomic numbers and bases

The set of partitions has a natural partial order, defined by $\lambda \subseteq \mu$ if and only if the (Ferrers) diagram of λ is contained in that of μ. When λ and μ are column 2-regular, $\lambda \subseteq \mu$ if and only if $A_2(\lambda) \subseteq A_2(\mu)$ for the corresponding Hopf subalgebras of A_2. We define a related partial order on the generators $Sq_t(2^s)$ of these subalgebras. The element $Sq_t(2^s)$ has degree $d = 2^s(2^t - 1)$. As d determines s and t uniquely, it is convenient to define the partial order on integers d of this form.

Definition 12.5.1 An integer is 2-**atomic** if it has the form $2^s(2^t - 1)$ for some $s \geq 0$ and $t \geq 1$. The **dominance order** on 2-atomic numbers is defined by $2^s(2^t - 1) \leq 2^{s'}(2^{t'} - 1)$ if and only if $s+t \leq s'+t'$ and $t \leq t'$.

The dominance order on the first few 2-atomic numbers is shown below.

$$
\begin{array}{ccccccccc}
\vdots \\
8 & \cdots \\
& \searrow \\
4 & \leftarrow & 12 & \cdots \\
& \searrow & & \searrow \\
2 & \leftarrow & 6 & \leftarrow & 14 & \leftarrow & \cdots \\
& \searrow & & \searrow & & \searrow \\
1 & \leftarrow & 3 & \leftarrow & 7 & \leftarrow & 15 & \leftarrow & \cdots
\end{array}
\tag{12.6}
$$

By (12.5), the Poincaré series $P(A_2(\lambda), t) = \prod_a (1 + t^a)$, where the product is over all 2-atomic numbers a such that $i < \lambda_j$ for $j \geq 1$. In particular, $P(A_2(n), t) = \prod_a (1 + t^a)$, where the product is over all 2-atomic numbers $a < 2^{n+1}$.

Definition 12.5.2 A linear order on 2-atomic numbers which extends the dominance order is called an **Arnon order**.

We give two examples of Arnon orders, the Y- and Z-orders. In terms of the diagram above, the Y-order reads along rows and then upwards, so that the sequence of 2-atomic numbers begins $1,3,7,15,\ldots,2,6,14,\ldots,4,12,\ldots,8,\ldots$, while the Z-order reads upwards on successive diagonals, so that the sequence begins $1,3,2,7,6,4,15,14,12,8,\ldots$.

Definition 12.5.3 Let $a = 2^s(2^t - 1)$ and $a' = 2^{s'}(2^{t'} - 1)$ be 2-atomic numbers. Then $a \leq_Y a'$ if and only if $(s,t) \leq_l (s',t')$, and $a \leq_Z a'$ if and only if $(s+t,s) \leq_l (s'+t',s')$.

Given an Arnon order $\leq_A$ and $d \geq 0$, we define a corresponding order on Milnor basis elements $Sq(R)$ of degree d as follows. We begin with the $(0,1)$-block whose rows are the reversed binary expansions of the terms of R. In particular, the element $Sq_t^{2^k} = Sq(0,\ldots,0,2^k)$ of degree $2^k(2^r - 1)$ is represented by a block with a single entry 1 in row r and column $k+1$. This block is rotated through $90°$ anticlockwise and each entry 1 is replaced by the corresponding 2-atomic number $2^k(2^r - 1)$.

We write these 2-atomic numbers as an increasing sequence $At(R)$ of modulus d in the chosen Arnon order $\leq_A$. We call this sequence the 2-**atomic sequence** of R. Given sequences $R,S \in$ Seq, we define $Sq(R) \leq_A Sq(S)$ if and only if $At(R) \leq_l At(S)$.

Example 12.5.4 Let $R = (5,4,0,2)$ and $S = (19,2,1,1)$, so that $Sq(R)$ and $Sq(S)$ are Milnor basis elements of degree $d = 47$. The corresponding blocks and 2-atomic numbers are

$$
\begin{array}{l}
1\ 0\ 1 \\
0\ 0\ 1 \\
0 \\
0\ 1
\end{array}
\quad \longrightarrow \quad
\begin{array}{cccc}
 & 4 & 12 & \\
0 & 0 & 0 & 30 \\
1 & 0 & 0 & 0
\end{array} ,
\qquad
\begin{array}{l}
1\ 1\ 0\ 0\ 1 \\
0\ 1 \\
1 \\
1
\end{array}
\quad \longrightarrow \quad
\begin{array}{cccc}
 & & 16 & \\
 & & 0 & \\
 & & 0 & \\
 & 2 & 6 & \\
1 & 0 & 7 & 15
\end{array} .
$$

For the Y-order, $At(R) = (1,30,4,12)$ and $At(S) = (1,7,15,2,6,16)$, and for the Z-order, $At(R) = (1,4,12,30)$ and $At(S) = (1,2,7,6,15,16)$. In both cases, $At(S) \leq_l At(R)$, and so $Sq(S) \leq_A Sq(R)$.

Here we have used binary blocks only as a step to the 2-atomic numbers representing the Milnor basis elements. The $90°$ rotation is made to fit the traditional array

$$\vdots$$

$$P_1^3 \quad \cdots$$

$$P_1^2 \quad P_2^2 \quad \cdots$$

$$P_1^1 \quad P_2^1 \quad P_3^1 \quad \cdots$$

$$P_1^0 \quad P_2^0 \quad P_3^0 \quad P_4^0 \quad \cdots$$

of the Milnor basis elements $P_t^s = Sq_t^{2^s} = Sq(0,\ldots,0,2^s)$ and the diagram (12.6).

The following result uses the dual algebra A_2^* to produce a large number of bases of A_2, which are triangularly related to the Milnor basis using an Arnon order $\leq_A$ on 2-atomic numbers. The basis elements can be regarded as square-free monomials in elements $\theta(d) \in \mathsf{A}_2^d$, one in each 2-atomic degree d, with the factors written in increasing $\leq_A$-order. We also write $\leq_A$ for the corresponding left lexicographic order on the basis elements.

Theorem 12.5.5 *Let $\leq_A$ be an Arnon order on the set of 2-atomic numbers $d = 2^s(2^t - 1)$, for $s \geq 0$ and $t \geq 1$, and for each such d let $\theta(d) \in \mathsf{A}_2^d$ be an element such that $P_t^s = Sq_t^{2^s}$ is a term in the expansion of $\theta(d)$ in the Milnor basis. Let Θ be the set of all products $\theta(d_1)\cdots\theta(d_r)$ such that $d_1 <_A \cdots <_A d_r$ for $r \geq 1$, together with $Sq^0 = 1$. Then Θ is a basis for A_2 and is triangularly related to the Milnor basis for the order $\leq_A$.*

Proof Let $\xi(d) = \xi_t^{2^s}$ denote the monomial in A_2^* dual to P_t^s. We shall prove that for all pairs of elements $\theta(d_1)\cdots\theta(d_\ell), \theta(e_1)\cdots\theta(e_m) \in \Theta$,

$$\langle \xi(d_1)\cdots\xi(d_\ell), \theta(e_1)\cdots\theta(e_m)\rangle = \begin{cases} 1, & \text{if } (e_1,\ldots,e_m) = (d_1,\ldots,d_\ell), \\ 0, & \text{if } (e_1,\ldots,e_m) <_A (d_1,\ldots,d_\ell). \end{cases} \quad (12.7)$$

The result follows immediately, since every monomial of positive degree in A_2^* is uniquely expressible as a product of distinct elements of the form $\xi(d)$, taken in a fixed but arbitrary order. If this order is an Arnon order $\leq_A$, then the monomial basis is dual to the Milnor basis for the order $\leq_A$.

By Proposition 11.5.2(i), $\phi(\xi_t) = \sum_{u+v=t}\xi_u^{2^v} \otimes \xi_v$. Hence $\phi(\xi_t^{2^s}) = \phi(\xi_t)^{2^s} = \sum_{u+v=t}\xi_u^{2^{v+s}} \otimes \xi_v^{2^s}$. We may write this as $\phi(\xi(d)) = \sum\phi(\xi(d')) \otimes \phi(\xi(d''))$, where $d' = 2^{s+v}(2^u - 1)$, $d'' = 2^s(2^v - 1)$ and the sum is over all $u,v \geq 0$ such that $u + v = t$, where $\xi(0) = 1$. Thus $d' = 0$ if $u = 0$, and since $d' = 2^{s+t} - 2^{s+v}$ and $d = 2^{s+t} - 2^s$, $d' \geq_A d$ if $u > 0$. Then

$$\langle \xi(d_1)\cdots\xi(d_\ell), \theta(e_1)\cdots\theta(e_m)\rangle = \langle \xi(d_1)\cdots\xi(d_\ell), \mu(\theta(e_1) \otimes \theta(e_2)\cdots\theta(e_m))\rangle,$$

where $\mu : A_2 \otimes A_2 \to A_2$ is the product map. Since the coproduct ϕ in A_2^* is dual to μ, this is $\langle \phi(\xi(d_1) \cdots \xi(d_\ell)), \theta(e_1) \otimes \theta(e_2) \cdots \theta(e_m) \rangle$. We have

$$\phi(\xi(d_1) \cdots \xi(d_\ell)) = \phi(\xi(d_1)) \cdots \phi(\xi(d_\ell))$$

$$= \sum (\xi(d_1') \otimes \xi(d_1'')) \cdots (\xi(d_\ell') \otimes \xi(d_\ell''))$$

$$= \sum (\xi(d_1') \cdots \xi(d_\ell')) \otimes (\xi(d_1'') \cdots \xi(d_\ell'')),$$

where the sum is over all d_i', d_i'' corresponding to d_i as above for $1 \le i \le \ell$. When this sum is substituted in $\langle \phi(\xi(d_1) \cdots \xi(d_\ell)), \theta(e_1) \otimes \theta(e_2) \cdots \theta(e_m) \rangle$, a nonzero term can arise only if $d_1' + \cdots + d_\ell' = e_1$, where all the numbers involved are 2-atomic.

If $(e_1, \ldots, e_m) \le_A (d_1, \ldots, d_\ell)$, then in particular $e_1 \le_A d_1$, and since $d_1 <_A \cdots <_A d_\ell$, $e_1 \le_A d_i$ for $1 \le i \le \ell$. But, as observed above, $d_i \le_A d_i'$ if $d_i' \ne 0$. Hence all the nonzero terms d_i' are $\ge_A e_1$. Since these 2-atomic numbers d_i' are also $\le e_1$, they must lie on the same diagonal of (12.6) as e_1, and in the same or higher rows. This is possible only if $e_1 = d_1'$ and $d_i' = 0$ for $i > 1$. Since $e_1 \le_A d_1 \le_A d_1'$, $e_1 = d_1'$ implies that $e_1 = d_1$. To summarize, we have proved that if $\langle \xi(d_1) \cdots \xi(d_\ell), \theta(e_1) \cdots \theta(e_m) \rangle = 1$, then $d_1 = e_1$. It follows by induction on ℓ that $\langle \xi(d_1) \cdots \xi(d_\ell), \theta(e_1) \cdots \theta(e_m) \rangle = 0$ if $(e_1, \ldots, e_m) <_A (d_1, \ldots, d_\ell)$.

It remains to prove that $\langle \xi(d_1) \cdots \xi(d_\ell), \theta(d_1) \cdots \theta(d_\ell) \rangle = 1$. In this case the argument above shows that

$$\langle \xi(d_1) \cdots \xi(d_\ell), \theta(d_1) \cdots \theta(d_\ell) \rangle = \langle \xi(d_1), \theta(d_1) \rangle \langle \xi(d_2) \cdots \xi(d_\ell), \theta(d_2) \cdots \theta(d_\ell) \rangle,$$

and the result follows by induction on ℓ, since the hypothesis on $\theta(d)$ is equivalent to $\langle \xi(d), \theta(d) \rangle = 1$. $\qquad \square$

Example 12.5.6 When $\theta(d) = P_t^s = Sq_t^{2^s}$, the resulting bases are called P_t^s-**bases**.

Example 12.5.7 When $\theta(d) \in A_2^d$ is the element

$$X_s^{s+t-1} = Sq^{2^{s+t-1}} \cdots Sq^{2^{s+1}} Sq^{2^s}$$

and the Arnon order is the Z-order, we obtain the **Arnon A basis**. This is a monomial basis in the Steenrod squares Sq^k. We use induction on t to prove that the expansion of $\theta(d)$ in the Milnor basis involves P_t^s. Since $P_1^s = Sq^{2^s}$ in the case $t = 1$, we assume as inductive hypothesis that the result holds for $t - 1$. We have $\langle \xi_t^{2^s}, X_s^{s+t-1} \rangle = \langle \xi_t^{2^s}, \mu(Sq^{2^{s+t-1}} \otimes X_t^{s+t-2}) \rangle = \langle \phi(\xi_t^{2^s}), Sq^{2^{s+t-1}} \otimes X_t^{s+t-2} \rangle$. With notation as for 12.5.5, $\phi(\xi(d)) = \sum \xi(d') \otimes \xi(d'')$, where $d' = 2^s(2^t - 2^v)$. Thus $d' = 2^{s+t-1}$ if and only if $v = t - 1$, $u = 1$, and so $\langle \phi(\xi_t^{2^s}), Sq^{2^{s+t-1}} \otimes X_t^{s+t-2} \rangle = \langle \xi_1^{2^{s+t-1}} \otimes \xi_{t-1}^{2^s}, Sq^{2^{s+t-1}} \otimes X_t^{s+t-2} \rangle = \langle \xi_1^{2^{s+t-1}}, Sq^{2^{s+t-1}} \rangle \langle \xi_{t-1}^{2^s}, X_t^{s+t-2} \rangle$. Since $\langle \xi_1^{2^{s+t-1}}, Sq^{2^{s+t-1}} \rangle = 1$, this proves the inductive step.

Example 12.5.8 When $\theta(d) = Xq^d = \chi(Sq^d)$ and the Arnon order is the Y-order, the Y-**basis** is obtained by applying χ to all elements of Θ, and the Z-**basis** is defined similarly using the Z-order. The Y- and Z-bases are monomial bases in the Steenrod squares with exponents in decreasing Y- or Z-order. For example, the top element of the subalgebra $A_2(2)$ is written as $Sq^4 Sq^6 Sq^2 Sq^7 Sq^3 Sq^1$ in the Y-basis and as $Sq^4 Sq^6 Sq^7 Sq^2 Sq^3 Sq^1$ in the Z-basis. The hypothesis on $\theta(d)$ needed to apply Theorem 12.5.5 follows from Proposition 3.5.11, since $\chi(Sq^d)$ is the sum of all $Sq(R)$ in degree d, and so in particular it involves P_t^s.

Example 12.5.9 If $\theta(d) = \chi(Z_s^{s+t-1})$, where

$$Z_s^{s+t-1} = Sq^{2^s} Sq^{2^{s+1}} \cdots Sq^{2^{s+t-1}}$$

and the Arnon order is the Z-order, the corresponding basis obtained by applying χ to all elements of Θ is the **Wall basis** (or **Arnon B-basis**). We shall show that P_t^s is a term in the Milnor basis expansion of $\theta(d) = Xq^{2^{s+t-1}} \cdots Xq^{2^{s+1}} Xq^{2^s}$.

We begin by using Proposition 3.5.11 to write the factors as sums of all $Sq(R)$ in degrees $2^{s+t-1}, \cdots, 2^{s=1}, 2^s$. A Milnor matrix X which produces a term $P_t^s = Sq(0, \ldots, 0, 2^s)$ in a product $Sq(R)Sq(S)$ must have all entries 0 except those on the tth diagonal. If the tth diagonal has more than one nonzero entry, then the coefficient $b_X = 0$ since a nontrivial partition of 2^s into 2-powers must have the smallest part repeated. Hence X has only one nonzero entry $x_{i,j} = 2^s$, where $i+j = t$. The initial Milnor matrix corresponding to X must then be

	0	$\cdots$	0	$x_{0,j} = 2^s$
0	0	$\cdots$	0	0
$\vdots$	$\vdots$	$\ddots$	$\vdots$	$\vdots$
$x_{i,0} = 2^{s+j}$	0	$\cdots$	0	0

Thus P_t^s is a term in the product $Sq(R)Sq(S)$ if and only if $Sq(R) = P_i^{s+j}$ and $Sq(S) = P_j^s$ for some i,j such that $i+j = t$. It follows by induction on t that P_t^s is a term in the Milnor basis expansion of $Xq^{2^{s+t-1}} \cdots Xq^{2^{s+1}} Xq^{2^s}$.

The following result follows immediately from Theorem 12.5.5 by counting the basis elements which lie in the subalgebra $A_2(n)$. Since all elements of A_2 of degree $< 2^{n+1}$ are in $A_2(n)$, all elements $\theta(d)$ for 2-atomic numbers $d < 2^{n+1}$ are in $A_2(n)$. There are $\binom{n+2}{2}$ such numbers, and so the number of basis elements $\theta(d_1) \cdots \theta(d_\ell)$ which lie in $A_2(n)$ is $\leq 2^{\binom{n+2}{2}} = \dim A_2(n)$.

Proposition 12.5.10 *The bases for* A_2 *given by Theorem 12.5.5 restrict to bases of* $A_2(n)$ *for all* $n \geq 0$. *The basis for* $A_2(n)$ *is given by substrings of the product representing the top element of* $A_2(n)$. □

Remark 12.5.11 The P_t^s-bases given by Example 12.5.6 belong to a larger family. Fix a linear ordering of the set of 2-atomic numbers, and consider all finite products $P_{t_1}^{s_1} \cdots P_{t_k}^{s_k}$ of P_t^s elements with factors in increasing order, together with 1. It follows that the products of this form which lie in $A_2(n)$ form a basis for $A_2(n)$, since the error terms introduced by multiplying Milnor basis elements $Sq(R)$ and $Sq(S)$ for which $At(R)$ and $At(S)$ are disjoint are $<_r Sq(R + S)$. Hence all these bases are triangular with respect to the Milnor basis using the $\leq_r$ order.

12.6 The Arnon C basis

The Arnon C basis of A_2 consists of monomials Sq^C in the Steenrod squares whose last factor Sq^k gives the excess k of Sq^C.

Definition 12.6.1 An **Arnon C-monomial** is an element of A_2 of the form $Sq^{[T]} = Sq^{2^{m-1}t_m} \cdots Sq^{2t_2} Sq^{t_1}$, where $T = (t_1, t_2, \ldots, t_m)$ is a decreasing sequence.

Proposition 12.6.2 *For* $d \geq 0$, *the Arnon C-monomials* $Sq^{[T]}$ *indexed by* $T \in$ Dec *with* $\deg_2 T = d$ *form a basis of* A_2^d, *which is triangularly related to the Milnor basis for the right order* $<_r$.

Proof We follow the same method as for the admissible basis, Theorem 3.4.4. Let $k \geq t_1$. Then in block notation, there is a term in $Sq^{[T]}(x_1 \cdots x_k)$ where Sq^{t_1} moves t_1 1s from column 1 to column 2, Sq^{t_1} moves t_2 of these 1s from column 2 to column 3, and so on. For example, in the case $T = (5, 3, 2)$, $k = 6$, so that $Sq^{[T]} = Sq^8 Sq^6 Sq^5$, the diagram below shows the leading terms of the $\leq_r$-minimal Cartan symmetric functions obtained at each stage:

$$
\begin{array}{cccc}
1 & 0\,1 & 0\,0\,1 & 0\,0\,0\,1 \\
1 & 0\,1 & 0\,0\,1 & 0\,0\,0\,1 \\
1 \xrightarrow{Sq^5} & 0\,1 \xrightarrow{Sq^6} & 0\,0\,1 \xrightarrow{Sq^8} & 0\,0\,1 \\
1 & 0\,1 & 0\,1 & 0\,1 \\
1 & 0\,1 & 0\,1 & 0\,1 \\
1 & 1 & 1 & 1
\end{array}
$$

This term cannot be obtained by any other sequence of choices of terms in the Cartan formula for the application of successive factors $Sq^{2^i t_{i+1}}$ of the Arnon C-monomial $Sq^{[T]}$ to $Sq^{2^{i-1} t_i} \cdots Sq^{t_1}(x_1 \cdots x_k)$, and it gives the leading term of the $\leq_r$-minimal Cartan symmetric function in $Sq^{[T]}(x_1 \cdots x_k)$, namely $c(k - t_1, t_1 - t_2, \ldots, t_{m-1} - t_m, t_m)$. Hence $Sq^{[T]} = Sq(R) + \sum_S Sq(S)$, where $R = (t_1 - t_2, \ldots, t_{m-1} - t_m, t_m)$ and $R \leq_r S$ for all S in the sum. $\qquad\square$

Example 12.6.3 The table below gives the conversion from the Arnon C-basis to the Milnor basis in degree 9. By Theorem 3.5.1, the Arnon C-basis is also triangularly related to the admissible basis for the right order.

	$Sq(2,0,1)$	$Sq(0,3)$	$Sq(3,2)$	$Sq(6,1)$	$Sq(9)$
$Sq^4 Sq^2 Sq^3$	1	0	1	1	1
$Sq^6 Sq^3$	0	1	0	0	0
$Sq^4 Sq^5$	0	0	1	1	1
$Sq^2 Sq^7$	0	0	0	1	1
Sq^9	0	0	0	0	1

Proposition 12.6.4 *For* $T = (t_1, t_2, \ldots) \in \mathrm{Dec}$, *the Arnon C-monomial* $Sq^{[T]}$ *has excess* t_1.

Proof The previous proof shows that the excess is $\leq t_1$. It cannot be $< t_1$, because Sq^{t_1} has excess t_1. $\qquad\square$

12.7 Remarks

For the relations of Proposition 12.1.7, see [15]. In [64], it is shown that the $d+1$ elements $Sq^a Xq^b$ with $a+b=d$ span a vector space of dimension $d+1-\beta(d)$, with basis given by the elements with $\beta(d) \leq a \leq d$. Any subset of these relations obtained using $\beta(d)$ consecutive values of m are linearly independent. M. G. Barratt and H. R. Miller proved these relations by an inductive argument on $\alpha_2(j)$, using the Bullett–Macdonald form of the Adem relations to start the induction. In [40], M. C. Crabb, M. D. Crossley and J. R. Hubbuck used ideas from K-theory to reformulate and simplify their argument. Linear dependence between the relations is studied further in [64].

Proposition 12.1.6 is due to C. T. C. Wall [222]. By choosing an explicit polynomial expression in $Sq^1, Sq^2, \ldots, Sq^{2^{r-1}}$ for the elements of $A_2(r-1)$ expressed here in the Milnor basis, the relations of Proposition 12.1.6 become a set of relations in the generators Sq^{2^r}, $r \geq 0$, of A_2. Wall proved that this is a minimal set of defining relations for A_2 in terms of these generators, and obtained the additive basis of A_2 in monomials in the elements Sq^{2^r} given by Example 12.5.9.

The importance of the finite Hopf subalgebras $A_2(n)$ of A_2 and of the elements $P_t^s = Sq_t^{2^s} = Sq(0, \ldots, 0\ 2^s)$ has been recognised since the earliest studies of the Steenrod algebra. The bases for A_2 constructed in Section 12.4 by using compositions of the elements Sq_t^k appear in Section 5 of the influential paper of J. F. Adams [1], where they were overshadowed by the application of A_2 to homotopy theory via the famous spectral sequence introduced by Adams in the same work. Adams defined a 2-parameter family of Hopf subalgebras and proved Proposition 12.4.7. The sub-Hopf algebras of A_2 were classified by Adams and H. R. Margolis [3, 129]. The Steenrod algebra A_q over the Galois field of q elements appears in [190, Chapter 11].

K. G. Monks [146] introduced the P_t^s bases of Remark 12.5.11 and showed that the top class of $A_2(\lambda)$ is an admissible monomial. The Arnon A, B and C bases were introduced in [9]. The B basis coincides with Wall's basis. The Y- and Z-bases of Example 12.5.8 appear in [231]. Proposition 12.4.8 goes back to Steenrod-Epstein [196, II, Lemma 3.1].

The collection of interesting bases of the Steenrod algebra continues to grow. In [157], J. H. Palmieri and J. J. Zhang construct a family of bases for A_2 whose elements are products of iterated commutators, in 2-atomic degrees, of the generators Sq^{2^j}, $j \geq 0$. These bases restrict to bases of $A_2(n)$.

13

Stripping and nilpotence in A_2

13.0 Introduction

In this chapter we introduce a technique for obtaining new relations in A_2 from old ones, which we call **stripping**. Formally the method consists of acting on a relation $X = 0$ in A_2 with an element ξ of the dual algebra A_2^*. By analogy with the construction of the cap product of a cohomology class ξ and a homology class X in algebraic topology, we write the resulting relation as $\xi \cap X = 0$. The degree of the relation is lowered by $\deg \xi$. Since $\xi \eta \cap X = \xi \cap (\eta \cap X)$ for all $\xi, \eta \in A_2^*$ and all $X \in A_2$, all such operations are compositions of stripping by the generators ξ_i of $A_2^* = \mathbb{F}_2[\xi_1, \xi_2, \xi_3, \ldots]$.

In practice, it is more efficient to treat all elements $\xi_i^{2^j}$ for $i \geq 1$ and $j \geq 0$ as the basic stripping operators, one in each 2-atomic degree $2^j(2^i - 1)$. Proposition 13.2.3 explains how to apply these operations to a monomial $Sq^A = Sq^{a_1} \cdots Sq^{a_r}$ in A_2. The result is the sum of all monomials which are obtained by subtracting the integers $2^{i+j-1}, \ldots, 2^{j+1}, 2^j$, in this order, from i of the exponents $a_1, \ldots, a_r$. We describe this as 'stripping Sq^A by $(2^{i+j-1}, \ldots, 2^{j+1}, 2^j)$'.

For example, by stripping the Adem relation $Sq^{13} Sq^7 = 0$ by $(2, 1)$ we obtain the Adem relation $Sq^{11} Sq^6 = 0$. We can also strip the relation $Sq^{13} Sq^7 = 0$ by (1), (2), (4) and (8) to obtain the Adem relations $Sq^{12} Sq^7 = Sq^{13} Sq^6$, $Sq^{11} Sq^7 = Sq^{13} Sq^5$, $Sq^9 Sq^7 = Sq^{13} Sq^3$ and $Sq^5 Sq^7 = 0$. Stripping these in turn we obtain many more relations. For example, by stripping $Sq^5 Sq^7 = 0$ by $(2, 1)$, (1) and (2) we obtain $Sq^3 Sq^6 = 0$, $Sq^4 Sq^7 = Sq^5 Sq^6$ and $Sq^3 Sq^7 = Sq^5 Sq^5$.

It follows from Proposition 12.4.7 that every element of A_2^+ is nilpotent. Given $\theta \in A_2^+$, the minimum h such that $\theta^h = 0$ is the **nilpotence height** of θ. One application of the stripping method is to obtain a relation of the form $\theta^m = 0$ for a given element $\theta \in A_2^+$, thus giving an upper bound m on the nilpotence height of θ. One such upper bound can be obtained by considering the highest degree of the subalgebra $A_2(n)$ of Section 12.4 containing θ. For example, $Sq^5 \in A_2(2)$ and the top class $Sq(7,3,1)$ of $A_2(2)$ has degree 23, so $(Sq^5)^5 = 0$. However, by premultiplying the relation $Sq^3Sq^7 = Sq^5Sq^5$ by Sq^5 and using the basic Adem relation $Sq^5Sq^3 = 0$ we obtain $(Sq^5)^3 = 0$. Since $Sq^5Sq^5 = Sq^9Sq^1$ is admissible, we conclude that the nilpotence height of Sq^5 is 3.

In Section 13.1 we consider stripping operations in the general context of the action of a Hopf algebra on its dual, and in Section 13.2 we apply this to the Steenrod algebra. The fact that the dual algebra A_2^* is commutative is needed here. In Section 13.3 we introduce the self-map V of A_2 known as the **halving map** or **Verschiebung**. The map V sends a monomial Sq^A with at least one odd exponent to 0, and divides all exponents by 2 otherwise.

In Section 13.4 we introduce the subalgebra Od of A_2 which has $\mathbb{F}_2$-basis given by elements $Sq(R)$ such that all entries of R are odd. Although Od is not a Hopf subalgebra, it can be applied to study the nilpotence order of its basis elements $Sq(R)$. In Section 13.5 we show that Sq^{2^k} has nilpotence order $2k + 2$ for $k \geq 0$. In Section 13.7 this result is generalized to give the nilpotence order of the elements $P_t^s = Sq_t^{2^s}$ of Section 12.5, using relations in the subalgebra A_q, $q = 2^t$, of Section 12.3 which we develop in Section 13.6.

13.1 Cap products for Hopf algebras

Given a Hopf algebra H over a field F with coproduct $\phi : H \to H \otimes H$ and counit $\varepsilon : H \to F$, a left action of the dual algebra H^* on H is defined by the composition

$$H^* \otimes H \xrightarrow{\text{id} \otimes \phi} H^* \otimes H \otimes H \xrightarrow{\langle \ \rangle \otimes \text{id}} H$$

where $\langle \ \rangle : H^* \otimes H \to F$ is the Kronecker evaluation $\xi \otimes \theta \mapsto \langle \xi, \theta \rangle = \xi(\theta)$. A right action of H^* on H is defined in a similar way. These coincide when H is cocommutative, and we write the action as a cap-product, associating to each element $\xi \in H^*$ the F-linear map $\xi \cap : H \to H$ defined by

$$\xi \cap \theta = \sum_\theta \langle \xi, \theta' \rangle \theta'' = \sum_\theta \theta' \langle \xi, \theta'' \rangle \tag{13.1}$$

where $\phi(\theta) = \sum_\theta \theta' \otimes \theta''$ as in (11.1). These cap-product operators obey the following rules. Since $\varepsilon : H \to F$ is the identity element of H^*, (i) states that $\xi \cap \theta$ defines an action of H^* on H, (ii) and (iv) relate this action to the products in H and H^*, and (iii) states that it commutes with the antipodes χ of H and χ^* of H^*.

Proposition 13.1.1 *Let H be a cocommutative Hopf algebra with dual algebra H^*. Then for θ, θ_1, $\theta_2 \in H$ and ξ, $\eta \in H^*$*

(i) $\xi \eta \cap \theta = \xi \cap (\eta \cap \theta)$ *and* $\varepsilon \cap \theta = \theta$,

(ii) $\xi \cap \theta_1 \theta_2 = \sum_\xi (\xi' \cap \theta_1)(\xi'' \cap \theta_2)$,

(iii) $\chi(\xi \cap \theta) = \chi^*(\xi) \cap \chi(\theta)$,

(iv) $\langle \xi \eta, \theta \rangle = \langle \xi, \eta \cap \theta \rangle$.

Proof Using the Kronecker product $H^* \otimes H \to F$ to express the duality between the product in H^* and the coproduct in H, we have

$$\langle \xi \eta, \theta \rangle = \sum_\theta \langle \xi, \theta' \rangle \langle \eta, \theta'' \rangle, \tag{13.2}$$

while the compatibility of the product and coproduct in H gives

$$\sum_{\theta_1 \theta_2} (\theta_1 \theta_2)' \otimes (\theta_1 \theta_2)'' = \sum_{\theta_1} \sum_{\theta_2} \theta_1' \theta_2' \otimes \theta_1'' \theta_2''. \tag{13.3}$$

To prove (i) we expand the left and right sides and compare them using the coassociativity of H. Thus

$$\xi \eta \cap \theta = \sum_\theta \langle \xi \eta, \theta' \rangle \theta'' = \sum_\theta \sum_{\theta'} \langle \xi, (\theta')' \rangle \langle \eta, (\theta')'' \rangle \theta'' = \sum_\theta \langle \xi, \theta' \rangle \langle \eta, \theta'' \rangle \theta''',$$

using Sweedler's notational convention (11.1) for coassociativity. But

$$\xi \cap (\eta \cap \theta) = \xi \cap \sum_\theta \langle \eta, \theta' \rangle \theta'' = \sum_\theta \langle \eta, \theta' \rangle (\xi \cap \theta'') = \sum_\theta \langle \eta, \theta' \rangle \sum_{\theta''} \langle \xi, \theta'' \rangle \theta''',$$

which is the same sum, since H^* is commutative. Further $\varepsilon \cap \theta = \sum_\theta \langle \varepsilon, \theta' \rangle \theta'' = \sum_\theta \varepsilon(\theta') \theta'' = \theta$.

To prove (ii), we have

$$\xi \cap \theta_1 \theta_2 = \sum_{\theta_1 \theta_2} \langle \xi, (\theta_1 \theta_2)' \rangle (\theta_1 \theta_2)'' = \sum_{\theta_1} \sum_{\theta_2} \langle \xi, \theta_1' \theta_2' \rangle \theta_1'' \theta_2''$$

by (13.3), while

$$\sum_\xi (\xi' \cap \theta_1)(\xi'' \cap \theta_2) = \sum_\xi \sum_{\theta_1} \langle \xi', \theta_1' \rangle \theta_1'' \sum_{\theta_2} \langle \xi'', \theta_2' \rangle \theta_2''$$

$$= \sum_{\theta_1} \sum_{\theta_2} \sum_\xi \langle \xi', \theta_1' \rangle \langle \xi'', \theta_2' \rangle \theta_1'' \theta_2'',$$

and these are equal since $\langle \xi, \theta_1' \theta_2' \rangle = \sum_\xi \langle \xi', \theta_1' \rangle \langle \xi'', \theta_2' \rangle$ by the dual statement to (13.2).

To prove (iii), we use the duality $\langle \chi^*(\xi), \theta \rangle = \langle \xi, \chi(\theta) \rangle$ and the fact that χ is a coalgebra map, i.e. $(\chi \otimes \chi)(\phi(\theta)) = \phi(\chi(\theta))$, so that $\sum_{\chi(\theta)}(\chi(\theta))' \otimes (\chi(\theta))'' = \sum_\theta \chi(\theta') \otimes \chi(\theta'')$. It follows that $\chi(\xi) \cap \theta = \chi(\sum_\theta \langle \xi, \theta' \rangle \theta'') = \sum_\theta \langle \xi, \theta' \rangle \chi(\theta'')$, while $\chi^*(\xi) \cap \chi(\theta) = \sum_{\chi(\theta)} \langle \chi^*(\xi), \chi(\theta)' \rangle \chi(\theta)''$ and (iii) follows since these are equal.

To prove (iv), we have $\langle \xi, \eta \cap \theta \rangle = \langle \xi, \sum_\theta \langle \eta, \theta' \rangle \theta'' \rangle = \sum_\theta \langle \eta, \theta' \rangle \langle \xi, \theta'' \rangle = \langle \eta \xi, \theta \rangle = \langle \xi \eta, \theta \rangle$, since H^* is commutative. $\qquad \square$

13.2 The action of A_2^* on A_2

By Theorem 11.3.7, A_2 is a cocommutative Hopf algebra. Hence by Proposition 13.1.1 a left action of the dual algebra A_2^* on A_2 is defined by the composition $A_2^* \otimes A_2 \xrightarrow{\mathrm{id} \otimes \phi} A_2^* \otimes A_2 \otimes A_2 \xrightarrow{\langle\ \rangle \otimes \mathrm{id}} A_2$, where $\langle\ \rangle : A_2^* \otimes A_2 \to \mathbb{F}_2$ is the Kronecker evaluation $\xi \otimes \theta \mapsto \langle \xi, \theta \rangle$. A right action of A_2^* on A_2 is defined in a similar way. As in Section 13.1, we associate to each element $\xi \in A_2^*$ the $\mathbb{F}_2$-linear map $\xi \cap : A_2 \to A_2$ defined by (13.1). We describe the process of mapping θ to $\xi \cap \theta$ as **stripping** θ by $\xi \in A_2^*$.

This is an important technique for deriving new relations in A_2 from existing ones. By Proposition 13.1.1(ii), $\xi \cap : H \to H$ is a derivation of the Hopf algebra H when ξ is a primitive element of H^*, i.e. $\phi(\xi) = 1 \otimes \xi + \xi \otimes 1$. By Proposition 11.5.4 the elements $\xi_1^{2^j}, j \geq 0$ are primitive in A_2^*. In particular, $\xi_1 \cap$ is a derivation of A_2, and $\xi_1 \cap Sq^r = Sq^{r-1}$ for all $r > 0$. Hence stripping the relation $Sq^{2k-1}Sq^k = 0$ by ξ_1 leads to the relation $Sq^{2k-2}Sq^k + Sq^{2k-1}Sq^{k-1} = 0$.

The effect of stripping on Milnor basis elements is easy to describe, and follows directly from the definition using Proposition 11.2.4.

Proposition 13.2.1 *Let* $Sq(R) = Sq(r_1, r_2, \ldots)$ *and let* $\xi^S = \xi_1^{s_1} \xi_2^{s_2} \cdots$. *Then* $\xi^S \cap Sq(R) = Sq(R - S) = Sq(r_1 - s_1, r_2 - s_2, \ldots)$ *if* $r_i \geq s_i$ *for all* $i \geq 1$, *and* $\xi^S \cap Sq(R) = 0$ *otherwise.* $\qquad \square$

The next result allows us to strip monomials in A_2 by the generators of A_2^*.

Proposition 13.2.2 (i) $\langle \xi_k, Sq^{2^{k-1}} \cdots Sq^2 Sq^1 \rangle = 1$, *and*
(ii) $\langle \xi_k, Sq^A \rangle = 0$, *where* Sq^A *is a monomial of degree* $2^k - 1$ *and length* $< k$.

Proof By definition of ξ_k, $\langle \xi_k, Sq(R) \rangle = 1$ when $R = (0, \ldots, 0, 1)$ of length k, and $\langle \xi_k, Sq(S) \rangle = 0$ for all sequences $S \neq R$. For $d = 2^k - 1$, the Milnor basis element $Sq(R) = Q_k$ and the admissible monomial $Sq^{2^{k-1}} \cdots Sq^2 Sq^1$ have length k, and all other elements of both bases have length $< k$. Since the correspondence between admissible and Milnor sequences preserves length,

$Sq^{2^{k-1}} \cdots Sq^2 Sq^1 = Q_k +$ an element of $Ad_{<k}$ (Definition 3.5.6). This proves (i), and also (ii) if A is admissible.

More generally, if $A \in Seq^d$ has length $< k$ then we can express Sq^A as a sum of admissible monomials of length $< k$ by using the Adem relations (3.1). Thus $Sq^A \in Ad_{<k}$, and (ii) follows. $\qquad\square$

More generally, we wish to describe the action of $\xi \cap : A_2 \to A_2$ for a monomial $\xi \in A_2^*$ in combinatorial terms. Although in principle we need consider only the cases $\xi = \xi_i$ for $i \geq 1$, in practice it is more efficient to factor a monomial in A_2^* as a product of monomials $\xi_i^{2^k}$, $k \geq 0$. We associate with this monomial the **allowable sequence** $(2^{k+i-1}, \ldots, 2^k)$ of length i, whose components are consecutive descending powers of 2, ending with 2^k.

Proposition 13.2.3 *For any monomial* $Sq^A \in A_2$, $\xi_i^{2^k} \cap Sq^A = \sum_B Sq^B$, *the sum of all monomials such that the nonzero entries of $A - B$, taken in order, are* $2^{k+i-1}, \ldots, 2^k$.

Proof First consider the case $k = 0$, i.e. the operation $\xi_i \cap$. Since $\phi(Sq^k) = \sum_{i+j=k} Sq^i \otimes Sq^j$, we have $\phi(Sq^A) = \sum_{B+C=A} Sq^B \otimes Sq^C$ by Proposition 11.3.2. By Proposition 13.2.2, $\langle \xi_i, Sq^B \rangle$ is nonzero only when $Sq^B = Sq^{2^{i-1}} \cdots Sq^2 Sq^1$. This occurs if and only if B is obtained by interspersing 0s in the allowable sequence $(2^{i-1}, \ldots, 2, 1)$ to make its length the same as that of A. The complementary sequences $C = A - B$ give the terms of $\xi_i \cap Sq^A$.

In the case $k = 1$, the operation $\xi_i^2 \cap$ is carried out by repeating this process twice. However, the terms obtained from decompositions $A = B_1 + (B_2 + C)$ and $A = B_2 + (B_1 + C)$ are identical, and cancel unless $B_1 = B_2$. The result is thus the same as that of applying the allowable sequence $(2^i, \ldots, 2)$. The case $k > 1$ follows by iteration. $\qquad\square$

Example 13.2.4 Starting with the Adem relation $Sq^9 Sq^5 = 0$, we can strip by (8) to get $Sq^1 Sq^5 = 0$, or strip by (4) to get $Sq^5 Sq^5 + Sq^9 Sq^1 = 0$. Hence $Sq^5 Sq^5 Sq^5 = Sq^9 Sq^1 Sq^5 = 0$. Stripping this relation by $(2, 1)$, we obtain

$$Sq^5 Sq^3 Sq^4 + Sq^3 Sq^5 Sq^4 + Sq^3 Sq^4 Sq^5 = 0. \qquad (13.4)$$

Theorem 13.1.1(iii) relates stripping operations with conjugation. The next result provides a combinatorial procedure for stripping a monomial Sq^A by $\chi^*(\xi_i)$.

Example 13.2.5 Since $\chi^*(\xi_2) = \xi_2 + \xi_1^3$, stripping by $\chi^*(\xi_2)$ is the sum of stripping by $(2, 1)$ and $(2)(1)$. Applying this operation to the relation

$Sq^5 Sq^5 Sq^5 = 0$, the terms which appear in (13.4) cancel, and so we obtain the relation $Sq^2 Sq^5 Sq^5 + Sq^5 Sq^2 Sq^5 + Sq^5 Sq^5 Sq^2 + Sq^4 Sq^3 Sq^5 + Sq^4 Sq^5 Sq^3 + Sq^5 Sq^4 Sq^3 = 0$. This is equivalent to stripping by $(1,2)$ and (3) and adding the results, but $(1,2)$ and (3) are not allowable sequences, and can only be applied in combination.

Proposition 13.2.6 *For any monomial* $Sq^A \in A_2$, $\chi^*(\xi_i^{2^k}) \cap Sq^A = \sum_B Sq(B)$, *the sum of all monomials such that the nonzero entries of* $A - B$, *taken in order, are sums of consecutive terms of* $2^k, \ldots, 2^{k+i-1}$, *each being used once.*

For example, when $k = 0$ and $i = 3$, we add the results of stripping by the sequences $(1,2,4)$, $(1,6)$, $(3,4)$ and (7).

Proof It suffices to consider the case $k = 0$, as the general case follows by iteration as in Proposition 13.2.3. By Proposition 11.5.2(ii), $\chi^*(\xi_i)$ contains a term corresponding to each of the 2^{i-1} ways of partitioning the allowable sequence $(2^{i-1}, \ldots, 2, 1)$ into subsequences. For example, stripping by $\chi^*(\xi_3)$ is effected by adding the results of stripping by $(4,2,1)$, $(4,2)(1)$, $(4)(2,1)$ and $(4)(2)(1)$.

The result of stripping by $(2^{i-1}) \cdots (2)(1)$ is the sum of the results of stripping by all the sequences $(v_1, \ldots v_r)$ whose terms are obtained by adding terms of $(2^{i-1}, \ldots, 2, 1)$, using each once. If these terms are not used in increasing order, then there will be an even number of ways to partition $(2^{i-1}, \ldots, 2, 1)$ into subsequences so that the corresponding operation includes stripping by $(v_1, \ldots v_r)$.

For example, in the case $i = 3$ the sequence $(2,5)$ is produced by stripping either by $(4)(2)(1)$ or by $(4)(2,1)$. Thus the terms produced by $\chi^*(\xi_1^{2^i-1})$ which are not cancelled by terms arising from other summands of $\chi^*(\xi_i)$ are as stated. $\qquad\square$

Example 13.2.7 We illustrate the stripping technique by proving that

$$\sum_{i+j=k} Sq^i Sq^j = \begin{cases} Sq(0, k/3), & \text{if } k = 0 \bmod 3, \\ 0, & \text{otherwise.} \end{cases}$$

Since $\xi_1 \cap$ is a derivation and $\xi_1 \cap Sq^r = Sq^{r-1}$,

$$\xi_1 \cap \sum_{i+j=k} Sq^i Sq^j = \sum_{i+j=k} (Sq^{i-1} Sq^j + Sq^i Sq^{j-1}) = 0.$$

From Proposition 13.2.1, the kernel of $\xi_1 \cap$ is the subspace spanned by Milnor basis elements $Sq(R)$ with $r_1 = 0$. By Milnor's product formula, Theorem 4.1.2, $Sq^i Sq^j$ is a sum of elements $Sq(R)$ of length ≤ 2. Since $Sq(0, r)$ has degree $3r$, the result follows for degrees $k \neq 0$ mod 3. For $k = 3r$, the result follows using $\xi_2 \cap$ by induction on r.

13.3 The halving map of A_2

For a commutative algebra A over $\mathbb{F}_2$, the **Frobenius map** F_A is defined by $F_A(x) = x^2$ for $x \in A$. It is an algebra map $A \to A$, and if $f : A \to B$ is a map of commutative algebras over $\mathbb{F}_2$, then $f \circ F_A = F_B \circ f$. In particular, if A is a Hopf algebra with coproduct $\phi : A \to A \otimes A$ and conjugation $\chi : A \to A$, then $\phi \circ F_A = (F_A \otimes F_A) \circ \phi$ and $\chi \circ F_A = F_A \circ \chi$, and so F_A is a map of Hopf algebras.

When $A = A_2^*$ is the dual of the Steenrod algebra, the Frobenius map doubles exponents of monomials in the generators ξ_i, i.e. $F_A(\xi_1^{a_1} \cdots \xi_r^{a_r}) = \xi_1^{2a_1} \cdots \xi_r^{2a_r}$ for $r \geq 1$ and $a_1, \ldots, a_r \geq 0$. Since F_A is a map of Hopf algebras, the subalgebra $F_A(A_2^*)$ generated by ξ_i^2 for $i \geq 1$ is a Hopf subalgebra of A_2^*, and is isomorphic to A_2^* as a Hopf algebra. As a graded Hopf algebra, $F_A(A_2^*)$ is isomorphic to the graded Hopf algebra obtained from A_2^* by doubling the grading.

Let $V : A_2 \to A_2$ be the linear dual of F_A. Since F_A is an injective map of Hopf algebras which doubles the grading, V is a surjective map of Hopf algebras which maps elements of even degree d to elements of degree $d/2$ and elements of odd degree to 0. The map V is the **halving map** (or **Verschiebung**) of A_2.

Proposition 13.3.1 *The halving map* $V : A_2 \to A_2$ *satisfies*

  (i) *for a Milnor basis element* $Sq(R)$, $V(Sq(R)) = Sq(S)$ *if* $R = 2S$, *and* $V(Sq(R)) = 0$ *otherwise;*
 (ii) *for a monomial* Sq^A, $V(Sq^A) = Sq^B$ *if* $A = 2B$, *and* $V(Sq^A) = 0$ *otherwise;*
(iii) *the kernel of* V *is the Hopf ideal* $I(1) = A_2 Sq^1 A_2$ *generated by* Sq^1, *and hence* V *induces an isomorphism of Hopf algebras* $A_2/I(1) \to A_2$ *which halves the grading;*
(iv) V *commutes with the conjugation* χ *of* A_2.

Proof (i) follows from the duality between the Milnor basis and the monomial basis in A_2^* and the action of the Frobenius map F_A on such monomials. In particular, this proves (ii) for the case Sq^k for $k \geq 1$, since $Sq^k = Sq(R)$ when $R = (k, 0, \ldots)$. Since V is an algebra map, (ii) holds in general. For the same reason, the fact that $V(Sq^1) = 0$ implies that the ideal $I(1)$ is in the kernel of V.

For the converse, let $V(\theta) = 0$ where $\theta \in A_2$. By writing θ as a sum of admissible monomials, and noting that by (ii) V maps linearly independent admissible monomials Sq^A with $A = 2B$ to linearly independent admissible monomials Sq^J, and maps admissible monomials Sq^J with an odd exponent to 0, all admissible monomials appearing in θ must have an odd factor Sq^{2k+1}. The result then follows from the Adem relation $Sq^1 Sq^{2k} = Sq^{2k+1}$. Finally (iv) holds since V is a map of Hopf algebras. $\qquad \Box$

Proposition 13.3.2 *The k-fold iterate* $V^k : A_2 \to A_2$ *of V satisfies*

(i) *for a Milnor basis element $Sq(R)$, $V^k(Sq(R)) = Sq(S)$ if $R = 2^k S$, and $V^k(Sq(R)) = 0$ otherwise;*

(ii) *for a monomial Sq^A, $V^k(Sq^A) = Sq^B$, if $A = 2^k B$, and $V^k(Sq^A) = 0$ otherwise;*

(iii) *the kernel of V^k is the Hopf ideal $I(k)$ generated by $Sq^1, Sq^2, \ldots, Sq^{2^{k-1}}$, and hence V^k induces an isomorphism of Hopf algebras $A_2/I(k) \to A_2$ which divides the degree by 2^k.*

Proof Parts (i) and (ii) follow by induction on k from the corresponding properties of V. Arguing as in the proof of (iii) for V, we see that $Sq^j \in \text{Ker}(V^k)$ for $j < 2^k$, so that $I(k) \subset \text{Ker}(V^k)$, and that conversely every element $\theta \in \text{Ker}(V^k)$ is expressible as a sum of admissible monomials each having a factor Sq^j with j not divisible by 2^k. Thus $j = 2^r(2s+1)$ where $r \leq k-1$. As in the proof of Proposition 3.2.4, the Adem relation

$$Sq^{2^r} Sq^{2^{r+1}s} = Sq^j + \sum_{t=1}^{2^r-1} \binom{2^{r+1}s - t - 1}{2^r - 2t} Sq^{j-t} Sq^t \qquad (13.5)$$

shows that $Sq^j \in I(k)$. Hence $\text{Ker}(V^k) \subseteq I(k)$. $\qquad \Box$

It follows from Proposition 13.3.1(ii) that any relation in A_2^{2d} expressed in terms of monomials Sq^A gives a relation in degree d by cancelling all terms with an odd exponent and halving the exponents of the remaining terms. For example the Adem relation $Sq^4 Sq^8 = Sq^{12} + Sq^{11} Sq^1 + Sq^{10} Sq^2$ yields the Adem relation $Sq^2 Sq^4 = Sq^6 + Sq^5 Sq^1$. Conversely, we can start with a relation in degree d and obtain a relation in degree $2d$ which holds modulo terms with an odd exponent.

For example, from the relation $Xq^9 = Sq^6 Sq^2 Sq^1 + Sq^8 Sq^1$ of Example 4.3.9 we obtain $Xq^{18} = Sq^{12} Sq^4 Sq^2 + Sq^{16} Sq^2 +$ admissible monomials with an odd exponent. In practice, information obtained in this way appears to have limited usefulness: for example this approximation to Xq^{18} has excess 6, whereas by Proposition 3.5.11 Xq^{18} has excess $\mu(18) = 2$.

13.4 The odd subalgebra Od

In this section and the next we apply stripping operations to study nilpotence in A_2. We begin with monomials Sq^A and Milnor basis elements $Sq(R)$ in which all terms of the sequences A or R are odd.

Proposition 13.4.1 *Let* Od *denote the* $\mathbb{F}_2$-*subspace of* A_2 *spanned by Milnor basis elements* $Sq(R)$, *where all terms of* R *are odd. Then* Od *is a subalgebra of* A_2, *and the set of admissible monomials* Sq^A *where all terms of* A *are odd is also a basis for* Od. *All monomials* $Sq^B = Sq^{b_1} \cdots Sq^{b_\ell}$ *with* $b_1, \ldots, b_\ell$ *odd are in* Od, *and are sums of admissible monomials of length* ℓ.

Proof The Milnor product formula, Theorem 4.1.2, implies that Od is closed under multiplication. In particular all terms in the product $Sq^{b_1} Sq(r_2, \ldots, r_\ell)$ are in Od if b_1 and $r_2, \ldots, r_\ell$ are all odd. It follows by induction on ℓ that all monomials $Sq^B = Sq^{b_1} \cdots Sq^{b_\ell}$ with $b_1, \ldots, b_\ell$ odd are in Od. The correspondence $r_i = a_i - 2a_{i+1}$ of Definition 3.4.1 pairs sequences A and R with all odd terms, and so the admissible monomials with all odd terms span Od.

The last statement follows from the Adem relations. If a and b are odd, then $a - 2j$ is odd and so $b - j - 1$ is also odd for all terms $Sq^{a+b-j} Sq^j$ appearing in the relation for $Sq^a Sq^b$ with coefficient 1. In particular, the term with $j = 0$, which reduces the length, has coefficient 0. $\square$

Since $\phi(Sq^k) = \sum_{i+j=k} Sq^i \otimes Sq^j$, Od is not a Hopf subalgebra of A_2.

Let $T^k = Sq(1, 1, \ldots, 1) = Sq^{2^k-1} Sq^{2^{k-1}-1} \cdots Sq^1$ be the top element of the Hopf subalgebra $A_2(\lambda)$, where $\lambda = \{1^k\}$. This is the exterior algebra generated by the primitive elements $P_t^0 = Sq(0, \ldots, 0, 1)$ in degrees $1, 3, \ldots, 2^k - 1$. Thus $T^k = \prod_{t=1}^k P_t^0$ and $\deg T^k = 2^{k+1} - 2 - k$.

Proposition 13.4.2 *Let* $Sq^B = Sq^{b_1} \cdots Sq^{b_\ell}$ *with at least one of* $b_1, \ldots, b_\ell$ *odd. Then* (i) $Sq^B = \sum_T Sq(T)$, *where each sequence* T *in the sum has length* $\leq \ell$ *and has at least one odd term, and* (ii) $T^\ell Sq^B = 0$.

Proof For (i), we argue by induction on ℓ using Milnor's product formula. If b_1 is odd and X is a Milnor matrix for the product $Sq(b)Sq(S)$ which produces $Sq(T)$, then $x_{1,0} = b = 1 \bmod 2$, and hence t_1 is odd. On the other hand, if b_1 is even but S has length $< \ell$ and has an odd term s_i, then either $x_{0,i}$ or $x_{1,i}$ is odd, and so at least one of t_i, t_{i+1} is odd.

For (ii), since $T^\ell = Sq(1, \ldots, 1)$ of length ℓ, $T^\ell Sq(T) = 0$ for all T of length $\leq \ell$ with at least one odd term. $\square$

The following result shows the effect of multiplying a monomial in even squares by T^ℓ.

Proposition 13.4.3 $T^\ell Sq^{2a_1} \cdots Sq^{2a_\ell} = Sq^{2a_1+2^\ell-1} Sq^{2a_2+2^{\ell-1}-1} \cdots Sq^{2a_\ell-1}$.

Proof Writing $T^\ell = Sq(1,\ldots,1)$, two applications of the Milnor product formula show that $T^\ell Sq(2a) = Sq(2a+1,1,\ldots,1) = Sq^{2a+2^\ell-1} T^{\ell-1}$, since only the initial Milnor matrix arises in the first product, and only the Milnor matrix

$$X = \frac{\begin{array}{cccc} & 0 & 0 & \cdots & 0 \end{array}}{\begin{array}{c|cccc} 2a+1 & 1 & 1 & \cdots & 1 \end{array}}$$

has coefficient $b_X = 1$ in the second product. The result follows by induction on ℓ.   □

Recall from Definition 3.5.6 that for $\ell \geq 0$, $\mathrm{Ad}_{\leq \ell}$ is the $\mathbb{F}_2$-subspace of $\mathbf{A}_2$ spanned by admissible monomials of length $\leq \ell$. By Proposition 13.3.1(iii), the kernel of the halving map $V : \mathbf{A}_2 \to \mathbf{A}_2$ is $\mathbf{A}_2 Sq^1 \mathbf{A}_2$, and so it is spanned by monomials Sq^A with at least one odd exponent. Thus formal duplication $Sq^{a_1} \cdots Sq^{a_r} \to Sq^{2a_1} \cdots Sq^{2a_r}$ is not a well defined map on $\mathbf{A}_2$. However, if we restrict V to the subspace $\mathrm{Ad}_{< \ell}$, then its kernel is spanned by monomials of length $\leq \ell$ with at least one odd exponent, and these elements are also in the kernel of premultiplication by T^ℓ. It follows that the composition $T^\ell \circ V^{-1}$ is well defined on the subspace $\mathrm{Ad}_{\leq \ell}$, so that we have a commutative diagram

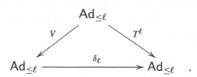

Proposition 13.4.4 *The composition* $\delta_\ell = T^\ell \circ V^{-1}$ *of formal duplication and premultiplication by* T^ℓ *is an injective linear map* $\delta_\ell : \mathrm{Ad}_{\leq \ell} \to \mathrm{Ad}_{\leq \ell}$, *and is given for* $1 \leq r \leq \ell$ *by*

$$\delta_\ell(Sq^{a_1} \cdots Sq^{a_r}) = Sq^{2a_1+2^\ell-1} \cdots Sq^{2a_r+2^{\ell-r+1}-1} Sq^{2^{\ell-r}-1} \cdots Sq^3 Sq^1$$

on a monomial in $\mathbf{A}_2$, *and by*

$$\delta_\ell(Sq(s_1,\ldots s_r)) = Sq(2s_1+1,\ldots,2s_r+1,1,\ldots,1)$$

on the Milnor basis.

Proof By the comments above, δ_ℓ is well defined. The formula for evaluating it on a monomial Sq^A follows from Proposition 13.4.3. Since δ_ℓ maps admissibles to admissibles, it is injective. The formula for Milnor basis elements follows from $T^\ell Sq(2s_1,\ldots,2s_r) = Sq(2s_1+1,\ldots,2s_r+1,1,\ldots,1)$. $\qquad\square$

By combining the stripping operation $\xi_\ell \cap$ and the map δ_ℓ, we can use a relation in A_2 between monomials of length $\leq \ell$ to generate further relations of the same type.

Example 13.4.5 By applying δ_2, the Adem relation $Sq^2 Sq^2 = Sq^3 Sq^1$ implies that $Sq^7 Sq^5 = Sq^9 Sq^3$. Stripping this by ξ_2, we obtain the further Adem relations $Sq^5 Sq^4 = Sq^7 Sq^2$, $Sq^3 Sq^3 = Sq^5 Sq^1$ and $Sq^1 Sq^2 = Sq^3$.

We can apply this method to study nilpotence in Od.

Example 13.4.6 Starting from the relation $Sq^5 Sq^5 Sq^5 = 0$ obtained in Example 13.2.4, we apply δ_3 to obtain $Sq^{17} Sq^{13} Sq^{11} = 0$, and then strip by ξ_3^2 to obtain $Sq^9 Sq^9 Sq^9 = 0$. Iterating this procedure shows that the nilpotence height of Sq^{2^k+1} is 3 for all $k \geq 1$.

The effect of the stripping operation $\xi_\ell^2 \cap$ on a monomial Sq^A of length ℓ is to subtract the sequence $(2^\ell, 2^{l-1}, \ldots, 2)$ from the sequence A, giving a single monomial Sq^B. Thus by following the map δ_ℓ on Ad_ℓ with the stripping operation $\xi_\ell^2 \cap$, we have $\xi_\ell^2 \cap \delta_\ell(Sq^A) = \xi_\ell^2 \cap (Sq^{2a_1+2^\ell-1} \cdots Sq^{2a_\ell-1+3} Sq^{2a_\ell+1}) = Sq^{2a_1-1} \cdots Sq^{2a_\ell-1-1} Sq^{2a_\ell-1} = Sq^B$, where $b_i = 2a_i - 1$ for $1 \leq i \leq \ell$. Thus monomials in Od of length ℓ are mapped to monomials in Od of length ℓ. We have also seen that using the Adem relations a monomial in Od of length ℓ can be written as a sum of admissible monomials in Od of length ℓ. This leads to the following definition.

Proposition 13.4.7 *Let* $\ell \geq 0$ *and let* $(a_1, \ldots a_\ell)$ *be a sequence of odd numbers of length* ℓ *in* Seq. *The formula*

$$\lambda(Sq^{a_1} \cdots Sq^{a_\ell}) = Sq^{2a_1-1} \cdots Sq^{2a_\ell-1}$$

gives a well defined map $\lambda : \mathrm{Od} \to \mathrm{Od}$. *The map* λ *is an injective algebra map.*

Proof Note that the case $\ell = 0$ gives $\lambda(1) = 1$. On sequences A of length ℓ, λ can be regarded as the map $Sq^A \to \xi_\ell^2 \cap \delta_\ell(Sq^A)$. We verify directly that λ is well defined, by showing that it is compatible with the Adem relations. We wish to show that we obtain the Adem relation

$$Sq^{2a-1} Sq^{2b-1} = \sum_j \binom{2b-2-j}{2a-1-2j} Sq^{2a+2b-2-j} Sq^j \qquad (13.6)$$

by applying λ to the relation (3.1). The coefficient in (13.6) is 0 if j is even, so let $j = 2k - 1$. Then $\binom{2b-2-j}{2a-1-2j} = \binom{2b-2-2k+1}{2a-4k+1} = \binom{b-k-1}{a-2k}$, as required.

Since λ is formally multiplicative and is well defined, it is actually multiplicative. It is also injective, because it sends distinct admissible monomials to distinct admissible monomials.   □

The effect of λ on Milnor basis elements is easily calculated using Propositions 13.2.1 and 13.4.4, since $\lambda(Sq(R)) = \xi_\ell^2 \cap \delta_\ell(Sq(R))$ where $\ell = \text{len}(R)$. We obtain

$$\lambda(Sq(r_1,\ldots,r_{\ell-1},r_\ell)) = Sq(2r_1 + 1,\ldots,2r_{\ell-1} + 1, 2r_\ell - 1). \qquad (13.7)$$

Proposition 13.4.8 *Let $\ell \geq 0$ and let $(a_1,\ldots a_\ell)$ be a sequence of odd numbers of length ℓ in* Seq. *Then the monomials $Sq^{a_1}\cdots Sq^{a_\ell}$ and $Sq^{2a_1-1}\cdots Sq^{2a_\ell-1}$ have the same nilpotence height. The same is true for the Milnor basis elements $Sq(r_1,\ldots,r_{\ell-1},r_\ell)$ and $Sq(2r_1 + 1,\ldots,2r_{\ell-1} + 1, 2r_\ell - 1)$.*

Proof This follows from the fact that λ is an injective algebra map.   □

Example 13.4.9 Since $Sq^3 Sq^3 = Sq^5 Sq^1$ is the highest degree element of $A_2(1)$, Sq^3 has nilpotence height 3. It follows that Sq^{2^k+1} has nilpotence height 3 for all $k \geq 1$. As in Example 13.2.4, this can also be seen by stripping the Adem relation $Sq^{2^{k+1}+1} Sq^{2^k+1} = 0$ by (2^k) to obtain the Adem relation $Sq^{2^k+1} Sq^{2^k+1} = Sq^{2^{k+1}+1} Sq^1$.

Example 13.4.10 The product formula gives $Sq(1,7)^2 = Sq(1,7,1,1)$ and $Sq(1,7)^3 = 0$, so $Sq(1,7)$ has nilpotence height 3. Hence $Sq(2^k - 1, 3 \cdot 2^k + 1)$ has nilpotence height 3 for all $k \geq 1$. For the corresponding admissible monomial $Sq^{15} Sq^7$, the Adem relations give $(Sq^{15} Sq^7)^2 = Sq^{27} Sq^{13} Sq^3 Sq^1$ and $(Sq^{15} Sq^7)^3 = 0$, so $Sq^{7 \cdot 2^k+1} Sq^{3 \cdot 2^k+1}$ has nilpotence height 3 for all $k \geq 1$.

We have seen that when a monomial Sq^B of length ℓ with m odd exponents is expressed in the admissible basis, the number of odd exponents in each admissible monomial Sq^A is $\geq m$ and $\text{len}(A) = \ell$. It follows that if $Sq^B \in$ Od then each admissible monomial in $(Sq^B)^k$ has $k\ell$ odd exponents. Since the admissible monomial of minimal degree with m odd exponents is $Sq^{2^m-1}\cdots Sq^3 Sq^1$, of degree $2^{m+1} - m - 2$, it follows that $(Sq^B)^k = 0$ if $k|B| < 2^{k\ell+1} - k\ell - 2$.

For example, $(Sq^{15} Sq^7)^3 = 0$ since $66 < 120$. Similarly, if R has a odd terms and S has b odd terms, the Milnor product formula implies that every $Sq(T)$ in $Sq(R)Sq(S)$ has $\geq a + b$ odd terms. Since $Sq^{2^m-1}\cdots Sq^3 Sq^1 = Sq(1,\ldots,1)$ is the Milnor basis element of minimal degree with m odd exponents, $Sq(R)^k = 0$ if $k \deg Sq(R) < (2^{k+1} - k - 2)\text{len}(R)$. For example, $Sq(1,7)^3 = 0$ since $66 < 120$.

Proposition 13.4.11 *The nilpotence height of* Sq^{2^n-1} *is* $n+1$.

Proof First we show more generally that $(Sq^{2k-1})^{n+1} = 0$ when $k \leq 2^{n-1}$. In this case, $Sq^{2k-1} \in \mathsf{A}_2(n-1)$. By the observations above, when $(Sq^{2k-1})^{n+1}$ is expressed as a sum of elements $Sq(R)$, every sequence R has length $n+1$. Since $\mathsf{A}_2(n-1)$ contains no elements $Sq(R)$ of length $> n$, $(Sq^{2k-1})^{n+1} = 0$.

Next we show that $(Sq^{2^n-1})^n \neq 0$. If we strip $(Sq^{2^n-1})^n$ successively by the sequences $(2^{n-1}, 2^{n-2}, \ldots, 2, 1)$, $(2^{n-1}, 2^{n-2}, \ldots, 2)$, $\ldots$, $(2^{n-1}, 2^{n-2})$, (2^{n-1}), then at each step we obtain a single monomial, and at the last step we obtain the admissible monomial $Sq^{2^{n-1}-1} Sq^{2^{n-2}-1} \cdots Sq^3 Sq^1$. For example, when $n = 3$ we have

$$Sq^7 Sq^7 Sq^7 \overset{(4,2,1)}{\mapsto} Sq^3 Sq^5 Sq^6 \overset{(4,2)}{\mapsto} Sq^3 Sq^1 Sq^4 \overset{(4)}{\mapsto} Sq^3 Sq^1.$$

Since the result of stripping is nonzero, the element $(Sq^{2^n-1})^n$ is also nonzero. □

13.5 Nilpotence of Sq^{2^s}

In this section we determine the nilpotence height of Sq^{2^s}. A simple stripping argument shows that this is even for all n.

Proposition 13.5.1 *If* $(Sq^{2^s})^{2k+1} = 0$, *then* $(Sq^{2^s})^{2k} = 0$.

Proof Strip the relation $(Sq^{2^s})^{2k+1} = 0$ by (2^s). This gives $(2k+1)(Sq^{2^s})^{2k} = 0$, and so $(Sq^{2^s})^{2k} = 0$. □

A further stripping argument yields the lower bound that we require. We first illustrate the argument with the case $n = 3$. To show that $(Sq^8)^6 \neq 0$, we strip by the sequences $(8, 4, 2, 1)$, $(8, 4, 2)$ and $(8, 4)$, and show that the result is the monomial $Sq^4 Sq^2 Sq^1$. This monomial arises as a result of the stripping sequence

8	8	8	8	8	8
8	4	0	2	0	1
0	0	8	4	0	2
0	0	0	0	8	4
0	4	0	2	0	1

(13.8)

However, in each row we have to take into account all possible choices of columns in which the 0 entries occur, for example the stripping sequences

$$
\begin{array}{cccccc|cccccc}
8 & 8 & 8 & 8 & 8 & 8 & 8 & 8 & 8 & 8 & 8 & 8 \\
\hline
8 & 0 & 4 & 0 & 2 & 1 & 8 & 0 & 4 & 0 & 2 & 1 \\
0 & 8 & 4 & 0 & 0 & 2 & 0 & 0 & 0 & 8 & 4 & 2 \\
0 & 0 & 0 & 8 & 1 & 0 & 0 & 8 & 0 & 0 & 0 & 4 \\
\hline
0 & 0 & 0 & 0 & 2 & 5 & 0 & 0 & 4 & 0 & 2 & 1
\end{array}
\qquad (13.9)
$$

give further terms $Sq^2 Sq^5$ and $Sq^4 Sq^2 Sq^1$. However, if the final sequence contains two consecutive 0s, then the same sequence is obtained by exchanging the corresponding columns of the array, and so it occurs an even number of times as a result of the whole stripping process. For example, the arrays obtained from (13.9) by exchanging the first two columns also give $Sq^2 Sq^5$ and $Sq^4 Sq^2 Sq^1$.

Proposition 13.5.2 $(Sq^{2^s})^{2s} \neq 0.$

Proof We assume that $(Sq^{2^s})^{2s} = 0$ and obtain a contradiction by stripping this relation by the sequences $S_k = (2^s, 2^{s-1} \ldots, 2^k)$ for $0 \leq k \leq s-1$. Since $|S_k| = 2^{s+1} - 2^k$, $\sum_{k=0}^{s-1} |S_k| = s \cdot 2^{s+1} - (2^s - 1) = 2s \cdot 2^s - (2^s - 1)$, and so the result of stripping is an element in A_2 of degree $2^s - 1$. We shall prove that this element is the admissible monomial $Sq^{2^{s-1}} Sq^{2^{s-2}} \cdots Sq^2 Sq^1$.

Each stripping sequence can be written as an array with s rows and $2s$ columns, as in (13.8) and (13.9). If the sum of the entries in any column is $> 2^s$, then the stripping sequence gives a zero term, so the s entries 2^s in the array must occur in different columns, and they must be the only nonzero entries in these columns. If two of these columns are consecutive, then a cancelling term is produced by the stripping array obtained by exchanging them. Hence we may assume that no two of these s columns are consecutive. The last column cannot contain an entry 2^s, since we do not strip by $S_s = (2^s)$, and so the entries 2^n occur in the odd numbered columns. Taking the stripping operations in the order of increasing k, the only stripping array with this property is the one generalizing (13.8), and this produces the monomial $Sq^{2^{s-1}} Sq^{2^{s-2}} \cdots Sq^2 Sq^1$. $\qquad\square$

The next result completes the proof that the nilpotence height of Sq^{2^s} is $2s+2$. Note that by Straffin's formula 12.1.2(ii), Xq^{2^s} and Sq^{2^s} both give $Sq^0 = 1$ when stripped by (2^s).

Proposition 13.5.3 $(Sq^{2^s})^{2s+2} = 0.$

The proof is recursive, and we use the example $(Sq^8)^8 = 0$ to illustrate it. First we show that it is sufficient to show that $Xq^8(Sq^8)^7 = 0$. Given this relation, stripping by the sequence (8) gives $(Sq^8)^7 + 7 \cdot Xq^8(Sq^8)^6 = 0$, i.e. $(Sq^8)^7 = Xq^8(Sq^8)^6$, and so $(Sq^8)^8 = Xq^8(Sq^8)^7 = 0$, as required. The point of this step is that the relation $Xq^8(Sq^8)^7 = 0$ can be manipulated using Proposition 12.1.2(i), since $Xq^8Sq^8 = Sq^{12}Xq^4$. Hence it is sufficient to show that $Xq^4(Sq^8)^6 = 0$.

We can repeat this argument to simplify further. At the next step, we show that it suffices to prove that $Xq^4Xq^8(Sq^8)^5 = 0$. Given this relation, stripping by (8) gives $Xq^4(Sq^8)^5 = 5 \cdot Xq^4Xq^8(Sq^8)^4 = Xq^4Xq^8(Sq^8)^4$, and so $Xq^4(Sq^8)^6 = Xq^4Xq^8(Sq^8)^5 = 0$, as required. As $Xq^4Xq^8Sq^8 = Xq^4Sq^{12}Xq^4 = Sq^{14}Xq^2Xq^4$ by Proposition 12.1.2(i), it suffices to prove that $Xq^2Xq^4(Sq^8)^4 = 0$.

We next prove that it suffices to show that $Xq^2Xq^4Xq^8(Sq^8)^3 = 0$. Given this relation, stripping by (8) gives $Xq^2Xq^4(Sq^8)^3 = 3 \cdot Xq^2Xq^4Xq^8(Sq^8)^2 = Xq^2Xq^4Xq^8(Sq^8)^2$, and so $Xq^2Xq^4(Sq^8)^4 = Xq^2Xq^4Xq^8(Sq^8)^3 = 0$. We have $Xq^2Xq^4Xq^8Sq^8 = Xq^2Xq^4Sq^{12}Xq^4 = Xq^2Sq^{14}Xq^2Xq^4 = Sq^{15}Xq^1Xq^2Xq^4$ by Proposition 12.1.2(i). Hence it is sufficient to show that $Xq^1Xq^2Xq^4(Sq^8)^2 = 0$.

We next prove that it is sufficient to show that $Xq^1Xq^2Xq^4Xq^8Sq^8 = 0$. Given this relation, stripping by (8) gives $Xq^1Xq^2Xq^4Sq^8 = Xq^1Xq^2Xq^4Xq^8$ and so $Xq^1Xq^2Xq^4(Sq^8)^2 = Xq^1Xq^2Xq^4Xq^8Sq^8 = 0$. Now $Xq^1Xq^2Xq^4Xq^8 = \chi(Sq^8Sq^4Sq^2Sq^1) = Sq^{15}$, so the result follows from the Adem relation $Sq^{15}Sq^8 = 0$.

Proof of Proposition 13.5.3 As in Example 12.5.7, let $X_k^s = Sq^{2^s}Sq^{2^{s-1}} \cdots Sq^{2^k}$ for $0 \le k \le s$, and let $X_s^{s-1} = 1$. We prove the statements

$$\text{(i)} \quad \chi(X_k^s)(Sq^{2^s})^{2k+1} = 0, \quad \text{(ii)} \quad \chi(X_k^{s-1})(Sq^{2^s})^{2k+2} = 0,$$

by induction on k for $0 \le k \le s$.

By Davis's formula 4.4.2, $Xq^{2^{s+1}-1} = Sq^{2^s}Sq^{2^{s-1}} \cdots Sq^2Sq^1 = X_0^s$, so $\chi(X_0^s) = Sq^{2^{s+1}-1}$ and so the Adem relation $Sq^{2^{s+1}-1}Sq^{2^s} = 0$ gives $\chi(X_0^s)Sq^{2^s} = 0$. This is the case $k = 0$ of (i), which starts the induction. We shall show that (i) for k implies (ii) for k, and that (ii) for k implies (i) for $k+1$. The case $k = s$ of (ii) is the required relation $(Sq^{2^s})^{2s+2} = 0$.

Statement (i) for k can be written as $\chi(X_k^{s-1})Xq^{2^s}(Sq^{2^s})^{2k+1} = 0$. Since $\deg X_k^{s-1} = 2^s - 2^k < 2^s$, stripping X_k^{s-1} by the sequence (2^s) gives 0, and by Straffin's formula 12.1.2(ii), stripping Xq^{2^s} by (2^s) gives 1. Hence stripping the relation $\chi(X_k^{s-1})Xq^{2^s}(Sq^{2^s})^{2k+1} = 0$ by (2^s) gives

$$\chi(X_k^{s-1})(Sq^{2^s})^{2k+1} + (2k+1)\chi(X_k^{s-1})Xq^{2^s}(Sq^{2^s})^{2k} = 0.$$

Hence $\chi(X_k^{s-1})(Sq^{2^s})^{2k+2} = \chi(X_k^{s-1})Xq^{2^s}(Sq^{2^s})^{2k+1} = 0$, which is statement (ii) for k.

Next we assume statement (ii) for k. By iterating Proposition 12.1.2(i),

$$
\begin{aligned}
\chi(X_{k+1}^s)Sq^{2^s} &= Xq^{2^{k+1}}\cdots Xq^{2^{s-1}}Xq^{2^s}Sq^{2^s} \\
&= Xq^{2^{k+1}}\cdots Xq^{2^{s-1}}Sq^{2^{s+1}-2^{s-1}}Xq^{2^{s-1}} \\
&= Xq^{2^{k+1}}\cdots Xq^{2^{s-2}}Sq^{2^{s+1}-2^{s-2}}Xq^{2^{s-2}}Xq^{2^{s-1}} \\
&= \cdots \\
&= Sq^{2^{s+1}-2^k}Xq^{2^k}\cdots Xq^{2^{s-2}}Xq^{2^{s-1}} \\
&= Sq^{2^{s+1}-2^k}\chi(X_k^{s-1}),
\end{aligned}
$$

and so we obtain

$$
\chi(X_{k+1}^s)(Sq^{2^s})^{2k+3} = \chi(X_{k+1}^s)Sq^{2^s}(Sq^{2^s})^{2k+2} = Sq^{2^{s+1}-2^k}\chi(X_k^{s-1})(Sq^{2^s})^{2k+2} = 0,
$$

which is statement (i) for $k+1$. This completes the induction. $\square$

13.6 Relations in A_q

In this section we generalize some of the relations proved for A_2 in Chapters 3, 4 and 12 to the subalgebra A_q of A_2 defined in Section 12.3, where $q = 2^t$ is a 2-power. By Proposition 1.2.5 the left action of A_q on $P(n)$ commutes with the right action of $\mathbb{F}_2 M(n)$ by linear substitutions. We begin by extending Definition 12.2.1 by considering polynomials over the field of q elements $\mathbb{F}_q$.

Definition 13.6.1 Let $q = 2^t$ where $t \geq 1$ and let $P(n, \mathbb{F}_q)$ be the polynomial algebra $\mathbb{F}_q[x_1, \ldots, x_n]$. The **total Steenrod qth power** $Sq_t : P(n, \mathbb{F}_q) \to P(n, \mathbb{F}_q)$ is the algebra map defined by $Sq_t(1) = 1$ and $Sq_t(x_i) = x_i + x_i^q$ for $1 \leq i \leq n$. For $k, d \geq 0$, the **kth Steenrod qth power** $Sq_t^k : P^d(n, \mathbb{F}_q) \to P^{d+k(q-1)}(n, \mathbb{F}_q)$ is the linear map given by restricting Sq_t to $P^d(n, \mathbb{F}_q)$ and projecting on to $P^{d+k(q-1)}(n, \mathbb{F}_q)$.

Example 13.6.2 Let α be a generator of the cyclic group $\mathbb{F}_q^* \cong \mathbb{Z}/(q-1)$ and let x_i be one of the generators of $P(n, \mathbb{F}_q)$. Then $Sq_t^1(\alpha x_i) = \alpha Sq_t^1(x_i) = \alpha x_i^q$. Since $\alpha^q = \alpha$, $Sq_t^1(x) = x^q$ where $x = \alpha x_i$. It follows by linearity over $\mathbb{F}_2$ that $Sq_t^1(x) = x^q$ for all $x \in P^1(n, \mathbb{F}_q)$.

The next result generalizes Proposition 1.2.5, and explains why we regard A_q as the algebra of Steenrod operations over $\mathbb{F}_q$. By taking the graded parts of

Sq_t, it follows from Proposition 13.6.3 that $Sq_t^k : \mathrm{P}^d(n, \mathbb{F}_q) \to \mathrm{P}^{d+(q-1)k}(n, \mathbb{F}_q)$ is a map of $\mathbb{F}_q(\mathrm{M}(n, \mathbb{F}_q))$-modules.

Proposition 13.6.3 *The right action of* $\mathbb{F}_q\mathrm{M}(n, \mathbb{F}_q)$ *by linear substitution on the polynomial algebra* $\mathrm{P}(n, \mathbb{F}_q) = \mathbb{F}_q[x_1, \ldots, x_n]$ *commutes with the left action of* Sq_t^k, *i.e.* $\mathrm{Sq}_t(f) \cdot A = \mathrm{Sq}_t(f \cdot A)$ *for* $f \in \mathrm{P}(n, \mathbb{F}_q)$ *and* $A \in \mathrm{M}(n, \mathbb{F}_q)$.

Proof Since Sq_t and $f \mapsto f \cdot A$ are algebra maps of $\mathrm{P}(n, \mathbb{F}_q)$, we can reduce to the case where f is one of the variables x_i. Let $A = (a_{i,j})$. Then $\mathrm{Sq}_t(x_i) \cdot A = (x_i + x_i^q) \cdot A = x_i \cdot A + (x_i \cdot A)^q$, while $\mathrm{Sq}_t(x_i \cdot A) = \mathrm{Sq}_t(\sum_{j=1}^n a_{i,j}x_j) = \sum_{j=1}^n a_{i,j}\mathrm{Sq}_t(x_j) = \sum_{j=1}^n a_{i,j}(x_j + x_j^q)$. But $\sum_{j=1}^n a_{i,j}x_j^q = (\sum_{j=1}^n a_{i,j}x_j)^q = (x_i \cdot A)^q$, so $\mathrm{Sq}(x_i) \cdot A = \mathrm{Sq}(x_i \cdot A)$. $\square$

In the rest of this section, we return to the action on $\mathrm{P}(n)$. Since $\phi(Sq_t^k) = \sum_{i+j=k} Sq_t^i \otimes Sq_t^j$ by Proposition 12.3.6(i), $\sum_{i+j=k} Sq_t^i \chi(Sq_t^j) = 0$ for $k \geq 1$. This gives a recursive formula for $\chi(Sq_t^k)$ as a sum of monomials $\sum_B Sq_t^B$ in A_q. Li's formula 11.6.1 provides a means to express $\chi(Sq_t^k)$ as a sum of Milnor basis elements $\sum_R Sq_t(R)$, since for all the Milnor matrices $X = (x_{i,j})$ which occur we have $x_{i,j} = 0$ unless t divides i and j.

Definition 13.6.4 For $q = 2^t$ and $t \geq 1$, the **total conjugate Steenrod** qth **power** $\mathrm{Xq}_t : \mathrm{P}(n) \to \mathrm{P}(n)$ is the algebra map defined by

$$\mathrm{Xq}_t(1) = 1, \quad \mathrm{Xq}_t(x_i) = \sum_{j \geq 0} x_i^{q^j}, \text{ for } 1 \leq i \leq n.$$

For $k, d \geq 0$, the **conjugate Steenrod** qth **power** $Xq_t^k : \mathrm{P}^d(n) \to \mathrm{P}^{d+(q-1)k}(n)$ is defined by restricting Xq_t to $\mathrm{P}^d(n)$ and projecting on to $\mathrm{P}^{d+(q-1)k}(n)$, so that $\mathrm{Xq}_t = \sum_{k \geq 0} Xq_t^k$ is the formal sum of its graded parts.

As in Section 2.3, this leads to the following result relation between the operations Xq_t^k and Sq_t^k on $\mathrm{P}(n)$. The last statement is clear from the formula $\sum_{i+j=k} Sq_t^i \chi(Sq_t^j) = 0$.

Proposition 13.6.5 *Let* $t, k \geq 1$ *and let* $f \in \mathrm{P}(n)$. *Then*

$$\text{(i)} \sum_{i+j=k} Sq_t^i \circ Xq_t^j(f) = 0, \quad \text{(ii)} \sum_{i+j=k} Xq_t^i \circ Sq_t^j(f) = 0,$$

and $\chi(Sq_t^k) = Xq_t^k$ *for all* t *and* k. $\square$

The next result generalizes Proposition 3.5.11.

Proposition 13.6.6 *For* $k \geq 0$, Xq_t^k *is the sum of all Milnor basis elements* $Sq_t(R)$ *of degree* $k(q-1)$.

Proof Since the total conjugate square Xq_t is multiplicative,

$$\mathsf{Xq}_t(c(n)) = \mathsf{Xq}_t(x_1)\mathsf{Xq}_t(x_2)\cdots \mathsf{Xq}_t(x_n) = \prod_{i=1}^{n}(x_i + x_i^q + x_i^{q^2} + \cdots),$$

and this is the sum of all monomials in $P(n)$ with q-powers as exponents. Thus the terms of degree $n + k(q-1)$ in $\mathsf{Xq}_t(c(n))$ give the sum of all Cartan symmetric functions of degree $n + k(q-1)$ with all exponents q-powers. This is $\sum_{R_t} c(R_t^+)$, where the sum is over all R such that $Sq_t(R) = Sq(R_t) \in A_2^{k(q-1)}$. Taking $n \geq k(q-1)$, the result follows from Proposition 3.4.5. ☐

Example 13.6.7 Let $t = 2$ and $k = 10$. Then $Xq_2^{10} = Sq_2(10) + Sq_2(5,1) + Sq_2(0,2)$, or $\chi(Sq(0,10)) = Sq(0,10) + Sq(0,5,0,1) + Sq(0,0,0,2)$ in the usual notation.

We next extend the formulae of Davis and Silverman (Proposition 4.4.1) to A_q. Given a sequence $R = (r_1, r_2, \ldots) \in$ Seq, we define $\deg_q R = \sum_i q^{i-1} r_i$.

Proposition 13.6.8 *Let* $a, b \geq 0$. *Then*

$$(i)\ Sq_t^a Xq_t^b = \sum_R \binom{\deg_q R}{a} Sq_t(R), \quad (ii)\ Xq_t^a Sq_t^b = \sum_R \binom{|R|}{b} Sq_t(R),$$

where the sums are over all Milnor basis elements $Sq_t(R)$ *of degree* $(a+b)$ $(q-1)$.

Proof (i) Let $Sq_t^a Xq_t^b = \sum_R m_R Sq_t(R)$, where $m_R \in \mathbb{F}_2$. By Proposition 3.5.11, Xq_t^b is the sum of all Milnor basis elements in A_q of degree $b(q-1)$. Since $Sq_t^a = Sq_t(a)$, it follows from Theorem 4.1.2 that m_R is the sum of the coefficients $b(X)$ for Milnor matrices $X = (x_{i,j})$ (shown for $t = 2$, $q = 2^t = 4$)

$$X = \begin{array}{c|ccccc} & 0 & r_1 - a_1 & 0 & r_2 - a_2 & \cdots \\ \hline 0 & 0 & 0 & 0 & 0 & \cdots \\ a_1 & 0 & a_2 & 0 & a_3 & \cdots \end{array},$$

where $x_{ij} = 0$ unless t divides i and j, and $a = \sum_{j\geq 1} q^{j-1} a_j$. Every sequence $A = (a_1, a_2, \ldots)$ of q-degree a such that $a_j \leq r_j$ for $j \geq 1$ gives rise to one such matrix X, and $b(X) = \prod_{j\geq 1} \binom{r_j}{a_j}$.

By Proposition 1.4.11, $\binom{r_j}{a_j} = \binom{q^j r_j}{q^j a_j}$ mod 2, and $\binom{q^j r_j}{b_j} = 0$ mod 2 if b_j is not divisible by q^j. Hence

$$m_R = \sum_A \prod_{j \geq 1} \binom{r_j}{a_j} = \sum_A \prod_{j \geq 1} \binom{q^j r_j}{q^j a_j} = \sum_B \prod_{j \geq 1} \binom{q^j r_j}{b_j}, \qquad (13.10)$$

where the sum is over all $B = (b_1, b_2, \ldots)$ such that $\sum_{j \geq 1} b_j = \sum_{j \geq 1} q^j a_j = qa$. But (4.3) is the coefficient of x^{qa} in $\prod_{j \geq 1} (1+x)^{q^j r_j}$. Since $\deg_q R = \sum_{j \geq 1} q^{j-1} r_j$,

$$\prod_{j \geq 1} (1+x)^{q^j r_j} = (1+x)^{q \deg_q R} = (1+x^q)^{\deg_q R}.$$

Comparing coefficients of x^{qa}, (13.10) reduces to $m_R = \binom{\deg_q R}{a}$.

(ii) By Proposition 3.5.11, $Xq^a Sq^b = \sum_S Sq(S)Sq(b)$, where the sum is over all Milnor basis elements $Sq(S)$ of degree a. A term $b(X)Sq(R)$ in the product $Sq(S)Sq(b)$ arises from each Milnor matrix (again shown for $t = 2$, $q = 2^t = 4$)

$$Y = \begin{array}{c|cc}
 & 0 & b_1 \\
\hline
0 & 0 & 0 \\
r_1 - b_1 & 0 & b_2 \\
0 & 0 & 0 \\
r_2 - b_2 & 0 & b_3 \\
\vdots & \vdots &
\end{array}.$$

Every sequence $B = (b_1, b_2, \ldots)$ such that $|B| = b = \sum_{j \geq 1} b_j$ and $b_j \leq r_j$ for $j \geq 1$ gives rise to one such Y, with coefficient $b(Y) = \prod_{j \geq 1} \binom{r_j}{b_j}$. Comparing coefficients of x^b in the identity $\prod_{j \geq 1} (1+x)^{r_j} = (1+x)^{|R|}$, we have $\sum_B \prod_{j \geq 1} \binom{r_j}{b_j} = \binom{|R|}{b}$. $\square$

We can now generalize the commutation formulae of Proposition 12.1.2.

Proposition 13.6.9 (i) *For* $r, s, t \geq 1$ *with* $r \geq t$,

$$Xq_t^{2^r} Sq_t^{2^r(2^s - 1)} = Sq_t^{2^{r-t}(2^{s+t} - 1)} Xq_t^{2^{r-t}},$$

(ii) **(Straffin's formula)** *For* $s > 0$,

$$Xq_t^{2^s} = Sq_t^{2^s} + Sq_t^{2^{s-1}} Xq_t^{2^{s-1}}.$$

Proof (i) Let $a = 2^{r+s} - 2^{r-t}$ and $b = 2^{r-t}$ in Proposition 13.6.8(i). The Milnor basis element $Sq_t(R)$ of degree $(q-1)(a+b)$ with maximum excess is Sq_t^{a+b}, and so $|R| \leq a+b = 2^{r+s} \leq 2^{r+s}(q-1)$, with equality only in the case $q=2$, $R = (2^{r+s})$. The coefficient α of $Sq_t(R)$ in 13.6.8(i) is

$$\binom{\deg_q R}{a} = \binom{q \deg_q R}{qa} = \binom{|R|+\deg Sq_t(R)}{qa} = \binom{|R|+2^{r+s+t}-2^{r+s}}{2^{r+s+t}-2^r}.$$

On the other hand, setting $a = 2^r$ and $b = 2^{r+s} - 2^r$ in Proposition 13.6.8(ii), the coefficient β of $Sq_t(R)$ is

$$\binom{|R|}{2^{r+s}-2^r} = \binom{|R|+2^{r+s}(2^t-1)}{2^{r+s}-2^r+2^{r+s}(2^t-1)}.$$

Thus $\alpha = \beta$ if $|R| < 2^{r+s}(q-1)$, and in the exceptional case $q=2$, $R = (2^{r+s})$, it is easily checked that $\alpha = \beta = 0$.

(ii) Let $a = b = 2^{r-1}$ in Proposition 13.6.8(i). The coefficient α of $Sq_t(R)$ is $\alpha = \binom{|R|+(q-1)2^r}{q \cdot 2^{r-1}} = \binom{|R|+2^{t+r}-2^r}{2^{t+r-1}}$. Hence $\alpha = 1$ unless $|R| + 2^{t+r} - 2^r \geq 2^{t+r}$, i.e. $|R| \geq 2^r$, and so $R = (2^r)$. Hence $Sq_t^{2^{r-1}} Xq_t^{2^{r-1}}$ is the sum of all elements $Sq_t(R)$ of degree $(q-1)2^r$ except $Sq_t^{2^r}$. By Proposition 13.6.6, this is $Xq_t^{2^r} + Sq_t^{2^r}$. □

Remark 13.6.10 In the case $t=1$, this gives an alternative proof of Proposition 12.1.2 which uses the Milnor basis rather than the Bullett–Macdonald identity.

The next result generalizes the formula $Xq^{2^{u+1}-1} = Sq^{2^u} Sq^{2^{u-1}} \cdots Sq^2 Sq^1$ of Proposition 4.4.2(i).

Proposition 13.6.11 Let $t \geq 1$, $q = 2^t$. For $s \geq 1$, let $s = ut + v$ where $0 \leq v < t$. Then

$$Xq_t^{2^s + 2^{s-t} + \cdots + 2^{s-ut}} = Sq_t^{2^s} Xq_t^{2^{s-t} + \cdots + 2^{s-ut}} = Sq_t^{2^s} Sq_t^{2^{s-t}} \cdots Sq_t^{2^{s-ut}}.$$

Proof Let $a = 2^s$, $b = 2^{s-t} + \cdots + 2^{s-ut}$. By Proposition 13.6.8(i) $Sq_t^a Xq_t^b = \sum_R \binom{\deg_q R}{a} Sq_t(R)$, where the sum is over all sequences R such that $\deg Sq_t(R) = (a+b)(q-1)$. The mod 2 binomial coefficient

$$\binom{\deg_q R}{a} = \binom{q \deg_q R}{qa} = \binom{\deg Sq_t(R) + |R|}{qa} = \binom{(a+b)(q-1)+|R|}{qa}.$$

Since $(a+b)(q-1) = (2^s + 2^{s-t} + \cdots + 2^{s-ut})(2^t - 1) = 2^{s+t} - 2^v$ and $qa = 2^{s+t}$, this binomial coefficient is even for $2^{s-t} \leq 2^{s+t} - 2^v + |R| < 2^{s-t+1}$. This is true for all R with $\deg Sq_t(R) = 2^{s+t} - 2^v$ since (by considering the right order $\leq_r$) the element of maximal excess $|R|$ is $Sq_t(a+b)$ with $|R| = a+b$, and the element of minimal excess $|R|$ is $Sq_t(0,\ldots,0,2^v)$, where 2^v is term $u+1$ of the sequence. Hence $Sq_t^a Xq_t^b$ is the sum of all elements $Sq_t(R)$ of degree

$(a+b)(q-1)$, and it follows from Proposition 13.6.6 that $Sq_t^a Xq_t^b = Xq_t^{a+b}$. The second statement follows by iteration on u. $\qquad\square$

The next result generalizes the Adem relation $Sq^{2^{s+1}-1} Sq^{2^s} = 0$. Note that $(2^{s+1}-1) - 2^s = 2^s - 1 = 2^{s-1} + 2^{s-2} + \cdots + 1$ is the degree in A_2 of the monomial $Sq^{2^{s-1}} Sq^{2^{s-2}} \cdots Sq^2 Sq^1$ produced by stripping $(Sq^{2^s})^{2s}$ in Proposition 13.5.2. The corresponding admissible monomial produced by stripping $(Sq_t^{2^s})^{2u}$ in Proposition 13.7.4 has degree $2^{s-t} + 2^{s-2t} + \cdots + 2^{s-ut}$ as a monomial in the elements Sq_t^k, $k \geq 0$.

Proposition 13.6.12 *Let $t \geq 1$, $s \geq 0$ and $s = ut + v$ where $0 \leq v < t$. Then*

$$Sq_t^{2^s + 2^{s-t} + \cdots + 2^{s-ut}} Sq_t^{2^s} = 0.$$

Proof We prove the conjugate of the relation using Proposition 13.6.11. We have

$$
\begin{aligned}
Xq_t^{2^s} Xq_t^{2^s + 2^{s-t} + \cdots + 2^{s-ut}} &= Xq_t^{2^s} Sq_t^{2^s} Sq_t^{2^{s-t}} \cdots Sq_t^{2^{s-ut}} \\
&= Sq_t^{2^{s-t}(2^{t+1}-1)} Xq_t^{2^{s-t}} Sq_t^{2^{s-t}} \cdots Sq_t^{2^{s-ut}} \\
&= Sq_t^{2^{s-t}(2^{t+1}-1)} Sq_t^{2^{s-2t}(2^{t+1}-1)} Xq_t^{2^{s-2t}} \cdots Sq_t^{2^{s-ut}} \\
&= \cdots \\
&= Sq_t^{2^{s-t}(2^{t+1}-1)} Sq_t^{2^{s-2t}(2^{t+1}-1)} \cdots Xq_t^{2^{s-ut}} Sq_t^{2^{s-ut}}.
\end{aligned}
$$

Since $s - ut = v$, it suffices to show that $Xq_t^{2^v} Sq_t^{2^v} = 0$. This follows from Proposition 13.6.8(ii). Since $v \leq t-1$, $2^{v+1}(q-1) \leq q(q-1)$. The only element $Sq_t(R)$ of degree $2^{v+1}(q-1)$ is $Sq_t^{2^{v+1}}$, and $\binom{2^{v+1}}{2^v} = 0$. $\qquad\square$

Example 13.6.13 To illustrate these relations, we tabulate $2^s + 2^{s-t} + \cdots + 2^{s-ut}$ for $s, t \leq 6$ and $u = [s/t]$. For example, the third column gives the relations $Sq^7 Sq^4 = 0$, $Sq_2^5 Sq_2^4 = 0$ (i.e. $Sq(0,5)Sq(0,4) = 0$) and $Sq_t^4 Sq_t^4 = 0$ for $t \geq 3$.

$t \backslash 2^s$	1	2	4	8	16	32	64
1	1	3	7	15	31	63	127
2	1	2	5	10	21	42	85
3	1	2	4	9	18	36	73
4	1	2	4	8	17	34	68
5	1	2	4	8	16	33	66
6	1	2	4	8	16	32	64

13.7 Nilpotence of P_t^s

In this section we determine the nilpotence height of P_t^s in A_2. Recall that $P_t^s = Sq(0,\dots,0,2^s)$, where 2^s is in the tth place, and is also denoted by $Sq_t(2^s)$ or by $Sq_t^{2^s}$. This element is the analogue of Sq^{2^s} for the subalgebra A_q, where $q = 2^t$. We adapt the notation for monomials in the squares Sq^k and for Milnor basis elements to A_q, writing $Sq_t^A = Sq_t^{a_1} \cdots Sq_t^{a_r}$ for a sequence $A = (a_1,\dots,a_r)$ and $Sq_t(R) = Sq(0,\dots,0,r_1,0,\dots,0,r_2,\dots,0,\dots,0,r_\ell)$ for a sequence $R = (r_1, r_2, \dots, r_\ell)$, where r_i is in the tr_ith position. The admissible basis for A_q is given by the monomials Sq_t^A such that $a_i \geq qa_{i+1}$ for all i.

Proposition 13.7.1 *For $s \geq 0$ and $t \geq 1$, let $s = ut + v$ where $0 \leq v < t$, so that $u = [s/t]$. Then the nilpotence height of $Sq_t^{2^s}$ is $2u + 2$.*

Stripping operations generalize in a straightforward way to A_q. Recall that the dual algebra $A_q^* = \mathbb{F}_2[\xi_t, \xi_{2t}, \dots]$ is a subalgebra (but a quotient Hopf algebra) of A_2^*. In particular, we have the analogue of Proposition 13.2.3.

Proposition 13.7.2 *For any monomial $Sq_t^A \in A_q$, $\xi_{it}^{q^k} \cap Sq_t^A = \sum_B Sq_t^B$, the sum of all monomials such that the nonzero entries of $A - B$, taken in order, are $q^{k+i-1}, \dots, q^k$.* $\qquad\square$

We refer to this operation as stripping by the sequence $(q^{k+i-1}, \dots, q^k)$. In fact, a more general stripping operation is available, since we can use 2-powers of the generators rather than q-powers. To strip by $\xi_{it}^{2^r}$, let $r = tk + \ell$ where $k \geq 0$ and $0 \leq \ell < t$. As in the proof of Proposition 13.2.3, it follows that the corresponding stripping sequence is obtained by multiplying the sequence $(q^{k+i-1}, \dots, q^k)$ by 2^ℓ, giving the sequence $(2^{r+(i-1)t}, \dots, 2^{r+t}, 2^r)$. In particular, we can strip by (2^r) in A_q for all $r \geq 0$.

As for Sq^{2^s}, we begin the proof of Proposition 13.7.1 by showing that the nilpotence height is even.

Proposition 13.7.3 *If $(Sq_t^{2^s})^{2k+1} = 0$, then $(Sq_t^{2^s})^{2k} = 0$.*

Proof Strip the relation $(Sq_t^{2^s})^{2k+1} = 0$ by the sequence (2^s). This gives $(2k + 1)(Sq_t^{2^s})^{2k} = 0$, and so $(Sq_t^{2^s})^{2k} = 0$. $\qquad\square$

A further stripping argument gives the lower bound.

Proposition 13.7.4 *$(Sq_t^{2^s})^{2u} \neq 0$, where $u = [s/t]$.*

Proof We assume that $(Sq_t^{2^s})^{2u} = 0$ and obtain a contradiction by stripping this relation in two stages to obtain an admissible monomial in A_q. As for Proposition 13.5.2, we strip using the sequences $S_j = (2^s, 2^{s-t} \dots, 2^{s-jt})$ for $1 \leq$

$j \leq u$. We find that all stripping arrays which have terms 2^s in consecutive columns cancel in pairs, and the only remaining array which gives a nonzero monomial is

2^s	2^s	2^s	2^s	$\cdots$	2^s	2^s
2^s	2^{s-t}	0	2^{s-2t}	$\cdots$	0	2^{s-ut}
0	0	2^s	2^{s-t}	$\cdots$	0	$2^{s-(u-1)t}$
0	0	0	0	$\cdots$	0	$2^{s-(u-2)t}$
$\vdots$	$\vdots$	$\vdots$	$\vdots$	$\vdots$	$\vdots$	$\vdots$
0	0	0	0	$\cdots$	0	2^{s-2t}
0	0	0	0	$\cdots$	2^s	2^{s-t}

$$(13.11)$$

Hence the result of the first stage of stripping is the monomial

$$Sq_t^{2^s-2^{s-t}} Sq_t^{2^s-2^{s-t}-2^{s-2t}} \cdots Sq_t^{2^s-2^{s-t}-\cdots-2^{s-ut}}.$$

This monomial is not admissible in A_q, so we carry out the second stage of stripping, using the sequences $(2^t - 2)T_j$ where $T_j = (2^{s-t}, 2^{s-2t}, \ldots, 2^{s-jt})$, $1 \leq j \leq u$. This is a legitimate stripping sequence, since only one nonzero monomial (of length u) will result at each stage for degree reasons. The result is the admissible monomial $Sq_t^{2^{s-t}} Sq_t^{2^{s-2t}} \cdots Sq_t^{2^{s-ut}}$. $\qquad\square$

Example 13.7.5 Let $t = 2$ and $s = 6$, so that $u = 3$ and we are stripping $(Sq_2^{64})^6$ in A_4. The arrays

64	64	64	64	64	64		0	48	0	44	0	43
64	16	0	4	0	1		0	32	0	8	0	2
0	0	64	16	0	4 ,		0	0	0	32	0	8 ,
0	0	0	0	64	16		0	0	0	0	0	32
0	48	0	44	0	43		0	16	0	4	0	1

are the array (13.11) and the array for the second stage, which gives the admissible monomial $Sq_2^{16} Sq_2^4 Sq_2^1$.

We turn next to the upper bound, with the aim of generalizing the method of Proposition 13.5.3. A preliminary result reduces the cases to be considered.

Proposition 13.7.6 *If Proposition 13.7.1 is true when $s = -1 \bmod t$, then it is true for all s.*

Proof Let $s = -1 \bmod t$. Then the result follows by applying the halving map V, since the required upper bound is the same for s and $s - 1$, and the lower bound obtained from Proposition 13.7.4 is also the same for s and $s - 1$. □

The following result completes the proof of Proposition 13.7.1.

Proposition 13.7.7 *Let $t \geq 1$, $s \geq 0$ and $s = ut + v$ where $0 \leq v < t$. Then $(Sq_t^{2^s})^{2u+2} = 0$.*

Proof We prove the statements

$$\text{(i) } \chi(X_{t,k}^u)(Sq_t^{2^s})^{2k+1} = 0, \text{ (ii) } \chi(X_{t,k}^{u-1})(Sq_t^{2^s})^{2k+2} = 0,$$

by induction on k for $0 \leq k \leq s$, where $X_{t,k}^u = Sq_t^{2^s}Sq_t^{2^{s-t}}\cdots Sq_t^{2^{s-(u-k)t}}$ for $0 \leq k \leq u$, as in Example 12.5.7, and $X_{t,u}^{u-1} = 1$.

By Proposition 13.6.8(i), $Xq_t^{2^s+2^{s-t}+\cdots+2^{s-ut}} = Sq_t^{2^s}Sq_t^{2^{s-t}}\cdots Sq_t^{2^{s-ut}} = X_{t,0}^u$, so $\chi(X_{t,0}^u) = Sq_t^{2^s+2^{s-t}+\cdots+2^{s-ut}}$ and so the relation $Sq_t^{2^s+2^{s-t}+\cdots+2^{s-ut}}Sq_t^{2^s} = 0$ of Proposition 13.6.12 gives $\chi(X_{t,0}^u)Sq_t^{2^s} = 0$. This is the base case $k = 0$ for (i). We shall show that (i) for k implies (ii) for k, and that (ii) for k implies (i) for $k+1$. The case $k = s$ of (ii) is the required relation $(Sq_t^{2^s})^{2u+2} = 0$.

Statement (i) for k can be written as $\chi(X_{t,k}^{u-1})Xq_t^{2^s}(Sq_t^{2^s})^{2k+1} = 0$. Since $\deg X_{t,k}^{u-1} = 2^{s-t} + 2^{s-2t} + \cdots + 2^{s-(u-k)t} < 2^s$, stripping $X_{t,k}^{u-1}$ by (2^s) gives 0. By Straffin's formula 12.1.2(ii), stripping $Xq_t^{2^s}$ by (2^s) gives 1. Hence stripping the relation $\chi(X_{t,k}^{u-1})Xq_t^{2^s}(Sq_t^{2^s})^{2k+1} = 0$ by (2^s) gives $\chi(X_{t,k}^{u-1})(Sq_t^{2^s})^{2k+1} + (2k+1)\chi(X_{t,k}^{u-1})Xq_t^{2^s}(Sq_t^{2^s})^{2k} = 0$. We conclude that $\chi(X_{t,k}^{u-1})(Sq_t^{2^s})^{2k+2} = \chi(X_{t,k}^{u-1})Xq_t^{2^s}(Sq_t^{2^s})^{2k+1} = 0$, which is statement (ii) for k.

Next we assume statement (ii) for k. By iterating Proposition 12.1.2(i),

$$\begin{aligned}
\chi(X_{t,k+1}^u)Sq_t^{2^s} &= Xq_t^{2^{s-(u-k)t}}\cdots Xq_t^{2^{s-t}}Xq_t^{2^s}Sq_t^{2^s}\\
&= Xq_t^{2^{s-(u-k)t}}\cdots Xq_t^{2^{s-t}}Sq_t^{2^{s+1}-2^{s-t}}Xq_t^{2^{s-t}}\\
&= Xq_t^{2^{s-(u-k)t}}\cdots Xq_t^{2^{s-2t}}Sq_t^{2^{s+1}-2^{s-2t}}Xq_t^{2^{s-2t}}Xq_t^{2^{s-t}}\\
&= \cdots\\
&= Sq_t^{2^{s+1}-2^{s-kt}}Xq_t^{2^{s-(u-k)t}}\cdots Xq_t^{2^{s-2t}}Xq_t^{2^{s-t}}\\
&= Sq_t^{2^{s+1}-2^{s-kt}}\chi(X_{t,k}^{u-1}),
\end{aligned}$$

and so we obtain

$$\begin{aligned}
\chi(X_{t,k+1}^u)(Sq_t^{2^s})^{2k+3} &= \chi(X_{t,k+1}^u)Sq_t^{2^s}(Sq_t^{2^s})^{2k+2}\\
&= Sq_t^{2^{s+1}-2^{s-kt}}\chi(X_{t,k}^{u-1})(Sq_t^{2^s})^{2k+2}\\
&= 0,
\end{aligned}$$

which is statement (i) for $k+1$. This completes the induction. □

13.8 Remarks

Section 13.2 is based on the treatment of stripping in [233, Section 5]. This method of studying relations in the Steenrod algebra was introduced by L. Kristensen [115], as a natural counterpart of the cap product relating the homology and cohomology groups of a topological space. In particular, Kristensen introduced the operator $\xi_1 \cap$ under the name of 'differentiation', and used it to obtain all the Adem relations from the relations $Sq^{2k-1} Sq^k = 0$, $k \geq 1$. A treatment of the Adem relations by Kristensen's method can be found in Gray [70].

The formula of Example 13.2.7 goes back to [142]. Proposition 13.2.2 can be found in Milnor [134, Lemma 7]. The halving map (or Verschiebung) V of A_2 is discussed by Steenrod–Epstein [196] and Li [123].

The nilpotence height of an element of A_2^+ is known only in special cases [123, 143, 216]. The odd subalgebra Od was introduced by K. G. Monks [143], who proved that the nilpotence height of $Sq^{2^m(2^k-1)+1}$ is $k + 2$ for all $m \geq 1$ and $k \geq 0$, generalizing Proposition 13.4.11. The main results of Section 13.7 are also due to Monks [145]. Proposition 13.6.6 first appeared in [61]. Proposition 13.5.2 is due to Davis 'On the height of Sq^{2^n}', preprint, Lehigh University, 1985.

14

The 2-dominance theorem

14.0 Introduction

The Peterson conjecture, Theorem 2.5.5, states that a monomial $f \in P^d(n)$ is hit if $\mu(d) > n$, so that there are no spikes in $P^d(n)$. In Theorem 6.3.12, it was shown that if $\mu(d) \leq n$ and $\omega(f) <_l \omega^{\min}(d)$, so that $\omega(f)$ is lower in the left order than the ω-sequence of every spike in $P^d(n)$, then f is again hit. The main result of this chapter, Theorem 14.1.3, strengthens this result by showing that a monomial $f \in P^d(n)$ is hit if $\omega(f)$ is not greater than or equal to $\omega^{\min}(d)$ in the 2-dominance order (see Section 5.4).

As this is only a partial order, there are many situations where $\omega(f) >_l \omega^{\min}(d)$ and yet $\omega(f) \not\geq_2 \omega^{\min}(d)$. For example, $\omega^{\min}(13) = (3,3,1)$, and so if $\omega(f) = (5,0,2)$, $\omega(f) >_l \omega^{\min}(13)$, but $\omega(f) \not\geq_2 \omega^{\min}(13)$, since $5 + 2 \cdot 0 = 5 < 3 + 2 \cdot 3 = 9$. Thus a monomial such as $f = x_1^5 x_2^5 x_3 x_4 x_5$ is hit.

The Silverman–Singer criterion 14.1.2 generalizes the μ-function criterion of Proposition 2.5.3 so as to apply to this example. In this case, $\omega(f) <_r \omega^{\min}(d)$. Since an element of Seq_d which is $<_r \omega^{\min}(d)$ cannot 2-dominate $\omega^{\min}(d)$, the 2-dominance theorem implies that the analogue of Theorem 6.3.12 is true for the right order, i.e. a monomial $f \in P^d(n)$ is hit if $\omega(f) <_r \omega^{\min}(d)$. The example of $\omega(f) = (5,0,5)$ and $\omega^{\min}(25) = (3,3,2,1)$ shows that it is possible that $\omega(f) >_l \omega^{\min}(d)$ and $\omega(f) >_r \omega^{\min}(d)$ but $\omega(f) \not\geq_2 \omega^{\min}(d)$.

The key result used in proving the Peterson conjecture is the χ-trick. This explains the appearance of the function $\mu(d)$ in connection with the hit problem, as it is the excess of $Xq^d = \chi(Sq^d)$. The proof of the 2-dominance theorem follows the same pattern. The iterated squaring operation $Sq^{[k;d]} = Sq^{2^{k-1}d} \cdots Sq^{2d} Sq^d$ maps a monomial $h \in P^d(n)$ to h^{2^k}, and the essential input, the excess theorem 14.1.1, states that its conjugate $Xq^{[k;d]}$ has excess $(2^k - 1)\mu(d)$.

In Section 14.1, we show how the 2-dominance theorem follows from the excess theorem. In Section 14.2, we prove Proposition 14.2.4 and collect preliminary results required for Proposition 14.4.4. These involve properties of the Adem relations and of the stripping operations introduced in Section 13.2. Section 14.3 introduces the numerical function γ, which is inverse to $\beta(d) = (d+\mu(d))/2$ in the sense that $q \geq \beta(d)$ if and only if $\gamma(q) \geq d$. We prove Proposition 14.4.4 in Section 14.4, and use it in Section 14.5 to prove the excess theorem 14.1.1. In Section 14.6 we relate these results to the Milnor basis.

14.1 The excess theorem

Theorem 14.1.1 (The excess theorem) *Let* $Sq^{[k;d]} = Sq^{2^{k-1}d} \cdots Sq^{2d} Sq^d$ *for* $d \geq 0$ *and* $k \geq 1$, *and let* $Xq^{[k;d]} = \chi(Sq^{[k;d]})$. *Then the excess of* $Xq^{[k;d]}$ *is* $(2^k - 1)\mu(d)$.

In the case $k = 1$, the excess theorem reduces to Proposition 2.4.10, and in the case $d = 1$, it follows from Proposition 4.4.2(i). For $d = 2$, a similar argument shows that $\chi(Sq^{2^{k-1}} \cdots Sq^2) = Sq^{2^k-2}$ (see Example 14.2.7), so that $\mathrm{ex}(Xq^{[k;2]}) = 2^k - 2 = (2^{k-1} - 1)\mu(2)$. This simple pattern soon breaks down, since $\chi(Sq^d)$ is not generally a monomial in A_2. For example, when $j \geq 2$, $Xq^{2^j} = Sq^{2^j} +$ other terms, and this is not a monomial since Sq^{2^j} is indecomposable.

However, Judith Silverman found a remarkable generalization of Davis's result 4.4.2(i), by showing that conjugation χ acts on the family of admissible monomials of the form $Sq^{2^{k-1}d} \cdots Sq^{2d} Sq^d$, $d = 2^j - 1$ by exchanging j and k (Proposition 14.2.4). For example, the conjugate of $Sq^{14} Sq^7$ is $Sq^{12} Sq^6 Sq^3$ and that of $Sq^{30} Sq^{15}$ is $Sq^{24} Sq^{12} Sq^6 Sq^3$, while $Sq^1, Sq^6 Sq^3, Sq^{28} Sq^{14} Sq^7$ are invariant under conjugation. The case $\mu(d) = 1$ of the excess theorem follows, since $\mu(d) = 1$ for $d = 2^j - 1$, and the conjugate of $Sq^{2^{k-1}d} \cdots Sq^{2d} Sq^d$ is $Sq^{2^{j-1}d'} \cdots Sq^{2d'} Sq^{d'}$, where $d' = 2^k - 1$, and so it has excess $2^k - 1$.

By Proposition 5.8.4, the excess of an element of A_2 is the excess of the $\leq_l$-minimal term in its admissible basis expansion. The proof of Theorem 14.1.1 depends on the fact that the $\leq_l$-minimal term in the admissible basis expansion of $Xq^{[k;d]}$ is obtained by multiplying all the exponents of $Sq^{A(d)}$ by $2^k - 1$, where $Sq^{A(d)}$ is the minimal admissible monomial of degree d (Definition 5.7.5). This monomial $Sq^{(2^k-1)A(d)}$ has excess $(2^k - 1)\mu(d)$; it appears in the expansion of $Xq^{[k;d]}$ in the admissible basis, and no admissible monomials of lower excess appear.

In fact, we shall prove that $Sq^{(2^k-1)A(d)}$ is minimal for both the left and right orders on the admissible basis, and thus obtain a corresponding result for the

Milnor basis expansion. Note that $A(d)$ has length j when $2^j - 1 \le d \le 2^{j+1} - 2$, since $Sq^{[j;1]} = Sq^{2^{j-1}} \cdots Sq^2 Sq^1$ is the minimum admissible monomial in degree $2^j - 1$, and there is no admissible sequence of length j in degrees $< 2^j - 1$. We shall see that for $k \ge 1$ the length of $Xq^{[k;d]}$, when expressed in the admissible basis, is the length of $A(d)$. Our main technical tool, Proposition 14.4.4, is a recursive formula which expresses $Xq^{[k;d]}$ in terms of stripping operations.

Theorem 14.1.1 has important implications for the hit problem. In particular, we obtain a sufficient condition for a monomial in $P(n)$ to be hit.

Theorem 14.1.2 (The Silverman–Singer criterion) *Let $f = gh^{2^k}$ be a homogeneous polynomial. If $\deg g < (2^k - 1)\mu(\deg h)$, then f is hit.*

Proof In the notation above, $h^{2^k} = Sq^{[k;d]}h$, where $d = \deg h$. Hence $gh^{2^k} \sim Xq^{[k;d]}(g)h$ by the χ-trick 2.5.2. The excess of $Xq^{[k;d]}$ is $(2^k - 1)\mu(d)$ by Theorem 14.1.1. Thus $Xq^{[k;d]}(g) = 0$ by the definition of excess, and hence gh^{2^k} is hit. $\qquad\square$

In turn this result leads to a criterion in terms of ω-sequences.

Theorem 14.1.3 (The 2-dominance theorem) *Let $f \in P^d(n)$ be a monomial such that $\omega(f) \not\ge_2 \omega^{\min}(d)$. Then f is hit.*

Proof Given $k \ge 1$, we can factor f uniquely as $f = gh^{2^k}$, where g and h are monomials and all exponents of g are $< 2^k$. Assume that f is not hit. It follows by Theorem 14.1.2 that $\deg g \ge (2^k - 1)\mu(\deg h)$. Let $d' = \deg g - (2^k - 1)\mu(\deg h)$, so that $d' \ge 0$. Let

$$R = \begin{array}{c|c} A & B \\ \hline C & 0 \end{array}$$

be a spike block of degree d, constructed as follows: B is a minimal spike block of degree $\deg h$, with $\mu(\deg h)$ rows, A is a $(\mu(\deg h), k)$-block with all entries 1, C is a spike block of degree d' with k columns and any number of rows and 0 is a zero block. We regard R as the concatenation $S|T$ of spike n-blocks S of degree $\deg g$ and T of degree $\deg h$. Since $\omega(R) \ge_l \omega^{\min}(d)$, $\sum_{i=1}^k 2^{i-1}\omega_i(R) \ge \sum_{i=1}^k 2^{i-1}\omega_i^{\min}(d)$. But $\sum_{i=1}^k 2^{i-1}\omega_i(R) = \sum_{i=1}^k 2^{i-1}\omega_i(S) = \deg g = \sum_{i=1}^k 2^{i-1}\omega_i(f)$. Hence $\sum_{i=1}^k 2^{i-1}\omega_i(f) \ge \sum_{i=1}^k 2^{i-1}\omega_i^{\min}(d)$. Since this holds for all $k \ge 1$, $\omega(f) \ge_2 \omega^{\min}(d)$. $\qquad\square$

Theorem 6.3.12 states that $Q^\omega(n) = 0$ for all $\omega <_l \omega^{\min}(d)$. We can use Theorem 14.1.3 to obtain a corresponding result for the right order $\le_r$.

Proposition 14.1.4 *Let* $\omega <_r \omega^{\min}(d)$ *where* $\deg_2 \omega = d$. *Then* $Q^\omega(n) = 0$, *where the filtration on* $Q^d(n)$ *is defined with respect to* $<_r$.

Proof By Proposition 5.2.3, $\omega \not\geq_2 \omega^{\min}(d)$. Hence by Theorem 14.1.3 all monomials f with $\omega(f) = \omega$ are hit. $\qquad\square$

14.2 Stripping and conjugation for $\mu(d) = 1$

In this section we begin the proof of Theorem 14.1.1 with some preliminary results on the Adem relations in A_2 and the stripping operations of Section 13.2. In particular, Proposition 14.2.4 establishes the case $\mu(d) = 1$ of Theorem 14.1.1.

Proposition 14.2.1 *For* $a < 2^k$, *the coefficients in the Adem relation*

$$Sq^a Sq^{b+2^k c} = \sum_{j=0}^{[a/2]} \binom{b+2^k c - 1 - j}{a - 2j} Sq^{a+b+2^k c - j} Sq^j$$

are independent of c, *and so can be evaluated by setting* $c = 0$.

Proof This follows immediately from Proposition 1.4.11, as the parity of the binomial coefficient depends only on $\mathrm{bin}(a - 2j)$, and $a - 2j \leq a < 2^k$. $\qquad\square$

Note that we do not assume that $a < 2b$. If $a \geq 2b$, then $Sq^a Sq^b$ is admissible, but the Adem relation remains valid. For example when $a = 4$, $b = 1$ we obtain $Sq^4 Sq^9 = Sq^{12} Sq^1 + Sq^{11} Sq^2$ when $k = 3$ and $c = 1$, and $Sq^4 Sq^1 = Sq^4 Sq^1 + Sq^3 Sq^2$ when $k = 3$ and $c = 0$, so that the coefficients in the two relations agree.

In particular, the identity $Sq^{2u} Sq^u = Sq^{2u} Sq^u$ and the relations $Sq^{2u-1} Sq^u = 0$ and $Sq^{2^a-1} Sq^{2^a} = Sq^{2^{a+1}-1}$ lead to the following relations, which we use in the proof of Proposition 14.2.3.

Proposition 14.2.2 (i) *For* $u < 2^{k-1}$, $Sq^{2u} Sq^{u+2^k c} = Sq^{2u+2^k c} Sq^u$.
(ii) *For* $u < 2^{k-1}$, $Sq^{2u-1} Sq^{u+2^k c} = 0$.
(iii) *For* $b \geq a \geq 0$, $Sq^{2^a-1} Sq^{2^b} = Sq^{2^a+2^b-1}$. $\qquad\square$

Proposition 14.2.3 *For* $a, k \geq 1$, $Xq^{[k; 2^{a+1}-1]} = Sq^{2^a(2^k-1)} Xq^{[k; 2^a-1]}$.

Proof The proof is by induction on k. The case $k = 1$ corresponds to formula 4.4.2(i). By conjugation, we wish to prove that $Sq^{[k; 2^{a+1}-1]} = Sq^{[k; 2^a-1]} Xq^{2^a(2^k-1)}$. Using Bausum's formula 4.4.3, we can expand $Xq^{2^a(2^k-1)}$ as

$$Xq^{2^a(2^k-1)} = Sq^{2^{a+k-1}} Xq^{2^a(2^{k-1}-1)} + Sq^{2^{a-1}(2^k-1)} Xq^{2^{a-1}(2^k-1)}.$$

By Proposition 14.2.2(ii), $Sq^{2^a-1}Sq^{2^{a-1}(2^k-1)} = 0$. Hence $Sq^{[k;2^a-1]}Xq^{2^a(2^k-1)} = Sq^{[k;2^a-1]}Sq^{2^{a+k-1}}Xq^{2^a(2^{k-1}-1)}$. As $Sq^{[k;2^a-1]} = Sq^{2^{k-1}(2^a-1)}\cdots Sq^{2(2^a-1)}Sq^{2^a-1}$, we can use Proposition 14.2.2(iii) with $b = k - 1$ to absorb the factor Sq^{2^a-1}, and using Proposition 14.2.2(i)

$$Sq^{2^l(2^a-1)}Sq^{2^{l-1}(2^a-1)+2^{a+k-1}} = Sq^{2^l(2^a-1)+2^{a+k-1}}Sq^{2^{l-1}(2^a-1)}$$

for $1 < l < k - 1$ to switch successive pairs of factors, we obtain

$$Sq^{[k;2^a-1]}Xq^{2^a(2^k-1)} = Sq^{2^{k-1}(2^a-1)+2^{a+k-1}}Sq^{[k-1;2^a-1]}Xq^{2^a(2^{k-1}-1)}. \qquad (14.1)$$

By applying χ to the induction hypothesis on k, we have

$$Sq^{[k-1;2^a-1]}Xq^{2^a(2^{k-1}-1)} = Sq^{[k-1;2^{a+1}-1]},$$

and so the right hand side of (14.1) is $Sq^{[k;2^{a+1}-1]}$, completing the induction. $\qquad \square$

Proposition 14.2.4 *For all $j, k \geq 1$, $Xq^{[k;2^j-1]} = Sq^{[j;2^k-1]}$.*

Proof This follows by iteration from Proposition 14.2.3. $\qquad \square$

Example 14.2.5 We illustrate the preceding argument by working through the case $j = 4$, $k = 3$. We wish to prove that $\chi(Sq^{60}Sq^{30}Sq^{15}) = Sq^{56}Sq^{28}Sq^{14}Sq^7$. We begin with the expansion $Xq^{56} = Sq^{32}Xq^{24} + Sq^{28}Xq^{28}$ of Proposition 4.4.3. Using the Adem relations $Sq^7Sq^{28} = 0$, $Sq^7Sq^{32} = Sq^{39}$, $Sq^{14}Sq^{39} = Sq^{46}Sq^7$ and $Sq^{28}Sq^{46} = Sq^{60}Sq^{14}$, we find $Sq^{28}Sq^{14}Sq^7Xq^{56} = Sq^{60}Sq^{14}Sq^7Xq^{24}$.

Again we use Bausum's formula to expand Xq^{24} as $Sq^{16}Xq^8 + Sq^{12}Xq^{12}$. Then the Adem relations $Sq^7Sq^{12} = 0$, $Sq^7Sq^{16} = Sq^{23}$ and $Sq^{14}Sq^{23} = Sq^{30}Sq^7$ give $Sq^{14}Sq^7Xq^{24} = Sq^{30}Sq^7Xq^8$. Finally Straffin's formula $Xq^8 = Sq^8 + Sq^4Xq^4$ and the Adem relations $Sq^7Sq^4 = 0$, $Sq^7Sq^8 = Sq^{15}$ give $Sq^7Xq^8 = Sq^{15}$.

Hence $Sq^{28}Sq^{14}Sq^7Xq^{56} = Sq^{60}Sq^{30}Sq^{15}$. Applying χ to this result gives $\chi(Sq^{60}Sq^{30}Sq^{15}) = Sq^{56}\chi(Sq^{28}Sq^{14}Sq^7)$. The argument is completed by using a similar method to show $\chi(Sq^{28}Sq^{14}Sq^7) = Sq^{28}Sq^{14}Sq^7$, an earlier instance of Proposition 14.2.4.

We next consider the stripping operations of A_2^* on A_2. As in Theorem 11.6.1, for $k \geq 1$ we write $\eta_k = \chi(\xi_k)$, the conjugate of the generator ξ_k of degree $2^k - 1$ of $A_2^* = \mathbb{F}_2[\xi_1, \xi_2, \ldots]$. We also write $\eta_0 = \xi_0 = 1$.

Proposition 14.2.6 *For all d and k, the stripping operators $\xi_k\cap$ and $\eta_k\cap$ satisfy:*

(i) $\eta_k \cap Sq^d = \eta_1^{2^k-1} \cap Sq^d = Sq^{d-(2^k-1)}$,

(ii) $\xi_k \cap Xq^d = \xi_1^{2^k-1} \cap Xq^d = Xq^{d-(2^k-1)}$,

(iii) $\xi_k \cap Sq^{[k;d]} = Sq^{[k;d-1]}$, and $\xi_k \cap Sq^{[\ell;d]} = 0$ for $k > \ell$,

(iv) $\eta_k \cap Xq^{[k;d]} = Xq^{[k;d-1]}$, and $\eta_k \cap Xq^{[\ell;d]} = 0$ for $k > \ell$.

Proof The generators ξ_k of A_2^* act on Sq^d by $\xi_1 \cap Sq^d = Sq^{d-1}$ and $\xi_k \cap Sq^d = 0$ for $k > 1$. Hence $\xi_1^r \cap Sq^d = Sq^{d-r}$ for all $r \geq 0$, and $\xi^A \cap Sq^d = 0$ for any monomial $\xi^A \in A_2^*$ involving ξ_j for $j > 1$. By Proposition 11.5.2, $\eta_k = \chi(\xi_k) = \xi_1^{2^k-1} +$ terms involving ξ_j for $j > 1$. Property (i) follows, and (ii) follows from (i) by conjugation.

For (iii) we use induction on k, using $Sq^{[k;d]} = Sq^{2^{k-1}d} Sq^{[k-1;d]}$ and noting that only the term $\xi_1^{2^{k-1}} \otimes \xi_{k-1}$ in $\phi(\xi_k)$ contributes to the result. Finally (iv) follows from (iii) by conjugation. $\qquad\square$

Example 14.2.7 Applying the operator $\xi_1 \cap$ to Davis's formula $Xq^{2^{k+1}-1} = Sq^{2^k} \cdots Sq^2 Sq^1$ (Proposition 4.4.2(i)), and using the Adem relations $Sq^{2^t-1} Sq^t = 0$, we find $Xq^{2^{k+1}-2} = Sq^{2^k} \cdots Sq^2$. Conjugation gives $Xq^{[k;2]} = Sq^{2^{k+1}-2}$.

The following relation between the operations of stripping by η_k and stripping by η_{k-1} will be useful in Section 14.4.

Proposition 14.2.8 *For all $r,k,a \geq 0$ and all $\theta \in A_2$,*

$$\eta_k^{2^r} \cap Sq^a \theta = Sq^a(\eta_k^{2^r} \cap \theta) + \eta_{k-1}^{2^{r+1}} \cap Sq^{a-2^r}\theta.$$

Proof Since $\phi(\xi_k) = \sum_{i+j=k} \xi_i^{2^j} \otimes \xi_j$ by Proposition 11.5.2, and $\eta_j = \chi(\xi_j)$, $\phi(\eta_k) = \sum_{i+j=k} \eta_j \otimes \eta_i^{2^j}$. Since ϕ is multiplicative, $\phi(\eta_k^{2^r}) = \sum_{i+j=k} \eta_j^{2^r} \otimes \eta_i^{2^{j+r}}$ for $r \geq 1$. Hence for $\theta_1, \theta_2 \in A_2$, $\eta_k^{2^r} \cap \theta_1\theta_2 = \sum_{i+j=k}(\eta_j^{2^r} \cap \theta_1)(\eta_i^{2^{j+r}} \cap \theta_2)$, and using Proposition 14.2.6 we obtain $\eta_k^{2^r} \cap Sq^a \theta = \sum_{i+j=k} Sq^{a-(2^j-1)2^r}(\eta_i^{2^{j+r}} \cap \theta)$, for all $\theta \in A_2$. The $j = 0$ term is $Sq^a(\eta_k^{2^r} \cap \theta)$, and, by shifting the summation index, the remaining terms give the corresponding expansion of $\eta_{k-1}^{2^{r+1}} \cap Sq^{a-2^r}\theta$. $\qquad\square$

When $\eta_k^{2^r} \cap \theta = 0$, this result can be used to rewrite $\eta_{k-1}^{2^{r+1}} \cap Sq^{a-2^r}\theta$ as $\eta_k^{2^r} \cap Sq^a \theta$. In particular, $\eta_k \cap Xq^{[k';b]} = 0$ when $k' < k$. We shall often apply this result in one of the following forms.

Proposition 14.2.9 (i) *For $r,k,a,b \geq 0$,*

$$\eta_k^{2^r} \cap Sq^{a-2^{r-1}} Xq^{[k;b]} = \eta_{k+1}^{2^{r-1}} \cap Sq^a Xq^{[k;b]}.$$

(ii) *For $j,k,u,v \geq 0$,*

$$\eta_{k+j}^2 \cap Sq^u Xq^{[k;v]} = \eta_{k+j+1} \cap Sq^{u+1} Xq^{[k;v]}). \qquad\square$$

14.3 The numerical function γ

In this section we introduce the numerical function γ, which is inverse to the function $\beta(d) = (d + \mu(d))/2$ of Proposition 5.7.6, in the sense that $\gamma(q) \geq d$ if and only if $q \geq \beta(d)$.

Definition 14.3.1 For an integer $q \geq 0$,

$$\gamma(q) = \sum_{j \geq 0} \left[\frac{q}{2^j} \right] = \nu_2((2q)!),$$

where $2^{\nu_2(m)}$ is the highest 2-power dividing m.

Clearly γ is a strictly increasing function. We tabulate its first few values.

q	0	1	2	3	4	5	6	7	8	9	10	11	12	13	14	15	16
$\gamma(q)$	0	1	3	4	7	8	10	11	15	16	18	19	22	23	25	26	31

In particular, $\gamma(2^r) = 2^{r+1} - 1$ for $r \geq 0$, and $\gamma(q) - \gamma(q-1) = 1 + \nu_2(q)$ for all q.

Proposition 14.3.2 *For $q \geq 0$, (i) $\gamma(q) = 2q - \alpha(q)$, (ii) $\gamma(q) - q = \gamma([q/2])$, and (iii) $\alpha(q) = \mu(\gamma(q))$.*

(iv) For $r > 0$, $r = \gamma(q)$ for some $q > 0$ if and only if r is a sum of distinct integers of the form $2^j - 1$.

Proof For (i), let $q = 2^{a_1} + 2^{a_2} + \cdots + 2^{a_m}$ where $a_1 > a_2 \cdots > a_m$, so that $m = \alpha(q)$. The formula $\gamma(q) = \sum_{j \geq 0}[q/2^j]$ can be evaluated directly to give

$$\gamma(q) = (2^{a_1+1} - 1) + (2^{a_2+1} - 1) + \cdots + (2^{a_m+1} - 1) = 2q - m. \quad (14.2)$$

The recursive formula (ii) follows easily from (i) by separating the cases where q is even and where q is odd. For (iii) and (iv), note that (14.2) is the minimal expression for $\gamma(q)$ as a sum of integers of the form $2^j - 1$, since $a_1 > a_2 \cdots > a_m$. Conversely, any sum r of distinct integers $2^j - 1$ can be written in the form (14.2), and so we can recover q such that $\gamma(q) = r$. $\qquad \square$

Recall from Definition 5.4.3 that $\sigma^{\min}(d)$ is the partition of d as the sum of a minimal number of integers of the form $2^k - 1$, and that either the parts of $\sigma^{\min}(d)$ are all distinct, or they are distinct except that the smallest part is repeated. It follows from (iv) that γ enumerates the integers d for which $\sigma^{\min}(d)$ has distinct parts. For such d, $\sigma^{\min}(d+1)$ is obtained by appending a 1 to $\sigma^{\min}(d)$, and hence $\mu(d+1) = \mu(d) + 1$. On the other hand, if the smallest part $2^j - 1$ of $\sigma^{\min}(d)$ is repeated, then the smallest part of $\sigma^{\min}(d+1)$ is $2(2^j - 1) + 1 = 2^{j+1} - 1$, and so $\mu(d+1) = \mu(d) - 1$.

Definition 14.3.3 An integer $d \geq 0$ is a μ^+**-number** if $\mu(d+1) = \mu(d) + 1$, and is a μ^-**-number** if $\mu(d+1) = \mu(d) - 1$.

Thus $d = \gamma(q)$ for some $q \geq 0$ if and only if d is a μ^+-number. The next statement follows from the fact that $\gamma(q) - \gamma(q-1) = 1$ if and only if q is odd, but we give a proof based on the definition of the minimal spike partition $\sigma^{\min}(d)$.

Proposition 14.3.4 μ^+*-numbers occur in pairs, separated by* μ^-*-numbers.*

Proof If d is a μ^+-number, we consider the smallest part of $\sigma^{\min}(d)$. If this is 1, then $d+1$ is a μ^--number, because we cannot have three parts equal to 1 in $\sigma^{\min}(d+2)$. In this case, $d-1$ is a μ^+-number, since $\sigma^{\min}(d-1)$ is obtained by dropping the 1 from $\sigma^{\min}(d)$. On the other hand, if the smallest part of $\sigma^{\min}(d)$ is > 1, then $d+1$ is also a μ^+-number, but $d+2$ is a μ^--number, as the smallest part of $\sigma^{\min}(d+1)$ is 1. $\qquad\square$

We next identify the inverse of γ, regarded as a bijection from the integers ≥ 0 to the μ^+-numbers, with the function $\beta(d) = (d + \mu(d))/2$. Recall from Section 5.8 and Proposition 5.7.6 that if Sq^A is an admissible monomial of minimum excess $\mu(d)$ in A_2^d, then the first term a_1 of A is $\beta(d)$, and that iteration of this calculation yields the $<_l$-minimal admissible monomial $A(d)$ of Definition 5.7.5.

Proposition 14.3.5 *For an integer $d \geq 0$, let $\beta(d) = (d + \mu(d))/2$. Then*

(i) $\beta(d+1) = \beta(d)$ *or* $\beta(d) + 1$ *according as d is a μ^- or a μ^+ number,*
(ii) $\beta(d)$ *is the number of μ^+ numbers $< d$,*
(iii) *if d is a μ^+-number, then $\beta(d) = \gamma^{-1}(d)$,*
(iv) *for all $q, d \geq 0$, $\gamma(q) \geq d$ if and only if $q \geq \beta(d)$.*

Proof (i) is clear from the definition of β, and (ii) follows from (i). For (iii), we reverse the argument in Proposition 14.3.2(i). Let $d = (2^{a_1+1} - 1) + (2^{a_2+1} - 1) + \cdots + (2^{a_m+1} - 1)$ where $a_1 > a_2 \cdots > a_m$, so that $m = \mu(d)$. Then $\beta(d) = (d+m)/2 = 2^{a_1} + 2^{a_2} + \cdots + 2^{a_m} = q$, say, and $\gamma(q) = d$.

For (iv), let $d' = d$ if d is a μ^+-number, and otherwise let d' be the next μ^+-number. Then $d' = \gamma(q')$ where $q' = \beta(d')$, and $\gamma(q) \geq d$ if and only if $\gamma(q) \geq d'$. Since γ is increasing, $\gamma(q) \geq \gamma(q')$ if and only if $q \geq q'$. But $q' = \beta(d') = \beta(d)$ by (ii). $\qquad\square$

The function β is interesting in its own right. We tabulate its first few values.

d	0	1	2	3	4	5	6	7	8	9	10	11	12	13	14	15	16
$\beta(d)$	0	1	2	2	3	4	4	4	5	6	6	7	8	8	8	8	9

Thus for all $q > 0$ there are $1 + v_2(q)$ values of d such that $\beta(d) = q$, namely $\gamma(q-1) + 1 \leq d \leq \gamma(q)$.

Proposition 14.3.6 *For $a \geq 0$ and $0 \leq d \leq 2^{a+1} - 1$,*

$$\beta(d) + \beta(2^{a+1} - 1 - a - d) = 2^a,$$

where $\beta(d') = 0$ if $d' < 0$.

Proof Let $e = 2^{a+1} - (a+1)$. Since $e = 1 + \sum_{j=0}^{a}(2^j - 1)$ and $e - 1 = \sum_{j=0}^{a}(2^j - 1)$ are minimal spike partitions, $\mu(e) = a + 1$ and $\mu(e-1) = a$. Since β is non-decreasing, β takes the value 2^a for the first time when $d = e$. For $d \leq e$, the equation $\beta(d) + \beta(e - d) = 2^a$ is equivalent to $\mu(d) + \mu(e - d) = 2^{a+1} - e = a + 1$, which is true since the minimal spike partitions of d and $e - d$ are complements in the multiset $\{2^a - 1, 2^{a-1} - 1, \ldots, 3, 1, 1\}$.

Since $\beta(d) = 2^a$ for $e \leq d \leq 2^{a+1} - 1$, and $\beta(e - d) = 0$ by definition for $e - d \leq 0$, the result holds in the whole range $0 \leq d \leq 2^{a+1} - 1$. □

14.4 Stripping and conjugation for general d

In this section we obtain a formula for evaluation of $Xq^{[k;d]}$ using stripping operators. This generalizes the results of Section 14.2 for the case $d = 2^j - 1$ and Bausum's formula 4.4.3 for Xq^d in the case $k = 1$. We begin by extending Proposition 14.2.3 to a range of similar cases.

Proposition 14.4.1 *For $1 \leq e \leq 1 + a$, $Xq^{[k;2^{a+1}-e]} = Sq^{2^a(2^k-1)}Xq^{[k;2^a-e]}$.*

Proof Proposition 14.2.3 gives the result for $e = 1$. We use induction on a. For $a = 0$, $e = 1$ is the only case. For $a > 0$, we argue by induction on e for $2 \leq e \leq a + 1$. Thus by the induction hypothesis, replacing e by $e - 1$,

$$Xq^{[k;2^{a+1}-(e-1)]} = Sq^{2^a(2^k-1)}Xq^{[k;2^a-(e-1)]}.$$

Stripping this by η_k, we have by Propositions 14.2.6 and 14.2.8

$$Xq^{[k;2^{a+1}-e]} = Sq^{2^a(2^k-1)}Xq^{[k;2^a-e]} + \eta_{k-1}^2 \cap Sq^{2^a(2^k-1)-1}Xq^{[k;2^a-(e-1)]}.$$

By the induction hypothesis on a, $Xq^{[k;2^a-(e-1)]} = Sq^{2^{a-1}(2^k-1)}Xq^{[k;2^a-1-(e-1)]}$, since $1 \leq e - 1 \leq a$. Hence $Sq^{2^a(2^k-1)-1}Xq^{[k;2^a-(e-1)]} = 0$, using the Adem relation $Sq^{2t-1}Sq^t = 0$ with $t = 2^{a-1}(2^k - 1)$. It follows that $Xq^{[k;2^{a+1}-e]} = Sq^{2^a(2^k-1)}Xq^{[k;2^a-e]}$, which completes the induction on e, and so also completes the induction on a. □

As a by-product of the preceding proof, we have the following relation, which we shall use in the proof of Proposition 14.4.4.

Proposition 14.4.2 *For* $1 \leq e \leq a$, $Sq^{2^a(2^k-1)-1}Xq^{[k;2^a-e]} = 0$. $\qquad\qquad$ □

Example 14.4.3 Let $a = 3$, so that $2^a - e = b = 5,6,7$. For $k = 1$, we have $Sq^7 Xq^b = 0$, and for $k = 2$, $Sq^{23}Xq^{[2;b]} = 0$, where $Xq^{[2;b]} = Xq^b Xq^{2b}$.

The aim of the remainder of this section is to prove Proposition 14.4.4, which is the main tool needed to prove the excess theorem. It involves the stripping operations explicitly, and reduces to Proposition 14.4.1 in the case $q = 0$.

Proposition 14.4.4 *Let* $d = 2^a + b = 2^{a+1} - e$ *where* $a \geq 0$ *and* $0 \leq b < 2^a$, *so that* $1 \leq e \leq 2^a$. *Then for* $k \geq 1$ *and* $q \geq 0$

$$Xq^{[k;d]} = \eta_k^q \cap Sq^{2^a(2^k-1)}Xq^{[k;b+q]},$$

for $1 + q \leq e \leq 1 + a + \gamma(q)$.

To set this result in context, we first relate it to the function β. Given d, by Proposition 14.3.5(iii) the minimum value of q such that $e \leq 1 + a + \gamma(q)$ is $q = \beta(e - 1 - a)$. By Proposition 14.3.6, this minimum value is $2^a - \beta(d)$, since $e = 2^{a+1} - d$.

Next we relate Proposition 14.4.4 to Bausum's formula 4.4.3. Using Propositions 14.2.6(iv) and 14.2.8, the formula of Proposition 14.4.4 can be expanded in the case $q = 2^c$ as

$$Xq^{[k;d]} = Sq^{2^a(2^k-1)}Xq^{[k;b]} + \eta_{k-1}^{2^{c+1}} \cap Sq^{2^a(2^k-1)-2^c}Xq^{[k;b+2^c]}. \qquad (14.3)$$

If we now set $k = 1$, the stripping operation is the identity since $\eta_0 = 1$, and the formula reduces to Proposition 4.4.3, valid in the range $2^c + 1 \leq e \leq 2^{c+1} + a$ rather than $2^c + 1 \leq e \leq 2^{c+1}$.

Before giving the proof of Proposition 14.4.4, we discuss its structure and illustrate it by an example. The proof is by induction on q, the base case $q = 0$ being given by Proposition 14.4.1. For the inductive step we split the range $1 + q \leq e \leq 1 + a + \gamma(q)$ into two cases as follows.

Case 1: $1 + q \leq e \leq 2 + a + \gamma(q - 1)$. By the induction hypothesis, the formula holds for $q' = q - 1$ with $e' = e - 1$. Thus $Xq^{[k;d+1]} = \eta_k^{q-1} \cap Sq^{2^a(2^k-1)}Xq^{[k;b+q]}$. Applying $\eta_k \cap$, we obtain $Xq^{[k;d]} = \eta_k^q \cap Sq^{2^a(2^k-1)}Xq^{[k;b+q]}$, as required. (Note that if $b = 2^a - 1$ and $e = 1$, then $q = 0$ and the base case Proposition 14.4.1 applies.)

Case 2: $3 + a + \gamma(q-1) \le e \le 1 + a + \gamma(q)$. Since $\gamma(q) - \gamma(q-1) = 1 + v_2(q)$, there are $v_2(q)$ values of e to consider. Thus we may assume that q is even and > 0. The proof in this case will consist of a sequence of reductions using several applications of the inductive hypothesis for 2-powers and the results of Section 14.2. We illustrate the argument by carrying out the case $q = 8$ in more detail.

Example 14.4.5 When $q = 8$, in Case 2 we have $e = a + 14$, $a + 15$ or $a + 16$. We shall use the inductive hypothesis for $q = 4, 2, 1$ and finally for $q = 0$. Since $a + 14 \le e \le 2^a$, we may assume that $a \ge 5$, and so $e \ge 19$.

We first apply the inductive hypothesis with $q' = 4$, $e' = e - 8 = a + 6$, $a + 7$ or $a + 8$. For this we have to check that $1 + q' \le e' \le 1 + a + \gamma(q')$. Since $\gamma(4) = 7$ this becomes $5 \le e' \le a + 8$, which is clear. Thus we have

$$Xq^{[k;d+8]} = \eta_k^4 \cap Sq^{2^a(2^k-1)}Xq^{[k;b+12]}.$$

Expanding this using Proposition 14.2.8 gives

$$Xq^{[k;d+8]} = Sq^{2^a(2^k-1)}Xq^{[k;b+8]} + \eta_{k-1}^8 \cap Sq^{2^a(2^k-1)-4}Xq^{[k;b+12]}.$$

Since $\eta_k^8 \cap Xq^{[k;d+8]} = Xq^{[k;d]}$ by Proposition 14.2.6(iv), the required formula holds if and only if the second term on the right is in the kernel of η_k^8. We shall prove that

$$\eta_k^8 \cap Sq^{2^a(2^k-1)-4}Xq^{[k;b+12]} = 0.$$

We use Proposition 14.2.9(i) to rewrite this as $\eta_{k+1}^4 \cap Sq^{2^a(2^k-1)}Xq^{[k;b+12]} = 0$. Next we apply the inductive hypothesis again, with $q' = 2$ and $e' = e - 12 = a + 2$, $a + 3$ or $a + 4$, checking that $3 \le e' \le a + 4$. This gives

$$Xq^{[k;d+12]} = \eta_k^2 \cap Sq^{2^a(2^k-1)}Xq^{[k;b+14]}$$
$$= Sq^{2^a(2^k-1)}Xq^{[k;b+12]} + \eta_{k-1}^4 \cap Sq^{2^a(2^k-1)-2}Xq^{[k;b+14]}.$$

Since $\eta_{k+1} \cap Xq^{[k;d+12]}) = 0$, it follows that

$$\eta_{k+1}^4 \cap Sq^{2^a(2^k-1)}Xq^{[k;b+12]} = \eta_{k+1}^4 \eta_{k-1}^4 \cap Sq^{2^a(2^k-1)-2}Xq^{[k;b+14]}.$$

Ignoring the factor η_{k-1}^4, we prove that $\eta_{k+1}^4 \cap Sq^{2^a(2^k-1)-2}Xq^{[k;b+14]} = 0$. By Proposition 14.2.9(i), we can rewrite this as $\eta_{k+2}^2 \cap Sq^{2^a(2^k-1)}Xq^{[k;b+14]} = 0$. Next we apply the inductive hypothesis with $q' = 1$ and $e' = e - 14 = a$, $a + 1$

or $a+2$, checking that $2 \le e' \le a+2$. This gives

$$Xq^{[k;d+14]} = \eta_k \cap Sq^{2^a(2^k-1)}Xq^{[k;b+15]}$$
$$= Sq^{2^a(2^k-1)}Xq^{[k;b+14]} + \eta_{k-1}^2 \cap Sq^{2^a(2^k-1)-1}Xq^{[k;b+15]}.$$

Hence $\eta_{k+2}^2 \cap Sq^{2^a(2^k-1)}Xq^{[k;b+14]} = \eta_{k+2}^2\eta_{k-1}^2 \cap Sq^{2^a(2^k-1)-1}Xq^{[k;b+15]}$. Again ignoring the factor η_{k-1}^2, we prove that $\eta_{k+2}^2 \cap Sq^{2^a(2^k-1)-1}Xq^{[k;b+15]} = 0$. By Proposition 14.2.9, the left hand side is $\eta_{k+3} \cap Sq^{2^a(2^k-1)}Xq^{[k;b+15]}$. Finally we apply the inductive hypothesis with $q' = 0$ and $e' = e - 15 = a - 1$, a or $a+1$, checking that $1 \le e' \le a+1$. This gives $Xq^{[k;d+15]} = Sq^{2^a(2^k-1)}Xq^{[k;b+15]}$, and hence $\eta_{k+3} \cap Sq^{2^a(2^k-1)}Xq^{[k;b+15]} = 0$, which completes the argument.

As in the example above, the proof consists of a sequence of steps of two kinds: **raising** steps, which use Proposition 14.2.9 to raise the exponent of the Sq^* factor while keeping the exponent of the Xq^* factor fixed, and **lowering** steps, which use the induction hypothesis for q of the form $q = 2^c$ with $c < r$ to lower the exponent of the Sq^* factor and to raise the exponent of the Xq^* factor. Although powers of η_{k-1} are introduced into the stripping operator, the resulting terms play no part in the argument. The exponent of the Sq^* factor remains $\le 2^a(2^k - 1)$ throughout.

Proof of Proposition 14.4.4   The discussion so far may be summarized as follows. The proof is by induction on q, with base $q = 0$ given by Proposition 14.4.1. Recall from the discussion above that we divide the inductive step into *Case 1*: $1 + q \le e \le 2 + a + \gamma(q - 1)$ and *Case 2*: $3 + a + \gamma(q - 1) \le e \le 1 + a + \gamma(q)$. We have seen that *Case 1* requires no further discussion. Thus we assume that we are in *Case 2* (which implies that q is even, since $\gamma(q) = \gamma(q-1)+1$ if q is odd) and that the inductive hypothesis holds (in both cases) for $q' < q$.

The inductive hypothesis will be used for every 2-power $q' = 2^{r_i}$ occurring in the binary expansion $\gamma(q) - q = \sum_{i=1}^{m} 2^{r_i}$, where $r_1 > \cdots > r_m \ge 0$, with corresponding values $e' = e - q - \sum_{j=1}^{i-1} 2^{r_j}$. For this, the inequalities $1 + q' \le e' \le 1 + a + \gamma(q')$ can be written as

$$a + 1 + \gamma(q) - q - (2^{r_1} + \cdots + 2^{r_{i-1}}) \le 1 + a + \gamma(2^{r_i}),$$
$$3 + a + \gamma(q - 1) - q - (2^{r_1} + \cdots + 2^{r_{i-1}}) \ge 1 + 2^{r_i}$$

for $1 \le i \le m$. Since $\gamma(2^{r_i}) = 2^{r_i+1} - 1$, the first inequality reduces to $2^{r_i} + \cdots + 2^{r_m} \le 2^{r_i+1} - 1$. Also $\gamma(q-1) = \gamma(q) - v_2(q) - 1$, so that the second inequality reduces to $1 + a - v_2(q) + (2^{r_{i+1}} + \cdots + 2^{r_m}) \ge 0$. This is clear if $i < m$, and it also holds for $i = m$, since $q \le e - 1 \le 2^a - 1$ and hence $v_2(q) \le a - 1$.

Between the lowering steps using the induction hypothesis, we use $\gamma(q) - q$ raising steps given by Proposition 14.2.9(ii). These have the form

$$\eta_{k+j}^2 \cap Sq^u Xq^{[k;v]} = \eta_{k+j+1} \cap Sq^{u+1} Xq^{[k;v]}, \tag{14.4}$$

and connect expressions indexed by integers s such that $q \leq s \leq \gamma(q)$. Writing $s = q + s_i - t$ where $s_i = 2^{r_1} + \cdots + 2^{r_i}$, $1 \leq t \leq 2^{r_i}$ and $1 \leq i \leq m$, the sth step has the form

$$\zeta(s) \eta_{k-1}^{2s_i} \cap Sq^{2^a(2^k-1)-t} Xq^{[k;b+s_i]} = \zeta(s+1) \eta_{k-1}^{2s_i} \cap Sq^{2^a(2^k-1)-t+1} Xq^{[k;b+s_i]}$$

where $\zeta(s)$ is a monomial in A_2^*. From (14.4), we see that each raising step increases the degree of $\zeta(s)$ by 1 and decreases the sum of its exponents by 1. Since $\zeta(q) = \eta_k^q$, $\zeta(s)$ has degree $(2^k - 1)q + (s - q) = (2^k - 2)q + s$ and exponent sum $q - (s - q) = 2q - s$ for all s. After the last raising step, $\zeta(\gamma(q))$ is the product of the elements η_{k+a_j} such that $2^{a_j} \in \text{bin}(q)$. For $q \leq s \leq \gamma(q)$, $\zeta(s)$ is a monomial of the form $\eta_k^{b_1} \eta_{k+1}^{b_2} \cdots \eta_{k+\ell-1}^{b_\ell}$, where $b_1 + b_2 + \cdots + b_\ell = 2q - s$.

Next we explain how the raising and lowering steps are fitted together to obtain the inductive step in Case 2 for a given q. We first apply the induction hypothesis with $q' = 2^{r_1}$ and $e' = e - q$. This gives

$$Xq^{[k;d+q]} = \eta_k^{2^{r_1}} \cap Sq^{2^a(2^k-1)} Xq^{[k;b+q+2^{r_1}]}$$
$$= Sq^{2^a(2^k-1)} Xq^{[k;b+q]} + \eta_{k-1}^{2^{r_1+1}} \cap Sq^{2^a(2^k-1)-2^{r_1}} Xq^{[k;b+q+2^{r_1}]}.$$

Applying $\eta_k^q \cap$, the first term on the right gives the required relation, and so it is sufficient to prove that $\eta_k^q \cap Sq^{2^a(2^k-1)-2^{r_1}} Xq^{[k;b+q+2^{r_1}]} = 0$.

We next apply 2^{r_1} raising steps to rewrite the left hand side of this equation as $\zeta(q + 2^{r_1}) \cap Sq^{2^a(2^k-1)} Xq^{[k;b+q+2^{r_1}]} = 0$ for some monomial $\zeta(q + 2^{r_1})$. Now the inductive hypothesis with $q' = 2^{r_2}$ and $e' = e - q - 2^{r_1}$ gives

$$Xq^{[k;d+q+2^{r_1}]} = \eta_k^{2^{r_2}} \cap Sq^{2^a(2^k-1)} Xq^{[k;b+q+2^{r_1}+2^{r_2}]}.$$

Expanding this gives

$$Xq^{[k;d+q+2^{r_1}]} = Sq^{2^a(2^k-1)} Xq^{[k;b+q+2^{r_1}]} + \eta_{k-1}^{2^{r_2+1}} \cap Sq^{2^a(2^k-1)-2^{r_2}} Xq^{[k;b+q+2^{r_1}+2^{r_2}]}.$$

Since $\zeta(q + 2^{r_1}) \cap Xq^{[k;d+q+2^{r_1}]} = 0$, it is therefore sufficient to prove that

$$\zeta(q + 2^{r_1}) \cap Sq^{2^a(2^k-1)-2^{r_2}} Xq^{[k;b+q+2^{r_1}+2^{r_2}]} = 0.$$

We next apply 2^{r_2} raising steps to rewrite the left hand side of this equation as $\zeta(q + 2^{r_1} + 2^{r_2}) \cap Sq^{2^a(2^k-1)} Xq^{[k;b+q+2^{r_1}+2^{r_2}]}$ for some monomial $\zeta(q + 2^{r_1} + 2^{r_2})$. The cycle of lowering and raising steps is then repeated a further $m - 2$

times, after which we have reduced the problem to proving that

$$\zeta(q+2^{r_1}+\cdots+2^{r_m}) \cap Sq^{2^a(2^k-1)} Xq^{[k;b+q+2^{r_1}+2^{r_2}+\cdots+2^{r_m}]} = 0.$$

Since $\gamma(q) - q = 2^{r_1} + 2^{r_2} + \cdots + 2^{r_m}$, this is $\zeta(\gamma(q)) \cap Sq^{2^a(2^k-1)} Xq^{[k;b+\gamma(q)]} = 0$. Finally we apply the induction hypothesis with $q' = 0$ and $e' = e - \gamma(q)$, checking the inequalities $a + 1 \le 1 + a + \gamma(q)$ and $3 + a + \gamma(q-1) - \gamma(q) \ge 1$. Hence $Sq^{2^a(2^k-1)} Xq^{[k;b+\gamma(q)]} = Xq^{[k;d+\gamma(q)]}$, which is in the kernel of each factor η_{k+i}, $i > 0$, of $\zeta(\gamma(q))$. This completes the proof. $\qquad\square$

14.5 Proof of the excess theorem

In this section we use the expansion of $Xq^{[k;d]}$ given by Proposition 14.4.4 to prove the excess theorem 14.1.1.

Proposition 14.5.1 (i) *For* $\theta_1, \theta_2 \in A_2$, $\mathrm{ex}(\theta_1\theta_2) \ge \mathrm{ex}(\theta_1) - \deg\theta_2$.
  (ii) *For* $\theta \in A_2$, $\mathrm{ex}(Sq^t\theta) \ge 2t - \deg Sq^t\theta$.

Proof Let f be a polynomial of degree $< \mathrm{ex}(\theta_1) - \deg\theta_2$. Then $\theta_2(f)$ has degree $< \mathrm{ex}(\theta_1)$, and hence $\theta_1\theta_2(f) = 0$. This proves (i), and (ii) follows by putting $\theta_1 = Sq^t$, $\theta_2 = \theta$ and noting that $\mathrm{ex}(Sq^t) = t$ and $\deg(Sq^t\theta) = t + \deg\theta$. $\qquad\square$

We can also see (ii) by noting that by Proposition 5.8.4 the bound is exact if $Sq^t\theta$ is an admissible monomial, and that if the Adem relations are used to convert a general monomial $Sq^A = Sq^t\theta$ to a sum of admissible monomials, these must have first exponent $\ge t$.

By the **last term** in the expansion of the formula

$$Xq^{[k;d]} = \eta_k^q \cap Sq^{2^a(2^k-1)} Xq^{[k;b+q]} \qquad (14.5)$$

we mean the term $(\eta_k^q \cap Sq^{2^a(2^k-1)}) Xq^{[k;b+q]}$, which is $Sq^t Xq^{[k;b+q]}$ by Proposition 14.2.6(i), where $t = (2^a - q)(2^k - 1)$. We next use Proposition 14.4.4 to obtain lower bounds for all terms in the expansion. For this purpose we use the minimum value of q for which Proposition 14.4.4 gives an expansion for $Xq^{[k;d]}$, namely $q = 2^a - \beta(d)$.

The following example shows how iteration of this procedure leads to the conclusion that $Xq^{[k;d]} = Sq^{(2^k-1)A(d)} + \sum_A Sq^A$, where $A(d)$ is the $\le_l$-minimal admissible sequence of modulus d, and $Sq^{(2^k-1)A(d)}$ is the maximal monomial in the sum in both the left and right orders. Given $d \ge 0$, we define decreasing sequences $(a_i), (b_i), (d_i)$ and (q_i) for $i \ge 1$ recursively as follows, starting with

$d_1 = d$: $d_i = 2^{a_i} + b_i$ where $0 \le b_i \le 2^{a_i} - 1$; $d_{i+1} = d_i - \beta(d_i) = (d_i - \mu(d_i))/2$; and $q_i = 2^{a_i} - \beta(d_i) = d_{i+1} - b_i$.

Example 14.5.2 We tabulate these sequences for $d = 47$ below, along with the sequences $(\mu(d_i))$ and $(\beta(d_i))$.

i	1	2	3	4	5	6
d_i	47	22	10	4	1	0
$\mu(d_i)$	3	2	2	2	1	0
$\beta(d_i)$	25	12	6	3	1	0
2^{a_i}	32	16	8	4	1	0
b_i	15	6	2	0	0	0
q_i	7	4	2	1	0	0

The sequence $(\beta(d_i))$ is $A(d)$, as was shown in Proposition 5.7.6. The sequences (d_i), (2^{a_i}) and (q_i) appear in the successive applications of Proposition 14.4.4, i.e. $Xq^{[k;d_i]} = \eta_k^{q_i} \cap Sq^{2^{a_i}(2^k-1)} Xq^{[k;d_{i+1}]}$. At each stage of the expansion, all terms have the form $Sq^u Xq^{(v_1,\dots,v_k)}$, and all terms other than the last term satisfy $u > (2^k - 1)\beta(d_i)$, since the stripping operation $\eta_k^{q_i} \cap$ is not then concentrated on the first factor $Sq^{2^{a_i}(2^k-1)}$.

Recall from Section 14.1 that the sequence $A(d)$ has length j for $2^j - 1 \le d \le 2^{j+1} - 2$. Since Proposition 14.2.4 deals with the case $d = 2^j - 1$, we may assume that $2^j \le d \le 2^{j+1} - 2$, so that $a = j$ and $0 \le b \le 2^a - 2$. In the notation above, the sequences (d_i) and (2^{a_i}) have the same length as $(\beta(d_i)) = A(d)$. Since $d_{i+1} < d_i/2 < 2^{a_i}$ when $d_i > 0$, (2^{a_i}) is a strictly decreasing sequence of 2-powers of length a with $a_1 = a$, so that exactly one 2-power is missing from the sequence, as in Example 14.5.2 above. The calculation of $Xq^{[k;d]}$ as a sum of monomials Sq^A using Proposition 14.4.4 requires a iterations.

Proposition 14.5.3 *Let $d = 2^a + b$ where $0 \le b \le 2^a - 1$, and let $q = 2^a - \beta(d)$. Then, for $k \ge 1$ and $\eta = \eta_k^q$, the last term $Sq^t Xq^{[k;b+q]}$ in the expansion*

$$\eta \cap Sq^{2^a(2^k-1)} Xq^{[k;b+q]} = \sum_\eta (\eta' \cap Sq^{2^a(2^k-1)})(\eta'' \cap Xq^{[k;b+q]})$$

has excess $\ge (2^k - 1)\mu(d)$, and all other terms have excess $> (2^k - 1)\mu(d)$.

Proof Since $t = (2^a - q)(2^k - 1)$ and $\deg Xq^{[k;b+q]} = (2^k - 1)(b + q)$, Proposition 14.5.1 gives $\mathrm{ex}(Sq^t Xq^{[k;b+q]}) \ge (2^k - 1)(2^a - b - 2q)$. For $q =$

$2^a - \beta(d)$, $2q = 2^{a+1} - (d + \mu(d)) = e - \mu(d)$, by definition of β. Thus $2^a - b - 2q = e - 2q = \mu(d)$, and so $\text{ex}(Sq^t Xq^{[k;b+q]}) \geq (2^k - 1)\mu(d)$.

For any other nonzero term in the sum, $\eta' \cap Sq^{2^a(2^k-1)} = Sq^{t'}$, where $t' > t$. Since all the terms have the same degree $(2^k - 1)d$, by Proposition 14.5.1 the excess of such a term is $> 2t - (2^k - 1)d = (2^k - 1)\mu(d)$. $\square$

Since $Sq^{A(d)}$ is the minimum admissible monomial in A_2^d in the 2-dominance order, it is both $\leq_l$- and $\leq_r$-minimal. Recall that the sequence $A(d)$ is defined recursively by $A(d) = (\beta(d), A(d - \beta(d)))$. Thus we have $\text{ex}(Sq^{(2^k-1)A(d)}) = (2^k - 1)\mu(d)$ for $k \geq 1$, and so Theorem 14.1.1 follows at once from the next result. When $d = 2^j - 1$, the next two results are immediate from Proposition 14.2.4.

Proposition 14.5.4 *For all $d \geq 0$ and $k \geq 1$, $Sq^{(2^k-1)A(d)}$ appears in the expansion of $Xq^{[k;d]}$ in the admissible basis, and is minimal for the left order $\leq_l$.*

Proof We use induction on d. As above, for $q = 2^a - \beta(d)$, the last term $Sq^t \theta$ in the expansion of $Xq^{[k;d]}$ is $Sq^t \theta$, where $t = (2^a - q)(2^k - 1) = (2^k - 1)\beta(d)$ and $b + q = b + 2^a - \beta(d) = d - \beta(d)$. Hence this term is $Sq^{(2^k-1)\beta(d)} Xq^{[k;d-\beta(d)]}$. By the inductive hypothesis, we may assume that $Sq^{(2^k-1)A(d-\beta(d))}$ appears in the expansion of $Xq^{[k;d-\beta(d)]}$ in the admissible basis, and that if Sq^B is any other admissible monomial in this expansion, then $Sq^B >_l Sq^{(2^k-1)A(d-\beta(d))}$. Since $A(d) = (\beta(d), A(d - \beta(d)))$, the term $Sq^t Sq^{(2^k-1)A(d-\beta(d))}$ in the sum for $Xq^{[k;d]}$ is $Sq^{(2^k-1)A(d)}$. Since Adem relations cannot lower monomials in the left order, any term Sq^C in the expansion of $Sq^t Sq^B$ in the admissible basis satisfies $Sq^C >_l Sq^{(2^k-1)A(d)}$. Thus $Sq^{(2^k-1)A(d)}$ cannot be cancelled in the admissible basis expansion of $Xq^{[k;d]}$, and it is the minimal term in the left order, completing the induction. $\square$

14.6 $Xq^{[k;d]}$ in the Milnor basis

Recall (Definition 5.7.5) that $Sq(R(d))$ is the $\leq_r$-minimal Milnor basis element in A_2^d, corresponding to the $\leq_r$-minimal admissible monomial $Sq^{A(d)}$. Since the admissible basis is triangularly related to the Milnor basis for the $\leq_r$ order, the second statement below follows from the first.

Proposition 14.6.1 *For all $d \geq 0$ and $k \geq 1$, $Sq^{(2^k-1)A(d)}$ is $\leq_r$-minimal in the admissible basis expansion of $Xq^{[k;d]}$, and $Sq((2^k - 1)R(d))$ is $\leq_r$-minimal in the Milnor basis expansion of $Xq^{[k;d]}$.*

Proof Since Proposition 14.2.4 deals with the case $d = 2^j - 1, j \geq 1$, we may assume that $2^j \leq d \leq 2^{j+1} - 2$. We define the sequences (a_i), (b_i), (d_i) and (q_i) as above. Note that $d_i - d_{i+1} = \beta(d_i) = 2^{a_i} - q_i$ is the ith entry of the sequence $A(d)$. We have seen that (a_i) and (d_i) have length j, (b_i) and (q_i) have length $\leq j$, and that by using the expansion

$$Xq^{[k;d_i]} = \eta_k^{q_i} \cap Sq^{2^{a_i}(2^k-1)} Xq^{[k;d_{i+1}]} \tag{14.6}$$

for $1 \leq i \leq j$, we can express $Xq^{[k;d]}$ as a sum of monomials Sq^C, $C = (c_1,\dots,c_j)$. These monomials need not be admissible, but, as Adem relations raise monomials in the right order, it suffices to prove that $Sq^{(2^k-1)A(d)}$ is the $\leq_r$ minimal monomial in the sum $\sum_C Sq^C$ using the $\leq_r$ order on the sequences C.

In order to compare the monomials Sq^C for the $\leq_r$ order, we consider (14.6) for $i = j, j - 1, \dots, 1$. For $i = j$ we have $Xq^{[k;d_j]} = \eta_k^{q_j} \cap Sq^{2^{a_j}(2^k-1)} Xq^{[k;d_{j+1}]}$. Since $d_{j+1} = 0$, this reduces to $Xq^{[k;d_j]} = Sq^{(2^{a_j}-q_j)(2^k-1)}$ using Proposition 14.2.6(i). As noted above, this is $Sq^{(2^k-1)\beta(d_j)}$. It is no surprise that we obtain a single squaring operation in this case, because $d_j = 1$ or 2 and so we recover the results $Xq^{[k;1]} = Sq^{2^k-1}$ and $Xq^{[k;2]} = Sq^{2^{k+1}-2}$.

Setting $i = j - 1$ in (14.6), $Xq^{[k;d_{j-1}]} = \eta_k^{q_{j-1}} \cap Sq^{2^{a_{j-1}}(2^k-1)} Sq^{(2^k-1)\beta(d_j)}$. Since all terms except the last term

$$(\eta_k^{q_{j-1}} \cap Sq^{2^{a_{j-1}}(2^k-1)}) Sq^{(2^k-1)\beta(d_j)} = Sq^{(2^{a_{j-1}}-q_{j-1})(2^k-1)} Sq^{(2^k-1)\beta(d_j)}$$
$$= Sq^{(2^k-1)\beta(d_{j-1})} Sq^{(2^k-1)\beta(d_j)}$$

lower the exponent of the second factor, this term gives the $\leq_r$-minimal exponent sequence, i.e. $Xq^{[k;d_{j-1}]} = Sq^{(2^k-1)\beta(d_{j-1})} Sq^{(2^k-1)\beta(d_j)} + \sum Sq^{c_{j-1}} Sq^{c_j}$, where $(c_{j-1}, c_j) >_r (2^k - 1)(\beta(d_{j-1}), \beta(d_j))$ for all terms in the sum.

Setting $i = j - 2$ in (14.6),

$$Xq^{[k;d_{j-2}]} = \eta_k^{q_{j-2}} \cap Sq^{2^{a_{j-2}}(2^k-1)} \left(Sq^{(2^k-1)\beta(d_{j-1})} Sq^{(2^k-1)\beta(d_j)} + \sum Sq^{c_{j-1}} Sq^{c_j} \right).$$

Here all terms except the last term lower the exponent of the second or third factor, and the last term is

$$Sq^{(2^k-1)\beta(d_{j-2})} \left(Sq^{(2^k-1)\beta(d_{j-1})} Sq^{(2^k-1)\beta(d_j)} + \sum Sq^{c_{j-1}} Sq^{c_j} \right).$$

Thus we can write $Xq^{[k;d_{j-2}]} = Sq^{(2^k-1)(\beta(d_{j-2}),\beta(d_{j-1}),\beta(d_j))} + \sum Sq^{(c_{j-2},c_{j-1},c_j)}$, with all $(c_{j-2}, c_{j-1}, c_j) >_r (2^k - 1)(\beta(d_{j-2}), \beta(d_{j-1}), \beta(d_j))$. By repeating the argument down to $i = 1$, we obtain $Xq^{[k;d]} = Sq^{(2^k-1)A(d)} + \sum Sq^C$, where $C >_r (2^k - 1)A(d)$ for all terms in the sum. $\square$

14.7 Remarks

The study of deeper properties of the conjugation χ of A_2 goes back to [49]. The work of David Bausum [16] is also a significant step. The main results in this chapter are due to Judith Silverman and William Singer. In [185], Theorems 14.1.2 and 14.1.3 were stated as conjectures, which were shown to be equivalent and which were proved in the case $k = 2$, the case $k = 1$ being already known from [230].

The excess theorem was first formulated as a conjecture in [180], where Proposition 14.2.4 was proved. Further cases of the theorem were proved in [183], where the inequality $\text{ex}(Xq^{[k;d]}) \leq (2^k - 1)\mu(d)$ was proved for all k and d. The proof of the excess theorem in the general case was completed by Silverman and Dagmar Meyer [184, 133] using the Milnor basis for A_2.

The proof offered in this chapter uses the admissible basis and relies on the recursive stripping formula of Section 14.4. It would be desirable to have a more direct proof of Proposition 14.1.4 analogous to that of Singer's Theorem 6.3.12 for the left order, but we do not know of one.

We take this opportunity to correct Lemma 4.2 of [233]: the statement about excess should be replaced by the corresponding statement about the right order, as in Theorem 3.5.1. Error terms can have the same excess as the corresponding admissible, for example $Sq^6 Sq^2 Sq^1$ in Example 3.5.4.

15

Invariants and the hit problem

15.0 Introduction

The **Dickson algebra** $D(n)$ is the subalgebra of $GL(n)$-invariant polynomials in $P(n)$. It is a polynomial algebra on n generators $d_{n,0}, d_{n,1}, \ldots, d_{n,n-1}$, where the ith **Dickson invariant** $d_{n,i}$ has degree $2^n - 2^i$. We shall usually write $d_{n,i}$ as d_i. For example, elements of $D(2)$ can be written uniquely as polynomials in $d_0 = x^2y + xy^2$ and $d_1 = x^2 + xy + y^2$. We call a monomial in the Dickson invariants a **Dickson monomial**, and a general element of $D(n)$ a **Dickson polynomial**.

If $f \in P(n)$ is invariant under the action of a subgroup $G \subseteq GL(n)$, then $\theta(f)$ is also G-invariant for any Steenrod operation $\theta \in A_2$. Hence $D(n)$ is an A_2-module, and we can study the action of A_2 on $D(n)$ in the context of the hit problem. In fact, there are two problems: we can ask whether an invariant polynomial $f \in D(n)$ is hit as an element of $P(n)$, i.e. $f = \sum_{i>0} Sq^i(g_i)$ where $g_i \in P(n)$, and we can ask whether f is hit as an element of $D(n)$, i.e. $f = \sum_{i>0} Sq^i(g_i)$ where $g_i \in D(n)$. We call the first question the 'relative' hit problem for $D(n)$, and the second the 'absolute' hit problem.

In this chapter we discuss the solution of the relative hit problem by Nguyen H. V. Hung and Tran Ngoc Nam, who proved that for $n > 2$ all Dickson monomials $\neq 1$ are hit in $P(n)$. We sketch their argument for the case $n = 3$ in Section 15.6, and give details for the general case in Sections 15.7 and 15.8. The case $n = 2$ is an exception (as is $n = 1$, since $D(1) = P(1)$), since d_1 contains the spike xy. More generally, $d_1^{2^k-1}$ is not hit in $P(2)$ for $k \geq 1$, as it contains the spike $x^{2^k-1}y^{2^k-1}$.

In Section 15.1 we introduce the Dickson invariant $d_{n,i} \in P(n)$ as the quotient $\Delta_{n,i}/\Delta_n$ of Vandermonde determinants whose entries are 2-powers of the variables. We prove Dickson's theorem, and show that $d_{n,i}$ is the sum of all monomials of degree $2^n - 2^i$ in $P(n)$ whose exponents are 2-powers.

297

In Section 15.2 we solve the hit problem for D(2), which is a case where the absolute and relative hit problems have the same solution. In Section 15.3 we give formulae for the action of the Steenrod squares Sq^k and the Milnor primitives $Q_k = Sq_k^1$ on the Dickson invariants. Sections 15.4 and 15.5 are concerned with the invariants $P(2)^{H_2}$ and $P(3)^{H_3}$, where H_2 and H_3 are maximal subgroups of odd order in GL(2) and GL(3) respectively. The absolute and relative hit problems again have the same solution in these special cases. A generalization of the case of $P(2)^G$ to all n is described in Chapter 27.

15.1 Dickson invariants

We begin by considering Vandermonde determinants in $P(n)$ of the form

$$\Delta(2^{j_1},\ldots,2^{j_n}) = \begin{vmatrix} x_1^{2^{j_1}} & x_2^{2^{j_1}} & \cdots & x_n^{2^{j_1}} \\ x_1^{2^{j_2}} & x_2^{2^{j_2}} & \cdots & x_n^{2^{j_2}} \\ \vdots & \vdots & \ddots & \vdots \\ x_1^{2^{j_n}} & x_2^{2^{j_n}} & \cdots & x_n^{2^{j_n}} \end{vmatrix}, \quad 0 \leq j_1 < j_2 < \cdots < j_n. \quad (15.1)$$

Since the exponents are 2-powers, we see by using column operations that this determinant is a GL(n)-invariant. It is the monomial symmetric function in $P(n)$ with exponents $2^{j_1},\ldots,2^{j_n}$, and in the notation of Section 3.3, it is the Cartan symmetric function $c(R)$ where the ith entry of the sequence R is 1 if $i = j_1 + 1,\ldots,j_n + 1$ and 0 otherwise. Since $\Delta(2^{j_1},\ldots,2^{j_n})$ is divisible by x_1 and is GL(n)-invariant, it is divisible by all nonzero elements of $P^1(n)$. Since

$$\Delta_n = \Delta(1,2,\ldots,2^{n-1}) = \begin{vmatrix} x_1 & x_2 & \cdots & x_n \\ x_1^2 & x_2^2 & \cdots & x_n^2 \\ \vdots & \vdots & \ddots & \vdots \\ x_1^{2^{n-1}} & x_2^{2^{n-1}} & \cdots & x_n^{2^{n-1}} \end{vmatrix}$$

has degree $2^n - 1$, it is the product of all nonzero elements of $P^1(n)$. Hence the quotient $\Delta(2^{j_1},\ldots,2^{j_n})/\Delta(1,2,\ldots,2^{n-1})$ is a GL(n)-invariant polynomial.

Definition 15.1.1 For $0 \leq i \leq n$, the *i*th **Dickson invariant** $d_{n,i} = \Delta_{n,i}/\Delta_n$, where $\Delta_{n,i} = \Delta(1,2,\ldots,2^{i-1},2^{i+1},\ldots,2^n)$. In particular, $\Delta_{n,n} = \Delta_n$ and $\Delta_{n,0} = \Delta_n^2$, and so $d_{n,n} = 1$ and $d_{n,0} = \Delta_n$. The **Dickson algebra** D(n) is the subalgebra of $P(n)$ generated by $d_{n,0},\ldots,d_{n,n-1}$.

Thus $d_{n,i}$ is a GL(n)-invariant polynomial of degree $2^n - 2^i$. Except in the case of $d_{n,0}$, Definition 15.1.1 does not give an explicit formula for $d_{n,i}$ as a sum of monomials. In Proposition 15.1.7 we shall show that $d_{n,i}$ is the sum of all monomials in P(n) of degree $2^n - 2^i$ whose exponents are 2-powers.

Example 15.1.2 When the value of n is clear from the context, we write $d_{n,i}$ as d_i. Thus for $n = 2$, $d_0 = x^2y + xy^2$, $d_1 = x^2 + xy + y^2$, and for $n = 3$,

$$d_0 = x^4y^2z + x^4yz^2 + x^2y^4z + xy^4z^2 + x^2yz^4 + xy^2z^4,$$
$$d_1 = x^4y^2 + x^2y^4 + x^4z^2 + x^2z^4 + y^4z^2 + y^2z^4 + x^4yz + xy^4z + xyz^4 + x^2y^2z^2,$$
$$d_2 = x^4 + y^4 + z^4 + x^2yz + xy^2z + xyz^2 + x^2y^2 + x^2z^2 + y^2z^2.$$

The next result shows that the polynomials $d_{n,i}$ are the nonzero elementary symmetric functions in the $2^n - 1$ nonzero elements of $P^1(n)$. When $n = 2$, for example, $x + y + (x + y) = 0$, $xy + x(x + y) + y(x + y) = d_1$ and $xy(x + y) = d_0$.

Proposition 15.1.3 *Let* $f_n(t) = \prod_{x \in P^1(n)}(t + x)$, *a polynomial of degree 2^n in t with coefficients in* P(n). *Then* $f_n(t) = \sum_{i=0}^{n} d_{n,i}t^{2^i}$, *and* $x^{2^n} = \sum_{i=0}^{n-1} d_{n,i}x^{2^i}$ *for all* $x \in P^1(n)$.

Proof The determinant

$$\Delta = \begin{vmatrix} x_1 & x_2 & \cdots & x_n & t \\ x_1^2 & x_2^2 & \cdots & x_n^2 & t^2 \\ \vdots & \vdots & \ddots & \vdots & \vdots \\ x_1^{2^n} & x_2^{2^n} & \cdots & x_n^{2^n} & t^{2^n} \end{vmatrix}$$

is the product of all $2^{n+1} - 1$ nonzero linear polynomials in the variables $x_1, \ldots, x_n$ and t. The 2^n factors which involve t have product $f_n(t)$, while the remaining $2^n - 1$ factors have product Δ_n. Hence $\Delta = f_n(t)\Delta_n$. Expanding Δ by the last column, $\Delta = \sum_{i=0}^{n} \Delta_{n,i}t^{2^i}$. Hence $f_n(t) = \sum_{i=0}^{n}(\Delta_{n,i}/\Delta_n)t^{2^i} = \sum_{i=0}^{n} d_{n,i}t^{2^i}$. The second statement follows, since $f_n(x) = 0$ for all $x \in P^1(n)$ and $d_{n,n} = 1$. $\quad\square$

The next result is the first part of Dickson's theorem.

Theorem 15.1.4 *The Dickson invariants* $d_{n,0}, \ldots, d_{n,n-1}$ *are algebraically independent over* $\mathbb{F}_2$, *so that* D(n) *is the polynomial algebra* $\mathbb{F}_2[d_{n,0}, \ldots, d_{n,n-1}]$.

Proof Let $d_0, \ldots, d_{n-1}$ be algebraically independent indeterminates over $\mathbb{F}_2$, and let $g_n(t) = t^{2^n} + \sum_{i=0}^{n-1} d_i t^{2^i}$, a monic polynomial of degree 2^n in t with coefficients in $\mathbb{F}_2[d_0, \ldots, d_{n-1}]$. Let F be a splitting field for $g_n(t)$. Since the derivative $g'_n(t) = d_0 \neq 0$, $g_n(t)$ has 2^n distinct roots in F, and since $g_n(a+b) = g_n(a) + g_n(b)$, the roots form a vector space of dimension n over $\mathbb{F}_2$.

Let $a_1, \ldots, a_n$ be a $\mathbb{F}_2$-basis for the roots of $g_n(t)$ in F. Define an algebra map $\theta : P(n) \to \mathbb{F}_2[a_1, \ldots, a_n]$ by $\theta(x_i) = a_i$ for $1 \leq i \leq n$. Let $f_n(t) = \prod_{x \in P^1(n)}(t+x)$, as in Proposition 15.1.3. Then θ maps the roots of $f_n(t)$ to the roots of $g_n(t)$, and so it maps the coefficient $d_{n,i}$ of t^{2^i} in $f_n(t)$ to the coefficient d_i of t^{2^i} in $g_n(t)$ for $0 \leq i \leq n-1$. Since $d_0, \ldots, d_{n-1}$ are algebraically independent over $\mathbb{F}_2$, the Dickson invariants $d_0, \ldots, d_{n-1}$ are also algebraically independent over $\mathbb{F}_2$. $\square$

Since $D(n)$ acts on $P(n)$ by multiplication, $P(n)$ is a $D(n)$-module. The next result describes its structure. Recall that $|GL(n)| = \prod_{i=0}^{n-1}(2^n - 2^i)$.

Proposition 15.1.5 *The polynomial algebra* $P(n)$ *is a free module of rank* $|GL(n)|$ *over the Dickson algebra* $D(n)$, *with a basis given by the monomials* $x_1^{a_0} \cdots x_n^{a_{n-1}}$ *such that* $0 \leq a_i < 2^n - 2^i$ *for* $0 \leq i \leq n-1$.

Proof For $0 \leq i \leq n-1$, let $g_{n,i}(t) = f_n(t)/f_i(t)$, where $f_n(t) = \prod_{x \in P^1(n)}(t+x)$ as in Proposition 15.1.3. Thus $g_{n,i}(t)$ is the product of the linear polynomials $t + x$ where $x \in P^1(n)$ has a term x_j with $j > i$. Hence $g_{n,i}(t)$ is monic of degree $2^n - 2^i$, with x_{i+1} as a root, and by Proposition 15.1.3, its numerator $f_n(t)$ has coefficients $d_{n,0}, \ldots, d_{n,n-1}$ and its denominator $f_i(t)$ has coefficients $d_{i,0}, \ldots, d_{i,i-1}$.

The coefficients of $g_{n,i}(t)$ in $P(n)$ can be expressed as polynomials in $x_1, \ldots, x_i$ and the Dickson invariants $d_{n,0}, \ldots, d_{n,n-1}$. Thus they lie in the subring of $P(n)$ generated by $P(i)$ and $D(n)$. Using the division algorithm in the subring of $P(n)[t]$ generated by t and both sets of Dickson invariants, we can write $f_n(t) = q(t)f_i(t) + r(t)$.

Here $q(t)$ and $r(t)$ are polynomials in t with coefficients in the two sets of Dickson invariants, and the remainder $r(t)$ has degree $< \deg f_i(t) = 2^i - 1$ in t. Since the division is exact in $P(n)[t]$, $q(t) = g_{n,i}(t)$ and $r(t) = 0$. In particular, $g_{n,i}(t)$ can be written as a polynomial in t whose coefficients are polynomials in the $n + i$ variables $x_1, \ldots, x_i$ and $d_{n,0}, \ldots, d_{n,n-1}$. The coefficients of powers of t in $r(t)$ give identities relating the two sets of Dickson invariants.

We wish to express a given polynomial $f \in P(n)$ as a polynomial in $x_1, \ldots, x_n$, with coefficients in $D(n)$, such that the degree of x_i in each term is $< 2^n - 2^i$. We use the polynomials $g_{n,i}(t)$ to do this. First, since $g_{n,n-1}(x_n) = 0$, we can express $x_n^{2^n - 2^{n-1}}$ as a polynomial of degree $< 2^n - 2^{n-1}$ in x_n, with coefficients which are polynomials in $x_1, \ldots, x_{n-1}$ and the Dickson invariants. By using this equation recursively, we can write f as a polynomial of degree $< 2^n - 2^{n-1}$ in x_n with coefficients in this form.

Next, since $g_{n,n-2}(x_{n-1}) = 0$, we can express $x_{n-1}^{2^n-2^{n-2}}$ as a polynomial of degree $< 2^n - 2^{n-2}$ in x_{n-1} with coefficients which are polynomials in $x_1, \ldots, x_{n-2}$ and the Dickson invariants. Again, using this equation recursively, we can write f as a polynomial of degree $< 2^n - 2^{n-1}$ in x_n and degree $< 2^n - 2^{n-2}$ in x_{n-1} with coefficients in this form. Continuing in this way, by using $g_{n,i}(t)$ for $n - 3 \geq i \geq 1$ we can reduce the degree of f in x_i below $2^n - 2^i$ for $1 \leq i \leq n$, with coefficients in $D(n)$.

Hence $f \in P(n)$ can be written as $f = \sum_A f_A x^A$, where x^A denotes a monomial in $P(n)$ in which x_i has exponent $< 2^n - 2^i$ for $1 \leq i \leq n$. We use Galois theory to show that this expression is unique. Let $R(n) = \mathbb{F}_2(x_1, \ldots, x_n)$ be the field of rational functions, and let $R(n)^{\mathrm{GL}(n)}$ be the subfield of $\mathrm{GL}(n)$-invariants in $R(n)$. Since $\mathrm{GL}(n)$ acts as a group of automorphisms of $R(n)$, $R(n)$ is a Galois extension of $R(n)^{\mathrm{GL}(n)}$ of degree $|\mathrm{GL}(n)|$.

If $f_1/f_2 \in R(n)$ then, by multiplying $f_2 \in P(n)$ by its conjugates under $\mathrm{GL}(n)$, $f_1/f_2 = f/g$, where $f \in P(n)$ and g is $\mathrm{GL}(n)$-invariant in $P(n)$. Thus $f/g = \sum_A (f_A/g) x^A$, where the coefficients f_A/g are in $R(n)^{\mathrm{GL}(n)}$. Since $R(n)$ is a vector space of dimension $|\mathrm{GL}(n)|$ over $R(n)^{\mathrm{GL}(n)}$, the monomials x^A form a $R(n)^{\mathrm{GL}(n)}$-basis for $R(n)$. In particular, $f = \sum_A f_A x^A$ is the unique expansion of f as a linear combination of the basis monomials x^A with coefficients in $D(n)$. $\quad\square$

The proof of Dickson's theorem can now be completed as follows.

Theorem 15.1.6 $D(n) = P(n)^{\mathrm{GL}(n)}$, *the subalgebra of $\mathrm{GL}(n)$-invariants in $P(n)$.*

Proof Given $f \in P(n)$, the expansion $f = \sum_A f_A x^A$ constructed in the proof of Proposition 15.1.5 is the unique expansion of f as a linear combination of the basis monomials x^A for $R(n)$ as a vector space over the invariant subfield $R(n)^{\mathrm{GL}(n)}$. If f itself is invariant under $\mathrm{GL}(n)$, then it has a trivial expansion of this form, given by setting f as the coefficient of the constant monomial 1 and setting all other coefficients as 0. By uniqueness, this expansion must be the same as $f = \sum_A f_A x^A$, so by comparing coefficients of 1 we obtain $f \in D(n)$. $\quad\square$

Proposition 15.1.7 *Let $d \geq 0$ be an integer such that $\alpha(d) \leq n$, and let $c_n(d)$ be the sum of all monomials in $P^d(n)$ whose exponents are 2-powers. Then $c_n(d)$ is in $D(n)$ if and only if $d = 0 \bmod 2^{n-\alpha(d)}$. In particular, $d_{n,i} = c_n(2^n - 2^i)$ for $0 \leq i \leq n - 1$.*

Proof Since $d_{n,i}$ is the only Dickson monomial in degree $2^n - 2^i$, the second statement follows from the first. The polynomial $c_n(d)$ is symmetric for all d, so $c_n(d) \in D(n)$ if and only if it is fixed by the standard transvection U, which maps x_1 to $x_1 + x_2$ and fixes $x_2, \ldots, x_n$. Let $f = x_1^{a_1} x_2^{a_2} \cdots x_n^{a_n}$ be a monomial in $c_n(d)$, so that a_i is a 2-power or 0 for $1 \le i \le n$. Then

(i) if $a_1 = 0$, then $f \cdot U = f$,

(ii) if $a_1 \ne a_2$ and both are > 0, then $(f + g) \cdot U = f + g$, where $g = x_1^{a_2} x_2^{a_1} x_3^{a_3} \cdots x_n^{a_n}$, and

(iii) if $a_1 = 2a > 1$ and $a_2 = 0$, then $(f + h) \cdot U = f + h$, where $h = x_1^a x_2^a x_3^{a_3} \cdots x_n^{a_n}$.

In the case $\alpha(d) = n$, all the variables x_i appear in every monomial f in $c_n(d)$ and have different exponents, and so by (ii) we can match up pairs of monomials f and g to obtain $c_n(d) \cdot U = c_n(d)$. Thus let $\mathrm{bin}(d) = \{2^{d_1}, \ldots, 2^{d_r}\}$, where the terms are in increasing order and $\alpha(d) = r \le n - 1$. Then by (i) U fixes monomials in $c_n(d)$ with $a_1 = 0$, and by (ii) and (iii) the only monomials with $a_1 > 0$ which are not paired with another monomial to give a sum fixed by U are those in which x_1 has exponent 1 and x_2 does not appear.

Since $1 + 1 + 2 + 4 + \cdots + 2^{d_1 - 1}$ is the partition of 2^{d_1} of minimal length containing 1, the minimal length of a partition of d containing 1 is $d_1 + r$. If x_2 does not appear in f, then $d_1 + r \le n - 1$. If $d = 0 \bmod 2^{n-r}$, then $d_1 \ge n - r$ and no such monomials f arise, so $c_n(d) \in D(n)$. On the other hand, if $d \ne 0 \bmod 2^{n-r}$, then, by considering $f = x_1 x_3 x_4^2 \cdots x_{d_1}^{2^{d_1 - 1}} x_{d_1 + 1}^{2^{d_2}} \cdots x_{d_1 + r - 1}^{2^{d_r}}$, it follows that $c_n(d)$ is not fixed by U, and so $c_n(d) \notin D(n)$. $\square$

The following result will be used in Section 21.6.

Proposition 15.1.8 *Let* $p_d(n) = \sum_{x \ne 0} x^d$, *where the sum is over all nonzero elements* $x \in P^1(n)$. *Then* $p_d(n) = 0$ *if* $0 < d < 2^n - 1$, *and* $p_d(n) = \Delta_n$ *if* $d = 2^n - 1$.

Proof Putting $x_1 = 0$ reduces $p_d(n)$ to a sum of terms, each appearing twice, so that the sum is 0 mod 2. Hence $p_d(n)$ is divisible by x_1. Since $p_d(n)$ is $GL(n)$-invariant, it is divisible by all $x \ne 0 \in P^1(n)$. Hence $p_d(n) = 0$ if $0 < d < 2^n - 1$, and $p_d(n)$ is a scalar multiple of Δ_n if $d = 2^n - 1$. By writing $d = \sum_{i=0}^{n-1} 2^i$, we see that the monomial $\prod_{i=0}^{n-1} x_{i+1}^{2^i}$ appears only once in $p_d(n)$, in the expansion of $(x_1 + \cdots + x_n)^d = \prod_{i=0}^{n-1} (x_1^{2^i} + \cdots + x_n^{2^i})$. Hence the scalar multiple is 1. $\square$

15.2 The hit problem for $D(2)$

The Dickson algebra $D(2) = \mathbb{F}_2[d_0, d_1]$, where $d_1 = x^2 + xy + y^2$ and $d_0 = x^2 y + xy^2$. It is easy to verify that $Sq^1(d_1) = d_0$, $Sq^1(d_0) = 0$ and $Sq^2(d_0) = d_1 d_0$. Since d_1 contains the spike xy, it is not hit in $P(2)$. More generally, since $d_1^{2^k - 1} =$

$\prod_{i=1}^{k}(x^{2^i}+x^{2^{i-1}}y^{2^{i-1}}+y^{2^i})$, we see by induction on k that $d_1^{2^k-1}$ contains the spike $x^{2^k-1}y^{2^k-1}$, and so it is not hit in P(2) for all $k \geq 1$. We shall show that all other Dickson monomials f are hit, not only in P(2) but in D(2) itself. In other words, there is a hit equation of the form $f = \sum_{i\geq 1} Sq^i(g_i)$ where $g_i \in D(2)$.

We use the following preliminary result.

Proposition 15.2.1 *If* $f \in D(2)$ *is hit in* D(2), *then* $d_1 f^2$ *is hit in* D(2).

Proof Let $f = \sum_{i\geq 1} Sq^i(g_i)$ where $g_i \in D(2)$. Then using Proposition 1.3.1

$$\sum_{i\geq 1} Sq^{2i}(d_1 g_i^2) = \sum_{i\geq 1} d_1(Sq^i(g_i))^2 + \sum_{i\geq 1} d_1^2(Sq^{i-1}(g_i))^2 = d_1 f^2 + g^2$$

where $g = \sum_{i\geq 1} d_1 Sq^{i-1}(g_i)$. Since every square is hit in D(2), $d_1 f^2$ is hit. □

Proposition 15.2.2 *The Dickson monomials* $d_1^{2^k-1}$, $k \geq 0$, *form a basis for* $D(2)/A_2^+ D(2)$. *All other Dickson monomials are hit in* D(2).

Proof We show that $d_0^a d_1^b$ is hit in D(2) unless $a = 0$ and $b = 2^k - 1$ for some $k \geq 0$.

Case 1: b is even. If a and b are both even, then $d_0^a d_1^b$ is a square and is therefore hit in D(2). If $a > 0$, then as $Sq^1(d_1) = d_0$ and $Sq^1(d_0) = 0$, the Cartan formula gives $Sq^1(d_0^{a-1} d_1^{b+1}) = (b+1)d_0^a d_1^b = d_0^a d_1^b$.

Case 2: a and b are odd. Let $a = 2a' + 1$ and $b = 2b' + 1$, and let $f = d_0^{a'} d_1^{b'}$. Then $d_0^a d_1^b = f^2 d_0 d_1 = f^2 Sq^2(d_0) = Sq^2(f^2 d_0) + (Sq^1(f))^2 d_0 \sim (Sq^1(f))^2 d_0 = (Sq^1(f))^2 Sq^1(d_1) = Sq^1((Sq^1(f))^2 d_1) \sim 0$. Hence $d_0^a d_1^b$ is hit in D(2).

Case 3: a is even and b is odd. We assume as inductive hypothesis that if a' is even and b' is odd then $d_0^{a'} d_1^{b'}$ is hit in D(2) if $a' + b' < a + b$, unless $a' = 0$ and b' is of the form $2^k - 1$. Let $a = 2a'$ and $b = 2b' + 1$. Then $d_0^a d_1^b = d_1(d_0^{a'} d_1^{b'})^2$. By Cases 1 and 2, $d_0^{a'} d_1^{b'}$ is hit in D(2) unless a' is even and b' is odd, and so $d_0^a d_1^b$ is hit in D(2) by Proposition 15.2.1. Using the inductive hypothesis, the same argument applies unless $b' = 2^k - 1$ for some $k \geq 0$ and $a' = 0$. But in this case $a = 0$ and $b = 2^{k+1} - 1$. This completes the inductive step. □

15.3 The action of A$_2$ on Dickson invariants

Since D(n) is the polynomial algebra $\mathbb{F}_2[d_0, d_1, \ldots, d_{n-1}]$ where $d_i = d_{n,i}$ for $0 \leq i \leq n - 1$, the action of A$_2$ on D(n) is determined by the action of the operations Sq^k on the Dickson invariants d_i and the Cartan formula. Note that $d_n = d_{n,n} = 1$.

Proposition 15.3.1 *For $n \geq 1$ and $0 \leq i \leq n-1$, the action of the Steenrod squares on the Dickson invariant $d_i \in P^{2^n - 2^i}(n)$ is given by*

$$Sq^k(d_i) = \begin{cases} d_j, & \text{if } k = 2^i - 2^j, \quad 0 \leq j \leq i, \\ d_j d_\ell, & \text{if } k = 2^i - 2^j + 2^n - 2^\ell, \quad 0 \leq j \leq i < \ell \leq n, \\ d_i^2, & \text{if } k = 2^n - 2^i, \\ 0, & \text{otherwise.} \end{cases}$$

In particular, when $k < 2^{n-1}$, $Sq^k(d_i) = 0$ unless $k = 2^i - 2^j$ where $0 \leq j \leq i$.

Example 15.3.2 The tables below show $Sq^k(d_i)$ in the cases $n = 2$ and $n = 3$.

	Sq^1	Sq^2	Sq^3
d_0	0	$d_0 d_1$	d_0^2
d_1	d_0	d_1^2	0

	Sq^1	Sq^2	Sq^3	Sq^4	Sq^5	Sq^6	Sq^7
d_0	0	0	0	$d_0 d_2$	0	$d_0 d_1$	d_0^2
d_1	d_0	0	0	$d_1 d_2$	$d_0 d_2$	d_1^2	0
d_2	0	d_1	d_0	d_2^2	0	0	0

Proof of Proposition 15.3.1 For $x \in P^1(n)$, $Sq^k(x^{2^j}) = 0$ unless $k = 0$ or $k = 2^j$. Hence by the Cartan formula, the effect of Sq^k on a Vandermonde determinant of the form

$$\Delta(2^{j_1}, \ldots, 2^{j_n}) = \begin{vmatrix} x_1^{2^{j_1}} & x_2^{2^{j_1}} & \cdots & x_n^{2^{j_1}} \\ x_1^{2^{j_2}} & x_2^{2^{j_2}} & \cdots & x_n^{2^{j_2}} \\ \vdots & \vdots & \ddots & \vdots \\ x_1^{2^{j_n}} & x_2^{2^{j_n}} & \cdots & x_n^{2^{j_n}} \end{vmatrix}, \quad 0 \leq j_1 < \cdots < j_n, \qquad (15.2)$$

is to produce a sum of similar determinants in which the entries in each row are either squared or unchanged. If $Sq^k(\Delta(2^{j_1}, \ldots, 2^{j_n})) \neq 0$, it follows that $\mathrm{bin}(k) \subseteq \{2^{j_1}, \ldots, 2^{j_n}\}$. Further, if $j_{r+1} = j_r + 1$ for some r such that $1 \leq r \leq n-1$, and if $\mathrm{bin}(k)$ contains 2^r but not 2^{r+1}, then $Sq^k(\Delta(2^{j_1}, \ldots, 2^{j_n})) = 0$, since the resulting determinant has two identical rows.

For $0 \leq i \leq n$, $d_i = \Delta_{n,i}/\Delta_{n,n}$, where $\Delta_{n,i} = \Delta(1, 2, \ldots, 2^{i-1}, 2^{i+1}, \ldots, 2^n)$. We consider $Sq^k(\Delta_{n,i})$ where $k < 2^n$, so that $2^n \notin \mathrm{bin}(k)$. It follows from the remarks above that $Sq^k(\Delta_{n,i}) = 0$ unless $\mathrm{bin}(k) = \{2^j, \ldots, 2^{i-1}\}$ for some j such that $0 \leq j \leq i$. In fact we have shown that

$$Sq^k(\Delta_{n,i}) = \begin{cases} \Delta_{n,j} & \text{if } k = 2^i - 2^j, \text{ for } 0 \leq j \leq i, \\ 0 & \text{otherwise, for } k < 2^n. \end{cases} \quad (15.3)$$

In particular, for $i < n$ we have $Sq^k(\Delta_{n,i}) = 0$ for $2^{n-1} \leq k \leq 2^n - 1$. Recall that $d_0 = \Delta_{n,n}$ and $d_0 d_j = \Delta_{n,j}$ for $0 \leq j \leq n$, so that $d_n = 1$. Thus putting $i = n$ in (15.3) gives

$$Sq^k(d_0) = \begin{cases} d_0 d_j, & \text{if } k = 2^n - 2^j, \text{ for } 0 \leq j \leq n, \\ 0, & \text{otherwise.} \end{cases}$$

By the Cartan formula, it follows that for $0 \leq i \leq n$ and $k < 2^n$

$$Sq^k(d_0 d_i) = \sum_{j=0}^{n} d_0 d_j Sq^{k - 2^n + 2^j}(d_i). \quad (15.4)$$

For $k < 2^{n-1}$ the terms on the right hand side for $j < n$ are 0, and since $d_n = 1$, we have $Sq^k(\Delta_{n,i}) = Sq^k(d_0 d_i) = d_0 Sq^k(d_i)$, so using (15.3) and cancelling a factor d_0, we obtain

$$Sq^k(d_i) = \begin{cases} d_j, & \text{if } k = 2^i - 2^j, \text{ for } 0 \leq j \leq i, \\ 0, & \text{otherwise, for } k < 2^{n-1}. \end{cases} \quad (15.5)$$

Next let $2^{n-1} \leq k \leq 2^n - 1$ and $i < n$. In this range $Sq^k(d_0 d_i) = Sq^k(\Delta_{n,i}) = 0$ as noted above following (15.3). Again cancelling the factor d_0, we can rewrite (15.4) as

$$Sq^k(d_i) = \sum_{j=0}^{n-1} d_j Sq^{k - 2^n + 2^j}(d_i). \quad (15.6)$$

We divide the range of k into n subintervals $2^n - 2^\ell \leq k < 2^n - 2^{\ell-1}$, indexed by ℓ for $0 \leq \ell \leq n-1$. Recall that $\deg d_\ell = 2^n - 2^\ell$. Since $Sq^k(f) = 0$ for $k > d$ and $Sq^d(f) = f^2$ when $f \in P^d(n)$, for k in the ℓth subinterval we may assume that $i < \ell$. In particular, we may assume that $\ell > 0$, since there is nothing more to prove if $k = 2^n - 1$. We shall show that when k is in the ℓth subinterval, all terms with $j \neq \ell$ on the right hand side of (15.6) are 0. In the case $j < \ell$, $k - 2^n + 2^j < 0$ and so $Sq^{k - 2^n + 2^j} = 0$. On the other hand, if $j > \ell$ then $k - 2^n + 2^j \geq 2^\ell$, and since $i < \ell$, $k - 2^n + 2^j$ is not of the form $2^i - 2^j$. It therefore follows from (15.5) that $Sq^{k - 2^n + 2^j}(d_i) = 0$.

Thus (15.6) reduces to $Sq^k(d_i) = d_\ell Sq^{k - 2^n + 2^\ell}(d_i)$ when k is in the ℓth subinterval $2^n - 2^\ell \leq k < 2^n - 2^{\ell-1}$. Using (15.5), we obtain $Sq^k(d_i) = d_\ell d_j$ if $k - 2^n + 2^\ell = 2^i - 2^j$ for $0 \leq j \leq i$, and $Sq^k(d_i) = 0$ for other k in the ℓth subinterval. $\qquad \square$

There are also convenient formulae for the action on the Dickson invariants of the Milnor basis elements $Q_k = Sq_k^1 = Sq(R)$, where $R = (0,\ldots,0,1)$ has length $k \geq 1$.

Proposition 15.3.3 *Let $1 \leq k \leq n$. Then for $0 \leq i \leq n-1$,*

$$
Q_k(\mathsf{d}_i) = \begin{cases} \mathsf{d}_0, & \text{if } k=i, \\ \mathsf{d}_i\mathsf{d}_0, & \text{if } k=n, \\ 0, & \text{if } k \neq i,n. \end{cases}
$$

Proof By Proposition 3.5.10, $Q_k = \sum_{j=1}^{n} x_j^{2^k} \partial/\partial x_j$ as an operator on $\mathsf{P}(n)$, and Q_k is a derivation on $\mathsf{P}(n)$. In particular, $Q_k(x_j) = x_j^{2^k}$ and $Q_k(x_j^{2^i}) = 0$ if $i \geq 1$. It follows that $Q_k(\Delta_{n,i}) = \Delta(2^k, 2, \ldots, 2^{i-1}, 2^{i+1}, \ldots, 2^n)$ for $k \geq 1$ by the Cartan formula. Since $\Delta_{n,n} = \mathsf{d}_0$ we obtain

$$
Q_k(\mathsf{d}_0) = \Delta(2^k, 2, \ldots, 2^{n-1}) = \begin{cases} 0, & \text{if } 1 \leq k \leq n-1, \\ \mathsf{d}_0^2, & \text{if } k=n, \end{cases}
$$

and since $\Delta_{n,i} = \mathsf{d}_i\mathsf{d}_0$ we obtain

$$
Q_k(\mathsf{d}_i\mathsf{d}_0) = \begin{cases} 0, & \text{if } 1 \leq k \leq n, \ k \neq i, \\ \mathsf{d}_0^2, & \text{if } k=i. \end{cases}
$$

Since $Q_k(\mathsf{d}_i\mathsf{d}_0) = Q_k(\mathsf{d}_i)\mathsf{d}_0 + \mathsf{d}_i Q_k(\mathsf{d}_0)$, the result for $i > 0$ now follows from that for $i = 0$. $\qquad\square$

The following formula gives a means to evaluate the operations Q_k on $\mathsf{D}(n)$ recursively for $k \geq n$. Alongside the Cartan formula 3.5.7, this gives a means to evaluate Milnor basis elements $Sq(R)$ on $\mathsf{D}(n)$.

Proposition 15.3.4 *For $k > n$, the operation Q_k on $\mathsf{P}(n)$ is given by*

$$
Q_k = \mathsf{d}_0^{2^{k-n}} Q_{k-n} + \mathsf{d}_1^{2^{k-n}} Q_{k-n+1} + \cdots + \mathsf{d}_{n-1}^{2^{k-n}} Q_{k-1}.
$$

Consequently, if $f \in \mathsf{P}(n)$ and $Q_k(f) = 0$ for $1 \leq k \leq n$, then $Q_k(f) = 0$ for all $k \geq 1$.

Proof By Proposition 15.1.3(ii), $x_j^{2^n} = \sum_{i=0}^{n-1} \mathsf{d}_i x_j^{2^i}$ for $1 \leq j \leq n$. Taking the 2^{k-n}th power of this equation and multiplying by $\partial f/\partial x_j$ gives $x_j^{2^k} \partial f/\partial x_j = \sum_{i=0}^{n-1} \mathsf{d}_i^{2^{k-n}} x_j^{2^{k-n+i}} \partial f/\partial x_j$. Summing over j and using Proposition 3.5.10(iv) yields the required formula. The last statement is proved by induction on i, starting with $i = n+1$. $\qquad\square$

Remark 15.3.5 Proposition 15.3.4 is also true for $k = 0$, where Q_0 is the derivation of $P(n)$ defined by $Q_0 = \sum_{j=1}^{n} x_j \, \partial/\partial x_j$. Thus $Q_0(f) = df$ if $f \in P^d(n)$, i.e. $Q_0(f) = f$ if d is odd and $Q_0(f) = 0$ if d is even. Note that Q_0 is not a Steenrod operation.

The actions of A_2 on $D(n)$ and on $D(n+1)$ are related by a degree-doubling map. Note that $\deg d_{n+1,i+1} = 2 \deg d_{n,i}$ and that the quotient $D(n+1)/\langle d_{n+1,0} \rangle$ of $D(n+1)$ is the polynomial algebra generated by $d_{n+1,1}, \ldots, d_{n+1,n}$. We define $\zeta : D(n) \to D(n+1)/\langle d_{n+1,0} \rangle$ to be the linear map defined on Dickson monomials by $\zeta(d_{n,0}^{a_0} d_{n,1}^{a_1} \cdots d_{n,n-1}^{a_{n-1}}) = d_{n+1,1}^{a_0} d_{n+1,2}^{a_1} \cdots d_{n+1,n}^{a_{n-1}} + \langle d_{n+1,0} \rangle$.

Proposition 15.3.6 *The map* $\zeta : D(n) \to D(n+1)/\langle d_{n+1,0} \rangle$ *is an isomorphism of* A_2-*modules via the halving map* V *on* A_2, *i.e.* $\zeta \circ Sq^k = Sq^{2k} \circ \zeta$ *for all* $k \geq 0$.

Proof For V, see Section 13.3. The result follows immediately from Proposition 15.3.1, since the formula

$$Sq(d_i) = \sum_{j \leq i} d_j + \sum_{j \leq i < k} d_j d_k + d_i^2$$

in $D(n)$ maps to the corresponding formula in $D(n-1)$ by mapping d_i in $D(n)$ to d_{i-1} in $D(n-1)$ for $i > 0$ and mapping d_0 in $D(n)$ to 0. $\qquad\square$

We use this result to obtain a partial calculation of the action of the conjugate Steenrod square $Xq^k = \chi(Sq^k)$ on the Dickson invariants.

Proposition 15.3.7 *In* $D(n)$, $Xq(d_i) = d_i + d_{i-1} + d_i d_{n-1} + d_{i-1} d_{n-1} + d_{i-1}^2 +$ *terms of degree* $> 2^{n+1} - 2^i$.

Proof By induction on n using Proposition 15.3.6, the formula holds modulo terms divisible by d_0. Since we are considering only terms of degree $\leq 2^{n+1} - 1$, all such terms are of the form $d_0 d_j$, where $0 \leq j \leq n$ and $d_n = 1$. Thus we must prove that none of these terms occur when $i > 1$, that only d_0, $d_0 d_{n-1}$ and d_0^2 occur when $i = 1$ and that only d_0 and $d_0 d_{n-1}$ occur when $i = 0$.

The Adem relation $Sq^1 Sq^{2k} = Sq^{2k+1}$ gives $Xq^{2k+1} = Xq^{2k} Xq^1 = Xq^{2k} Sq^1$. Since $Sq^1(d_i) = 0$ for $i \neq 1$, it follows that all terms in $Xq(d_i)$ are of even degree when $i > 1$, while all terms in $Xq(d_0)$ are of odd degree. Since $d_0 d_j$ has odd degree for $j > 0$, this completes the proof in the case $i > 1$.

Next consider the case $i = 0$. Since $Sq(d_0) = d_0(1 + d_{n-1} + \cdots + d_1 + d_0)$ and $Xq(Sq(d_0)) = d_0$, we have $Xq(d_0) Xq(1 + d_{n-1} + \cdots + d_1 + d_0) = d_0$. The results already proved for $i > 1$, together with $Xq^1(d_1) = Sq^1(d_1) = d_0$, show that

$$Xq(1 + d_{n-1} + \cdots + d_1 + d_0) = 1 + (d_{n-1} + d_{n-2} + d_{n-1}^2) + (d_{n-2} + d_{n-3}) +$$
$$\cdots + (d_1 + d_0) + d_0) + \text{terms of degree} > 2^n$$
$$= 1 + d_{n-1} + d_{n-1}^2 + \text{terms of degree} > 2^n,$$

and it follows that $Xq(d_0) = d_0 + d_0 d_{n-1} + \text{terms of degree} > 2^{n+1} - 1$.

Finally let $i = 1$. Since $Xq^{2k+1}(d_1) = Xq^{2k}Sq^1(d_1) = Xq^{2k}(d_0)$, it follows from the result for d_0 that $d_0 d_{n-1}$ occurs in $Xq(d_1)$ but $d_0 d_j$ does not for $1 \le j \le n-2$.

It remains to prove that $Xq^{2^n}(d_1) = d_0^2$. This follows from Straffin's formula 12.1.2(ii) and the preceding results, as follows: $Xq^{2^n}(d_1) = Sq^{2^{n-1}}Xq^{2^{n-1}}(d_1) = Sq^{2^{n-1}}(d_1 d_{n-1}) = Sq^{2^{n-1}}(d_1)d_{n-1} + Sq^1(d_1)Sq^{2^{n-1}-1}(d_{n-1}) + d_1 Sq^{2^{n-1}}(d_{n-1}) = d_1 d_{n-1}d_{n-1} + d_0 d_0 + d_1 d_{n-1}^2 = d_0^2.$ $\square$

15.4 The hit problem for $P(2)^{H_2}$

For any subgroup H of $GL(n)$, the ring of invariants $P(n)^H$ is an A_2-submodule of $P(n)$, since if $f \cdot A = f$ where $A \in GL(n)$ then $Sq^k(f) \cdot A = Sq^k(f \cdot A) = Sq^k(f)$. Hence we may consider the absolute and relative hit problems for $P(n)^H$. We have seen in Section 15.2 that in the case of $D(2)$ the cohits are the same in both cases, but the main result of this chapter, Theorem 15.8.2, shows that this is an exception: all elements $\ne 1$ in $D(n)$ are hit in $P(n)$.

However, the absolute and relative hit problems for $P(n)^H$ coincide when H is a subgroup of odd order in $GL(n)$. We prove this below, and study the case where H is the cyclic subgroup of $GL(2)$ of order 3. In Section 15.5 we consider the more complicated case where H is a maximal odd order subgroup of $GL(3)$.

Proposition 15.4.1 *Let* $e_H = \sum_{A \in H} A \in \mathbb{F}_2 GL(n)$, *where* H *is a subgroup of* $GL(n)$ *of odd order* $|H|$. *Then* e_H *is idempotent, i.e.* $e_H^2 = e_H$, *and* $P(n) \cdot e_H = P(n)^H$, *the ring of invariants of* H *in* $P(n)$.

Proof If $A \in H$ then $Ae_H = e_H = e_H A$, and so $e_H^2 = |H|e_H = e_H$ since $|H|$ is odd. If $f \in P(n)^H$, then $f \cdot e_H = \sum_{A \in H} f \cdot A = \sum_{A \in H} f = |H|f = f$, so $f \in P(n) \cdot e_H$. Conversely, if $f \in P(n) \cdot e_H$ then $f = g \cdot e_H$ for some $g \in P(n)$, and so $f \cdot A = (g \cdot e_H) \cdot A = g \cdot (e_H A) = g \cdot e_H = f$, so $f \in P(n)^H$. $\square$

Proposition 15.4.2 *Let* $e \in \mathbb{F}_2 GL(n)$ *be idempotent, so that* $P(n) \cdot e$ *is an* A_2-*module, and let* $f \in P(n) \cdot e$. *Then* f *is hit in* $P(n) \cdot e$ *if and only if* f *is hit in* $P(n)$.

Proof Since $f \in P(n) \cdot e$, $f = g \cdot e$ for some $g \in P(n)$, and so $f \cdot e = g \cdot e^2 = g \cdot e = f$. Let f be hit in $P(n)$, so that $f = \sum_{i \geq 1} Sq^i(h_i)$ where $h_i \in P(n)$. Then $f = f \cdot e = (\sum_{i \geq 1} Sq^i(h_i)) \cdot e = \sum_{i \geq 1} Sq^i(h_i \cdot e)$. Hence f is hit in $P(n) \cdot e$.   □

Let $H = H_2$ be the cyclic subgroup of $GL(2)$ of order 3, generated by

$$C = \begin{pmatrix} 0 & 1 \\ 1 & 1 \end{pmatrix}.$$

Thus $f \in P(2)^H$ if and only if $f \cdot C = f$. Clearly $D(2) \subseteq P(2)^H$, and we check that $P(2)^H$ also contains $g = x^3 + y^3 + x^2y$. Hence $gD(2) \subseteq P(2)^H$. To see that $D(2) \cap gD(2) = 0$, suppose that f and gf are in $D(2)$. Since $g \cdot S = g + d_0$ where $S = \begin{pmatrix} 0 & 1 \\ 1 & 0 \end{pmatrix}$, $gf = (gf) \cdot S = (g \cdot S)(f \cdot S) = (g + d_0)f$, so $d_0f = 0$ and hence f $= 0$. Hence $D(2) \oplus gD(2) \subseteq P(2)^H$, and we note that $g^2 = g \cdot d_0 + d_0^2 + d_1^3$.

By Proposition 15.4.1, $P(2)^H = P(2) \cdot e_H$, where $e_H = I_2 + C + C^2$ is idempotent. By Proposition 15.1.5, $P(2)$ is the free $D(2)$-module of rank 6 with basis $\{1, x, y, x^2, xy, x^2y\}$. Since $f \cdot e_H = 0$ for $f = 1$, x, y or x^2, $xy \cdot e_H = d_1$ and $x^2y \cdot e_H = g$, it follows that d_1 and g span $P(2)e_H$ as a $D(2)$-module. Hence $P(2)^H = D(2) \oplus gD(2)$.

Proposition 15.4.3 *Let* $H_2 \subseteq GL(2)$ *be the cyclic subgroup of order 3, and let* $d = (2^{s+t} - 1) + (2^t - 1)$ *where* $s, t \geq 0$. *Let* $d_0 = x^2y + xy^2$, $d_1 = x^2 + xy + y^2$ *and* $g = x^3 + y^3 + x^2y$. *Then the polynomials*

$$\text{(i) } d_1^{2^t - 1}, \; t \geq 0, \quad \text{(ii) } g \cdot d_1^{2^{s+t-1} - 2^t - 1} d_0^{2^t - 1}, \; s \geq 2, \; t \geq 0,$$

are a basis for $Q^d(P(2)^{H_2})$.

The following table shows these '$P(2)^{H_2}$-monomials' for $s, t \leq 3$.

Cohit $P(2)^{H_2}$-monomials

	$s = 0$	$s = 1$	$s = 2$	$s = 3$
$t = 0$	1	0	g	$g \cdot d_1^2$
$t = 1$	d_1	0	gd_1d_0	$gd_1^5d_0$
$t = 2$	d_1^3	0	$g \cdot d_1^3d_0^3$	$g \cdot d_1^{11}d_0^3$
$t = 3$	d_1^7	0	$g \cdot d_1^7d_0^7$	$g \cdot d_1^{23}d_0^7$

Proof The polynomials listed are in $P(2)^H$, where $H = H_2$. We have seen that $d_1^{2^t-1}$ contains the spike $(xy)^{2^t-1}$. The term in $g \cdot d_1^{2^{s+t-1}-2^t-1}d_0^{2^t-1}$ with highest degree in x is the spike $x^3 \cdot (x^2)^{2^{s+t-1}-2^t-1} \cdot (x^2)^{2^t-1}y^{2^t-1} = x^{2^{s+t}-1}y^{2^t-1}$. Hence the polynomials listed are not hit in $P(2)$, and so also in $P(2)^H$. Hence $Q^d(P(2)^H) \neq 0$ for all $s, t \geq 0$ with $s \neq 1$. In the case $s = 0$, $\dim Q^d(P(2)^H) = 0$ or 1 since $\dim Q^d(2) = 1$, and there is nothing more to prove.

For $s > 0$ we consider the polynomials $f \cdot e_H$ where f runs through a basis for $Q^d(2)$, since these polynomials span $P^d(2)^H$ mod hit polynomials. First let f be the spike $x^{2^{s+t}-1}y^{2^t-1}$. Then $f \cdot e_H = f + f \cdot C + f \cdot C^2$ where $f \cdot C = y^{2^{s+t}-1}(x+y)^{2^t-1}$ and $f \cdot C^2 = (x+y)^{2^{s+t}-1}x^{2^t-1}$. Since $s > 0$ there are two distinct spikes f and $f \cdot S$ in $P^d(2)$, where $S = \begin{pmatrix} 0 & 1 \\ 1 & 0 \end{pmatrix}$. Since $f \cdot C$ contains $f \cdot S$ but not f, while $f \cdot C^2$ contains both spikes, neither f nor $f \cdot S$ is a term of $f \cdot e_H$. Since e_H commutes with S, $(f \cdot S) \cdot e_H$ also contains no spikes. In the case $s = 1$, the spikes f and $f \cdot S$ span $Q^d(2)$, and we conclude that all elements of $P^d(2)^H$ are hit.

It remains to prove that $\dim(P^d(2)^H / A_2^+ P(2)^H) = 1$ when $s \geq 2$. In this case $\dim Q^d(2) = 3$, so that we must apply e_H to a further cohit monomial. This may be taken to be any non-spike monomial m with $\omega(m) = (2, \ldots, 2, 1, \ldots, 1)$, all choices being equivalent mod $H^d(2)$. By Theorem 6.3.12, all monomials with ω-sequences $<_l \omega$ are hit. We choose $m = x^{2^{s+t-1}-1}y^{2^{s+t-1}+2^t-1}$, corresponding to the block

$$
\begin{array}{l}
1 \cdots 1\ 1 \cdots 1\ 0 \\
1 \cdots 1\ 0 \cdots 0\ 1
\end{array}.
$$

While $m \cdot C = y^{2^{s+t-1}-1}(x+y)^{2^{s+t-1}+2^t-1}$ has two terms

$$
\begin{array}{l}
1 \cdots 1\ 0 \cdots 0\ 1 \\
1 \cdots 1\ 1 \cdots 1\ 0
\end{array},
\quad
\begin{array}{l}
1 \cdots 1\ 0 \cdots 0\ 1 \\
0 \cdots 1\ 1 \cdots 1\ 1
\end{array}
$$

with ω-sequence ω, $m \cdot C^2 = (x+y)^{2^{s+t-1}-1}x^{2^{s+t-1}+2^t-1}$ has 2^{s-1} such terms of the form

$$
\begin{array}{l}
1 \cdots 1\ * \cdots *\ 1 \\
1 \cdots 1\ * \cdots *\ 0
\end{array},
$$

where the stars represent 0s or 1s. Hence $m \cdot e_H$ contains both spikes and a sum of $2^{s-1} + 1$ other monomials with ω-sequence ω, so that $m \cdot e_H = f + f \cdot S + m \neq 0$ in $Q^d(2)$. Hence $\dim(P^d(2)^H / A_2^+ P(2)^H) = 1$ for $s \geq 2$. □

Remark 15.4.4 We shall see in Chapter 19 that $\dim P^d(2)^{H_2}$ is the number of $\mathbb{F}_2 GL(2)$-composition factors of $P^d(2)$ isomorphic to the trivial module $I(2)$. For example $g \in P^3(2)$ generates an indecomposable submodule of dimension 2 containing $d_0 = x^2y + xy^2$. This is a direct summand, with complementary summand generated by x^3 and isomorphic to the defining module $V(2)$.

We shall show that $\mathbb{F}_2$-basis elements of $Q(P(2)^{H_2}) = P(2)^{H_2}/A_2^+P(2)^{H_2}$ correspond to occurrences of $I(2)$ in $Q(2)$. By Theorem 1.8.2 and the remarks above, $Q^d(2) \cong I(2)$, $V(2)$ or $Q^3(2) \cong I(2) \oplus V(2)$ for $d = (2^{s+t} - 1) + (2^t - 1)$ where $t \geq 0$ and $s = 0$, $s = 1$ or $s \geq 2$ respectively, and $Q^d(2) = 0$ otherwise. Thus we wish to show that $P(2)^{H_2}/A_2^+P(2)^{H_2}$ has dimension 1 if $s = 0$ or $s \geq 2$ and 0 if $s = 1$.

It follows from Proposition 15.4.2 that $Q(P(2)^{H_2}) = P(2)^{H_2}/A_2^+P(2)^{H_2}$ is the quotient of $Q(2) \cdot e_{H_2}$ by elements which are hit in $P(2)$. Comparison of Proposition 15.4.3 with the diagram following Theorem 1.8.2 shows that the $GL(2)$-invariant elements of $Q(2)$ are given by $Q(P(2)^{H_2})$.

15.5 The hit problem for P(3)^{H_3}

In this section we consider the hit problem for the ring $P(3)^{H_3}$ of polynomials in $P(3)$ which are invariants of the subgroup $H = H_3$ of $GL(3)$ generated by the matrices

$$A = \begin{pmatrix} 1 & 0 & 0 \\ 0 & 0 & 1 \\ 0 & 1 & 1 \end{pmatrix}, \quad B = \begin{pmatrix} 0 & 1 & 0 \\ 0 & 0 & 1 \\ 1 & 1 & 0 \end{pmatrix}.$$

Thus $f \in P(3)$ is in $P(3)^H$ if and only if $f \cdot A = f \cdot B = f$. We have $A^3 = I_3$, $B^7 = I_3$ and $AB = B^2A$, and it follows that $H = \{B^iA^j : 0 \leq i \leq 6, 0 \leq j \leq 2\}$, so $|H| = 21$. Since H has odd order, $P(n)^H = P(3)e_H$ by Proposition 15.4.1, where the idempotent $e_H \in \mathbb{F}_2 GL(3)$ is the sum of the elements of H.

We shall show that H is a maximal subgroup of $GL(3)$. Recall that $GL(3)$ is a simple group, because its conjugacy classes C_1, C_2, C_3, C_4, C_7, C_7' have order $1, 21, 56, 42, 24, 24$ respectively, and so no subset of $GL(3)$ containing I_3 and closed under conjugacy has order dividing $|GL(3)| = 168$. If K is a proper subgroup of $GL(3)$ containing H, then K has index 2 or 4 in $GL(3)$. Hence there is a nonzero homomorphism from $GL(3)$ to the group of permutations of the cosets of K, whose kernel is a proper normal subgroup of GL(3). This is a contradiction, so H is maximal.

Clearly $P(3)^H$ contains $D(3) = \mathbb{F}_2[d_2, d_1, d_0]$, where the Dickson invariants d_2, d_1, d_0 are given in Example 15.1.2. We check that the polynomial $h = x^3 + y^3 + z^3 + xyz + xy^2 + y^2z + xz^2$ is invariant under H, that $Sq^1(h) = d_2$, and that $k = Sq^2(h)$ is not in the subalgebra of $P(3)$ generated by $D(3)$ and h. The action of the Steenrod algebra A_2 on $D(3)$ is given in Example 15.3.2, and it follows that $Sq^1(k) = h^2$, $Sq^2(k) = Sq^2Sq^2(h) = Sq^3Sq^1(h) = Sq^3(d_2) = d_0$, $Sq^3(k) = 0$ and $Sq^4(k) = Sq^4Sq^2(h) = Sq^6(h) + Sq^5Sq^1(h) = 0$. The following result describes the structure of $P(3)^H$, where $H = H_3$.

Proposition 15.5.1 (i) *The ring of invariants* $P(3)^{H_3}$ *is generated by the Dickson invariants* d_2, d_1, d_0 *and the polynomials*

$$h = x^3 + y^3 + z^3 + xyz + xy^2 + y^2z + xz^2,$$
$$k = x^5 + y^5 + z^5 + x^2y^2z + x^2yz^2 + xy^2z^2 + xy^4 + y^4z + xz^4,$$

subject to the relations

$$k^2 = h^2 \cdot d_2 + h \cdot d_0 + d_2 d_1, \quad h^4 = d_2^3 + d_1^2 + k \cdot d_0 + h^2 \cdot d_1. \qquad (15.7)$$

(ii) *As a module over the Dickson algebra* $D(3)$, $P(3)^{H_3}$ *is free of dimension* 8 *with basis elements* 1, h, k, h^2, hk, h^3, h^2k, h^3k *of degrees* 1, 3, 5, 6, 8, 9, 11 *and* 14 *respectively.*

Proof Let E be the subalgebra of $P(3)$ generated by h, k, d_2, d_1 and d_0, so that $D(3) \subset E \subseteq P(3)^H \subset P(3)$. The relations 15.7 are verified by direct calculation, and it follows from them that every element of E is a linear combination of the eight elements 1, h, k, h^2, hk, h^3, h^2k and h^3k with coefficients in $D(3)$. We wish to show that these elements are linearly independent over $D(3)$.

To prove this, we use Galois theory. Given an integral domain D, let $F(D)$ denote its field of fractions. Then the inclusions above give corresponding inclusions $F(D(3)) \subset F(E) \subseteq F(P(3)^H) \subset F(P(3))$. Since $F(P(3)) = \mathbb{F}_2(x, y, z)$ is the field of rational functions over $\mathbb{F}_2$ in the variables x, y, z, $F(D(3))$ is the subfield fixed by GL(n), and $F(P(3)^H)$ is the subfield fixed by H, the Galois correspondence implies that $F(E)$ is the subfield of $F(P(3))$ fixed by some group G such that $H \subseteq G \subset$ GL(n). However, we have shown above that H is a maximal subgroup of GL(3). Hence $G = H$ and $F(E) = F(P(3))^H$. Thus $F(E)$ is a vector space of dimension 8 over $F(D(3))$.

Proposition 15.5.2 shows that the eight elements $1, h, k, h^2, hk, h^3, h^2k$ and h^3k span $F(E)$ as a vector space over $F(D(3))$. It follows that the eight elements are linearly independent over $F(D(3))$, and hence also over $D(3)$. $\square$

Proposition 15.5.2 *Every element of $F(E)$ is a linear combination of $1, h, k, h^2, hk, h^3, h^2k$ and h^3k with coefficients in $F(D(3))$.*

Proof Every element $e \in E$ is a linear combination of the eight elements $1, h, k, h^2, hk, h^3, h^2k$ and h^3k with coefficients in $D(3)$. Since $k \cdot d_0 = h^4 + d_2^3 + d_1^2 + h^2 \cdot d_1$, we can write ed_0 as a linear combination of the eight elements h^i, $0 \le i \le 7$ with coefficients in $D(3)$. Conversely, it suffices to prove that every element of $F(E)$ is a linear combination of the elements h^i, $0 \le i \le 7$, with coefficients in $F(D(3))$.

Relations (15.7) imply that $h^8 = h^4 \cdot d_1^2 + h^2 \cdot d_2 d_0^2 + h \cdot d_0^3 + (d_2 d_1 d_0^2 + d_2^6 + d_1^4)$. Hence every element of $F(E)$ can be written in the form f/g, where $f, g \in E$ are polynomials of degree ≤ 7 in h with coefficients in $D(3)$.

It suffices to prove that $1/g$ can be written in the required form. This follows by induction on $\deg g$, using the observation that $1/g = f/fg$ and $f \in E$ can be chosen so that $\deg fg < \deg g$. $\square$

The next two results give representatives g for the cohits in P(3)^{H_3} as 'P(3)^{H_3}-monomials', i.e. monomials in the variables h, k, d_2, d_1, d_0. The cohits $Q(\text{P}(3)^{H_3})$ correspond to occurrences of I(3) in Q(3), as given in Theorem 10.6.2. We tabulate these according to their ω-sequences, i.e. the $\le_l$-maximal ω-sequences of their terms. In Proposition 15.5.3 we deal with the case '$u = 0$' where these elements, regarded as polynomials in P(3), have terms with ω-sequences $(2, \ldots, 2, 1, \ldots, 1)$ with t 2s and s 1s. In the case $t = 0$, these polynomials contain spikes of both the spike types which occur in degree $2^s - 1$. In Proposition 15.5.4 we deal with the case '$u > 0$' where these elements, regarded as polynomials in P(3), have terms with ω-sequences $(3, \ldots, 2, \ldots, 2, 1, \ldots, 1)$ with $u > 0$ 3s, t 2s and s 1s.

Proposition 15.5.3 *For $u = 0$, the following* P(3)^{H_3}*-monomials span the cohits* $Q^\omega(\text{P}(3)^{H_3})$, *where $h_1 = h$, $h_2 = h \cdot Sq^2(h)$ and $h_3 = h^3 \cdot Sq^2(h)$:*

(i) *1 in degree 0 and h_2 in degree 8;*
(ii) $h_1 \cdot d_2^{2^{s-2}-1}$ *for $s \ge 3$, $t = 0$;*
(iii) $h_3 \cdot d_2^{2^{t+1}-4}$ *for $s = 0$, $t \ge 3$;*
(iv) $h_3 \cdot d_2^{2^{s+t-2}-2^{t-1}-1} d_1^{2^{t-1}-2}$ *for $s \ge 2$, $t \ge 2$.*

We tabulate these elements for $s, t \leq 4$ below.

Cohit $P(3)^{H_3}$-monomials, $u = 0$

	$s = 0$	$s = 1$	$s = 2$	$s = 3$	$s = 4$
$t = 0$	1	0	0	$h_1 \cdot d_2$	$h_1 \cdot d_2^3$
$t = 1$	0	0	h_2	0	0
$t = 2$	0	0	$h_3 \cdot d_2$	$h_3 \cdot d_2^5$	$h_3 \cdot d_2^{13}$
$t = 3$	h_3	0	$h_3 \cdot d_2^3 d_1^2$	$h_3 \cdot d_2^{11} d_1^2$	$h_3 \cdot d_2^{27} d_1^2$
$t = 4$	$h_3 \cdot d_2^4$	0	$h_3 \cdot d_2^7 d_1^6$	$h_3 \cdot d_2^{23} d_1^6$	$h_3 \cdot d_2^{55} d_1^6$

Proof For (i), we observe that $x^6 yz$, $xy^6 z$ and xyz^6 are terms of h_2, and that all other terms which are not squares occur in pairs which are equivalent mod hits by one-back splicing. Hence $h_2 \cong x^6 yz + xy^6 z + xyz^6$ mod hits.

For (ii), (iii) and (iv) we show that the stated elements contain spikes. In case (ii), it is clear that $h_1 \cdot d_2^{2^{s-2}-1}$ contains the spike $x^3 \cdot (x^4)^{2^{s-2}-1} = x^{2^s-1}$.

For (iii), we observe that $h_3 = (x^6 + y^6 + x^2 y^4)(x^3 + y^3 + xy^2)(x^5 + y^5 + xy^4) +$ terms involving z. As the spike $x^7 y^7$ occurs three times when this product is multiplied out without cancellation, it is a term in h_3. Since $d_2^4 = x^{16} + y^{16} + x^8 y^8 +$ terms involving z, $h_3 \cdot d_2^4$ contains $x^{15} y^{15}$. The cases $t \geq 5$ are similar.

For (iv), we begin with the case $t = 2$. The product above shows that h_3 has a term $x^{11} y^3 = x^6 \cdot y^3 \cdot x^5$, but no term $x^{13} y$. Hence $x^{15} y^3 = x^{11} y^3 \cdot x^4$ is a term in $h_3 \cdot d_2$ for $s = 2$, $x^{31} y^3 = x^{11} y^3 \cdot x^{20}$ is a term in $h_3 \cdot d_2^5$ for $s = 3$, and in general $x^{2^{s+2}-1} y^3 = x^{11} y^3 \cdot x^{2^{s+2}-12}$ is a term in $h_3 \cdot d_2^{2^s-3}$. For $s = 2$ and $t = 3$, we find similarly that $h_3 \cdot d_2^3 d_1^2$ has a term $x^{31} y^7 = x^{11} y^3 \cdot x^{12} \cdot x^8 y^4$. In the general case, we find similarly that $h_3 \cdot d_2^{2^{s+t-2}-2^{t-1}-1} d_1^{2^{t-1}-2}$ has a term $x^{2^{s+t}-1} y^{2^t-1} = x^{11} y^3 \cdot (x^4)^{2^{s+t-2}-2^{t-1}-1} \cdot (x^4 y^2)^{2^{t-1}-2}$.   □

The next result gives representatives for $Q(P(3)^{H_3})$ which (as polynomials in $P(3)$) have terms with ω-sequences $(3, \ldots, 3, 2, \ldots, 2, 1, \ldots, 1)$ with $u > 0$ 3s, t 2s and s 1s. In degree $2^s - 1$, these contain spikes of type $(u, t, s) = (1, s - 2, 0)$

only. Hence they are linearly independent of the elements in Proposition 15.5.3 mod hits, and $\dim Q^d(P(3)^{H_3}) = 2$ in degrees $d = 2^s - 1$, $s \geq 5$.

Proposition 15.5.4 *For $u > 0$, the following $P(3)^{H_3}$-monomials span the cohits $Q^\omega(P(3)^{H_3})$, where $h_1 = h$, $h_2 = h \cdot Sq^2(h)$ and $h_3 = h^3 \cdot Sq^2(h)$:*

(i) $h_1 \cdot d_1^{2^{u-1}-1}$ *in degree* $3(2^u - 1)$ *and* $h_2 \cdot (d_2 d_0)^{2^u-1}$ *in degree* $11 \cdot 2^u - 3$;

(ii) $h_1 \cdot d_2^{2^{s+u-2}-2^u} d_1^{2^u-1}$ *for* $s \geq 3$, $t = 0$;

(iii) $h_3 \cdot d_2^{2^{u+t+1}-2^{u+2}} (d_2 d_1 d_0)^{2^u-1}$ *for* $s = 0$, $t \geq 3$;

(iv) $h_3 \cdot d_2^{2^{u+s+t-2}-2^{u+t-1}-2^u} d_1^{2^{u+t-1}-2^{u+1}} (d_2 d_1 d_0)^{2^u-1}$ *for* $s \geq 2$, $t \geq 2$.

We tabulate these elements for $u = 1$ and $s, t \leq 4$ below.

Cohit $P(3)^{H_3}$-monomials, $u = 1$

	$s = 0$	$s = 1$	$s = 2$	$s = 3$	$s = 4$
$t = 0$	h_1	0	0	$h_1 \cdot d_2^2 d_1$	$h_1 \cdot d_2^6 d_1$
$t = 1$	0	0	$h_2 \cdot d_2 d_0$	0	0
$t = 2$	0	0	$h_3 \cdot d_2^3 d_1 d_0$	$h_3 \cdot d_2^{11} d_1 d_0$	$h_3 \cdot d_2^{27} d_1 d_0$
$t = 3$	$h_3 \cdot d_2 d_1 d_0$	0	$h_3 \cdot d_2^7 d_1^5 d_0$	$h_3 \cdot d_2^{23} d_1^5 d_0$	$h_3 \cdot d_2^{55} d_1^5 d_0$
$t = 4$	$h_3 \cdot d_2^9 d_1 d_0$	0	$h_3 \cdot d_2^{15} d_1^{13} d_0$	$h_3 \cdot d_2^{47} d_1^{13} d_0$	$h_3 \cdot d_2^{111} d_1^{13} d_0$

Proof We use the down Kameko map $\kappa : P^{2d+3}(3) \to P^d(3)$. The map κ preserves the invariants $P(3)^{H_3}$, since if $c(n)f^2 \cdot A = c(n)f^2$ then $f \cdot A = \kappa(c(n)f^2) \cdot A = \kappa(c(n)f^2 \cdot A) = \kappa(c(n)f^2) = f$, since κ is a $\mathbb{F}_2 GL(3)$-module map. Further κ induces a map $Q^{2d+3}(n) \to Q^d(n)$ for all d.

By direct calculation $\kappa(h_1 \cdot d_1) = h_1$, $\kappa(h_2 \cdot d_2 d_0) = h_2 + d_2^2$, $\kappa(h_3 \cdot d_2 d_1 d_0) = h_3 + h_1^2 \cdot d_2^2 + h_2 \cdot d_1$. Since d_2^2 and $h_1^2 \cdot d_2^2$ are squares and $h_2 \cdot d_1 = hk \cdot d_1 = Sq^2(hk \cdot d_2 + d_2^3)$, the induced map $\kappa : Q^{2d+3}(P(3)^{H_3}) \to Q^d(P(3)^{H_3})$ satisfies $\kappa(h_1 \cdot d_1) = h_1$, $\kappa(h_2 \cdot d_2 d_0) = h_2$ and $\kappa(h_3 \cdot d_2 d_1 d_0) = h_3$. Using the fact that $\kappa(fg^2) = \kappa(f)g$ for any polynomials f and g, we can check that κ maps the given generators for $(u+1, s, t)$ to the given generators for (u, s, t) in all cases. $\square$

Since the generators g of the cohits $Q^d(P(3)^{H_3})$ in degrees $d > 0$ tabulated in Section 15.5 are not symmetric, the submodule of $P(3)^{H_3}$ generated by such an element g is not 1-dimensional, and so g cannot be replaced by a Dickson invariant. It follows that all elements of $D(3)$ except 1 are hit in $P(3)^{H_3}$, and hence also in $P(3)$. This solves the relative hit problem for $D(3)$. However, it seems unlikely that this approach to the relative hit problem for $D(3)$ can be extended to the case $n > 3$, as the groups $GL(2)$ and $GL(3)$ are exceptional in having subgroups whose order is the odd part of the order of the group.

15.6 The relative hit problem for $D(3)$

In the remaining sections of this chapter, we follow the proof by Hung and Nam that for $n \geq 3$ all elements of positive degree in $D(n)$ are hit in $P(n)$. In this section, we sketch this argument in the case $n = 3$. We extend the block notation for monomials to Dickson monomials $d \neq 1$ by representing $d = d_0^{a_0} d_1^{a_1} \cdots d_{n-1}^{a_{n-1}} \in D(n)$ by the n-block whose ith row is the reversed base 2 expansion of a_{i-1}, for $1 \leq i \leq n$. The columns are indexed by $1 \leq j \leq \ell$, so that the (i,j)th entry is 1 if $2^{j-1} \in \text{bin}(a_{i-1})$ and is 0 otherwise. For example, the blocks

$$
D_1 = \begin{matrix} 1\,0\,1 \\ 1\,1\,0 \\ 1\,1\,0 \end{matrix}, \quad
D_2 = \begin{matrix} 0\,1\,0 \\ 1\,0\,1 \\ 1\,1\,0 \end{matrix}, \quad
D_3 = \begin{matrix} 0\,0\,0 \\ 1\,1\,1 \\ 1\,1\,1 \end{matrix}
$$

represent $d_0^5 d_1^3 d_2^3$, $d_0^2 d_1^5 d_2^3$ and $d_1^7 d_2^7$ in $D(3)$ respectively. We refer to such blocks D as **Dickson blocks**. As in these examples, we retain trailing 0s in the rows and assume that the last (ℓth) column is nonzero.

Dickson blocks such as D_3, where all entries in rows other than the first row are 1s, are called **full**. The proof of Theorem 15.8.2 is by induction on ℓ, the number of columns in the block, and, for blocks which are not full, on the position (s,j) of the first 0 entry not in the first row, reading down the columns and from left to right. In the examples above, the first 0 in D_1 is in position $(2,3)$, the first 0 in D_2 is in position $(2,2)$, while $D(3)$ is full.

An $n \times \ell$ block D representing a Dickson monomial $d \neq 1$ can be regarded as a concatenation $D'|E$, where E is the ℓth column and is nonzero. At the ℓth stage of the induction, we consider blocks such that D' is full, and prove (i) if the entry in row 2 of E is 0, then df^{2^ℓ} is hit in $P(n)$ for all $f \in P(n)$, (ii) the same is true if the entry in row 2 of E is 1 and the entry in row 3 of E is 0 and (iii) if the entries in rows 2 and 3 of E are 1s (so that D is full), then df^{2^ℓ} is hit in $P(n)$ if f satisfies a restricted hit equation of the form $f = Sq^1(g_1) + Sq^2(g_2)$.

For illustrative purposes, we represent $df^{2^{\ell}}$ by the 'concatenation' $B = D|F$, where D is the Dickson block representing d and F is a sum of ordinary blocks representing $f \in P(n)$. We begin by carrying out the above three steps in the case $\ell = 1$.

(i) Let

$$B_1 = \begin{array}{c} * \\ 0 \\ * \end{array} \Bigg| \, F$$

where F is an ordinary 3-block and the stars represent 0 or 1. The block B_1 represents df^2, where $d = 1, d_0, d_2$ or $d_0 d_2$ and f is a monomial in $P(3)$. Since $Sq^1(d_s) = 0$ in $D(n)$ if $s \neq 1$, $Sq^1(df^2) = 0$ in all cases. Hence $df^2 = Sq^1(f_0)$ for some $f_0 \in P(n)$ by Proposition 1.3.5.

(ii) Let

$$B_2 = \begin{array}{c} * \\ 1 \\ 0 \end{array} \Bigg| \, F.$$

The block B_2 represents df^2, where $d = d_1$ or $d_0 d_1$ and again f is arbitrary. The two cases are similar, so let $d = d_0 d_1$. Then $df^2 = d_0 Sq^2(d_2)f^2 = Sq^2(d_0 d_2 f^2) + d_0 d_2 (Sq^1(f))^2$. By (i) the second term is hit, and so df^2 is hit.

(iii) Let

$$B_3 = \begin{array}{c} * \\ 1 \\ 1 \end{array} \Bigg| \, F.$$

The block B_3 represents df^2, where $d = d_1 d_2$ or $d_0 d_1 d_2$. In this case, F is a formal sum of blocks representing a polynomial $f \in P(3)$, and we show that df^2 is hit when $f = Sq^1(g_1) + Sq^2(g_2)$. Taking the star in B_3 as 1 and $g = g_1$, we have $d_0 d_1 d_2 (Sq^1(g))^2 = Sq^2(d_0 d_1 d_2 g^2) + Sq^2(d_0 d_1 d_2)g^2$ and $Sq^2(d_0 d_1 d_2) = d_0 d_1^2$, so $d_0 d_1 d_2 (Sq^1(g))^2$ is hit using (i). Taking $g = g_2$, we have $d_0 d_1 d_2 (Sq^2(g))^2 = Sq^4(d_0 d_1 d_2 g^2) + Sq^2(d_0 d_1 d_2)(Sq^1(g))^2 + Sq^4(d_0 d_1 d_2)g^2$. The second term on the right is hit using (i). Since $Sq^4(d_0 d_1 d_2) = d_0 d_1 d_2^2 + d_0^3$, the third term on the right is hit using (i) and (ii). Hence df^2 is hit when f is hit using only Sq^1 and Sq^2.

The next three steps of the induction follow a similar pattern for $\ell = 2$. We consider blocks of the form

$$B_4 = \begin{array}{|cc|} * & * \\ 1 & 0 \\ 1 & * \end{array} F, \quad B_5 = \begin{array}{|cc|} * & * \\ 1 & 1 \\ 1 & 0 \end{array} F, \quad B_6 = \begin{array}{|cc|} * & * \\ 1 & 1 \\ 1 & 1 \end{array} F,$$

representing df^4, where d is a Dickson monomial and, in B_6, F is a formal sum of blocks representing a polynomial $f \in P(3)$ which is hit using only Sq^1 and Sq^2. With the stars taken as 1s, B_4 represents $df^4 = d_0 d_1 d_2 (d_0 d_2 f^2)^2 = d_0 d_1 d_2 (Sq^1(g))^2$ by (i) for some g, and so df^4 is hit by (iii). In the same way, B_5 represents $df^4 = d_0 d_1 d_2 (d_0 d_1 f^2)^2$. By (i) and (ii), $d_0 d_1 f^2$ is hit using Sq^1 and Sq^2, and so by (iii) df^4 is hit. For B_6 the argument is similar to (iii), taking $f = Sq^1(g_1) + Sq^2(g_2)$. The induction then proceeds to the cases where $\ell = 3$.

In order to use this inductive argument to prove Theorem 15.8.2 for $n = 3$, we separate the cases where $d \neq 1$ is full from the cases where d is not full. If it is not full, then d is represented by a block with (i, ℓ)th entry 0 for some ℓ and $i = 2$ or 3. By taking $f = 1$ in these cases, we show that d is hit in $P(3)$. If d is full, then $d = d'e^{2^\ell}$ where d' is full and $e = d_1 d_2$ or $d_0 d_1 d_2$.

For a general n, the next step in the argument is provided by Proposition 15.7.1. In the case $n = 3$, we can apply the results of Section 15.5, as follows. Let $h, k \in P(3)$ be as in Proposition 15.5.1. Then $Sq^2(hk) = Sq^2(h)k + Sq^1(h)Sq^1(k) + hSq^2(k) = k^2 + d_2 h^2 + hd_0 = d_1 d_2$, and since $Sq^1(d_0) = 0$ and $Sq^2(d_0) = 0$ it follows that $Sq^2(d_0 hk) = d_0 d_1 d_2$. Thus e is hit using only Sq^1 and Sq^2, and so by taking $f = e$ in (iii), we conclude that d is hit.

15.7 Elementary Dickson monomials

The results of this section will be used in the solution of the relative hit problem for $D(n)$. We assume throughout that $n \geq 3$. We begin by finding some Dickson monomials which are hit in $P(n)$ using only Sq^1 and Sq^2.

Proposition 15.7.1 *For $n \geq 3$, let $d \in D(n)$ be a Dickson monomial in which d_1 and d_2 have exponent 1, i.e. $d = cd_1 d_2$ where c is a monomial in $d_0, d_3, \ldots, d_{n-1}$. Then $d = Sq^1(f_1) + Sq^2(f_2)$ where $f_1, f_2 \in P(n)$.*

Proof First consider the case $c = 1$. Since $Sq^1(d_2) = 0$, $d_2 = Sq^1(f)$ for some $f \in P(n)$ by Proposition 1.3.5. Then $d = d_1 d_2 = Sq^2(d_2)d_2 = Sq^2 Sq^1(f)Sq^1(f)$. Using the Cartan formula and the relations $Sq^1 Sq^2 Sq^1 = Sq^3 Sq^1 = Sq^2 Sq^2$, this implies that $Sq^1(Sq^2 Sq^1(f)f) = Sq^2 Sq^2(f)f + d_1 d_2$. We also have $Sq^2(Sq^2(f)f) = Sq^2 Sq^2(f)f + g$, where $g = Sq^2(f)Sq^2(f) +$

$Sq^1 Sq^2(f)Sq^1(f)$. Since $Sq^1 Sq^1 = 0$ and Sq^1 annihilates squares of polynomials, $Sq^1(g) = 0$, so $g = Sq^1(h)$ for some $h \in P(n)$, by Proposition 1.3.5. Collecting terms, $d = d_1 d_2 = Sq^1(f_1) + Sq^2(f_2)$ where $f_1 = Sq^2 Sq^1(f)f + h$ and $f_2 = Sq^2(f)f$.

In general, $d = cd_1 d_2$ where d_1 and d_2 do not divide the Dickson monomial c. Since $Sq^1(d_i) = 0$ and $Sq^2(d_i) = 0$ for $i \neq 1, 2$, $Sq^1(c) = 0$ and $Sq^2(c) = 0$ by the Cartan formula. Thus $d = c(Sq^1(f_1) + Sq^2(f_2)) = Sq^1(cf_1) + Sq^2(cf_2)$. $\quad\square$

We call a Dickson monomial $e \in D(n)$ **elementary** if e is a product of distinct Dickson invariants d_i, $0 \leq i \leq n - 1$. If $e \neq 1$, then e is represented by a 1-column Dickson block E.

Proposition 15.7.2 *For $1 < s \leq n - 1$, let $e \in D(n)$ be an elementary Dickson monomial divisible by d_s but not d_{s-1}, so that $e = e'd_s e''$ where e' and e'' are elementary Dickson monomials such that e' divides $d_1 \cdots d_{s-2}$ and e'' divides $d_0 d_{s+1} \cdots d_{n-1}$. Then $Sq^k(e) = g_k d_s e''$ for $0 \leq k < 2^{s-1}$, where g_k is a polynomial in $d_0, d_1, \ldots, d_{s-2}$, and $Sq^{2^{s-1}}(e) = e'd_{s-1}e''$.*

Proof By Proposition 15.3.1, nonzero values of $Sq^k(d_i)$ for $k < 2^{n-1}$ occur only when $k = 0$ or $2^{i-1} \leq k \leq 2^i - 1$. In particular, $Sq^k(d_i) = 0$ for $0 < k \leq 2^{s-1}$ when $i > s$ or $i = 0$. Thus the Cartan formula gives $Sq^k(e'') = 0$. Since these nonzero values of $Sq^k(d_i)$ are d_j where $0 \leq j \leq i$, the Cartan formula shows that $Sq^k(e') = g_k$ is a polynomial in $d_0, d_1, \ldots, d_{s-2}$, and $Sq^k(e') = 0$ if $2^{s-1} \geq k > \sum_{i=1}^{s-2}(2^i - 1) = 2^{s-1} - s$. Finally $Sq^k(d_s) = 0$ if $0 < k < 2^{s-1}$ and $Sq^{2^{s-1}}(d_s) = d_{s-1}$. The result follows by applying the Cartan formula again to the product $e = e'd_s e''$. $\quad\square$

This result leads to the following reduction formula.

Proposition 15.7.3 *For $n \geq 3$ and $1 < s < n$, let $e \in D(n)$ be an elementary Dickson monomial which is divisible by d_i for $1 \leq i < s$ but not by d_s. Then for any polynomial $f \in P(n)$, $ef^2 = Sq^{2^{s-1}}(f_0) + \sum_i e_i f_i^2$, where $f_i \in P(n)$ for $i \geq 0$, and each $e_i \in D(n)$ is an elementary Dickson monomial which is divisible by d_s but not by d_{s-1}.*

Proof Using the notation of Proposition 15.7.2, let $e = e'd_{s-1}e''$, where $e' = d_1 \cdots d_{s-2}$ and e'' divides $d_{s+1} \cdots d_{n-1}d_0$. Let $\tilde{e} = e'd_s e''$. Applying Proposition 15.7.2 to $\tilde{e}$,

$$ef^2 = Sq^{2^{s-1}}(\tilde{e})f^2 = Sq^{2^{s-1}}(f_0) + \sum_{0 \leq j < 2^{s-2}} Sq^{2j}(\tilde{e})(Sq^{2^{s-2}-j}(f))^2,$$

where $f_0 = \tilde{e}f^2$ and each $Sq^{2j}(\tilde{e})$ is of the form $g_{2j}d_s e''$ for some polynomial g_{2j} in $d_0, d_1, \ldots, d_{s-2}$. By writing g_{2j} for each j as a sum of Dickson monomials

$e'_i(g'_{j,i})^2$, where e'_i divides $d_0 d_1 \cdots d_{s-2}$, and taking $e_i = e'_i d_s e''$ and f_i as a sum of terms of the form $(g'_{j,i} Sq^{2^{s-2}-j}(f))^2$, we obtain an expression of the required form for ef^2. $\qquad\square$

The following numerical result will be needed in Section 15.8. In terms of Dickson blocks, it states that the result of applying certain Steenrod squares to a full Dickson block is a sum of blocks which are not full. By a **term** of an element of $D(n)$, we mean a term in the expression for d as an irredundant sum of Dickson monomials.

Proposition 15.7.4 *For $n \geq 3$ and $\ell \geq 1$, let $d = d_0^r (d_1 \cdots d_{n-1})^{2^\ell - 1}$ where $0 \leq r \leq 2^\ell - 1$, and let the Dickson monomial c be a term of $Sq^{2^\ell j}(d)$ for some j such that $1 \leq j < 2^{n-1}$. Then the exponent of d_i in c is $\neq -1 \bmod 2^\ell$ for some i such that $1 \leq i \leq n - 1$.*

Proof We argue by contradiction. Suppose that $c = d_0^s (d_1 \cdots d_{n-1})^{2^\ell - 1} g^{2^\ell}$, where $0 \leq s \leq 2^\ell - 1$ and g is a Dickson monomial. Since $\deg d + 2^\ell j = \deg c$, $r \deg d_0 + 2^\ell j = s \deg d_0 + 2^\ell \deg g$. Since $\deg d_0 = 2^n - 1$, $r = s \bmod 2^\ell$. Since $0 \leq r, s \leq 2^\ell - 1$, it follows that $r = s$ and $\deg g = j$. This is a contradiction, since there is no Dickson monomial g of degree j when $1 \leq j < 2^{n-1}$. $\qquad\square$

15.8 The relative hit problem for $D(n)$

In this section we complete the proof that all Dickson monomials $d \neq 1$ are hit in $P(n)$ when $n \geq 3$. We define a Dickson monomial $d = d_0^{a_0} d_1^{a_1} \cdots d_{n-1}^{a_{n-1}}$ to have **size** $\ell \geq 0$ when the corresponding Dickson block D has ℓ columns, i.e. $a_i < 2^\ell$ for $0 \leq i \leq n - 1$ and $a_i \geq 2^{\ell-1}$ for some i. Thus an elementary Dickson monomial $e \neq 1$ has size 1 and the identity element 1 has size 0. We say that d is **full** if the block D is full, i.e. all entries of D except those in the first row, given by the binary expansion of a_0, are 1s.

The following result generalizes statements (i), (ii) and (iii) of Section 15.6. Note that the hit equation $f = Sq^1(g_1) + Sq^2(g_2)$ is satisfied by f in case (iii) when $n = 3$ is replaced by $f = \sum_{j=0}^{n-2} Sq^{2^j}(g_j)$, i.e. $f \in A_2(n-2)^+ P(n)$.

Proposition 15.8.1 *For $n \geq 3$ and $\ell \geq 1$, let $d = d' e^{2^{\ell-1}} \in D(n)$ be a Dickson monomial of size ℓ, where d' is a full Dickson monomial of size $\ell - 1$ and $e \neq 1$ is an elementary Dickson monomial. Then*

(i) *if e (and hence d) is not full, then df^{2^ℓ} is hit in $P(n)$ for all $f \in P(n)$,*
(ii) *if e (and hence d) is full, then df^{2^ℓ} is hit in $P(n)$ for $f \in A_2(n-2)^+ P(n)$.*

Proof We argue by induction on ℓ for $\ell \geq 1$. Thus we assume that both statements hold for $1, 2, \ldots, \ell - 1$. In Step 1, we prove (i) for a given ℓ by induction on the row containing the first 0 entry of the Dickson block D representing d. This is row $s + 1$, which corresponds to d_s for $1 \leq s \leq n - 1$. In Step 2, we prove that (i) for $1 \leq q \leq \ell$ implies (ii) for ℓ, and in Step 3 that (ii) for ℓ implies (i) for $\ell + 1$ when $s = 1$. This provides the base case for the induction on s in Step 1.

To start the whole argument, we prove (i) for $\ell = 1$ and $s = 1$. In this case $d = e$ is an elementary Dickson monomial not involving d_1. Since $Sq^1(d_i) = 0$ if $i \neq 1$, the Cartan formula implies that $Sq^1(d) = 0$, and so $Sq^1(df^2) = 0$ for all $f \in P(n)$. Hence $df^2 = Sq^1(f_0)$ for some $f_0 \in P(n)$ by Proposition 1.3.5.

Step 1: (i) for ℓ, s implies (i) for $\ell, s + 1$. We assume that $s > 1$, that (i) is true for ℓ and smaller values of s, and that (ii) is true for $\ell - 1$ if $\ell > 1$.

We have $d = d'e^{2^{\ell-1}}$, where $d' = d_0^r(d_1 \cdots d_{n-1})^{2^{\ell-1}-1}, 0 \leq r \leq 2^{\ell-1} - 1$, and e is an elementary Dickson monomial divisible by $d_1 \cdots d_{s-1}$ but not by d_s. If $\ell = 1$ then $d' = 1$ and $d = e$. Thus given $f \in P(n)$, $df^{2^\ell} = d'(ef^2)^{2^{\ell-1}}$. By Proposition 15.7.3, $ef^2 = Sq^{2^{s-1}}(f_0) + \sum_i e_i f_i^2$ for some polynomials f_i, where each e_i is an elementary Dickson monomial divisible by d_s but not by d_{s-1}. Hence $df^{2^\ell} = d'(ef^2)^{2^{\ell-1}} = d'(Sq^{2^{s-1}}(f_0))^{2^{\ell-1}} + \sum_i d'e_i^{2^{\ell-1}} f_i^{2^\ell}$. By (ii) for $\ell - 1$, $d'(Sq^{2^{s-1}}(f_0))^{2^{\ell-1}}$ is hit, since $s - 1 \leq n - 2$ and so $Sq^{2^{s-1}} \in A_2(n-2)$. On the other hand, the inductive hypothesis on s implies that $d'e_i^{2^{\ell-1}} f_i^{2^\ell}$ is hit for each e_i. This completes the inductive step on s.

Step 2: (i) for $1 \leq q \leq \ell$ implies (ii) for ℓ. Here d is full. By linearity and the fact that $A_2(n-2)$ is generated by Sq^{2^m}, $0 \leq m \leq n - 2$, it suffices to prove that df^{2^ℓ} is hit in P(n) when $f = Sq^{2^m}(f_0)$ and $0 \leq m \leq n - 2$. Then $f^{2^\ell} = Sq^{2^{m+\ell}}(f_0^{2^\ell})$. By the Cartan formula and Proposition 1.3.2, $Sq^{2^{m+\ell}}(df_0^{2^\ell}) = dSq^{2^{m+\ell}}(f_0^{2^\ell}) + \sum_{j=1}^{2^m} Sq^{2^\ell j}(d)f_j^{2^\ell}$, where $f_j = Sq^{2^m-j}(f_0)$.

Let the Dickson monomial c be a term of $Sq^{2^\ell j}(d)$, where $1 \leq j \leq 2^m$. Then it suffices to prove that $cf_j^{2^\ell}$ is hit in P(n). If c is a square then $cf_j^{2^\ell}$ is hit, so we may assume that the exponent b_i of d_i in c is odd for some i, where $0 \leq i \leq n - 1$.

Let $c = d_0^{b_0} d_1^{b_1} \cdots d_{n-1}^{b_{n-1}}$. Since $1 \leq j \leq 2^m < 2^{n-1}$, it follows from Proposition 15.7.4 that $b_i \neq -1 \bmod 2^\ell$ for some $i > 0$. Let $q \geq 1$ be minimal such that $2^{q-1} \notin \text{bin}(b_i)$ for some i such that $1 \leq i \leq n - 1$, and let $s \geq 1$ be minimal such that $2^{q-1} \notin \text{bin}(b_s)$, so that the first 0 in the Dickson block for c is in position $(s + 1, q)$. Then $1 \leq q \leq \ell$.

Let $b_i = b'_i \bmod 2^q$ for $0 \leq i \leq n - 1$, where $0 \leq b'_i \leq 2^q - 1$. Then $b'_i = 2^q - 1$ for $1 \leq i \leq s - 1$, $b'_s = 2^{q-1} - 1$ and $b'_i = 2^{q-1} - 1$ or $2^q - 1$ for $s + 1 \leq i \leq n - 1$, and $c = c_1 c_2^{2^q}$ where $c_1 = d_0^{b'_0} d_1^{b'_1} \cdots d_{n-1}^{b'_{n-1}}$ and c_2 is a Dickson monomial. Since

$q \leq \ell$, we may apply (i) to c_1. We conclude that $c_1 f^{2^q}$ is hit for all $f \in P(n)$. Taking $f = c_2 f_j^{2^{\ell-q}}$, it follows that $c f_j^{2^\ell}$ is hit.

Step 3: (ii) *for ℓ implies* (i) *for $\ell+1$ when $s = 1$.*   We have $\mathsf{d} = \mathsf{d}' e^{2^\ell}$, where $\mathsf{d}' = \mathsf{d}_0^r (\mathsf{d}_1 \cdots \mathsf{d}_{n-1})^{2^\ell - 1}$ and $0 \leq r \leq 2^\ell - 1$, and e is an elementary Dickson monomial which is not divisible by d_1.

Thus given $f \in P(n)$, $\mathsf{d} f^{2^{\ell+1}} = \mathsf{d}' (e f^2)^{2^\ell}$. Since $Sq^1(e f^2) = Sq^1(e) f^2 = 0$, $e f^2 = Sq^1(f_0)$ for some $f_0 \in P(n)$. Hence $\mathsf{d} f^{2^{\ell+1}} = \mathsf{d}'(Sq^1(f_0))^{2^\ell}$. We may apply (ii) to d' with the hit polynomial f of (ii) taken as $Sq^1(f_0)$. We conclude that $\mathsf{d}'(Sq^1(f_0))^{2^\ell}$ is hit. Thus (i) is true for $\ell+1$ when $s = 1$. $\square$

We can now prove the theorem of Hung and Nam. In fact, the proof of Proposition 15.8.1 is valid also for $n = 2$, the appeal to Proposition 15.7.4 in Step 2 being replaced by a direct calculation of $\deg \mathsf{c}$. The hypothesis $n \geq 3$ is needed for the next argument.

Theorem 15.8.2 *For $n \geq 3$, every Dickson monomial $\mathsf{d} \neq 1$ is hit in $P(n)$.*

Proof Let d be a Dickson monomial of size $\ell \geq 1$. If d is not full, let $\mathsf{d} = \mathsf{d}'\mathsf{d}''$, where the last column of d' contains the first 0 entry of d which is not in the first row. Then d' is not full, and the result follows by applying (i) of Proposition 15.8.1 with $f = \mathsf{d}''$. If d is full, then $\mathsf{d} = \mathsf{d}' e^{2^{\ell-1}}$ where d' is full (or $\mathsf{d}' = 1$) and e is an elementary Dickson monomial of the form $e = e' \mathsf{d}_1 \mathsf{d}_2$.

By Proposition 15.7.1, $e = Sq^1(f_1) + Sq^2(f_2)$ where $f_1, f_2 \in P(n)$. In the case $\ell = 1$, $\mathsf{d} = e$ and there is no more to prove. For $\ell > 1$ we apply Proposition 15.8.1(ii) to d', which has size $\ell - 1$. Since $n \geq 3$, $e \in A_2(n-2)^+ P(n)$, and so we may choose $f = e$ in (ii). Hence $\mathsf{d} = \mathsf{d}' e^{2^{\ell-1}}$ is hit. $\square$

15.9 Remarks

In 1911, L. E. Dickson proved that for every finite field $\mathbb{F}_q$ the invariants of the general linear group $GL(n, \mathbb{F}_q)$ acting by linear transformations on $\mathbb{F}_q[x_1, \ldots, x_n]$ form a polynomial algebra over $\mathbb{F}_q$ with generators $\mathsf{d}_{n,i}$ of degree $q^n - q^i$, $0 \leq i \leq n - 1$. Proofs can be found in Macdonald [128, Section I.2, Ex. 26, 27] and Smith [190, Sections 8.1 and 10.6] The argument presented in Section 15.1 for the case $q = 2$ follows [128], where it is attributed to Steinberg [199]. For the action of A_2 on the Dickson algebra (Section 15.3) see Wilkerson [225] and Hung [80].

The symmetric polynomials $c_n(\mathsf{d})$ were introduced by Hung and Peterson [93], where they are denoted by $\omega_n(\mathsf{d})$. They have been used by Arnon [10] to study the structure of $D(n)$ as an A_2-module. They are related to the elementary symmetric functions $e_k(n)$ in $x_1, \ldots, x_n$ by the formula $c_n(d) = \sum_{k=0}^{n} Xq^{d-k}(e_k(n))$. To prove this, we observe that $\sum_{d \geq 0} c_n(d) = \prod_{i=1}^{n}(1 + x_i + x_i^2 + x_i^4 + \cdots) = \prod_{i=1}^{n}(1 + Xq(x_i)) = Xq(\prod_{i=1}^{n}(1 + x_i)) = Xq(\sum_{k=0}^{n} e_k(n))$.

The interplay of the Steenrod operations and the action of matrices on $P(n)$ leads naturally to a study of algebras of invariants of subgroups of $GL(n)$. This broadens the scope of the hit problem, and justifies the purely algebraic approach to the study of modules over A_2 which are not in general realizable in topology as the cohomology of a space. The Dickson algebra $D(n)$ is realizable as the mod 2 cohomology of $\mathbb{R}P^\infty$, $BSO(3)$ and BG_2 respectively for $n = 1, 2$ and 3, but for $n \geq 6$ $D(n)$ is not realizable as the cohomology of any space [189, Section 3].

Sections 15.2, 15.4 and 15.5 are based in large part on the work of W. M. Singer [186]. In particular, Proposition 15.4.3 is the main step in the proof that Singer's transfer map is an isomorphism in homological degree 2 [186, Section 9], while the isomorphism $Q(P(3)^{H_3}) \cong Q(3)^{H_3}$ is used as a means of determining the $GL(3)$-invariant elements of $Q(3)$.

The relative hit problem for the Dickson algebra was solved by Hung and Nam in [91], in work motivated by the study of Singer's algebraic transfer map. We follow their argument in Sections 15.6 to 15.8. Theorem 15.8.2 implies that the only spherical classes in $Q_0 S^0$ which are detected by the algebraic transfer are the Hopf invariant 1 and the Kervaire invariant 1 elements [82].

As noted in Section 15.0, the algebra $D(n)$ is itself an A_2-module, and so there is an absolute problem of finding a minimal set of A_2 generators of $D(n)$. The absolute problem has been solved by Hung and Peterson [92] for $n \leq 4$ using detailed case by case calculations. Giambalvo and Peterson [67] obtain a number of general results on the A_2-module structure of $D(n)$, leading to a less computational solution and an extension to the case $n = 5$.

Bibliography

[1] J. F. Adams, On the structure and applications of the Steenrod algebra, Comment. Math. Helv. **32** (1958), 180–214.

[2] J. F. Adams, J. Gunawardena and H. Miller, The Segal conjecture for elementary abelian 2-groups, Topology **24** (1985), 435–460.

[3] J. F. Adams and H. R. Margolis, Sub-Hopf algebras of the Steenrod algebra, Math. Proc. Cambridge Philos. Soc. **76** (1974), 45–52.

[4] J. Adem, The iteration of Steenrod squares in algebraic topology, Proc. Nat. Acad. Sci. U.S.A. **38** (1952), 720–726.

[5] J. Adem, The relations on Steenrod powers of cohomology classes, in Algebraic Geometry and Topology, a symposium in honour of S. Lefschetz, 191–238, Princeton Univ. Press, Princeton, NJ, 1957.

[6] J. L. Alperin and Rowen B. Bell, Groups and Representations, Graduate Texts in Mathematics 162, Springer-Verlag, New York, 1995.

[7] M. A. Alghamdi, M. C. Crabb and J. R. Hubbuck, Representations of the homology of BV and the Steenrod algebra I, Adams Memorial Symposium on Algebraic Topology vol. 2, London Math. Soc. Lecture Note Ser. **176**, Cambridge Univ. Press 1992, 217–234.

[8] D. J. Anick and F. P. Peterson, A_2-annihilated elements in $H_*(\Omega\Sigma(\mathbb{R}P^2))$, Proc. Amer. Math. Soc. **117** (1993), 243–250.

[9] D. Arnon, Monomial bases in the Steenrod algebra, J. Pure App. Algebra **96** (1994), 215–223.

[10] D. Arnon, Generalized Dickson invariants, Israel J. Maths **118** (2000), 183–205.

[11] M. F. Atiyah and F. Hirzebruch, Cohomologie-Operationen und charakteristische Klassen, Math. Z. **77** (1961), 149–187.

[12] Shaun V. Ault, Relations among the kernels and images of Steenrod squares acting on right $\mathcal{A}$-modules, J. Pure. Appl. Algebra **216**, (2012), no. 6, 1428–1437.

[13] Shaun Ault, Bott periodicity in the hit problem, Math. Proc. Camb. Phil. Soc. **156** (2014), no. 3, 545–554.

[14] Shaun V. Ault and William Singer, On the homology of elementary Abelian groups as modules over the Steenrod algebra, J. Pure App. Algebra 215 (2011), 2847–2852.

[15] M. G. Barratt and H. Miller, On the anti-automorphism of the Steenrod algebra, Contemp. Math. **12** (1981), 47–52.

[16] David R. Bausum, An expression for $\chi(Sq^m)$, Preprint, Minnesota University (1975).

[17] D. J. Benson, Representations and cohomology II: Cohomology of groups and modules, Cambridge Studies in Advanced Mathematics **31**, Cambridge University Press (1991).

[18] D. J. Benson and V. Franjou, Séries de compositions de modules instables et injectivité de la cohomologie du groupe $\mathbb{Z}/2$, Math. Zeit **208** (1991), 389–399.

[19] P. C. P. Bhatt, An interesting way to partition a number, Information Processing Letters **71** (1999), 141–148.

[20] Anders Björner and Francesco Brenti, Combinatorics of Coxeter Groups, Graduate Texts in Mathematics 231, Springer-Verlag, 2005.

[21] J. M. Boardman, Modular representations on the homology of powers of real projective spaces, Algebraic Topology, Oaxtepec 1991, Contemp. Math. **146** (1993), 49–70.

[22] Kenneth S. Brown, Buildings, Springer-Verlag, New York, 1989.

[23] Robert R. Bruner, Lê M Hà, and Nguyen H. V. Hung, On the algebraic transfer, Trans. Amer. Math. Soc. **357** (2005), 473–487.

[24] S. R. Bullett and I. G. Macdonald, On the Adem relations, Topology **21** (1982), 329–332.

[25] H. E. A. Campbell and P. S. Selick, Polynomial algebras over the Steenrod algebra, Comment. Math. Helv. **65** (1990), 171–180.

[26] David P. Carlisle, The modular representation theory of $GL(n,p)$ and applications to topology, Ph.D. dissertation, University of Manchester, 1985.

[27] D. Carlisle, P. Eccles, S. Hilditch, N. Ray, L. Schwartz, G. Walker and R. Wood, Modular representations of $GL(n,p)$, splitting $\Sigma(\mathbb{C}P^\infty \times \ldots \times \mathbb{C}P^\infty)$, and the β-family as framed hypersurfaces, Math. Zeit. **189** (1985), 239–261.

[28] D. P. Carlisle and N. J. Kuhn, Subalgebras of the Steenrod algebra and the action of matrices on truncated polynomial algebras, Journal of Algebra **121** (1989), 370–387.

[29] D. P. Carlisle and N. J. Kuhn, Smash products of summands of $B(\mathbb{Z}/p)^n_+$, Contemp. Math. **96** (1989), 87–102.

[30] David P. Carlisle and Grant Walker, Poincaré series for the occurrence of certain modular representations of $GL(n,p)$ in the symmetric algebra, Proc. Roy. Soc. Edinburgh **113A** (1989), 27–41.

[31] D. P. Carlisle and R. M. W. Wood, The boundedness conjecture for the action of the Steenrod algebra on polynomials, Adams Memorial Symposium on Algebraic Topology, Vol. 2, London Math. Soc. Lecture Note Ser. **176**, Cambridge University Press, (1992), 203–216.

[32] D. P. Carlisle, G. Walker and R. M. W. Wood, The intersection of the admissible basis and the Milnor basis of the Steenrod algebra, J. Pure App. Algebra **128** (1998), 1–10.

[33] Séminaire Henri Cartan, 2 Espaces fibrés et homotopie (1949–50), 7 Algèbre d'Eilenberg-MacLane et homotopie (1954–55), 11 Invariant de Hopf et opérations cohomologiques secondaires (1958–59), available online at http://www.numdam.org

[34] H. Cartan, Une théorie axiomatique des carrés de Steenrod, C. R. Acad. Sci. Paris **230** (1950), 425–427.

[35] H. Cartan, Sur l'itération des opérations de Steenrod, Comment. Math. Helv. **29** (1955), 40–58.

[36] R. W. Carter, Representation theory of the 0-Hecke algebra, J. of Algebra **104** (1986), 89–103.

[37] R. W. Carter and G. Lusztig, Modular representations of finite groups of Lie type, Proc. London Math. Soc. (3) **32** (1976), 347–384.

[38] Chen Shengmin and Shen Xinyao, On the action of Steenrod powers on polynomial algebras, Proceedings of the Barcelona Conference on Algebraic Topology, Lecture Notes in Mathematics **1509**, Springer-Verlag (1991), 326–330.

[39] D. E. Cohen, On the Adem relations, Math. Proc. Camb. Phil. Soc. **57** (1961), 265–267.

[40] M. C. Crabb, M. D. Crossley and J. R. Hubbuck, K-theory and the anti-automorphism of the Steenrod algebra, Proc. Amer. Math. Soc. **124** (1996), 2275–2281.

[41] M. C. Crabb and J. R. Hubbuck, Representations of the homology of BV and the Steenrod algebra II, Algebraic Topology: new trends in localization and periodicity (Sant Feliu de Guixols, 1994) 143–154, Progr. Math. **136**, Birkhaüser, Basel, 1996.

[42] M. D. Crossley and J. R. Hubbuck, Not the Adem relations, Bol. Soc. Mat. Mexicana (2) **37** (1992), No. 1–2, 99–107.

[43] M. D. Crossley, $\mathcal{A}(p)$-annihilated elements of $H_*(\mathbb{C}P^\infty \times \mathbb{C}P^\infty)$, Math. Proc. Cambridge Philos. Soc. **120** (1996), 441–453.

[44] M. D. Crossley, H^*V is of bounded type over $\mathcal{A}(p)$, Group Representations: Cohomology, group actions, and topology (Seattle 1996), Proc. Sympos. Pure Math. **63**, Amer. Math. Soc. (1998), 183–190.

[45] M. D. Crossley, $\mathcal{A}(p)$ generators for H^*V and Singer's homological transfer, Math. Zeit. **230** (1999), No. 3, 401–411.

[46] M. D. Crossley, Monomial bases for $H^*(\mathbb{C}P^\infty \times \mathbb{C}P^\infty)$ over $\mathcal{A}(p)$, Trans. Amer. Math. Soc. **351** (1999), No. 1, 171–192.

[47] M. D. Crossley and Sarah Whitehouse, On conjugation invariants in the dual Steenrod algebra, Proc. Amer. Math. Soc. **128** (2000), 2809–2818.

[48] Charles W. Curtis and Irving Reiner, Representation theory of finite groups and associative algebras, Wiley, New York, 1962.

[49] D. M. Davis, The antiautomorphism of the Steenrod algebra, Proc. Amer. Math. Soc. **44** (1974), 235–236.

[50] D. M. Davis, Some quotients of the Steenrod algebra, Proc. Amer. Math. Soc. **83** (1981), 616–618.

[51] J. Dieudonné, A history of algebraic and differential topology 1900–1960, Birkhäuser, Basel, 1989.

[52] A. Dold, Über die Steenrodschen Kohomologieoperationen, Annals of Math. **73** (1961), 258–294.

[53] Stephen Donkin, On tilting modules for algebraic groups, Math. Zeitschrift **212** (1993), 39–60.

[54] Stephen Doty, Submodules of symmetric powers of the natural module for GL_n, Invariant Theory (Denton, TX 1986) 185–191, Contemp. Math. **88**, Amer. Math. Soc., Providence, RI, 1989.

[55] Stephen Doty and Grant Walker, The composition factors of $\mathbb{F}_p[x_1,x_2,x_3]$ as a $GL(3,\mathbb{F}_p,$-module, J. of Algebra **147** (1992), 411–441.

[56] Stephen Doty and Grant Walker, Modular symmetric functions and irreducible modular representations of general linear groups, J. Pure App. Algebra **82** (1992), 1–26.

[57] Stephen Doty and Grant Walker, Truncated symmetric powers and modular representations of GL_n, Math. Proc. Cambridge Philos. Soc. **119** (1996), 231–242.

[58] Jeanne Duflot, Lots of Hopf algebras, J. Algebra **204** (1998), No. 1, 69–94.

[59] V. Franjou and L. Schwartz, Reduced unstable A-modules and the modular representation theory of the symmetric groups, Ann. Scient. Ec. Norm. Sup. **23** (1990), 593–624.

[60] W. Fulton, Young Tableaux, London Math. Soc. Stud. Texts **35**, Cambridge Univ. Press, 1997.

[61] A. M. Gallant, Excess and conjugation in the Steenrod algebra, Proc. Amer. Math. Soc. **76** (1979), 161–166.

[62] L. Geissinger, Hopf algebras of symmetric functions and class functions, Springer Lecture Notes in Mathematics **579** (1977), 168–181.

[63] V. Giambalvo, Nguyen H. V. Hung and F. P. Peterson, $H^*(\mathbb{R}P^\infty \times \cdots \times \mathbb{R}P^\infty)$ as a module over the Steenrod algebra, Hilton Symposium 1993, Montreal, CRM Proc. Lecture Notes **6**, Amer. Math. Soc. Providence RI (1994), 133–140.

[64] V. Giambalvo and H. R. Miller, More on the anti-automorphism of the Steenrod algebra, Algebr. Geom. Topol. **11** (2011), No. 5, 2579–2585.

[65] V. Giambalvo and F. P. Peterson, On the height of Sq^{2^n}, Contemp. Math. **181** (1995), 183–186.

[66] V. Giambalvo and F. P. Peterson, The annihilator ideal of the action of the Steenrod algebra on $H^*(\mathbb{R}P^\infty)$, Topology Appl. **65** (1995), 105–122.

[67] V. Giambalvo and F. P. Peterson, $\mathcal{A}$-generators for ideals in the Dickson algebra, J. Pure Appl. Algebra **158** (2001), 161–182.

[68] D. J. Glover, A study of certain modular representations, J. Algebra **51** (1978), No. 2, 425–475.

[69] M. Y. Goh, P. Hitczenko and Ali Shokoufandeh, s-partitions, Information Processing Letters **82** (2002), 327–329.

[70] Brayton I. Gray, Homotopy Theory, Academic Press, New York, 1975.

[71] Lê Minh Hà, Sub-Hopf algebras of the Steenrod algebra and the Singer transfer, Proceedings of the school and conference on algebraic topology, Hanoi 2004, Geom. Topol. Publ. Coventry, **11** (2007), 81–105.

[72] Nguyen Dang Ho Hai, Generators for the mod 2 cohomology of the Steinberg summand of Thom spectra over $B(\mathbb{Z}/2)^n$, J. Algebra **381** (2013), 164–175.

[73] G. H. Hardy and E. M. Wright, An Introduction to the Theory of Numbers, Clarendon Press, Oxford, 1979.

[74] J. C. Harris and N. J. Kuhn, Stable decomposition of classifying spaces of finite abelian p-groups, Math. Proc. Cambridge Philos. Soc. **103** (1988), 427–449.

[75] J. C. Harris, T. J. Hunter and R. J. Shank, Steenrod algebra module maps from $H^*(B(\mathbf{Z}/p)^n$ to $H^*(B(\mathbf{Z}/p)^s$, Proc. Amer. Math. Soc. **112** (1991), 245–257.

[76] T. J. Hewett, Modular invariant theory of parabolic subgroups of $GL_n(\mathbf{F}_q)$ and the associated Steenrod modules, Duke Math. J. **82** (1996), 91–102.

[77] Florent Hivert and Nicolas M. Thiéry, The Hecke group algebra of a Coxeter group and its representation theory, J. Algebra **321**, No. 8 (2009), 2230–2258.

[78] Florent Hivert and Nicolas M. Thiéry, Deformation of symmetric functions and the rational Steenrod algebra, Invariant Theory in all Characteristics, CRM Proc. Lecture Notes **35**, Amer. Math. Soc, Providence, RI, 2004, 91–125.

[79] J. E. Humphreys, Modular Representations of Finite Groups of Lie Type, London Math. Soc. Lecture Note Ser. **326**, Cambridge Univ. Press, 2005.

[80] Nguyen H. V. Hung, The action of Steenrod squares on the modular invariants of linear groups, Proc. Amer. Math. Soc. **113** (1991), 1097–1104.

[81] Nguyen H. V. Hung, The action of the mod p Steenrod operations on the modular invariants of linear groups, Vietnam J. Math. **23** (1995), 39–56.

[82] Nguyen H. V. Hung, Spherical classes and the algebraic transfer, Trans. Amer. Math. Soc. **349** (1997), 3893–3910: Erratum, ibid. **355** (2003), 3841–3842.

[83] Nguyen H. V. Hung, The weak conjecture on spherical classes, Math. Z. **231** (1999), 727–743.

[84] Nguyen H. V. Hung, Spherical classes and the lambda algebra, Trans. Amer. Math. Soc. **353** (2001), 4447–4460.

[85] Nguyen H. V. Hung, On triviality of Dickson invariants in the homology of the Steenrod algebra, Math. Proc. Camb. Phil. Soc. **134** (2003), 103–113.

[86] Nguyen H. V. Hung, The cohomology of the Steenrod algebra and representations of the general linear groups, Trans. Amer. Math. Soc. **357** (2005), 4065–4089.

[87] Nguyen H. V. Hung, On A_2-generators for the cohomology of the symmetric and the alternating groups, Math Proc. Cambridge Philos. Soc. **139** (2005), 457–467.

[88] Nguyen H. V. Hung and Tran Dinh Luong, The smallest subgroup whose invariants are hit by the Steenrod algebra, Math. Proc. Cambridge Philos. Soc. **142** (2007), 63–71.

[89] Nguyen H. V. Hung and Pham Anh Minh, The action of the mod p Steenrod operations on the modular invariants of linear groups, Vietnam J. Math. **23** (1995), 39–56.

[90] Nguyen H. V. Hung and Tran Ngoc Nam, The hit problem for modular invariants of linear groups, J. Algebra **246** (2001), 367–384.

[91] Nguyen H. V. Hung and Tran Ngoc Nam, The hit problem for the Dickson algebra, Trans. Amer. Math. Soc. **353** (2001), 5029–5040.

[92] Nguyen H. V. Hung and F. P. Peterson, A_2-generators for the Dickson algebra, Trans. Amer. Math. Soc. **347** (1995), 4687–4728.

[93] Nguyen H. V. Hung and F. P. Peterson, Spherical classes and the Dickson algebra, Math. Proc. Cambridge Philos. Soc. **124** (1998), 253–264.

[94] Nguyen H. V. Hung and Võ T. N. Quynh, The image of Singer's fourth transfer, C. R. Acad. Sci. Paris, Ser I **347** (2009), 1415–1418.

[95] B. Huppert and N. Blackburn, Finite Groups II, Chapter VII, Springer-Verlag, Berlin, Heidelberg, 1982.

[96] Masateru Inoue, A_2-generators of the cohomology of the Steinberg summand M(n), Contemp. Math. **293** (2002), 125–139.

[97] Masateru Inoue, Generators of the cohomology of M(n) as a module over the odd primary Steenrod algebra, J. Lond. Math. Soc. **75**, No. 2 (2007), 317–329.

[98] G. D. James and A. Kerber, The representation theory of the symmetric group, Encyclopaedia of Mathematics, vol. **16**, Addison-Wesley, Reading, Mass., 1981.

[99] A. S. Janfada, The hit problem for symmetric polynomials over the Steenrod algebra, Ph.D. thesis, University of Manchester, 2000.

[100] A. S. Janfada, A criterion for a monomial in P(3) to be hit, Math. Proc. Cambridge Philos. Soc. **145** (2008), 587–599.

[101] A. S. Janfada, A note on the unstability conditions of the Steenrod squares on the polynomial algebra, J. Korean Math. Soc **46** (2009), No. 5, 907–918.

[102] A. S. Janfada, On a conjecture on the symmetric hit problem, Rend. Circ. Mat. Palermo, **60**, 2011, 403–408.

[103] A. S. Janfada, Criteria for a symmetrized monomial in $B(3)$ to be non-hit, Commun. Korean Math. Soc. **29** (2014), No. 3, 463–478.

[104] A. S. Janfada and R. M. W. Wood, The hit problem for symmetric polynomials over the Steenrod algebra, Math. Proc. Cambridge Philos. Soc. **133** (2002), 295–303.

[105] A. S. Janfada and R. M. W. Wood, Generating $H^*(BO(3), \mathbb{F}_2)$ as a module over the Steenrod algebra, Math. Proc. Camb. Phil. Soc. **134** (2003), 239–258.

[106] M. Kameko, Products of projective spaces as Steenrod modules, Ph.D. thesis, Johns Hopkins Univ., 1990.

[107] M. Kameko, Generators of the cohomology of BV_3, J. Math. Kyoto Univ. **38** (1998), 587–593.

[108] M. Kameko, Generators of the cohomology of BV_4, preprint, Toyama Univ., 2003.

[109] M. Kaneda, M. Shimada, M. Tezuka and N. Yagita, Representations of the Steenrod algebra, J. of Algebra **155** (1993), 435–454.

[110] Ismet Karaca, On the action of Steenrod operations on polynomial algebras, Turkish J. Math. **22** (1998), No. 2, 163–170.

[111] Ismet Karaca, Nilpotence relations in the mod p Steenrod algebra, J. Pure App. Algebra **171** (2002), No. 2–3, 257–264.

[112] C. Kassel, Quantum Groups, Graduate Texts in Mathematics **155**, Springer-Verlag, 1995.

[113] N. Kechagias, The Steenrod algebra action on generators of subgroups of GL($n, \mathbb{Z}/p\mathbb{Z}$), Proc. Amer. Math. Soc. **118** (1993), 943–952.

[114] D. Kraines, On excess in the Milnor basis, Bull. London Math. Soc. **3** (1971), 363–365.

[115] L. Kristensen, On a Cartan formula for secondary cohomology operations, Math. Scand. **16** (1965), 97–115.

[116] Nicholas J. Kuhn, The modular Hecke algebra and Steinberg representation of finite Chevalley groups, J. Algebra **91** (1984), 125–141.

[117] N. J. Kuhn, Generic representations of the finite general linear groups and the Steenrod algebra: I, Amer. J. Math. **116** (1994), 327–360; II, K-Theory **8** (1994), 395–428; III, K-theory **9** (1995), 273–303.

[118] N. J. Kuhn and S. A. Mitchell, The multiplicity of the Steinberg representation of $GL_n\mathbb{F}_q$ in the symmetric algebra, Proc. Amer. Math. Soc. **96** (1986), 1–6.

[119] J. Lannes and L. Schwartz, Sur la structure des $\mathcal{A}$-modules instables injectifs, Topology **28** (1989), 153–169.

[120] J. Lannes and S. Zarati, Sur les $\mathcal{U}$-injectifs, Ann. Scient. Ec. Norm. Sup. **19** (1986), 593–603.

[121] M. Latapy, Partitions of an integer into powers, in Discrete Mathematics and Theoretical Computer Science Proceedings, Paris, 2001, 215–228.

[122] Cristian Lenart, The combinatorics of Steenrod operations on the cohomology of Grassmannians, Adv. Math. **136** (1998), 251–283.

[123] Li Zaiqing, Product formulas for Steenrod operations, Proc. Edinburgh Math. Soc. **38** (1995), 207–232.

[124] Arunas Liulevicius, The factorization of cyclic reduced powers by secondary cohomology operations, Mem. Amer. Math. Soc. No. 42 (1962).

[125] Arunas Liulevicius, On characteristic classes, Lectures at the Nordic Summer School in Mathematics, Aarhus University, 1968.

[126] L. Lomonaco, A basis of admissible monomials for the universal Steenrod algebra, Ricer. Mat. **40** (1991), 137–147.

[127] L. Lomonaco, The iterated total squaring operation, Proc. Amer. Math. Soc. **115** (1992), 1149–1155.

[128] I. G. Macdonald, Symmetric Functions and Hall Polynomials (second edition), Oxford mathematical monographs, Clarendon Press, Oxford, 1995.

[129] Harvey Margolis, Spectra and the Steenrod algebra, North Holland Math Library, vol. 29, Elsevier, Amsterdam (1983).

[130] J. P. May, A general algebraic approach to Steenrod operations, The Steenrod Algebra and its Applications, Lecture Notes in Mathematics **168**, Springer-Verlag (1970), 153–231.

[131] Dagmar M. Meyer, Stripping and conjugation in the Steenrod algebra and its dual, Homology, Homotopy and Applications **2** (2000), 1–16.

[132] Dagmar M. Meyer, Hit polynomials and excess in the mod p Steenrod algebra, Proc. Edinburgh Math. Soc. (2) **44** (2001), 323–350.

[133] Dagmar M. Meyer and Judith H. Silverman, Corrigendum to 'Hit polynomials and conjugation in the dual Steenrod algebra', Math. Proc. Cambridge Philos. Soc. **129** (2000), 277–289.

[134] John Milnor, The Steenrod algebra and its dual, Annals of Math. **67** (1958), 150–171.

[135] J. Milnor and J. C. Moore, On the structure of Hopf algebras, Annals of Math. **81** (1965), 211–264.

[136] J. W. Milnor and J. D. Stasheff, Characteristic Classes, Princeton University Press, 1974.

[137] Pham Anh Minh and Ton That Tri, The first occurrence for the irreducible modules of the general linear groups in the polynomial algebra, Proc. Amer. Math. Soc. **128** (2000), 401–405.

[138] Pham Anh Minh and Grant Walker, Linking first occurrence polynomials over $\mathbb{F}_p$ by Steenrod operations, Algebr. Geom. Topol. **2** (2002), 563–590.

[139] S. A. Mitchell, Finite complexes with $A(n)$-free cohomology, Topology **24** (1985), 227–248.

[140] S. A. Mitchell, Splitting $B(\mathbb{Z}/p)^n$ and BT^n via modular representation theory, Math. Zeit. **189** (1985), 285–298.

[141] S. A. Mitchell and S. B. Priddy, Stable splittings derived from the Steinberg module, Topology **22** (1983), 285–298.

[142] K. Mizuno and Y. Saito, Note on the relations on Steenrod squares, Proc. Jap. Acad. **35** (1959), 557–564.

[143] K. G. Monks, Nilpotence in the Steenrod algebra, Bol. Soc. Mat. Mex. **37** (1992), 401–416.

[144] K. G. Monks, Polynomial modules over the Steenrod algebra and conjugation in the Milnor basis, Proc. Amer. Math. Soc. **122** (1994), 625–634.

[145] K. G. Monks, The nilpotence height of P_t^s, Proc. Amer. Math. Soc. **124** (1996), 1296–1303.

[146] K. G. Monks, Change of basis, monomial relations, and the P_t^s bases for the Steenrod algebra, J. Pure App. Algebra **125** (1998), 235–260.

[147] R. E. Mosher and M. C. Tangora, Cohomology operations and applications in homotopy theory, Harper and Row, New York, 1968.

[148] M. F. Mothebe, Generators of the polynomial algebra $\mathbb{F}_2[x_1,\ldots,x_n]$ as a module over the Steenrod algebra, Communications in Algebra **30** (2002), 2213–2228.

[149] M. F. Mothebe, Dimensions of subspaces of the polynomial algebra $\mathbb{F}_2[x_1,\ldots,x_n]$ generated by spikes, Far East J. Math. Sci. **28** (2008), 417–430.

[150] M. F. Mothebe, Admissible monomials and generating sets for the polynomial algebra as a module over the Steenrod algebra, Afr. Diaspora J. Math. **16** (2013), 18–27.

[151] M. F. Mothebe, Dimension result for the polynomial algebra $\mathbb{F}_2[x_1,\ldots,x_n]$ as a module over the Steenrod algebra, Int. J. Math. Math. Sci. (2013) Art. ID 150704, 6pp., MR3144989.

[152] Huynh Mui, Dickson invariants and Milnor basis of the Steenrod algebra, Topology, theory and application, Coll. Math. Soc. Janos Bolyai **41**, North Holland (1985), 345–355.

[153] Huynh Mui, Modular invariant theory and cohomology algebras of symmetric groups, J. Fac. Sci. Univ. Tokyo Sec. 1A **22** (1975), 319–369.

[154] Tran Ngoc Nam, A_2-générateurs génériques pour l'algèbre polynomiale, Adv. Math. **186** (2004), 334–362.

[155] Tran Ngoc Nam, Transfert algébrique et action du groupe linéaire sur les puissances divisées modulo 2, Ann. Inst. Fourier (Grenoble) **58** (2008), 1785–1837.

[156] P. N. Norton, 0-Hecke algebras, J. Austral. Math. Soc. (Ser. A) **27** (1979), 337–357.

[157] John H. Palmieri and James J. Zhang, Commutators in the Steenrod algebra, New York J. Math. **19** (2013), 23–37.

[158] S. Papastavridis, A formula for the obstruction to transversality, Topology **11** (1972), 415–416.

[159] David J. Pengelley, Franklin P. Peterson and Frank Williams, A global structure theorem for the mod 2 Dickson algebras, and unstable cyclic modules over the Steenrod and Kudo-Araki-May algebras, Math. Proc. Cambridge Philos. Soc. **129** (2000), 263–275.

[160] D. J. Pengelley and F. Williams, Sheared algebra maps and operation bialgebras for mod 2 homology and cohomology, Trans. Amer. Math. Soc. **352** (2000), No. 4, 1453–1492.

[161] D. J. Pengelley and F. Williams, Global Structure of the mod 2 symmetric algebra $H^*(BO, \mathbb{F}_2)$ over the Steenrod algebra, Algebr. Geom. Topol. **3** (2003), 1119–1138.

[162] D. J. Pengelley and F. Williams, The global structure of odd-primary Dickson algebras as algebras over the Steenrod algebra, Math. Proc. Cambridge Philos. Soc. **136** (2004), No. 1, 67–73.

[163] D. J. Pengelley and F. Williams, Beyond the hit problem: minimal presentations of odd-primary Steenrod modules, with application to $\mathbb{C}P^\infty$ and BU, Homology, Homotopy and Applications, **9**, No. 2 (2007), 363–395.

[164] D. J. Pengelley and F. Williams, A new action of the Kudo-Araki-May algebra on the dual of the symmetric algebras, with applications to the hit problem, Algebraic and Geometric Topology **11** (2011), 1767–1780.

[165] D. J. Pengelley and F. Williams, The hit problem for $H^*(BU(2); \mathbb{F}_p)$, Algebraic and Geometric Topology **13** (2013), 2061–2085.

[166] D. J. Pengelley and F. Williams, Sparseness for the symmetric hit problem at all primes, Math. Proc. Cambridge Philos. Soc. **158** (2015), No. 2, 269–274.

[167] F. P. Peterson, Some formulas in the Steenrod algebra, Proc. Amer. Math. Soc. **45** (1974), 291–294.

[168] F. P. Peterson, Generators of $\mathbf{H}^*(RP^\infty \wedge RP^\infty)$ as a module over the Steenrod algebra, Abstracts Amer. Math. Soc. (1987), 833-55-89.

[169] F. P. Peterson, $\mathcal{A}$-generators for certain polynomial algebras, Math. Proc. Camb. Phil. Soc. **105** (1989), 311–312.

[170] Dang Vo Phuc and Nguyen Sum, On the generators of the polynomial algebra as a module over the Steenrod algebra, C. R. Acad. Sci. Paris, Ser. 1 **353** (2015), 1035–1040.

[171] Dang Vo Phuc and Nguyen Sum, On a minimal set of generators for the polynomial algebra of five variables as a module over the Steenrod algebra, Acta Math. Vietnam. **42** (2017), 149–162.

[172] Geoffrey M. L. Powell, Embedding the flag representation in divided powers, J. of Homotopy and Related Structures **4**(1) (2009), 317–330.

[173] J. Repka and P. Selick, On the subalgebra of $H_*((\mathbb{R}P^\infty)^n; \mathbb{F}_2)$ annihilated by Steenrod operations, J. Pure Appl. Algebra **127** (1998), 273–288.

[174] J. Riordan, Combinatorial Identities, John Wiley & Sons, New York, 1968.

[175] B. E. Sagan, The Symmetric Group, Graduate Texts in Mathematics **203**, Springer (2001).

[176] Robert Sandling, The lattice of column 2-regular partitions in the Steenrod algebra, MIMS EPrint 2011.101, University of Manchester 2011, http://www.manchester.ac.uk/mims/eprints

[177] L. Schwartz, Unstable modules over the Steenrod algebra and Sullivan's fixed point set conjecture, Chicago Lectures in Mathematics, University of Chicago Press, 1994.

[178] J. Segal, Notes on invariant rings of divided powers, CRM Proceedings and Lecture Notes **35**, Invariant Theory in All Characteristics, ed. H. E. A. Campbell and D. L. Wehlau, Amer. Math. Soc. 2004, 229–239.

[179] J.-P. Serre, Cohomologie modulo 2 des complexes d'Eilenberg-MacLane, Comment. Math. Helv. **27** (1953), 198–232.

[180] Judith H. Silverman, Conjugation and excess in the Steenrod algebra, Proc. Amer. Math. Soc. **119** (1993), 657–661.

[181] Judith H. Silverman, Multiplication and combinatorics in the Steenrod algebra, J. Pure Appl. Algebra **111** (1996), 303–323.

[182] Judith H. Silverman, Hit polynomials and the canonical antiautomorphism of the Steenrod algebra, Proc. Amer. Math. Soc. **123** (1995), 627–637.

[183] Judith H. Silverman, Stripping and conjugation in the Steenrod algebra, J. Pure Appl. Algebra **121** (1997), 95–106.

[184] Judith H. Silverman, Hit polynomials and conjugation in the dual Steenrod algebra, Math. Proc. Cambridge Philos. Soc. **123** (1998), 531–547.

[185] Judith H. Silverman and William M. Singer, On the action of Steenrod squares on polynomial algebras II, J. Pure App. Algebra **98** (1995), 95–103.

[186] William M. Singer, The transfer in homological algebra, Math. Z. **202** (1989), 493–523.

[187] William M. Singer, On the action of Steenrod squares on polynomial algebras, Proc. Amer. Math. Soc. **111** (1991), 577–583.

[188] William M. Singer, Rings of symmetric functions as modules over the Steenrod algebra, Algebr. Geom. Topol. **8** (2008), 541–562.

[189] Larry Smith and R. M. Switzer, Realizability and nonrealizability of Dickson algebras as cohomology rings, Proc. Amer. Math. Soc. **89** (1983), 303–313.

[190] Larry Smith, Polynomial Invariants of Finite Groups, A. K. Peters, Wellesley, Mass., 1995.

[191] Larry Smith, An algebraic introduction to the Steenrod algebra, in: Proceedings of the School and Conference in Algebraic Topology, Hanoi, 2004, Geometry and Topology Monographs **11** (2007), 327–348.

[192] R. P. Stanley, Enumerative Combinatorics, vol. 2, Cambridge Studies in Advanced Mathematics **62**, Cambridge University Press (1999).

[193] N. E. Steenrod, Products of cocycles and extensions of mappings, Ann. of Math. **48** (1947), 290–320.

[194] N. E. Steenrod, Reduced powers of cohomology classes, Ann. of Math. **56** (1952), 47–67.

[195] N. E. Steenrod, Homology groups of symmetric groups and reduced power operations, Proc. Nat. Acad. Sci. U.S.A. **39** (1953), 213–217.

[196] N. E. Steenrod and D. B. A. Epstein, Cohomology Operations, Annals of Math. Studies 50, Princeton University Press (1962).

[197] R. Steinberg, Prime power representations of finite general linear groups II, Can. J. Math. **9** (1957), 347–351.

[198] R. Steinberg, Representations of algebraic groups, Nagoya Math. J. **22** (1963), 33–56.

[199] R. Steinberg, On Dickson's theorem on invariants, J. Fac. Sci. Univ. Tokyo, Sect. 1A Math. **34** (1987), No. 3, 699–707.

[200] P. D. Straffin, Identities for conjugation in the Steenrod algebra, Proc. Amer. Math. Soc. **49** (1975), 253–255.

[201] Nguyen Sum, On the action of the Steenrod-Milnor operations on the modular invariants of linear groups, Japan J. Math. **18** (1992), 115–137.

[202] Nguyen Sum, On the action of the Steenrod algebra on the modular invariants of special linear group, Acta Math. Vietnam **18** (1993), 203–213.

[203] Nguyen Sum, Steenrod operations on the modular invariants, Kodai Math. J. **17** (1994), 585–595.

[204] Nguyen Sum, The hit problem for the polynomial algebra of four variables, Quy Nhon University, Vietnam, Preprint 2007, 240pp. Available online at http://arxiv.org/abs/1412.1709.

[205] Nguyen Sum, The negative answer to Kameko's conjecture on the hit problem, C. R. Acad. Sci. Paris, Ser I **348** (2010), 669–672.

[206] Nguyen Sum, The negative answer to Kameko's conjecture on the hit problem, Adv. Math. **225** (2010), 2365–2390.

[207] Nguyen Sum, On the hit problem for the polynomial algebra, C. R. Acad. Sci. Paris, Ser I **351** (2013), 565–568.

[208] Nguyen Sum, On the Peterson hit problem of five variables and its application to the fifth Singer transfer, East-West J. Math. **16** (2014), 47–62.

[209] Nguyen Sum, On the Peterson hit problem, Adv. Math. **274** (2015), 432–489.

[210] René Thom, Une théorie intrinsèque des puissances de Steenrod, Colloque de Topologie de Strasbourg, Publication of the Math. Inst. University of Strasbourg (1951).

[211] René Thom, Espaces fibrés en sphères et carrés de Steenrod, Ann. Sci. Ec. Norm. Sup. **69** (1952), 109–182.

[212] René Thom, Quelque propriétés globales des variétés différentiables, Comment. Math. Helv. **28** (1954), 17–86.

[213] Ton That Tri, The irreducible modular representations of parabolic subgroups of general linear groups, Communications in Algebra **26** (1998), 41–47.

[214] Ton That Tri, On a conjecture of Grant Walker for the first occurrence of irreducible modular representations of general linear groups, Comm. Algebra **27** (1999), No. 11, 5435–5438.

[215] Neset Deniz Turgay, An alternative approach to the Adem relations in the mod p Steenrod algebra, Turkish J. Math. **38** (2014), No. 5, 924–934.

[216] G. Walker and R. M. W. Wood, The nilpotence height of Sq^{2^n}, Proc. Amer. Math. Soc. **124** (1996), 1291–1295.

[217] G. Walker and R. M. W. Wood, The nilpotence height of P^{p^n}, Math. Proc. Cambridge Philos. Soc. **123** (1998), 85–93.

[218] G. Walker and R. M. W. Wood, Linking first occurrence polynomials over $\mathbb{F}_2$ by Steenrod operations, J. Algebra **246** (2001), 739–760.

[219] G. Walker and R. M. W. Wood, Young tableaux and the Steenrod algebra, Proceedings of the School and Conference in Algebraic Topology, Hanoi 2004, Geometry and Topology Monographs **11** (2007), 379–397.

[220] G. Walker and R. M. W. Wood, Weyl modules and the mod 2 Steenrod Algebra, J. Algebra **311** (2007), 840–858.

[221] G. Walker and R. M. W. Wood, Flag modules and the hit problem for the Steenrod algebra, Math. Proc. Cambridge Philos. Soc. **147** (2009), 143–171.

[222] C. T. C. Wall, Generators and relations for the Steenrod algebra, Annals of Math. **72** (1960), 429–444.

[223] William C. Waterhouse, Two generators for the general linear groups over finite fields, Linear and Multilinear Algebra **24**, No. 4 (1989), 227–230.

[224] Helen Weaver, Ph.D. thesis, University of Manchester, 2006.

[225] C. Wilkerson, A primer on the Dickson invariants, Proc. of the Northwestern Homotopy Theory Conference, Contemp. Math. **19** (1983), 421–434.

[226] W. J. Wong, Irreducible modular representations of finite Chevalley groups, J. Algebra **20** (1972), 355–367.

[227] R. M. W. Wood, Modular representations of $GL(n, F_p)$ and homotopy theory, Algebraic Topology, Göttingen, 1984, Lecture Notes in Mathematics **1172**, Springer-Verlag (1985), 188–203.

[228] R. M. W. Wood, Splitting $\Sigma(\mathbb{C}P^\infty \times \ldots \times \mathbb{C}P^\infty)$ and the action of Steenrod squares on the polynomial ring $F_2[x_1, \ldots, x_n]$, Algebraic Topology Barcelona 1986, Lecture Notes in Mathematics **1298**, Springer-Verlag (1987), 237–255.

[229] R. M. W. Wood, Steenrod squares of Polynomials, Advances in homotopy theory, London Mathematical Society Lecture Notes 139, Cambridge University Press (1989), 173–177.

[230] R. M. W. Wood, Steenrod squares of polynomials and the Peterson conjecture, Math. Proc. Cambridge Philos. Soc. **105** (1989), 307–309.

[231] R. M. W. Wood, A note on bases and relations in the Steenrod algebra, Bull. London Math. Soc. **27** (1995), 380–386.

[232] R. M. W. Wood, Differential operators and the Steenrod algebra, Proc. London Math. Soc. **75** (1997), 194–220.

[233] R. M. W. Wood, Problems in the Steenrod algebra, Bull. London Math. Soc. **30** (1998), 194–220.

[234] R. M. W. Wood, Hit problems and the Steenrod algebra, Proceedings of the Summer School 'Interactions between Algebraic Topology and Invariant Theory', Ioannina University, Greece (2000), 65–103.

[235] R. M. W. Wood, Invariants of linear groups as modules over the Steenrod algebra, Ingo2003, Invariant Theory and its interactions with related fields, University of Göttingen (2003).

[236] R. M. W. Wood, The Peterson conjecture for algebras of invariants, Invariant Theory in all characteristics, CRM Proceedings and Lecture Notes **35**, Amer. Math. Soc., Providence R.I. (2004), 275–280.

[237] Wu Wen Tsün, Les i-carrés dans une variété grassmanniènne, C. R. Acad. Sci. Paris **230** (1950), 918–920.

[238] Wu Wen Tsün, Sur les puissances de Steenrod, Colloque de Topologie de Strasbourg, Publication of the Math. Inst. University of Strasbourg (1952).

[239] Hadi Zare, On the Bott periodicity, $\mathcal{A}$-annihilated classes in $H_*(QX)$, and the stable symmetric hit problem, submitted to Math. Proc. Cambridge Philos. Soc. 2015.

Index of Notation for Volume 1

Index for Volume 1

Index of Notation for Volume 2

Index for Volume 2